# Texts and Monographs in Physics

*Series Editors:* R. Balian   W. Beiglböck   H. Grosse   E. H. Lieb
N. Reshetikhin   H. Spohn   W. Thirring

*A selection of titles:*

**Quantum Mechanics:
Foundations and Applications**
3rd enlarged edition   By A. Böhm

**Operator Algebras and Quantum
Statistical Mechanics I + II**   2nd edition
By O. Bratteli and D. W. Robinson

**Geometry of the Standard Model
of Elementary Particles**
By. A. Derdzinski

**Quantum**
The Quantum Theory of Particles, Fields,
and Cosmology   By E. Elbaz

**Quantum Relativity**
A Synthesis of the Ideas of Einstein
and Heisenberg   By D. R. Finkelstein

**Quantum Mechanics I + II**
By A. Galindo and P. Pascual

**Local Quantum Physics**
Fields, Particles, Algebras
2nd revised and enlarged edition
By R. Haag

**Quantum Groups
and Their Representations**
By A. Klimyk and K. Schmüdgen

**Path Integral Approach
to Quantum Physics**   An Introduction
2nd printing   By G. Roepstorff

**Finite Quantum Electrodynamics**
The Causal Approach   2nd edition
By G. Scharf

**The Theory of Quark and Gluon
Interactions**   3rd revised and enlarged
edition   By F. J. Ynduráin

**Relativistic Quantum Mechanics
and Introduction to Field Theory**
By F. J. Ynduráin

**Renormalization**   An Introduction
By M. Salmhofer

**Quantum Field Theory in Condensed
Matter Physics**   By N. Nagaosa

**Quantum Field Theory in Strongly
Correlated Electronic Systems**
By N. Nagaosa

**Information Theory and Quantum
Physics**   Physical Foundations for Understanding the Conscious Process
By H.S. Green

**Quantum Non-linear Sigma Models**
From Quantum Field Theory to Supersymmetry, Conformal Field Theory, Black
Holes and Strings   By S. V. Ketov

**Perturbative Quantum
Electrodynamics and Axiomatic Field
Theory**
By O. Steinmann

Series homepage – http://www.springer.de/phys/books/tmp

Othmar Steinmann

# Perturbative Quantum Electrodynamics and Axiomatic Field Theory

With 37 Figures

 Springer

Professor Othmar Steinmann
Fakultät für Physik
Theoretische Physik
Universität Bielefeld
Postfach 10 01 31
33501 Bielefeld, Germany

*Editors*

Roger Balian
CEA
Service de Physique Théorique de Saclay
91191 Gif-sur-Yvette, France

Wolf Beiglböck
Institut für Angewandte Mathematik
Universität Heidelberg, INF 294
69120 Heidelberg, Germany

Harald Grosse
Institut für Theoretische Physik
Universität Wien
Boltzmanngasse 5
1090 Wien, Austria

Elliott H. Lieb
Jadwin Hall
Princeton University, P.O. Box 708
Princeton, NJ 08544-0708, USA

Nicolai Reshetikhin
Department of Mathematics
University of California
Berkeley, CA 94720-3840, USA

Herbert Spohn
Zentrum Mathematik
Technische Universität München
80290 München, Germany

Walter Thirring
Institut für Theoretische Physik
Universität Wien
Boltzmanngasse 5
1090 Wien, Austria

Library of Congress Cataloging-in-Publication Data. Steinmann, Othmar. Perturbative quantum electrodynamics and axiomatic field theory/Othmar Steinmann. p. cm. – (Texts and Monographs in physics, ISSN 0172-5998) Includes bibliographical references and index. ISBN 3540670246 (alk. paper) 1. Quantum electrodynamics. 2. Perturbation (Quantum dynamics) 3. Quantum field theory. I. Title. II. Series. QC680.S77 2000   530.14'33–dc21   00-030798

ISSN 0172-5998
ISBN 3-540-67024-6 Springer-Verlag Berlin Heidelberg New York

This work is subject to copyright. All rights are reserved, whether the whole or part of the material is concerned, specifically the rights of translation, reprinting, reuse of illustrations, recitation, broadcasting, reproduction on microfilm or in any other way, and storage in data banks. Duplication of this publication or parts thereof is permitted only under the provisions of the German Copyright Law of September 9, 1965, in its current version, and permission for use must always be obtained from Springer-Verlag. Violations are liable for prosecution under the German Copyright Law.

Springer-Verlag Berlin Heidelberg New York
a member of BertelsmannSpringer Science+Business Media GmbH

© Springer-Verlag Berlin Heidelberg 2000
Printed in Germany

The use of general descriptive names, registered names, trademarks, etc. in this publication does not imply, even in the absence of a specific statement, that such names are exempt from the relevant protective laws and regulations and therefore free for general use.

Data conversion: LE-TeX, Leipzig
Cover design: *design & production* GmbH, Heidelberg
Printed on acid-free paper
SPIN: 10756297      55/3141/ba - 5 4 3 2 1 0

# Preface

This book is concerned with relativistic quantum field theory, especially QED, its most successful example. It is set in the no-man's land between the mathematically rigorous but numerically barren general field theory of the mathematical physicist and the computationally fertile but mathematically sometimes adventurous field theory of the more phenomenologically inclined, and it aims at demonstrating that closer contact between these two disparate cultures may be of benefit to both. Perturbative QED serves as an example. It is shown how the rules of perturbative quantum field theory, one of the major tools of phenomenology, can be derived from well-defined general assumptions in a mathematically clean way, in particular, not using any regularizations. Special emphasis is placed on giving the infrared problem its full due. This leads, among other things, to an unorthodox method based on the local-observables approach to field theory, of describing particles and their reactions. The resulting scattering formalism is immediately applicable to the infraparticle situation of QED by directly yielding expressions for the observable-inclusive cross sections, dispensing with the notion of an S-matrix. Interestingly enough, these expressions differ somewhat from those of the conventional approach. This point will hopefully give rise to interesting and fruitful discussions.

I have been working on these sorts of problems on and off for a long time. During this time I have profited from discussions with and inspiration from a good many friends and colleagues, too many to remember them all or even single out those whose influence was most vital. Let me therefore extend my collective thanks to all of them. Let me also thank the numerous members of the physics department at Bielefeld University who supported me in my attempt to come to grips with the intricacies of computer composition. But first and foremost I wish to thank Klaus Hepp for suggesting to me that I write this book, for carefully reading the manuscript, and for the ensuing constructive criticism leading to definite improvements in the presentation.

Bielefeld, March 2000 *Othmar Steinmann*

# Contents

1 Introduction ................................................. 1

## Part I  What Is Quantum Electrodynamics?

2 Relativistic Covariance .................................... 7
   2.1 The Lorentz Group ...................................... 7
   2.2 Representations ......................................... 9
   2.3 Covariant Fields........................................ 16

3 Electrodynamics as a Classical Field Theory .............. 19

4 The Basic Principles of Relativistic Quantum Field Theory  25
   4.1 The Wightman Axioms .................................. 25
   4.2 The Wightman Functions................................ 32

5 Free Fields ................................................ 39
   5.1 The Hermitian Scalar Field ............................ 39
         Relation to the Canonical Formalism..................... 45
   5.2 Dirac Spinors ......................................... 47
   5.3 The Electromagnetic Field.............................. 53
   5.4 The Functions $\Delta_+$, $D_+$, and Related Functions............ 61

6 An Outline of Interacting QED ........................... 67
   6.1 General Description.................................... 67
   6.2 Solving the Field Equations............................ 73
   6.3 Renormalization and the UV Problem..................... 75

7 The Electric Charges ..................................... 85

## Part II  Perturbation Theory

**8 The Program of Perturbation Theory** .................... 95
   8.1 The Problem ............................................. 95
      (a) Equations of Motion ................................. 95
      (b) W-Properties ........................................ 97
      (c) Renormalization Conditions .......................... 97
   8.2 Perturbation Theory .................................... 98
   8.3 Uniqueness ............................................ 100
   8.4 Time-Ordered Products ................................. 103

**9 Unrenormalized Solution** ................................ 109
   9.1 Generalities .......................................... 109
   9.2 Special Examples in Low Orders ........................ 110
   9.3 The General Solution .................................. 115
      Configuration Space .................................... 115
      Momentum Space ......................................... 120
   9.4 Verifications ......................................... 123
   9.5 Current Conservation and the Charge Identity .......... 131

**10 Renormalization and the UV Problem** .................... 135
   10.1 Low Orders .......................................... 135
      The Photon 2-Point Function ........................... 135
      The Electron 2-Point Function ......................... 142
      The Vertex Function ................................... 147
   10.2 The General Case .................................... 151
   10.3 Renormalization in Configuration Space .............. 156

**11 The IR Problem for Wightman Functions** ................. 165
   11.1 The Problem ......................................... 165
   11.2 The Solution ........................................ 167
   11.3 The W-Properties and Other Conditions ............... 176
      The W-Properties ...................................... 176
      The Charge Identity ................................... 178
      Smoothness Conditions ................................. 180

**12 Physical States** ....................................... 187
   12.1 The General Strategy ................................ 187
   12.2 Unrenormalized Perturbation Theory .................. 192
   12.3 The UV Problem ...................................... 197
   12.4 The IR Problem and the Final Verifications .......... 201

## Part III  Particles and Their Reactions

**13 The Standard View** .................................................. 213
  13.1 Massive Theories .................................................. 214
  13.2 The IR Problem .................................................... 217

**14 Particle Probes** ..................................................... 221
  14.1 Mathematical Preliminaries ........................................ 222
  14.2 Free Particles .................................................... 229

**15 Interacting Particles** ............................................... 239
  15.1 Probes for Interacting Particles .................................. 239
  15.2 IR Cancellations .................................................. 246
      Proof of Lemma 15.4 ................................................ 259
      Proof of Theorems 15.1–3 ........................................... 265
  15.3 Photons ........................................................... 272
      Derivation of the Decomposition (15.87) ............................ 276

**16 Reactions** .......................................................... 283
  16.1 Massive Scalar Fields ............................................. 283
  16.2 Reactions in QED .................................................. 296
  16.3 Photons ........................................................... 300
  16.4 Electrons ......................................................... 304

**17 Cross Sections** ..................................................... 321
  17.1 Cross Sections: the Problem ....................................... 321
  17.2 A Solution ........................................................ 330
  17.3 Compton Scattering, an Example .................................... 336

**References** ........................................................... 347

**Index** ................................................................ 351

# Chapter 1

# Introduction

This is a frankly idiosyncratic book. It is not meant to be yet another general introduction to quantum electrodynamics (henceforth called QED), though it will indeed present, or so its author hopes, a self-contained description of QED and a systematic derivation of its perturbation theory. But this description and this derivation are given from a particular point of view. I do not aim at anything remotely approaching completeness concerning either results or methods. And I do not aim at a balanced exposition of all the facets of the problem even in the chosen approach. Rather I intend to stress some critical aspects that have, perhaps, not found sufficient attention in the existing literature, but are frequently treated in a somewhat haphazard and overly heuristic way, if at all. This is especially true for some points related to the "infrared problem", which is caused by the vanishing mass of the photon. Hence, though the book should be understandable to any reader with a good knowledge of the elements of classical electrodynamics and of the general principles of quantum mechanics, a previous encounter with QED or quantum field theory in general would be distinctly helpful.[1]

Our interest will be centred on studying the mathematical and physical structure of the theory, not on its comparison with experiment. In particular, we will use the best-developed approximation method, perturbation theory, mainly as a tool of investigating structural problems, not as a means of calculating numbers. The lucky fact that the electromagnetic interactions are quite weak, so that perturbation theory yields excellent numerical agreement with experiment, is therefore of little importance to us. This remark is especially relevant in view of possible generalizations of some of our results to other quantum field theories, including quantum chromodynamics, which describe strong interactions.

---

[1] A very brief and highly subjective selection of pertinent textbooks will be given in the bibliography at the end of this volume. These texts contain also more extensive bibliographies.

Heuristically speaking, QED is the quantum-mechanical description of the electromagnetic field and its interaction with charged particles. We will also demand this description to be fully relativistic. But particles are rather protean entities. On hitting one another they can completely change their character and their number. E.g. an electron and a positron can annihilate on meeting, turning into a number of photons, or vice versa. Now, what is of greatest interest to the physicist is the understanding of the dynamical processes underlying such transformations, i.e. of what goes on where the interaction takes place. But there the particles seem to enjoy at best a precarious existence, quite losing their individuality. In the region of interaction the system does not look like a collection of billiard-balls, but more like a jelly with infinitely many degrees of freedom. It is therefore hardly surprising that attempts at describing this interaction within a particle picture have not been successful, but had to be replaced by a field-theoretical formulation.

This is the point of view that we will take. The fundamental objects of the theory are quantized fields. Particles play no role in its basic formulation, but emerge as secondary objects, which are helpful for the phenomenological description of certain observed processes called scattering events. QED, or relativistic quantum field theory in general, is *not* based on the notion of "point particles", as one sees stated so often and yet so erroneously.

As is customary, we will only consider the special case of the electromagnetic field in interaction with a single charged field and its adjoint, which are associated with a particle of spin $\frac{1}{2}$, called the "electron", and its antiparticle called the "positron". The generalization to several fields of spin $\frac{1}{2}$ is immediate. The inclusion of charged scalar (spin 0) fields proceeds along similar lines and presents no essential new features. Fields of spin $\geq 1$ are more difficult to handle. This problem will not be considered.

We will work throughout in the Heisenberg picture, which is the most natural formulation of quantum mechanics for a relativistic theory.

Our presentation is largely inspired by the ideas and results of what is known under the unnecessarily forbidding name of "axiomatic field theory" [StW, J, BLOT]. QED is defined as a theory of certain quantized fields, which satisfy specific equations of motion and the Wightman axioms in a specially adapted version. Nothing more is needed. In particular, neither canonical commutation relations nor asymptotic conditions are assumed. From the more general algebraic "local observables" formalism [H] we take over its basic observation that all measurements and other experimental manipulations are carried out in bounded regions of space-time. The objects of direct physical relevance are therefore localized observables and their expectation values in physical states. This insistence on the local aspects of the theory motivates our preference for the operator formalism over the path integral method. Our introduction of particles will also be closely related to corresponding ideas in the algebraic approach, as developed especially by Buchholz [Bu 93]. In addition, the local-observables formalism has yielded vital insights into the superselection structure of QED (see e.g. [Bu 82]). These results will not be discussed here, however.

The essential role of locality and relativistic invariance in our approach enforces the use of the "Gupta-Bleuler formalism", a formulation of QED in which these two properties are manifest at all stages.

The book is organized as follows. Part I begins with a brief recapitulation of the relevant notions and theorems, without proofs, of (a) electrodynamics as a classical field theory, and (b) the Wightman formulation of general quantum field theory. These are the two ingredients whose combination into a coherent whole is the subject of this book. As an initial simple example, the theory of free quantized fields is then studied, concentrating on those aspects which are relevant to our purpose. Next, a description of what we understand under the name QED is arrived at in a partly heuristic way. The resulting definition is not as precise as one might wish. This reflects the present state of knowledge. Not only is it not known whether a rigorous theory deserving to be called QED exists, it is not even exactly known what "deserving to be called QED" means. For all we know, there may exist no rigorous QED, or a uniquely defined one, or several distinct versions having equal claim to authenticity. The final chapter of Part I (Chap.7) contains a collection of rigorous definitions and results concerning the electric charge that are pertinent to our objective.

Since it is not known how to fashion a coherent, exact, theory out of the principles enunciated in Part I, we set ourselves the humbler task of carrying out this program in perturbation theory, which is the most intensively studied and best understood approximation scheme in QED. This task is accomplished in Part II. Perturbation theory consists in expanding the quantities of interest into power series in a coupling constant, which measures the strength of the interaction. Such expansions will be derived for some basic objects of the theory: the Wightman functions (closely akin to the Green's functions of the more conventional treatments), from which in principle all physical quantities of interest can be calculated. We will work with what is called the Gupta-Bleuler formalism. Special attention will be paid to an awkward feature of this formulation: its mathematical state space contains, of necessity, vectors that cannot possibly represent physically admissible states of a physical system. The problem of extracting physically relevant information from this unphysical formulation is solved in a novel way.

The object of Part III is to establish a convincing scattering formalism in QED, which solves the notorious infrared problem in a natural way. Based on the results of Part II it is shown how particle-like objects emerge from our field theoretical formulation, and how their interactions can be described. The result is unexpected. It does not accord with the traditional picture. The notion of charged particles appears to be subtler than generally assumed. The problem is described and a solution is proposed in Chap. 17. But further discussions will be needed: the volume closes with a question, not with a solid result.

The perturbative construction of the Wightman functions developed in Part II can easily be generalized to other relativistic field theories, in particular to non-Abelian gauge theories. This is *not* true, however, for the construction of the physical states, which is a much more difficult and not yet satisfactorily solved problem in the non-Abelian case. Also, a generalization of the results of Part III to non-Abelian theories like quantum chromodynamics has not yet been found. Both these difficulties may be connected with the confinement problem: the widely held opinion that confinement is an entirely non-perturbative phenomenon may be erroneous.

Finally, let me mention some of the questions that will not be considered at all or only inadequately. The important subject of gauge invariance will be given rather short shrift, for reasons that will be explained in Chap. 5. The problem is much more complicated than in classical electrodynamics, and a rigorous general theory is still lacking. Other important topics that fall completely outside the scope of the present work are: path integrals (as already mentioned), approximation methods other than perturbation theory, numerical methods for the calculation of observable quantities, systems in external fields (e.g. the determination of the anomalous magnetic moment of the electron), and the bound-state problem (positronium, muonium, Lamb shift of the H-atom, etc.). For information on these points the reader should consult accounts written by more competent authors than the present one.

Throughout the book, except in a few places where the dependence on $\hbar$ is essential, units will be used in which $\hbar = c = 1$. The Minkowski tensor $g_{\mu\nu}$ is defined with signature $(1, -1, -1, -1)$.

No completeness concerning the cited literature is intended. The existing literature is vast. I quote mostly those books and articles that I personally found most useful and which are in my opinion the most helpful complements to the material presented here. References without year, e.g. [A], refer to books, those with years, e.g. [Aa 97], to articles or contributions to conferences or summer schools.

Lastly, a remark concerning the use of the pronouns *we* and *I* is indicated. "We" includes the reader. It is used in statements with which the reader is expected to agree. "I", on the other hand, is used for the expression of the author's opinions, with which the reader may or may not agree.

# Part I

# What Is Quantum Electrodynamics?

The first two chapters of this part contain recapitulations of background material that is expected to be known to the reader. They serve also to establish notations. In the third chapter we give a brief, mostly descriptive, non-deductive, introduction to the general principles of relativistic quantum field theory, as distilled in the axioms of "axiomatic field theory". The real work starts in Chap. 5, in which the theory of free fields is developed. Chap. 6 contains a heuristic overview of the basic definitions and the major problems of interacting QED. In Chap. 7 the definition of the electric charge operator(s) is discussed in a rigorous way.

# Chapter 2

# Relativistic Covariance

Useful references in addition to this chapter are, for example, Chap. 3 of [BLOT], Chaps. 7 and 15 of [ED], and [N].

## 2.1 The Lorentz Group

The space-time underlying QED is the Minkowski space $M$. This is a four-dimensional real vector space equipped with a scalar product

$$x, y \in M \quad \rightarrow \quad (x, y) \in \mathbb{R}, \tag{2.1}$$

which is bilinear and symmetric, but indefinite with signature $(+---)$. This means that in Cartesian coordinates $x^0 = ct, x^1, x^2, x^3$ the scalar product of $x, y \in M$ is given by

$$(x, y) = x^0 y^0 - \sum_{i=1}^{3} x^i y^i. \tag{2.2}$$

The following *notation* will be used. Minkowski indices running over the full range of values $0,\ldots,3$ will usually be denoted by lower-case Greek letters $\mu, \nu, \ldots$; those running only over the spatial values 1,2,3, by lower-case Roman letters $i, j, \ldots$ . Indices occurring twice in a term, once as a superscript, once as a subscript, must be summed over, unless explicitly stated otherwise. The spatial 3-vector $(x^1, x^2, x^3)$ will be denoted $\vec{x}$. The expression (2.2) then becomes $(x, y) = x^0 y^0 - \vec{x} \cdot \vec{y}$. The notation $x^2 = (x, x)$ is often used if there is no danger of confusion with the 2-component of $x$. Yet another way of writing the scalar product is

$$(x, y) = x^\mu g_{\mu\nu} y^\nu, \tag{2.3}$$

where the $g_{\mu\nu}$ are the components of the Minkowski tensor

$$g_{00} = 1, \quad g_{ii} = -1, \quad g_{\mu\nu} = 0 \text{ for } \mu \neq \nu. \tag{2.4}$$

## 2. Relativistic Covariance

The point $x \in M$ is said to be timelike, lightlike, or spacelike, relative to $y \in M$, if $(x-y)^2$ is positive, zero, or negative, respectively. The points $x$ with $(x-y)^2 > 0$ and $x^0 > y^0$ form the *open forward cone* $V_+$ with apex $y$, those with $(x-y)^2 \geq 0$ and $x^0 \geq y^0$ the *closed forward cone* $\bar{V}_+$ with apex $y$. If the apex is not explicitly mentioned, it is understood to be the origin. The backward cones $V_-$ and $\bar{V}_-$ are defined analogously with $x^0 < y^0$ or $x^0 \leq y^0$ respectively.

Two Cartesian systems of coordinates $\{x^\mu\}$ and $\{x'^\mu\}$ are related through a *Poincaré transformation*

$$x'^\mu = \Lambda^\mu{}_\nu x^\nu + a^\mu , \tag{2.5}$$

or in matrix notation

$$x' = \Lambda x + a , \tag{2.6}$$

where $x', x, a$, are $1 \times 4$ matrices, $\Lambda$ a $4 \times 4$ matrix with entries $\Lambda^\mu{}_\nu$. $\Lambda$ must satisfy

$$\Lambda^T G \Lambda = G \tag{2.7}$$

with $G$ the $4 \times 4$ matrix with entries $g_{\mu\nu}$. The homogeneous form of (2.6):

$$x' = \Lambda x , \tag{2.8}$$

which transforms between systems with the same origin, is called a *Lorentz transformation*. Condition (2.7) is equivalent to the requirement that Lorentz transformations leave the scalar product $(x, y)$ invariant.

The translations

$$x' = x + a \tag{2.9}$$

form an additive Abelian group, the *translation group* $\mathcal{T}$, the Lorentz transformations form the *Lorentz group* $L$, and the Poincaré transformations the *Poincaré group* $\mathcal{P}$. Any Poincaré transformation can be written as a Lorentz transformation followed by a translation. For our purposes it is sufficient to consider $\mathcal{T}$ and $L$ separately.

The Lorentz group $L$ can be decomposed into two mutually disconnected components in two different ways.

(a) From (2.7) it follows that

$$\det \Lambda = \pm 1 . \tag{2.10}$$

The $\Lambda \in L$ with $\det \Lambda = 1$ form the "proper Lorentz group" $L_+$, those with $\det \Lambda = -1$ the set $L_-$, which is not a group, because the product of two $L_-$ transformations lies in $L_+$. Obviously we have $L = L_+ \cup L_-$.

(b) A Lorentz transformation either leaves the forward cone $V_+$ invariant or maps it onto $V_-$. The former transformations form the "orthochronous Lorentz group" $L^\uparrow$, the latter the set $L^\downarrow$ which is not a group. Again we have $L = L^\uparrow \cup L^\downarrow$.

The intersection
$$L_+^\uparrow = L_+ \cap L^\uparrow \tag{2.11}$$
is called the "restricted Lorentz group". It is connected, i.e. any two transformations in $L_+^\uparrow$ can be connected by a continuous path within $L_+^\uparrow$. Continuity of a function $\Lambda(\lambda) \in L$ of a real parameter $\lambda$ means continuity of each matrix element $\Lambda^\mu{}_\nu$.

An important element of $L_-^\uparrow$ is the *parity transformation*
$$P \ : \ x^0 \to x^0, \quad x^i \to -x^i \ , \tag{2.12}$$
while the *time reversal*
$$T \ : \ x^0 \to -x^0, \quad x^i \to x^i \tag{2.13}$$
lies in $L_-^\downarrow$. Any transformation $\Lambda$ in $L_-^\uparrow$, $L_-^\downarrow$, or $L_+^\downarrow$ can be decomposed as
$$\Lambda = P\Lambda', \ T\Lambda', \ PT\Lambda' \ , \tag{2.14}$$
respectively, with $\Lambda' \in L_+^\uparrow$.

## 2.2 Representations

A *representation* of the group $G$ on the real or complex vector space $\mathcal{V}$ is a map $g \to R(g)$ of $G$ into the set $\mathcal{L}(\mathcal{V})$ of linear operators on $\mathcal{V}$, which respects the group operations:
$$R(g_1 g_2) = R(g_1) R(g_2) \ , \tag{2.15}$$
$$R(g^{-1}) = (R(g))^{-1} \ . \tag{2.16}$$

Representations of $L^\uparrow$ on infinite-dimensional vector spaces will be introduced in Chap. 4. But at present we confine our attention to finite-dimensional representations. Let $\mathcal{V}$ be $n$-dimensional and $e_1, \ldots, e_n$ a basis of $\mathcal{V}$. Then the operator $A \in \mathcal{L}(\mathcal{V})$ is uniquely determined by the $n \times n$ matrix $\hat{A}$ with elements $A_{ij}$ defined by
$$A e_i = \sum_j A_{ij} e_j \ . \tag{2.17}$$

$\hat{A}$ is called the matrix of $A$ in the basis $\{e_i\}$. The relations (2.15), (2.16) hold also for the matrices $\hat{R}(g)$ of $R(g)$ in a given basis. The mapping $g \to \hat{R}(g)$ of $G$ into the set of regular $n \times n$ matrices is called a matrix representation. The matrix representations in two different bases of $\mathcal{V}$ are related by a similarity transformation
$$\hat{R}_1(g) = V^{-1} \hat{R}_2(g) V \ , \tag{2.18}$$

with $V$ a regular matrix. The representations $\hat{R}_1$ and $\hat{R}_2$ are called *"equivalent"*. Two operator representations $R_1, R_2$ on the $n$-dimensional spaces $\mathcal{V}_1, \mathcal{V}_2$ (possibly identical) are also called equivalent if

$$R_1(g) = V^{-1} R_2(g) V , \qquad (2.19)$$

where $V$ is a linear bijective mapping of $\mathcal{V}_1$ onto $\mathcal{V}_2$. In the applications to quantum mechanics, equivalent representations are usually also physically equivalent, since the choice of basis in a quantum mechanical state space is physically irrelevant. We will therefore in general not clearly distinguish between equivalent representations, and also not between operator and matrix representations, as long as no harmful misunderstandings are to be feared.

An important class of representations are the *tensor representations* of the Lorentz group $L$. A *contravariant tensor $T$ of rank $n$* is an element of a $4^n$-dimensional vector space $\mathcal{V}^{(n,0)}$ with components $T^{\mu_1 \ldots \mu_n}$, with $\mu_i$ being Minkowski indices. The representation $R^{(n)}$ of $L$ on $\mathcal{V}^{(n)}$ is defined by

$$\left(R^{(n)}(\Lambda) T\right)^{\mu_1 \ldots \mu_n} = \prod_{i=1}^{n} \Lambda^{\mu_i}{}_{\nu_i} T^{\nu_1 \ldots \nu_n} . \qquad (2.20)$$

A tensor of rank 0 is called a *scalar*. Its representation space is 1-dimensional, and the scalar representation is the trivial representation $R^{(0)}(\Lambda) = 1$ for all $\Lambda$. A contravariant tensor of rank 1 is called a *contravariant vector*. The corresponding representation is the fundamental representation

$$R^{(1)}(\Lambda) = \Lambda. \qquad (2.21)$$

The scalars and the vectors define *irreducible representations*, meaning that $\mathcal{V}^{(n)}$ itself and its zero-dimensional subspace $\{0\}$ are the only subspaces which are mapped into themselves by all $R^{(n)}(\Lambda)$. The tensor representation of rank 2 is reducible. There exist proper subspaces of $\mathcal{V}^{(2)}$ which are invariant under $R^{(2)}$. Such subspaces are the symmetric subspace $\mathcal{V}^{(2)s}$ with $T^{\mu\nu} = T^{\nu\mu}$ for every pair $\mu, \nu$, and the antisymmetric subspace $\mathcal{V}^{(2)a}$ defined by $T^{\mu\nu} = -T^{\nu\mu}$. The restrictions of $R^{(2)}$ to these subspaces form lower-dimensional (10 and 6, respectively) representations $R^{(2)s}$, $R^{(2)a}$ of $L$. The former one can be reduced further; the latter is irreducible.

An important example of a contravariant symmetric tensor of rank 2 is the *Minkowski tensor* $g^{\mu\nu}$ defined by $g^{\mu\nu} = g_{\mu\nu}$. This tensor has the remarkable property of being translated into itself componentwise by all $R^{(2)}(\Lambda)$.

The tensors of rank $> 2$ are of course also highly reducible. We need not go into this any further.

The *contravariant pseudotensors* (or tensor densities) are distinguished from the genuine tensors by an additional factor $\det \Lambda$ in the transformation law (2.20). An important example of a pseudotensor of rank 4 is the totally antisymmetric $\varepsilon$-tensor

$$\varepsilon^{\mu_1\ldots\mu_4} = \begin{cases} 0 & \text{if two } \mu_i \text{ coincide,} \\ \pm 1 & \text{if } (\mu_1,\ldots,\mu_4) \text{ is an } \begin{Bmatrix} even \\ odd \end{Bmatrix} \text{ permutation of } (1,\ldots,4) \,. \end{cases} \quad (2.22)$$

Like $g^{\mu\nu}$, this tensor is componentwise invariant.

Let the matrix $\check{\Lambda}$ with elements $\check{\Lambda}_\mu{}^\nu =: \Lambda_\mu{}^\nu$ be defined as

$$\check{\Lambda} = \left(\Lambda^T\right)^{-1}, \quad (2.23)$$

so that

$$\Lambda_\rho{}^\mu \Lambda^\rho{}_\nu = \delta^\mu_\nu \,. \quad (2.24)$$

A *covariant tensor* $T$ of rank $n$ is an element of a $4^n$-dimensional vector space $\mathcal{V}_{(n)}$ with components $T_{\mu_1\ldots\mu_n}$. The representation $R_{(n)}$ of $L$ is defined on $\mathcal{V}_{(n)}$ by

$$\left(R_{(n)}(\Lambda)T\right)_{\mu_1\ldots\mu_n} = \prod_{i=1}^{n} \Lambda_{\mu_i}{}^{\nu_i}\, T_{\nu_1\ldots\nu_n} \,. \quad (2.25)$$

The corresponding definition of covariant pseudotensors contains again an additional factor $\det \check{\Lambda} = \det \Lambda$. A covariant (pseudo-)tensor of rank 1 is called a covariant (pseudo-)vector. The Minkowski tensor $g_{\mu\nu}$ is a covariant tensor of rank 2; the $\varepsilon$-tensor $\varepsilon_{\alpha\beta\gamma\delta} = \varepsilon^{\alpha\beta\gamma\delta}$ is a covariant pseudotensor of rank 4.

A *mixed tensor* of type $(n,m)$ and rank $n+m$, with components $T^{\mu_1\ldots\mu_n}_{\nu_1\ldots\nu_m}$, transforms according to

$$\left(R^{(n)}_{(m)}(\Lambda)T\right)^{\mu_1\ldots}_{\nu_1\ldots} = \prod_{i=1}^{n} \Lambda^{\mu_i}{}_{\mu'_i} \prod_{j=1}^{m} \Lambda_{\nu_j}{}^{\nu'_j}\, T^{\mu'_1\ldots}_{\nu'_1\ldots} \,. \quad (2.26)$$

Here we have used a somewhat ambiguous, simplified notation. A fixed ordering of the indices should be specified not only within the contravariant and the covariant index sets separately, but in the total index set $\mu_1,\ldots,\nu_m$. For example, the tensor $T^\mu{}_\nu$ is different from the tensor $T_\nu{}^\mu$.

The coefficients $\Lambda^\mu{}_\nu$ and $\Lambda_\mu{}^\nu$ of a Lorentz transformation refer to two different coordinate systems. They are *not* the components of a mixed tensor.

Note that the representation matrices of all tensor representations are real.

Let $\{\mu\}$, etc. stand for ordered sets of indices: $\{\mu\} = \{\mu_1\ldots\mu_n\}$, etc. Let $S^{\{\mu\}}_{\{\nu\}}$, $T^{\{\mu'\}}_{\{\nu'\}}$ be tensors of rank $r$ and $r'$, respectively. Then

$$U^{\{\mu,\mu'\}}_{\{\nu,\nu'\}} = S^{\{\mu\}}_{\{\nu\}} T^{\{\mu'\}}_{\{\nu'\}} \quad (2.27)$$

is a tensor of rank $r+r'$. $U = S \otimes T$ is called the *tensor product* of $S$ and $T$.

The *contraction*

$$S^{\mu_1...\mu_n}_{\nu_1...\nu_m} = T^{\mu_1...\mu_i \rho \mu_{i+1}...\mu_n}_{\nu_1...\nu_j \rho \nu_{j+1}...\nu_m} \qquad (2.28)$$

of a tensor of type $(n+1, m+1)$ over a contravariant and a covariant index is a tensor of type $(n, m)$.

These two operations, tensor product and contraction, can be combined to define the *raising and lowering* of indices. Let $\mu$ be a contravariant index of the tensor $T^{...\mu...}_{......}$ of type $(n, m)$. Then

$$T^{......}_{...\mu...} = g_{\mu\nu} T^{...\nu...}_{......} \qquad (2.29)$$

is a tensor of type $(n-1, m+1)$. We say that the index $\mu$ has been lowered. The original tensor is recovered by raising the index again by contraction with $g^{\mu\nu}$. This permits the identification of tensors of the same rank but different types, if they are related through raising and lowering of indices. For instance, we say that $\xi^\mu$ and $\xi_\mu = g_{\mu\nu}\xi^\nu$ are the contravariant and covariant components, respectively, of the same abstract vector $\xi$, and similarly for tensors of higher rank. Thus $g^{\mu\nu}$ and $g_{\mu\nu}$ are the contravariant and covariant components of *the* metric tensor $g$. Its mixed components are

$$g^\mu{}_\nu = g_\nu{}^\mu = \delta^\mu_\nu . \qquad (2.30)$$

Another representation that we need to discuss is the *spinor representation* $S(\Lambda)$ of the orthochronous group $L^\uparrow$. It is not a representation in the strict sense of the word, but what is known among physicists as a "2-valued representation".[1] This means that we associate to each element $\Lambda \in L^\uparrow$ a pair $\pm S(\Lambda)$ of matrices, more precisely complex $4 \times 4$ matrices, which satisfy the relation

$$S(\Lambda_1) S(\Lambda_2) = \pm S(\Lambda_1 \Lambda_2) . \qquad (2.31)$$

The representation is constructed in two steps. At first one finds a two-dimensional representation of the restricted Lorentz group $L^\uparrow_+$ as follows. The relation

$$X(x) = \begin{pmatrix} x^0 + x^3 & x^1 - ix^2 \\ x^1 + ix^2 & x^0 - x^3 \end{pmatrix} \qquad (2.32)$$

defines a linear bijective map of the Minkowski space onto the space of hermitian $2 \times 2$ matrices. We have

$$(x, x) = \det X . \qquad (2.33)$$

The expression (2.32) can also be written as

$$X = x^0 \mathbf{1} + x^i \sigma_i , \qquad (2.34)$$

---

[1] Mathematicians call it a representation of the universal covering group of $L^\uparrow$. But the physically relevant group is the Lorentz group itself, not its covering group.

where
$$\sigma_1 = \begin{pmatrix} 0 & 1 \\ 1 & 0 \end{pmatrix}, \quad \sigma_2 = \begin{pmatrix} 0 & -i \\ i & 0 \end{pmatrix}, \quad \sigma_3 = \begin{pmatrix} 1 & 0 \\ 0 & -1 \end{pmatrix} \quad (2.35)$$
are the Pauli matrices and $\mathbf{1}$ is the unit matrix.

Let $SL(2,C)$ be the group of complex $2 \times 2$ matrices with determinant 1. Choose $A \in SL(2,C)$ and let $X$ be an arbitrary hermitian matrix. Then
$$X' = AXA^* \quad (2.36)$$
is again hermitian and depends linearly on $X$ and therefore on the 4-vector $x$ represented by $X$. Moreover, $\det X' = \det X$. Hence the mapping
$$x \longrightarrow X \longrightarrow X' \longrightarrow x' = \Lambda x \quad (2.37)$$
defines a Lorentz transformation $\Lambda$. It can be shown that $\Lambda \in L_+^\uparrow$, and that any $\Lambda \in L_+^\uparrow$ can be obtained in this way by choosing a suitable $A$. The mapping $\Lambda \to A$ is clearly ambiguous, since $A$ can be replaced by $-A$ in (2.36) without changing $X'$. But this is the only ambiguity. The mapping $\Lambda \to A(\Lambda)$ is two-valued, and it can be shown to define a two-valued representation of $L_+^\uparrow$ in the sense of (2.31).

Let us mention two important special examples.

(a) Let $\Lambda$ be a rotation of 3-space by the angle $\varphi$ around an axis with direction $\vec{n}$, $|\vec{n}| = 1$. Then
$$A(\Lambda) = A_{\vec{n}}(\varphi) := \exp\left\{i\frac{\varphi}{2}(\vec{n},\vec{\sigma})\right\} = \mathbf{1}\cos\frac{\varphi}{2} + i(\vec{n},\vec{\sigma})\sin\frac{\varphi}{2}, \quad (2.38)$$
with $(\vec{n},\vec{\sigma}) = \sum_i n_i \sigma_i$. Note that these representatives of rotations are unitary.

(b) Let $\Lambda$ be a Lorentz boost with velocity $v$ in the direction of the 3-axis. Then $A(\Lambda)$ is of the form
$$A(\Lambda) = \begin{pmatrix} e^{\frac{\chi}{2}} & 0 \\ 0 & e^{-\frac{\chi}{2}} \end{pmatrix} \quad (2.39)$$
with $\cosh \chi = v$.

This representation must now be extended to a representation of the full $L^\uparrow$. This cannot be achieved on a 2-dimensional space; we must go over to 4-dimensional matrices. We first notice that if $A(\Lambda)$ is a 2-valued matrix representation, then the same is true for its complex conjugate $\overline{A(\Lambda)}$. Hence the $4 \times 4$ matrices
$$S(\Lambda) = \begin{pmatrix} A(\Lambda) & 0 \\ 0 & \overline{A(\Lambda)} \end{pmatrix} \quad (2.40)$$
define a representation of $L_+^\uparrow$, albeit a reducible one. Here each entry in the matrix on the right-hand side stands itself for a $2 \times 2$ matrix. The representation $S$ can be extended to an irreducible representation of $L^\uparrow$ by adding

the definition

$$S(P) = \begin{pmatrix} 0 & \varepsilon \\ -\varepsilon & 0 \end{pmatrix}, \quad \varepsilon = \begin{pmatrix} 0 & 1 \\ -1 & 0 \end{pmatrix} \qquad (2.41)$$

of the representative of the parity operation. Since every $\Lambda \in L^\uparrow_-$ can be written as $\Lambda = P\Lambda'$ with $\Lambda' \in L^\uparrow_+$, the definitions (2.40), (2.41) fix the value of $S(\Lambda)$ for every $\Lambda \in L^\uparrow$. The elements of the 4-dimensional vector space on which $S$ acts are called "4-spinors" or "*Dirac spinors*".

The special values of the $S$-matrices that we have given are by no means unique. If $V$ is a regular $4 \times 4$ matrix, then

$$S'(\Lambda) = V S(\Lambda) V^{-1} \qquad (2.42)$$

defines again an irreducible representation of $L^\uparrow$. If $V$ is unitary, then the desirable unitarity of the representations of rotations is retained. For the application to QED the choice of a particular representation is merely a matter of convenience. In this work we will use throughout the representation $S'$ obtained by inserting into (2.42) the matrix

$$V = \begin{pmatrix} 1 & 0 \\ 0 & \varepsilon \end{pmatrix},$$

and $S(\Lambda)$ will henceforth denote the resulting representation

$$S(\Lambda) = \begin{pmatrix} A & 0 \\ 0 & -\varepsilon \overline{A} \varepsilon \end{pmatrix} \quad \text{for} \quad \Lambda \in L^\uparrow_+ ,$$

$$S(P) = \begin{pmatrix} 0 & 1 \\ 1 & 0 \end{pmatrix}. \qquad (2.43)$$

The *infinitesimal generators* of the representation $R(\Lambda)$ of $L^\uparrow_+$, which is a Lie group, are defined as follows. Let $\Lambda'(\varphi)$ be the elements of a one-parameter subgroup of $L^\uparrow_+$, the parametrization being chosen such that $\Lambda'(\varphi_1 + \varphi_2) = \Lambda'(\varphi_1) \Lambda'(\varphi_2)$. An example of such a subgroup are the rotations around a fixed axis, with $\varphi$ the angle of rotation. Define $R'(\varphi) = R(\Lambda'(\varphi))$. Then the infinitesimal generator of this subgroup is defined as

$$\Sigma' = -i \left. \frac{dR'(\varphi)}{d\varphi} \right|_{\varphi=0}. \qquad (2.44)$$

This definition differs from the one current in the mathematical literature by the factor $-i$. In this way $\Sigma'$ is hermitian for unitary representations $R'$, which are of particular interest in quantum mechanics.

If $L_i$ is the group of rotations around the $i$-axis and $R_i$ a unitary representation of it, then its generator $\Sigma_i$ is a spin operator. To fix the ideas, consider the rotations around the 3-axis. Then

$$\Lambda_3(\varphi) = \begin{pmatrix} 1 & 0 & 0 & 0 \\ 0 & \cos\varphi & -\sin\varphi & 0 \\ 0 & \sin\varphi & \cos\varphi & 0 \\ 0 & 0 & 0 & 1 \end{pmatrix} \quad (2.45)$$

is the matrix representing the rotation by the angle $\varphi$. This $\Lambda_3(\varphi)$ is also the matrix representative of $\Lambda_3$ in the contravariant vector representation, and the corresponding infinitesimal generator is

$$\Sigma_3 = \begin{pmatrix} 0 & 0 & 0 & 0 \\ 0 & 0 & i & 0 \\ 0 & -i & 0 & 0 \\ 0 & 0 & 0 & 0 \end{pmatrix}. \quad (2.46a)$$

Analogously we find the other two spin operators in the vector representation

$$\Sigma_1 = \begin{pmatrix} 0 & 0 & 0 & 0 \\ 0 & 0 & 0 & 0 \\ 0 & 0 & 0 & i \\ 0 & 0 & -i & 0 \end{pmatrix}, \quad \Sigma_2 = \begin{pmatrix} 0 & 0 & 0 & 0 \\ 0 & 0 & 0 & -i \\ 0 & 0 & 0 & 0 \\ 0 & i & 0 & 0 \end{pmatrix}. \quad (2.46b)$$

For the spinor representation $S(\Lambda)$ in the form (2.43) one finds the simple result

$$\Sigma_i = \frac{1}{2} \begin{pmatrix} \sigma_i & 0 \\ 0 & \sigma_i \end{pmatrix}. \quad (2.47)$$

The matrices (2.46) have the eigenvalues 0 and $\pm 1$; the matrices (2.47) the eigenvalues $\pm 1/2$, and $\vec{\Sigma}^2$ the eigenvalues 2 and 3/4 respectively, corresponding to the spin values 1 and 1/2.

The "*Dirac matrices*" $\gamma^\mu$ are objects that combine vector and spinor properties. In our special spinor representation they are defined as

$$\gamma^0 = \begin{pmatrix} 0 & 1 \\ 1 & 0 \end{pmatrix}, \quad \gamma^i = \begin{pmatrix} 0 & -\sigma_i \\ \sigma_i & 0 \end{pmatrix}, \quad (2.48)$$

where again all the entries are $2\times 2$ matrices, so that the $\gamma^\mu$ are $4 \times 4$ matrices. They satisfy

$$S(\Lambda^{-1}) \gamma^\mu S(\Lambda) = \Lambda^\mu{}_\nu \gamma^\nu, \quad (2.49)$$

i.e. the $\gamma^\mu$ form a 4-vector with matrix-valued components. $\gamma^0$ is hermitian, the $\gamma^i$ are anti-hermitian:

$$\gamma^{0*} = \gamma^0, \quad \gamma^{i*} = -\gamma^i. \quad (2.50)$$

The anticommutators $\{\gamma^\mu, \gamma^\nu\} := \gamma^\mu\gamma^\nu + \gamma^\nu\gamma^\mu$ are

$$\{\gamma^\mu, \gamma^\nu\} = 2 g^{\mu\nu} \mathbf{1}. \quad (2.51)$$

Henceforth the factor **1** in this formula will not be written explicitly but will be understood tacitly. The same convention holds whenever formulas seemingly equating matrices or operators with numbers occur.

If going over to equivalent spinor representations as in (2.42), the $\gamma$-matrices must be transformed accordingly:

$$\gamma'^{\mu} = V \gamma^{\mu} V^{-1} . \tag{2.52}$$

The anticommutators (2.51) remain unchanged under this transformation, the hermiticity properties (2.50) only if $V$ is unitary.

## 2.3 Covariant Fields

Finally we need to define the notion of *covariant fields*. Let $R$ be one of the representations of $L$ or $L^{\uparrow}$ defined above, and $\mathcal{V}_R$ its representation space of dimension $n$. Let $\varphi(x)$ be a function on $M$ with values in $\mathcal{V}_R$. $\varphi(x)$ is called an *R-covariant field* if its components $\varphi'_\alpha(x')$ after the change of coordinates

$$x'^{\mu} = \Lambda^{\mu}{}_{\nu} x^{\nu} + a^{\mu}$$

are given by

$$\varphi'_\alpha(x') = \sum_{\beta=1}^{n} R_{\alpha\beta}(\Lambda) \varphi_\beta(x) . \tag{2.53}$$

Clearly this condition implies that the choice of basis in $\mathcal{V}_R$ must be related to the choice of coordinates in $M$. For the tensor representations there is an obvious canonical way of doing this: for the vectors we choose a basis $(e_0, \ldots, e_3)$ in $\mathcal{V}^{(1)}$ such that the differentials $dx^\mu$ are the contravariant components of the vector $dx \in \mathcal{V}^{(1)}$ relative to this basis. Bases of the higher tensor representations can be constructed from these $e_\mu$ by forming tensor products. For example, $e_{\mu\nu} = e_\mu \otimes e_\nu$ is a basis of $\mathcal{V}^{(2)}$.

For the spinors, such a canonical basis does not exist. This is related to the freedom (2.42) in the choice of representations. For given $\{x^\mu\}$ we may choose any basis in the spinor space $\mathcal{V}_S$ as the basis with respect to which the spinor field $\psi(x)$ is to be expanded. But we insist that, once a basis has been chosen for one coordinate system, the basis connected to another system $\{x'^\mu\}$ should be chosen such that (2.53) holds with $R = S$. This requirement is motivated as follows. Spinor fields by themselves have no direct observable meaning in quantum mechanics, in particular in QED. But let $\psi(x)$ be a spinor field, considered as a $1 \times 4$ matrix, so that

$$\psi'(x') = S(\Lambda) \psi(x) . \tag{2.54}$$

It can be shown that

$$S^*(\Lambda) \gamma^0 = \gamma^0 S^{-1}(\Lambda) . \tag{2.55}$$

## 2.3 Covariant Fields

We define the $4 \times 1$ matrix

$$\bar{\psi}(x) = \psi^*(x)\,\gamma^0 \tag{2.56}$$

and find

$$\bar{\psi}'(x') = \bar{\psi}(x)\,S^{-1}(\Lambda)\ . \tag{2.57}$$

This allows the construction of tensor fields out of spinor fields. For example, $\bar{\psi}(x)\psi(x)$ is a scalar field, and

$$j^\mu(x) = \bar{\psi}(x)\gamma^\mu\psi(x) \tag{2.58}$$

a vector field. Especially this latter field plays a vital role in QED, where the electric current density, an observable, is of this form.

Note that the word "covariant" as used in "covariant field" has a more general meaning than in "covariant tensor", denoting merely a collection of quantities transforming in a simple way under the Lorentz group. This notational confusion is deplorable but corresponds to current usage.

# Chapter 3

# Electrodynamics as a Classical Field Theory

The fundamental objects of a relativistic field theory are a finite set $\varphi^{(1)}, \ldots, \varphi^{(N)}$ of covariant fields, with components $\varphi_1^{(i)}, \ldots, \varphi_{d_i}^{(i)}$, each representing an independent degree of freedom, if for the moment we disregard the possible existence of subsidiary conditions. $d_i$ is the dimension of the corresponding representation $R^{(i)}$ of the Lorentz group. For ease of notation we at first number these $f = \sum d_i$ components consecutively: $\varphi_1, \ldots, \varphi_f$. They satisfy dynamical equations of motion, henceforth also called field equations, which are most simply characterized with the help of a Hamiltonian principle.

We use the symbols $\partial_\mu = \frac{\partial}{\partial x^\mu}$, $\varphi_{i,\mu} = \partial_\mu \varphi_i$. Let the Lagrangian $\mathcal{L}(x) = \mathcal{L}(\varphi_i(x), \varphi_{i,\mu}(x))$ be a function of the fields $\varphi_i$ and their first derivatives. The locality requirement, that only fields at the same point should interact, is the field theoretical expression of the relativistic prohibition of action at a distance.

Let $V$ be a bounded subset of the Minkowski space with a piecewise smooth boundary $\partial V$. Then the physically possible values of the fields are the values for which the action[1]

$$S_V = \int_V dx\, \mathcal{L}(x) \tag{3.1}$$

is stationary under infinitesimal variations $\delta \varphi_i$ which vanish at $\partial V$:

$$\delta S_V = \int_V dx\, \delta \mathcal{L}(x) = 0 \tag{3.2}$$

---

[1] Throughout this book the following notation will be used: in an integral over a 4-dimensional subset of Minkowski space the differential $\prod_{\mu=0}^{3} dx^\mu$ is denoted $dx$, but if we integrate only over 3-space we write $\prod_1^3 dx^i =: d^3 x$.

if $\delta\varphi_i = 0$ at $\partial V$ for all $i$. This requirement is equivalent, under suitable regularity conditions, to the Euler-Lagrange equations (= field equations),

$$\partial_\mu \frac{\partial \mathcal{L}}{\partial \varphi_{i,\mu}} - \frac{\partial \mathcal{L}}{\partial \varphi_i} = 0 , \quad i = 1, \ldots, f . \tag{3.3}$$

Lagrangians which differ only by a divergence,

$$\mathcal{L}_1(x) - \mathcal{L}_2(x) = \partial_\nu \Big( D^\nu(\varphi_i(x)) \Big) , \tag{3.4}$$

lead to the same field equations. We will take the point of view that the field equations are the real embodiment of the dynamical content of the theory, while the Lagrangian is merely a convenient bookkeeping device.

The theory is relativistically invariant if its field equations are invariant. This means the following. Let the fields $\varphi^{(1)}(x) \ldots \varphi^{(N)}(x)$ solve the field equations. Then their Poincaré transforms $\varphi'^{(i)}(x') = R^{(i)}(\Lambda) \varphi^{(i)}(x)$, with $x' = \Lambda x + a$, solve the same field equations. A sufficient condition for this to be the case is that the Lagrangian $\mathcal{L}(x)$ is a scalar field.

*Classical electrodynamics* is the theory of the electromagnetic field $F^{\mu\nu}(x)$, a real, antisymmetric tensor of rank 2. Its definition in terms of the electric and magnetic field strengths $\vec{E}(x)$ and $\vec{B}(x)$ is

$$(F^{\mu\nu}) = \begin{pmatrix} 0 & -E_1 & -E_2 & -E_3 \\ E_1 & 0 & -B_3 & B_2 \\ E_2 & B_3 & 0 & -B_1 \\ E_3 & -B_2 & B_1 & 0 \end{pmatrix} , \tag{3.5}$$

where $\mu$ numbers the lines, $\nu$ the columns of this matrix. This tensor satisfies the Maxwell equations

$$\partial_\nu F^{\nu\mu}(x) = j^\mu(x) , \tag{3.6}$$

$$\partial_\mu F_{\nu\rho}(x) + \partial_\nu F_{\rho\mu}(x) + \partial_\rho F_{\mu\nu}(x) = 0 . \tag{3.7}$$

The current density $j^\mu$ must satisfy the continuity equation

$$\partial_\mu j^\mu = 0, \tag{3.8}$$

so that the total charge

$$Q = \int_{x^0=t} d^3x \, j^0(x) \tag{3.9}$$

is conserved, i.e. independent of $t$. We assume that $j^\mu(x)$ vanishes sufficiently rapidly for $|\vec{x}| \to \infty$ to make this integral converge.

In truly classical electrodynamics the current $j^\mu$ describes the movement of charged matter, e.g. of charged particles. But as a "classical" starting point for developing QED we need a fully field theoretical formulation: a half-classical theory which treats the electromagnetic field classically, but the particles, in our case electrons, quantum mechanically, more exactly wave-mechanically. Since electrons carry spin 1/2, and since we need a relativistic theory, the wave functions of the particles are chosen to be a Dirac spinor $\psi(x)$ and its adjoint in the guise of the field $\bar{\psi}(x) = \psi^*(x)\gamma^0$. The asterisk denotes taking the adjoint matrix, as introduced in Chap. 2.

The field equations of the theory cannot be written down easily in terms of $\psi$, $\bar{\psi}$ and $F^{\mu\nu}$. We need to introduce the *electromagnetic potentials* $A^\mu(x)$. The homogeneous equation (3.7) implies that four real-valued functions $A^\mu(x)$ can be found such that

$$F^{\mu\nu}(x) = \partial^\mu A^\nu(x) - \partial^\nu A^\mu(x) \tag{3.10}$$

holds, with $\partial^\mu = g^{\mu\nu}\partial_\nu$. With this ansatz the inhomogeneous equation (3.6) becomes

$$\Box A^\mu - \partial^\mu \partial_\nu A^\nu = j^\mu , \tag{3.11}$$

with $\Box = \partial_\nu \partial^\nu$ the wave operator. The desired classical field equations are derived from the Lagrangian

$$\mathcal{L}(x) = \mathcal{L}_0^A(x) + \mathcal{L}_0^\psi(x) + \mathcal{L}_{\text{int}}(x) , \tag{3.12}$$

with

$$\begin{aligned}
\mathcal{L}_0^A(x) &= -\frac{1}{4} F_{\mu\nu}(x) F^{\mu\nu}(x) , \\
\mathcal{L}_0^\psi(x) &= \bar{\psi}(x)(i\slashed{\partial} - m)\psi(x) , \\
\mathcal{L}_{\text{int}}(x) &= -e\,\bar{\psi}(x)\,\slashed{A}(x)\,\psi(x) .
\end{aligned} \tag{3.13}$$

In these formulae $F^{\mu\nu}$ must be considered as an abbreviation of the expression (3.10), and for any 4-vector $v$ we define $\slashed{v} = v_\mu \gamma^\mu$. $m$ is the mass of the electron, $e$ the elementary charge quantum: the charge of the positron. In applying the variational principle (3.2), the $A^\mu$, not the $F^{\mu\nu}$, must be taken as the independent variables, and the components $\bar{\psi}_\rho$ must be varied independently of the $\psi_\sigma$, despite the fact that $\bar{\psi}$ is uniquely defined in terms of $\psi$. A cleaner way of proceeding would be to treat the real and imaginary parts of $\psi$ as fields to be varied independently. But the result would be the same, namely

$$\Box A^\mu(x) - \partial^\mu \partial_\nu A^\nu(x) = e\,\bar{\psi}(x)\gamma^\mu\psi(x) , \tag{3.14}$$

$$\begin{aligned}
\left(i\slashed{\partial} - e\,\slashed{A}(x) - m\right)\psi(x) &= 0 , \\
\bar{\psi}(x)\left(i\overleftarrow{\slashed{\partial}} + e\,\slashed{A}(x) + m\right) &= 0 .
\end{aligned} \tag{3.15}$$

The arrow $\leftarrow$ in $\overleftarrow{\slashed{\partial}}$ indicates that the derivation $\partial_\mu$ acts to the left, on the function $\bar{\psi}(x)$. Comparison of (3.14) with (3.11) shows that (3.14) is the Maxwell equation with

$$j^\mu(x) = e\,\bar{\psi}(x)\gamma^\mu\psi(x). \tag{3.16}$$

With the help of the Dirac equations (3.15) this current can easily be shown to satisfy the continuity equation.

Note that with the definition (3.16) the charge density $j^0$ is non-negative, hence cannot describe *two* particles with opposite charges. This difficulty is related to the well-known problem of the negative-energy solutions of the Dirac equation. A Dirac spinor as a classical field is a dubious object, physically speaking. It finds its proper place only in quantum field theory. But for the moment we need not worry about this problem, which will be duly solved in the quantum version of the theory.

The relation (3.10) does not define the potentials $A^\mu$ uniquely. Let $A^\mu$ be a solution of (3.10) for given $F^{\mu\nu}$, and let $G(x)$ be an arbitrary, real, twice differentiable function. Then

$$A'^\mu(x) = A^\mu(x) + \partial^\mu G(x) \tag{3.17}$$

produces the same fields $F^{\mu\nu}$ as $A^\mu$. The mapping $A^\mu \to A'^\mu$ is called a *gauge transformation*. The field equations (3.14), (3.15) remain satisfied if the fields $\psi, \bar\psi$ are changed along with $A^\mu$, to

$$\psi'(x) = \psi(x)\,e^{-ieG(x)}, \quad \bar\psi'(x) = \bar\psi(x)\,e^{ieG(x)}. \tag{3.18}$$

Since the potentials $A^\mu$ are only auxiliary quantities without a direct physical significance, gauge transformations must not change the observational content of the theory. This means in particular that only gauge invariant functionals of the fields $A, \psi, \bar\psi$ can be observable quantities. Examples of such invariants are $F^{\mu\nu}$ and $j^\mu$.

Because of the gauge ambiguity, the $A^\mu$ are not necessarily components of a covariant field, even if the $F^{\mu\nu}$ are. Also, the vast number of solutions which the field equations possess in consequence of their gauge invariance, makes them mathematically cumbersome. For instance, they are not deterministic: prescribing the values of $A, \psi, \bar\psi$ and a sufficient number of their derivatives at a given time $x^0 = t$ does not fix the solution uniquely, because a solution satisfying these initial conditions can always be changed into another one by a gauge transformation with a gauge function $G$ which vanishes in a neighbourhood of $\{x^0 = t\}$. It is therefore convenient for many purposes to restrict the set of possible solutions by imposing a *gauge condition*. This is an additional requirement on the fields, which is chosen such that it does not restrict physical generality, i.e. such that from any solution not satisfying the condition another one satisfying it can be constructed by applying a suitable gauge transformation.

Two popular choices of such gauge conditions are
(a) the *Coulomb condition*

$$\partial_i A^i(x) = 0, \tag{3.19}$$

(b) the *Lorentz condition*

$$\partial_\mu A^\mu(x) = 0. \tag{3.20}$$

The Coulomb condition is not Lorentz invariant, so that the corresponding $A^\mu$ do not form a covariant field. Such a choice leads to serious problems in the development of QED, especially in connection with renormalization (a notion that will be explained in due course). We will therefore work with the Lorentz condition (3.20). This condition does not yet fix the gauge uniquely: it still allows gauge transformations with gauge functions satisfying $\Box G(x) = 0$. As a consequence, the covariance of $A$ is not guaranteed. However, contravariant $A^\mu$'s satisfying condition (3.20) *can* be found if the $F^{\mu\nu}$ transform contravariantly. We will insist on using only such potentials until a rather late stage of our construction.

After adding condition (3.20) to the system of field equations, the Maxwell equation (3.14) can be replaced by the simpler form

$$\Box A^\mu = j^\mu , \tag{3.21}$$

or more generally by

$$\Box A^\mu - \lambda \partial^\mu \partial_\nu A^\nu = j^\mu , \tag{3.22}$$

with an arbitrary real number $\lambda$. This latter freedom does not seem to be very helpful in the present classical context. But it will make sense in the quantum version of the theory.

The theory described above is only covariant under the orthochronous Lorentz group. From the physical definitions of the measurable fields $\vec{E}, \vec{B}, j^\mu$ it is clear that they transform under the *time reversal* $T: x'^0 = -x^0, x'^i = x^i$ as

$$\begin{aligned} j'^0(x') &= j^0(x) , & j'^i(x') &= -j^i(x) , \\ F'^{ij}(x') &= -F^{ij}(x) , & F'^{0i}(x') &= F^{0i}(x) , \end{aligned} \tag{3.23}$$

i.e. they acquire an extra factor $-1$ relative to a proper covariant behaviour. But they are not pseudotensors either, because this additional sign is absent in the parity operation. The spinor representation $S$ is in any case only defined for $L^\uparrow$. Nevertheless, the $T$-operation can be represented inside the formalism as follows. Define $T$ on $A^\mu$ and $\psi$ by

$$A'^0(x') = A^0(x) , \quad A'^i(x') = -A^i(x) , \tag{3.24}$$

$$\psi'(x') = \gamma^1 \gamma^3 \psi^c(x) , \tag{3.25}$$

where the superscript $c$ denotes complex conjugation and the $\gamma$-matrices (2.48) are used. Using the special reality and symmetry properties of these matrices, it can easily be ascertained that $j^\mu = e\bar{\psi}\gamma^\mu \psi$ transform according to (3.23), and that the field equations (3.14), (3.15) are preserved. The special form of the action of $T$ on $\psi$ used here is not unique. Other, similar, prescriptions achieving the same purpose exist.

We conclude the chapter with a remark on possible alternatives to the choice of the Lagrangian (3.12). For practical reasons we want $\mathcal{L}$ to be a polynomial in the basic fields and their first derivatives, and we want it to be of the additive form (3.12). These are essentially the only Lagrangians that we know to handle with any confidence in quantum field theory.[2] $\mathcal{L}(x)$ should be a scalar in order to ensure relativistic invariance, and it should be gauge invariant. The free part $\mathcal{L}_0^A + \mathcal{L}_0^\psi$ is pretty much fixed, up to divergence terms, by the properties of the particles (photons and electrons) that are described by the corresponding free fields (see below in Chap. 5). The $\mathcal{L}_{\text{int}}$ given in (3.13) is then the simplest ansatz satisfying all the requirements. In fact, it is the only possibility leading to a "renormalizable" quantum field theory, i.e. a theory which depends only on a finite number of parameters (masses and coupling constants) that must be determined by measurement. The final justification of the choice (3.13) lies of course in its empirical success.

A possible generalization of $\mathcal{L}_{\text{int}}$ would consist in the addition of a "Pauli term" $ig\, F_{\mu\nu}\, \bar{\psi}[\gamma^\mu, \gamma^\nu]\psi$ to the expression (3.13), which would preserve Lorentz and gauge invariance. But it leads to a non-renormalizable generalization of QED. Nevertheless, this addition can be of some practical use, e.g. for describing in a summary way the effects of the anomalous magnetic moments of the nucleons, i.e. the parts of these moments that are due to the strong interaction and are therefore not explained by QED. But in such applications the Pauli term must be treated as an effective interaction, a rough approximation that should be taken into account only to lowest order in the coupling constant $g$.

---

[2] Exceptions are certain lower-dimensional theories like e.g. the sine-Gordon model. But here we are exclusively concerned with 4-dimensional space-time.

# Chapter 4

# The Basic Principles of Relativistic Quantum Field Theory

This chapter still provides background material, which will be described but not proved. Proofs can be found in the texts cited below, or are easy generalizations of proofs found there.

## 4.1 The Wightman Axioms

In the quantized form of relativistic field theory the fields $\varphi_\alpha^{(i)}(x)$ of Chap. 3 are operator-valued functions. The general properties of these functions and of the space on which they act have been codified in the postulates of "axiomatic field theory". These postulates are partly based on clear physical requirements, partly they are technical assumptions motivated by extensive practical experience in the handling of such theories. Our starting point is the list of postulates given by Wightman [StW, J, BLOT]. However, it has been realized from the start that QED, the paradigm of a successful quantum field theory, does not quite fit into the Wightman framework. We will therefore use the postulates in a generalized form adapted to the needs of QED. These postulates will now be enumerated and discussed.

**Postulate 1: Quantum mechanics.** *The observables and other quantities of physical interest, in particular the fields, are represented by linear operators on a complex vector space $\mathcal{V}$ which is equipped with a non-degenerate, hermitian, scalar product.*

The scalar product associates to every pair $v, w \in \mathcal{V}$ a complex number $(v, w)$ which depends linearly on $w$, antilinearly on $v$, and which satisfies

$$(v, w) = (w, v)^* , \qquad (4.1)$$

the asterisk denoting complex conjugation. Non-degeneracy means that $v = 0$ is the only element of $\mathcal{V}$ which is orthogonal to all of $\mathcal{V}$:

$$(v, w) = 0 \; \forall \; w \in \mathcal{V} \quad \Longrightarrow \quad v = 0 . \qquad (4.2)$$

Positivity of this scalar product is not assumed. There may exist vectors $v \in \mathcal{V}$ with $(v, v) < 0$, and $(v, v) = 0$ does not imply $v = 0$.

In $\mathcal{V}$ we introduce a weak notion of convergence. The sequence $w_1, w_2, \ldots, w_i, \ldots$, of elements of $\mathcal{V}$ is said to converge to $w \in \mathcal{V}$ if and only if

$$\lim_{i \to \infty} (v, w_i) = (v, w) \; \forall \, v \in \mathcal{V} . \qquad (4.3)$$

$\mathcal{V}$ is not assumed to be complete with respect to this notion of convergence.

Since positive observables, in particular the trivial "observable" $\mathbf{1}$ (=identity), must have positive expectation values in physical states, states with $(v, v) \leq 0$ cannot represent physically realizable states. The characterization of the physical states is one of the major problems of QED.

The fundamental fields of a field theory are not necessarily observables, hence the reference to "other quantities of physical interest". In QED the field strengths $F^{\mu\nu}$ and the currents $j^\mu$ are observables, the potentials $A^\mu$ and the spinors $\psi, \bar\psi$ are not.

**Postulate 2: Relativistic invariance.** *On $\mathcal{V}$ are defined a continuous unitary representation $T(a)$ of the translation group and a continuous unitary representation $U(\Lambda)$ of the orthochronous Lorentz group such that $U(\Lambda, a) = T(a) U(\Lambda)$ forms a representation of the orthochronous Poincaré group. The representation $U(\Lambda)$, and therefore $U(\Lambda, a)$, may be two-valued on a linear subspace $\mathcal{V}_-$ of $\mathcal{V}$.*

Unitarity of operators on $\mathcal{V}$ is defined with respect to the scalar product $(\cdot, \cdot)$. The operator $U$ is said to be unitary if it is invertible and if

$$(Uv, Uw) = (U^{-1}v, U^{-1}w) = (v, w) \qquad (4.4)$$

for all $v, w \in \mathcal{V}$. The adjoint $A^*$ of the operator $A$ is also defined relative to $(\cdot, \cdot)$:

$$(A^* v, w) = (v, Aw) \quad \forall \; v, w \in \mathcal{V} . \qquad (4.5)$$

Due to the defining relation (2.16) of a representation we have

$$T^{-1}(a) = T^*(a) = T(-a) , \quad U^{-1}(\Lambda) = U^*(\Lambda) = U(\Lambda^{-1}) . \qquad (4.6)$$

Convergence of a sequence $\{\Lambda_i\}$ of Lorentz transformations is defined as convergence of its coefficents $(\Lambda_i)^\mu{}_\nu$, and analogously for translations. Continuity of the representation $U(\Lambda)$ means that

$$\lim_{i \to \infty} U(\Lambda_i) v = U(\Lambda) v \quad \forall \; v \in \mathcal{V} \qquad (4.7)$$

if $\lim_{i \to \infty} \Lambda_i = \Lambda$, and analogously for translations.

The admission of two-valuedness for $U(\Lambda)$ is necessary in QED because of the two-valuedness of the spinor representation $S$. The space $\mathcal{V}_-$ will be the space of states with odd fermion number (see below).

It is well known that Lorentz transformations which reverse the time direction cannot be implemented by unitary operators. Henceforth the term "Lorentz transformation" will always denote an orthochronous transformation, and $L$ the orthochronous Lorentz group, unless noted otherwise.

**Postulate 3: Existence and uniqueness of the vacuum.** *In $\mathcal{V}$ there exists a state $\Omega$, unique up to multiplication with a complex number, which is invariant under translations:*

$$T(a)\,\Omega = \Omega \quad \forall\, a \in M \,. \tag{4.8}$$

*The scalar product of $\Omega$ with itself is positive, so that it can be normalized to 1,*

$$(\Omega, \Omega) = 1 \,. \tag{4.9}$$

*Moreover, for any test function $\chi(\vec{a}) \in \mathcal{S}$ and any state $\Phi \in \mathcal{V}$ we have*

$$\lim_{\lambda \to 0} \lambda^3 \int d^3 a \, T(\vec{a})\,\chi(\lambda\vec{a})\,\Phi = (\Omega, \Phi) \int d^3 a \, \chi(\vec{a})\,\Omega \tag{4.10}$$

*in every Lorentz frame.*

$\mathcal{S}$ is the Schwartz space of tempered test functions (see below Postulate 4 for its definition and motivation). This choice of admissible $\chi$ is unnecessarily restrictive, but it fits well into the general formalism, as we will see.

The condition (4.10) is called *"weak clustering"*. It is based on the assumption, which will become clearer after the formulation of the next postulate, that any state in $\mathcal{V}$ differs from the vacuum appreciably only in a finite region of space. I.e. every state is at a sufficiently large distance indistinguishable from the vacuum within a given experimental accuracy. But the $\lambda$-limit in (4.10) spreads this deviation from the vacuum over the whole space, thereby diluting it to unrecognizability: the state is locally no longer distinguishable from the vacuum, and only this local aspect determines its scalar product with an ordinary, undiluted state.

The requirement that the $U(\Lambda, a) = T(a)U(\Lambda)$ form a representation of the Poincaré group implies that $\Omega$ is also invariant under Lorentz transformations:

$$U(\Lambda)\,\Omega = \Omega \quad \forall \; \Lambda \in L \,. \tag{4.11}$$

Here we have assumed that $\Omega$ is orthogonal to $\mathcal{V}_-$.

**Postulate 4: Field theory.** *On $\mathcal{V}$ exist a finite number $\varphi^{(1)}(x), \ldots, \varphi^{(N)}(x)$ of covariant operator-valued fields which generate the full space $\mathcal{V}$ from the vacuum $\Omega$.*

This statement needs some explanation. Firstly, covariance of $\varphi^{(i)}$, with components $\varphi_\alpha^{(i)}$, means that

$$\varphi_\alpha^{(i)}(x+a) = T(a)\,\varphi_\alpha^{(i)}(x)\,T^*(a) \tag{4.12}$$

and that there is a finite-dimensional representation $R^{(i)}(\Lambda)$ of $L$ such that

$$\varphi_\alpha^{(i)}(\Lambda x) = \sum_\beta R_{\alpha\beta}^{(i)}(\Lambda)\,U(\Lambda)\,\varphi_\beta^{(i)}(x)\,U^*(\Lambda)\,, \tag{4.13}$$

where $T$ and $U$ are the representations introduced in Postulate 2.

Secondly it must be noted that the fields $\varphi_\alpha^{(i)}(x)$ are not functions in the familiar sense of the word, but distributions (= generalized functions). This means that the value of $\varphi_\alpha^{(i)}$ at a given point $x$ is not defined. Defined are only averages

$$\varphi_\alpha^{(i)}(f) = \int dx\,\varphi_\alpha^{(i)}(x)\,f(x) \tag{4.14}$$

over sufficiently well-behaved "test functions" $f(x)$. The exact characterization of the admissible test functions depends on the particular distribution in question. We demand that our fields be tempered distributions, i.e. they are defined on test functions $f$ from the Schwartz space $\mathcal{S}$ (see e.g. [StW] and [C]). $f(x)$ is an element of $\mathcal{S}$ if it is infinitely differentiable and if $x^\alpha D^\beta f(x)$ is bounded on $M$ for every monomial

$$x^\alpha = \prod_{\mu=0}^{3}(x^\mu)^{\alpha_\mu}\,, \quad \alpha_\mu \geq 0\,,$$

and every derivation

$$D^\beta = \prod \partial_\mu^{\beta_\mu}\,, \quad \beta_\mu \geq 0\,.$$

$\alpha = \{\alpha_0,\ldots,\alpha_3\}$ and $\beta = \{\beta_0,\ldots,\beta_3\}$ are called multi-indices.

That the fields cannot be ordinary functions will be seen in the next chapter, which describes the theory of free fields. The introduction of an interaction does not improve the situation. We insist on temperedness of the fields in order to have the well developed theory of Fourier transforms for tempered distributions at our disposal, which will play a fundamental role in our undertaking. The choice of $\mathcal{S}$ as the space of admissible test functions is not dictated by physical requirements. But $\mathcal{S}$ is mathematically very convenient, and we have learned from experience that it provides a suitable framework for field theories, especially QED. For ease of notation, we will formally write the fields and other distributions as functions: $\varphi(x)$, etc., with the understanding that really expressions of the type (4.14) should be considered. We will be careful not to abuse this sloppy notation by turning it into sloppy mathematics. Especially we will be careful to avoid multiplying distributions whose product does not exist, except for heuristic reasons.

## 4.1 The Wightman Axioms

Thirdly, that the fields generate $\mathcal{V}$ out of $\Omega$ means that every vector $v \in \mathcal{V}$ is a finite sum over "monomial states"

$$v_\alpha^I(f) = \int \prod_{h=1}^n dx_h\, f(x_1,\ldots,x_n) \prod_{h=1}^n \varphi_{\alpha_h}^{(i_h)}(x_h)\, \Omega\,, \tag{4.15}$$

with $f \in \mathcal{S}$. $I = \{i_h\}$ and $\alpha = \{\alpha_h\}$ are multi-indices. The factors in $\prod_h$ must occur in a prescribed order, since different fields do not necessarily commute. We demand that $v_\alpha^I$ depends continuously on $f$. This means that if $v_\alpha^I$ is formed with the test functions $f_\nu$, the fields being chosen independent of $\nu$, then

$$\lim_{\nu \to \infty} v_\alpha^I(f_\nu) = 0 \tag{4.16}$$

if

$$\lim_{\nu \to \infty} x^\beta D^\gamma f_\nu(x_1,\ldots,x_n) = 0 \tag{4.17}$$

uniformly in $(x_1,\ldots,x_n)$, for every choice of the multi-indices $\beta, \gamma$. $x^\beta$ and $D^\gamma$ are defined as

$$x^\beta = \prod_{h,\mu} (x_h{}^\mu)^{\beta_{h\mu}}\,,\ D^\gamma = \prod_{h,\mu} (\partial_{h,\mu})^{\gamma_{h\mu}}\,.$$

The fundamental fields of QED in the sense of Postulate 4 are the potentials $A$ and the spinors $\psi$, $\bar\psi$.

We define the Fourier transform $\tilde f(p)$ of the test function $f(x)$ by

$$\tilde f(p) = (2\pi)^{-\frac{3}{2}} \int dx\, e^{-ipx}\, f(x)\,, \tag{4.18}$$

with the inverse transformation

$$f(x) = (2\pi)^{-\frac{5}{2}} \int dp\, e^{ipx}\, \tilde f(p)\,. \tag{4.19}$$

If $f \in \mathcal{S}$, then $\tilde f \in \mathcal{S}$. The Fourier transform $\tilde\varphi_\alpha^{(i)}(p)$ of the field $\varphi_\alpha^{(i)}$ is formally written as

$$\tilde\varphi_\alpha^{(i)}(p) = (2\pi)^{-\frac{5}{2}} \int dx\, e^{ipx}\, \varphi_\alpha^{(i)}(x)\,. \tag{4.20}$$

This expression must be interpreted as Fourier transform in the sense of distributions. That is, the value of $\tilde\varphi$ on the test function $\tilde f(p) \in \mathcal{S}$ is defined as

$$\int dp\, \tilde\varphi(p)\, \tilde f(p) = \int dx\, \varphi(x)\, f(x)\,. \tag{4.21}$$

The inverse of (4.19) is

$$\varphi_\alpha^{(i)}(x) = (2\pi)^{-\frac{3}{2}} \int dp\, e^{-ipx}\, \tilde\varphi_\alpha^{(i)}(p)\,. \tag{4.22}$$

The monomial states (4.15) can also be written as

$$v_\alpha^I(f) = \int \prod_{h=1}^n dp_h \, \tilde{f}(p_1,\ldots,p_n) \prod_{h=1}^n \tilde{\varphi}_{\alpha_h}^{(i_h)}(p_h) \, \Omega \;, \qquad (4.23)$$

with

$$\tilde{f}(p_1,\ldots,p_n) = (2\pi)^{-\frac{3}{2}n} \int \prod_h \left(dx_h \, e^{-ip_h x_h}\right) f(x_1,\ldots,x_n) \;. \qquad (4.24)$$

We define energy-momentum operators $P_\mu$ on $\mathcal{V}$ by

$$P_\mu \Omega = 0 \;, \qquad \left[P_\mu, \tilde{\varphi}_\alpha^{(i)}(p)\right] = -p_\mu \, \tilde{\varphi}_\alpha^{(i)}(p), \qquad (4.25)$$

so that

$$P_\mu v_\alpha^I(f) = -\int \prod_h dp_h \, \tilde{f}(p_1,\ldots,p_n) \sum_h p_{h\mu} \prod_h \tilde{\varphi}_{\alpha_h}^{(i_h)}(p_h) \, \Omega, \qquad (4.26)$$

which definition can be extended to all of $\mathcal{V}$ by linearity. The $P_\mu$ are hermitian. The second equation in (4.25) reads in $x$-space

$$\left[P_\mu, \varphi_\alpha^{(i)}(x)\right] = -i\partial_\mu \varphi_\alpha^{(i)}(x) \;, \qquad (4.27)$$

which shows that $P_\mu$ generates infinitesimal translations.

A condition demanding positivity of the energy in every Lorentz frame can then be formulated as follows.

**Postulate 5: Spectral property.** *Define $v_\alpha^I(f)$ by (4.23). Then*

$$v_\alpha^I(f) = 0 \qquad (4.28)$$

*if $\tilde{f}(p_1,\ldots,p_n) \equiv 0$ for $\sum_i p_i \in V_-$, the open backward cone.*

Note that $\Omega$ is the only eigenstate of $P_\mu$ with eigenvalues 0, so that

$$v = (\Omega, v)\Omega \quad \text{if} \quad P_\mu v = 0 \;\forall\; \mu. \qquad (4.29)$$

Equation (4.10) implies

$$\lim_{\lambda \to \infty} \int \prod_h dp_h \, \tilde{f}(p_1,\ldots,p_n) \, \gamma\!\left(\lambda \sum \vec{p}_h\right) \prod \tilde{\varphi}_{\alpha_h}^{(i_h)}(p_h) \, \Omega = \gamma(0) \left(\Omega, v_\alpha^I(f)\right) \Omega \qquad (4.30)$$

for $\gamma \in \mathcal{S}$.

**Postulate 6: Locality.** *Fields at spacelike separated points either commute or anticommute,*

$$[\varphi_\alpha^{(i)}(x), \varphi_\beta^{(j)}(y)]_\pm = 0 \text{ if } (x-y)^2 < 0 \ . \tag{4.31}$$

*The upper sign applies if both $\varphi^{(i)}$ and $\varphi^{(j)}$ transform under two-valued representations of the Lorentz group, the lower sign in the other cases.*

We define $[A, B]_\pm = AB \pm BA$ for any two operators $A, B$. Alternative notations are $[A, B]_- = [A, B]$ for the commutator and $[A, B]_+ = \{A, B\}$ for the anticommutator.

The only two-valued representation of $L$ that has been discussed in Chap. 3 is the Dirac spinor representation $S$, because it is the only one relevant to QED. There are others, connected to half-integer spins higher than $1/2$. The irreducible uni-valued $R^{(i)}$ belong to integer spins. Fields satisfying local commutativity are called Bose fields, those satisfying local anticommutativity, Fermi fields. The subspace $\mathcal{V}_-$ of Postulate 2 is spanned by the monomial states with an odd number of Fermi fields.

The condition (4.31) of local commutativity has its origin in the canonical formulation of quantum field theory, an extension to systems with infinitely many degrees of freedom of the familiar canonical formalism of elementary quantum mechanics. But the condition can also be motivated more generally. Let $G \subset M$ be a domain, i.e. an open, connected subset of Minkowski space. A quantum mechanical observable $\mathcal{O}$ is said to be localized in $G$ if it can be measured by a procedure executed in $G$. The symbol $\mathcal{O}$ stands for both the physical observable and the operator representing it in the mathematical formalism. Because the fundamental fields $\varphi^{(i)}$ should describe the theory fully, we assume that $\mathcal{O}$ can be written as a function of the fields $\varphi_\alpha^{(i)}(x)$ and their derivatives at $x \in G$. For simplicity we restrict our attention to polynomials, because more general functions of operators are difficult to define. Let now $G_1, G_2$ be two domains which are causally disjoint, meaning that $(x-y)^2 < 0$ for all $x \in G_1$ and $y \in G_2$. Let $\mathcal{O}_1, \mathcal{O}_2$ be observables localized in $G_1$ and $G_2$ respectively. The measurement of $\mathcal{O}_1$ then cannot disturb the value of $\mathcal{O}_2$ in any way, and vice versa: $\mathcal{O}_1$ and $\mathcal{O}_2$ are compatible observables, so that

$$[\mathcal{O}_1, \mathcal{O}_2] = 0 \tag{4.32}$$

must hold. We call this condition the condition of *relativistic causality*, or shorter but less to the point: *locality*. For observable fields like $F^{\mu\nu}$ it implies at once the relation (4.31) with the lower sign. But Fermi fields cannot be observables for several reasons, so that an equally simple conclusion cannot be drawn for them. However, (4.32) is satisfied if condition (4.31) holds and observables are polynomials containing only terms of even degree in the Fermi fields. The reverse is not true: (4.31) is not a consequence of (4.32). In studying the physical states of QED we will actually have occasion to introduce fields not satisfying locality. But in establishing the basic mathematical formalism of the theory there is no need to forgo this simple way of satisfying relativistic causality.

That the signs in (4.31) must be chosen as indicated, assuming that relations of the form (4.31) hold at all, is the content of the celebrated *spin–statistics theorem*. It cannot be proved from the postulates as formulated here but only from a stricter version, to be discussed presently, assuming positivity of the scalar product $(\cdot,\cdot)$. But in our formulation we need locality in the stated form, because otherwise the construction of a meaningful physical state space cannot be accomplished.

The enumeration of the basic postulates is now complete. They are related to Wightman's original, more restrictive formulation as follows. Assume that the scalar product $(v,w)$ is positive definite,

$$(v,v) > 0 \quad \text{for all} \quad v \in \mathcal{V}, \; v \neq 0 \, . \tag{4.33}$$

We can then define on $\mathcal{V}$ a norm

$$||v|| = (v,v)^{\frac{1}{2}} \, , \tag{4.34}$$

with the usual properties. $\mathcal{V}$ can be completed in the topology induced by this norm to a Hilbert space $\mathcal{H}$ in which $\mathcal{V}$ is dense. The unitary operators $U(\Lambda, a)$ can be extended to unitary operators on $\mathcal{H}$ which still form a continuous representation of the Poincaré group. Its subrepresentation $T(a)$ of the translation group can be written by the SNAG theorem [StW], a generalization of Stone's theorem, as

$$T(a) = e^{ia^\mu P_\mu} \, , \tag{4.35}$$

with $P_\mu$ a self-adjoint extension of the original momentum operator defined on $\mathcal{V}$. The spectral property (4.28) implies that the joint spectrum $\{p_\mu\}$ of the four commuting operators $P_\mu$ is contained in the closed forward cone $\bar{V}_+$. Condition (4.10) becomes a theorem and need not be demanded separately. The field operators $\varphi_\alpha^{(i)}$ are defined on the dense subset $\mathcal{V}$. They can possibly be continued to larger domains of definition, such that their covariance and locality properties remain satisfied. In this way the original Wightman axioms are recovered: if the scalar product $(\cdot,\cdot)$ is positive definite, then our postulates are equivalent to Wightman's.

## 4.2 The Wightman Functions

We turn to the discussion of the *Wightman functions*, an essential tool of our formalism. First we need to introduce some notation. Together with $\varphi^{(i)}$ its adjoint $\varphi^{(i)*}$ is also one of the basic fields of the theory. Let it be numbered $\varphi^{(j)}$. We define $i^* = j$. For a hermitian field we have $i^* = i$. Like in (4.15) we associate to the field product $\varphi_{\alpha_1}^{(i_1)}(x_1) \cdots \varphi_{\alpha_n}^{(i_n)}(x_n)$ the multi-indices

$$I = (i_n, \ldots, i_n), \qquad \alpha = (\alpha_1, \ldots, \alpha_n) \, , \tag{4.36}$$

and define
$$I^* = (i_n^*, \ldots, i_1^*), \alpha^* = (\alpha_n, \ldots, \alpha_1) . \tag{4.37}$$

Note that the component indices are not necessarily simple numbers. In the case of tensor fields they may be sets of Minkowski indices, possibly including contravariant ones irrespective of $\alpha_i$ being written as a subscript. The order and position (up or down) of such internal indices inside $\alpha_i$ is not changed in going over to $\alpha^*$. Whenever the exact nature of $\alpha_i$ counts, it will be properly identified.

The Wightman function $W_\alpha^I$ is defined as
$$W_\alpha^I(x_1, \ldots, x_n) = \left(\Omega, \varphi_{\alpha_1}^{(i_1)}(x_1) \cdots \varphi_{\alpha_n}^{(i_n)}(x_n)\Omega\right) . \tag{4.38}$$

Of course, the $W_\alpha^I$ are not functions in the strict sense of the word, but tempered distributions.

From the postulates 1–6 the following properties, called "W-properties", can be derived.

**Property W1: Hermiticity.** *The relation*
$$\left(W_\alpha^I(x_1, \ldots, x_n)\right)^* = W_{\alpha^*}^{I^*}(x_n, \ldots, x_1) \tag{4.39}$$

*holds.*

In QED it is convenient to use $\bar\psi = \psi^* \gamma^0$ instead of $\psi^*$ as one of the basic fields. Written in these fields the relation (4.39) has a more complicated but obvious form (see Sect. 7.1).

**Property W2: Covariance.** $W_\alpha^I$ *transforms under the Poincaré transformation* $(\Lambda, a)$ *as*
$$W_\alpha^I(\Lambda x_1 + a, \ldots, \Lambda x_n + a) = \prod_{i=1}^n \sum_{\beta_i} R_{\alpha_i \beta_i}^{(i)}(\Lambda) W_\beta^I(x_1, \ldots, x_n). \tag{4.40}$$

As a consequence of translation invariance $W_\alpha^I$ can be written as a function of the $(n-1)$ 4-vectors $\xi_i = x_i - x_{i+1}$, $i = 1, \ldots, n-1$:
$$W_\alpha^I(x_1, \ldots, x_n) = w_\alpha^I(\xi_1, \ldots, \xi_{n-1}) . \tag{4.41}$$

The Fourier transforms
$$\tilde W_\alpha^I(p_1, \ldots, p_n) = \left(\Omega, \tilde\varphi_{\alpha_1}^{(i_1)}(p_1) \ldots \tilde\varphi_{\alpha_n}^{(i_n)}(p_n)\Omega\right)$$
$$= (2\pi)^{-\frac{5}{2}n} \int \prod_1^n dx_j \, \exp\left(i \sum p_j x_j\right) W_\alpha^I(x_1, \ldots, x_n) ,$$
$$\tilde w_\alpha^I(q_1, \ldots, q_{n-1}) = (2\pi)^{-\frac{5}{2}n+4} \int \prod_1^{n-1} d\xi_j \, \exp\left(i \sum q_j \xi_j\right) w_\alpha^I(\xi_1, \ldots, \xi_{n-1})$$
$$\tag{4.42}$$

are connected by

$$\tilde{W}^I_\alpha(p_1,\ldots,p_n) = \delta^4\left(\sum p_i\right) \tilde{w}^I_\alpha(q_1,\ldots,q_{n-1}), \quad (4.43)$$

with $q_k = \sum_{i=1}^k p_i$. Postulate 5 implies

**Property W3: Spectral property.** *The support of $\tilde{w}^I_\alpha(q_1,\ldots,q_{n-1})$ is contained in the set $\{q_i \in \bar{V}_+,\ i=1,\ldots,n-1\}$.*

**Property W4: Weak cluster property.** *Let $\gamma(\vec{p})$, $\tilde{f}(p_1,\ldots,p_n)$, $\tilde{g}(q_1,\ldots,q_m)$, be test functions. Then*

$$\lim_{\lambda\to\infty} \int \prod_1^n dp_i \prod_1^m dq_j\, \gamma(\lambda\sum \vec{p}_i)\, \tilde{f}(\ldots,p_i,\ldots)\, \tilde{g}(\ldots,q_j,\ldots)$$
$$\times \tilde{W}^{IJ}_{\alpha\beta}(p_1,\ldots,q_m)$$
$$= \gamma(0) \int \prod dp_i\, \tilde{f}(\ldots,p_i,\ldots)\, \tilde{W}^I_\alpha(\ldots,p_i,\ldots)$$
$$\times \int \prod dq_j\, \tilde{g}(\ldots,q_j,\ldots)\, \tilde{W}^J_\beta(\ldots,q_j,\ldots). \quad (4.44)$$

This is a direct consequence of (4.30).

**Property W5: Locality.**

$$W^{\ldots ij\ldots}_{\ldots\alpha_i\alpha_j\ldots}(\ldots,x_i,x_j,\ldots) = \pm W^{\ldots ji\ldots}_{\ldots\alpha_j\alpha_i\ldots}(\ldots,x_j,x_i,\ldots), \quad (4.45)$$

if $(x_i - x_j)^2 < 0$. The upper sign applies if $\varphi^{(i)}$ and $\varphi^{(j)}$ are Fermi fields, the lower sign if at least one of them is a Bose field.

The approach to perturbative QED that will be developed in Part II is based on the fact that the W-properties just enunciated are equivalent to the Postulates 1–6 of a field theory. This is the content of *Wightman's reconstruction theorem*:

**Theorem 4.1.** *Let $1,\ldots,r$, be a finite set of field indices, each associated with a Lorentz representation $R^{(i)}$, and let $\alpha_i$ be corresponding component indices. Assume that to every multi-index $I$ with entries from $1,\ldots,r$, and to every corresponding multi-index $\alpha$ a Wightman function $W^I_\alpha$ is given, such that the properties W1–W5 are satisfied. Then there exists a unique quantum field theory with fields $\varphi^{(1)},\ldots,\varphi^{(r)}$, which satisfies the postulates 1–6, and whose Wightman functions are the given ones.*

"Uniqueness" means uniqueness up to unitary equivalence.

The theorem is proved by explicit construction of the claimed field theory. Let $F = \{f^I_\alpha\}$ be an assignment of a test function to each pair $I,\alpha$, of multi-indices, such that only a finite number of $f^I_\alpha$'s are different from zero.

The empty index $I = \emptyset$ is included, the corresponding test function being a complex number. These $F$ form a complex vector space $\mathcal{V}'$, with addition and multiplication by a scalar being defined componentwise. A hermitian scalar product is defined on $\mathcal{V}'$ by

$$(F, G) = \sum_{I,J} \int \prod dx_i \prod dy_j \left( f_{\alpha^*}^{I^*}(x_n, \ldots, x_1) \right)^* g_\beta^J(y_1, \ldots, y_m)$$
$$\times W_{\alpha\beta}^{IJ}(x_1, \ldots, x_n, y_1, \ldots, y_m) \tag{4.46}$$

in obvious notation. This scalar product is degenerate. But the elements of $\mathcal{V}'$ which are orthogonal to all of $\mathcal{V}'$ form a linear subspace $\mathcal{V}_0$. The state space $\mathcal{V}$ of the desired field theory is defined as the quotient

$$\mathcal{V} = \mathcal{V}'/\mathcal{V}_0 . \tag{4.47}$$

This means that the elements of $\mathcal{V}$ are equivalence classes of elements of $\mathcal{V}'$, $F_1, F_2 \in \mathcal{V}'$ being equivalent if $(F_1 - F_2) \in \mathcal{V}_0$. The scalar product $(F, G)$ depends only on the classes of $F$ and $G$, so that it induces a scalar product on $\mathcal{V}$, which is non-degenerate.

The Poincaré representation $U(\Lambda, a)$ is defined on $\mathcal{V}'$ by

$$\left( U(\Lambda, a) F \right)_\alpha^I = \sum_\beta \prod_{i \in I} R_{\alpha_i \beta_i}^{(i)} f_\beta^I \left( \Lambda^{-1}(x_1 - a), \ldots, \Lambda^{-1}(x_n - a) \right) , \tag{4.48}$$

the field operator $\varphi_\alpha^{(i)}$ by

$$\left( \varphi_{\alpha_i}^{(i)}(f) F \right)_{\beta_j \alpha}^{jI} = \begin{cases} f(x_1) f_\alpha^I(x_2, \ldots, x_n) & \text{if } j = i, \; \beta_j = \alpha_i , \\ 0 & \text{otherwise} . \end{cases} \tag{4.49}$$

Here $I$ and $\alpha$ are multi-indices of order $n-1$, $j$ and $\beta_j$ are simple indices. Both $U(\Lambda, a)$ and $\varphi_\alpha^{(i)}(f)$ map equivalence classes of $\mathcal{V}'$ into equivalence classes, so that (4.48) and (4.49) define $U$ and $\varphi^{(i)}$ as operators on $\mathcal{V}$.

The vacuum state $\Omega \in \mathcal{V}$ is the equivalence class of $\Omega' \in \mathcal{V}'$ defined by

$$\Omega'^\emptyset = 1, \quad \Omega_\alpha'^I = 0 \quad \text{for } I \neq \emptyset. \tag{4.50}$$

It is not difficult to ascertain that these definitions satisfy the Wightman postulates in our general form. In establishing the uniqueness of the vacuum it is most convenient to use the criterion (4.29) in conjunction with (4.30).

For the stricter version of the Wightman postulates, which assumes a positive scalar product, the reconstruction theorem remains valid if the following condition is added to the list of W-properties.

**Property W6: Positivity.** *For any $F \in \mathcal{V}'$ we have*

$$(F, F) \geq 0 . \tag{4.51}$$

Both versions (general and strict) of the theorem still hold if locality and Lorentz covariance – but not translation invariance – are omitted from both the list of postulates and the list of W-properties. This fact is essential for the construction of physical states in QED that will be explained in Chap. 12.

We conclude the chapter with two pertinent remarks.

**Remark 1.** In the strict version of the Wightman formalism the vacuum condition W4 can be replaced by the physically more transparent *cluster property* which states that

$$\lim_{a\to\infty} W^{IJ}_{\alpha\beta}(x_1,\ldots,x_n,y_1+a,\ldots,y_m+a) = W^{I}_{\alpha}(x_1,\ldots,x_n)\, W^{J}_{\beta}(y_1,\ldots,y_m), \tag{4.52}$$

if the translation $a$ tends to infinity in a spacelike direction (see Sect. 7.2B of [BLOT]), $n$ and $m$ being the orders of $(I,\alpha)$ and $(J,\beta)$, respectively. In theories with an indefinite scalar product, especially in gauge theories, the cluster property is a rather more problematic requirement [MSt 80].

**Remark 2.** In the above description of the general principles of quantum field theory no mention was made of canonical commutation relations. This raises the question of the status of Planck's constant in the theory. The constant enters via the equation (4.27), which reads in a general system of units, with $\hbar \neq 1$,

$$\left[P_\mu, \varphi^{(i)}_\alpha(x)\right] = -i\hbar\, \partial_\mu \varphi^{(i)}_\alpha(x)\,. \tag{4.53}$$

In these general units we write the Fourier integral (4.20) as

$$\hat{\varphi}^{(i)}_\alpha(k) = (2\pi)^{-\frac{5}{2}} \int dx\, e^{-ikx}\, \varphi(x)\,, \tag{4.54}$$

where the wave vector $k$ is of dimension (length)$^{-1}$. The formal state

$$\Psi(k_1,\ldots,k_n) = \prod_i \hat{\varphi}^{(i)}_{\alpha_i}(k_i)\, \Omega \tag{4.55}$$

is then an improper eigenstate of $P_\mu$ with eigenvalue $-\hbar \sum_i k_{i\mu}$, meaning that

$$P_\mu \int \prod_i dk_i\, \hat{f}(k_1,\ldots,k_n)\, \Psi(k_1,\ldots,k_n) =$$
$$-\hbar \int \prod_i dk_i\, \sum_i k_{i\mu}\, \hat{f}(k_1,\ldots,k_n)\, \Psi(k_1,\ldots,k_n) \tag{4.56}$$

for $\hat{f} \in \mathcal{S}$. Choose $n=1$ and $\hat{f}(k) \in \mathcal{S}$ with a narrow support centered at the 4-vector $K$. Under suitable conditions the state

$$\Psi = \int dk\, \hat{f}(k)\, \hat{\varphi}^{(i)}_\alpha(-k)\, \Omega \tag{4.57}$$

may then be an excellent approximation to a physical one-particle state. In QED this turns out to be the case if $\varphi_\alpha^{(i)}$ is a field strength $F^{\mu\nu}$ and $K^2 = 0$, or if $\varphi^{(i)}$ is the spinor $\psi$ and $K^2 = \hbar^{-2} m^2$, with $m$ the electron mass. In these cases $\Psi$ is an approximate eigenstate of $P_\mu$ with eigenvalues

$$p_\mu = \hbar K_\mu , \qquad (4.58)$$

$p_\mu$ being the 4-momentum of the particle in question. Equation (4.58) is the famous Einstein–de Broglie relation, which is experimentally verifiable, thus allowing the experimental determination of $\hbar$. The 4-momentum $p_\mu$ can be measured with the classical standard methods, the wave vector $K$ by interference experiments, e.g. by scattering off crystals or other convenient gratings, as is well known from the history of quantum mechanics. Of course, at the present stage of our investigation these remarks are still heuristic, since the notion of a particle has not yet been introduced. This will be done in the next chapter for free fields, in Part III for interacting fields.

For later reference we note the relation

$$\tilde{\varphi}(p) = \hbar^{-4} \hat{\varphi}\left(\frac{p}{\hbar}\right) , \qquad (4.59)$$

with which we have

$$\varphi(x) = (2\pi)^{-\frac{3}{2}} \int dp \, \exp\left(-i\frac{px}{\hbar}\right) \tilde{\varphi}(p) . \qquad (4.60)$$

# Chapter 5

# Free Fields

In this chapter we construct free field theories by deriving their $W$-functions from the field equations and the properties W1–W5 enumerated in the previous chapter, but not assuming the canonical commutation relations of the traditional, canonical approach. According to Theorem 4.1 the specification of the $W$-functions fully determines the theory. This method will be extended to interacting QED in Part II, where a perturbative expansion of the $W$-functions will be constructed by induction with respect to the perturbative order. The methods and results of the present chapter will be needed there for establishing the uniqueness of the procedure. It is with a view to these later needs that we renounce the use of canonical commutation relations, because for interacting fields these relations are of a dubious status (see below after (5.33)).

We first illustrate the method with the simplest possible example, the theory of a free, hermitian, scalar field. The results obtained in this case are then generalized to the fields of relevance to QED, namely the Dirac spinors $\psi$, $\bar\psi$ and the electromagnetic field strengths $F^{\mu\nu}$ and 4-potentials $A^\mu$.

## 5.1 The Hermitian Scalar Field

We study a scalar field $\varphi(x)$ with $\varphi^*(x) = \varphi(x)$, which satisfies the *Klein–Gordon equation*

$$(\Box + m^2)\,\varphi(x) = 0 , \qquad (5.1)$$

with $m \geq 0$. This equation, considered as a classical equation, is the Euler–Lagrange equation of the Lagrangian density

$$\mathcal{L} = \frac{1}{2}\partial_\mu\varphi\,\partial^\mu\varphi - \frac{1}{2}\,m^2\,\varphi^2 . \qquad (5.2)$$

In quantum field theory the use of such a Lagrangian is more problematic on account of the distributional character of the field $\varphi$ (see below), which makes products like $(\varphi(x))^2$ at first meaningless. This problem does not concern us, since our starting point is the linear field equation (5.1), not the Lagrangian.

The function
$$W(x_1,\ldots,x_n) = (\Omega, \varphi(x_1)\cdots\varphi(x_n)\Omega) \tag{5.3}$$
must satisfy the system of differential equations
$$\left(\Box_i + m^2\right) W(\ldots, x_i, \ldots) = 0 \quad \text{for} \quad i = 1,\ldots,n, \tag{5.4}$$
with $\Box_i = \partial_{i\mu}\partial_i^\mu$, $\partial_{i\mu} = \partial/\partial x_i^\mu$.

For the one-point function $W(x_1)$ one finds from translational invariance and hermiticity (properties W1 and W2) that it must be a real constant, $W(x) = a \in R$. Equation (5.4) implies $a = 0$ if $m > 0$:
$$W(x) = 0. \tag{5.5}$$
We assume this to hold also in the case $m = 0$. This is no restriction of generality, because if the original field $\varphi'(x)$ does not satisfy this condition, then $\varphi(x) = \varphi'(x) - a$ does, and still satisfies (5.1) and properties W1–W5. The original $\varphi'$ can be reconstructed from $\varphi$ as $\varphi' = \varphi + a$. Important examples of scalar fields with non-vanishing vacuum expectation values are the Higgs fields of the standard electroweak theory and its generalizations.

Next we turn to the 2-point function $W(x_1, x_2)$. By (4.43) its Fourier transform is of the form
$$\tilde{W}(p_1, p_2) = \delta^4(p_1 + p_2)\,\tilde{w}(p_1). \tag{5.6}$$
Equation (5.4) becomes
$$\left(p_1{}^2 - m^2\right) \tilde{w}(p_1) = 0. \tag{5.7}$$
This together with the spectral property W3 and Lorentz invariance implies
$$\tilde{w}(p_1) = b\,\delta_+(p_1), \tag{5.8}$$
where $\delta_\pm(p) = \theta(\pm p_0)\,\delta(p^2 - m^2)$ is the Dirac measure of the positive mass shell.[1] A contribution of the form $b'\delta^4(p_1)$ which would also solve (5.7) if $m = 0$, is excluded by (5.5) and condition (4.44). $b$ is real by hermiticity. We postpone the discussion of how to choose the numerical value of $b$.

Transformed back into $x$-space, the expression (5.8) becomes
$$W(x_1, x_2) = -i\,b\,\Delta_+(x_1 - x_2), \tag{5.9}$$
with the definition
$$\Delta_\pm(\xi) = \pm\frac{i}{(2\pi)^3}\int dp\,\delta_\pm(p)\,e^{-ip\xi}. \tag{5.10}$$

---

[1] $\theta(u)$ denotes the step function $\theta(u) = \begin{cases} 1 & \text{for } u \gtrless 0. \\ 0 & \end{cases}$

These "invariant functions" will play an important part in perturbation theory, as will their combination

$$\Delta(\xi) = \Delta_+(\xi) - \Delta_+(-\xi) = \Delta_+(\xi) + \Delta_-(\xi)$$
$$= \frac{i}{(2\pi)^3} \int dp\, \varepsilon(p_0)\, \delta(p^2 - m^2)\, e^{-ip\xi}\,, \qquad (5.11)$$

with $\varepsilon(u) = \theta(u) - \theta(-u)$. The relevant properties of these functions will be discussed in the last section of this chapter.

As a side remark we note that $\Delta_+(\xi)$ clearly diverges at $\xi = 0$. Hence $W(x,x)$ diverges, hence the vector $\varphi(x)\Omega$ cannot be a function of $x$ in the strict sense of the word, and hence the field $\varphi(x)$, which must be defined on $\Omega$, cannot be a function. This is the reason in a nutshell why relativistic quantum fields are distributions, not functions.

For the higher functions with $n > 2$ we prove

**Theorem 5.1.**
*(a)*   *For $n$ odd we have*

$$W(x_1, \ldots, x_n) = 0\,. \qquad (5.12)$$

*(b)*   *For $n$ even, $W(x_1, \ldots, x_n)$ is given by the "cluster expansion"*

$$W(x_1, \ldots, x_n) = \sum \prod_\alpha W(x_{i_\alpha}, x_{j_\alpha})\,, \qquad (5.13)$$

*where the sum extends over all possible partitions of the set of variables into $n/2$ pairs $(x_{i_\alpha}, x_{j_\alpha})$ with $i_\alpha < j_\alpha$.*

For the proof of this theorem we need to know some facts about the Laplace transform of distributions (see [StW], Sects. 2-3 and 2-5). Let $F(\xi)$ be a tempered distribution in the 4-vector $\xi$ with support in the closed forward cone. Then its Laplace transform, or complex Fourier transform,

$$\hat{F}(z) = \int d\xi\, e^{iz\xi}\, F(\xi)\,, \qquad (5.14)$$

is a function of the complex 4-vector $z = p + iq$, which is analytic in the four variables $z^\mu$ in the *forward tube*

$$\mathcal{T}_+ = \{z = p + iq\,:\, q \in V_+\}\,. \qquad (5.15)$$

This is so because for these values of $z$ the function $e^{iz\xi}$ has in the support of $F$ all the properties of a test function: infinite differentiability and strong decrease at infinity. The same is true for the complex derivatives

$$\frac{\partial}{\partial z^\mu} \hat{F}(z) = i \int d\xi\, e^{iz\xi}\, \xi_\mu\, F(\xi).$$

The real Fourier transform $\tilde{F}(p)$ is a boundary value of the analytic function $\hat{F}(z)$. This means that

$$\int dp\, \tilde{F}(p)\, \tilde{f}(p) = \lim_{q \to 0} \int dp\, \hat{F}(p+iq)\, \tilde{f}(p) \qquad (5.16)$$

for all $\tilde{f} \in \mathcal{S}$, if $q$ tends to zero from inside the open cone $V_+$. If this boundary value $\tilde{F}(p)$ vanishes on an open set $D \subset R^4$, i.e. on all $\tilde{f}$ with support in $D$, then $\hat{F}(z) \equiv 0$, hence also $\tilde{F}(p) \equiv 0$ ([StW], Theorem 2.17).

We are now in a position to prove Theorem 5.1. The proof is somewhat involved, but it is indispensable, because the theorem is of vital importance to the perturbative constructions to be given in Part II.[2]

**Proof of Theorem 5.1.** The proof proceeds in several steps. We first note that (5.4) reads in $p$-space

$$\left(p_i^2 - m^2\right) \tilde{W}(p_1, \ldots, p_n) = 0, \qquad (5.17)$$

which implies that the support of $\tilde{W}$ is contained in the set $S_n = \{\{p_i\} : p_i^2 = m^2 \,\forall\, i\}$. Consider the function

$$\begin{aligned}D(\ldots, x_i, x_{i+1}, \ldots) &= W(\ldots, [x_i, x_{i+1}], \ldots) \\ &:= W(\ldots, x_i, x_{i+1}, \ldots) - W(\ldots, x_{i+1}, x_i, \ldots)\end{aligned} \qquad (5.18)$$

for a given $i$. The support of its Fourier transform $\tilde{D}$ is contained in the set $p_i^2 = p_{i+1}^2 = m^2$. We replace the variables $p_i, p_{i+1}$ by their linear combinations $P = \frac{1}{2}(p_i + p_{i+1})$, $Q = \frac{1}{2}(p_i - p_{i+1})$, which are conjugate to $X = x_i + x_{i+1}$, $\xi = x_i - x_{i+1}$. The support of $\tilde{D}$, as far as the variables $P, Q$ are concerned, consists then of four components defined by

$$P_0 + Q_0 = \pm \omega(\vec{P} + \vec{Q}), \qquad P_0 - Q_0 = \pm \omega(\vec{P} - \vec{Q}), \qquad (5.19)$$

with $\omega(\vec{p}) = \left(\vec{p}^2 + m^2\right)^{1/2}$. Adding the two conditions we find

$$2\, P_0 = \pm \omega(\vec{P} + \vec{Q}) \pm \omega(\vec{P} - \vec{Q}). \qquad (5.20)$$

Notice that $Q_0$ does not occur in this form of the conditions. For a fixed value of $P \neq 0$ none of the four conditions is satisfied identically in $Q$: there exists an open neighbourhood in $Q$-space in which $\tilde{D}$ vanishes identically.

Locality demands that $D(\ldots, X, \xi, \ldots)$ vanish for $\xi^2 < 0$. Hence the support of $D_\pm = \theta(\pm \xi_0)\, D$ is contained in the cone $\bar{V}_\pm$. The Fourier transforms

$$\tilde{D}_\pm = \pm i \int du \frac{1}{Q_0 - u \pm i\varepsilon}\, \tilde{D}(\ldots, P, u, \vec{Q}, \ldots) \qquad (5.21)$$

---

[2] The proof becomes simpler if positivity of the scalar product is assumed [FJ 60, J 61, P 69]. But in our future applications of the theorem this assumption is not satisfied.

are then, as functions of $Q$, boundary values of functions of complex $Q$'s which are analytic in the tubes $\mathcal{T}_\pm = \{Q : \Im Q \in V_\pm\}$, respectively. Since the support conditions (5.20) do not contain $Q_0$, they remain valid for $\tilde{D}_\pm$: these two functions vanish on open sets in $Q$-space, hence vanish identically, if $P \neq 0$. Hence also $\tilde{D} = \tilde{D}_+ + \tilde{D}_- \equiv 0$: the support of $\tilde{D}$ is contained in $\{P = 0\}$. This implies that $\tilde{D}$ is a finite sum of terms containing a factor $\delta^4(p_i + p_{i+1})$ or a derivative thereof. Transformed into $x$-space this means that $D(\ldots, x_i, x_{i+1}, \ldots)$ is a polynomial in $X$ with coefficients depending on the remaining variables $\xi$ and $x_j$, $j \neq i, i+1$.

Consider now vacuum expectation values containing double commutators,

$$W(\ldots, [[x_i, x_{i+1}], x_{i+2}], \ldots) = W(\ldots, [x_i, x_{i+1}], x_{i+2}, \ldots) \\ - W(\ldots, x_{i+2}, [x_i, x_{i+1}], \ldots) \ . \tag{5.22}$$

Because of locality this expression has its support contained in the set $S = \{(x_i - x_{i+2})^2 \geq 0 \text{ or } (x_{i+1} - x_{i+2})^2 \geq 0\}$. By the previous consideration $W(\ldots, [[\ldots].], \ldots)$ is a polynomial in $X$ if written as a function of $X, \xi$, and the remaining $x_j$. Hence its support is invariant under simultaneous translation of $x_i$ and $x_{i+1}$: $x_i \to x_i + a$, $x_{i+1} \to x_{i+1} + a$ for any $a \in R^4$, the remaining $x_j$ being unchanged. But no non-empty subset of $S$ is invariant under these operations. Therefore

$$W(\ldots, [[x_i, x_{i+1}], x_{i+2}], \ldots) \equiv 0 \ . \tag{5.23}$$

From this we find

$$\tilde{W}(\ldots, [p_i, p_{i+1}], \ldots) = \tilde{W}([p_i, p_{i+1}], p_1, \ldots, \hat{p}_i, \hat{p}_{i+1}, \ldots, p_n) \ , \tag{5.24}$$

where the careted variables must be omitted. Remembering the $\{p_i + p_{i+1} = 0\}$-support of this expression and using the vacuum condition (4.44) we obtain

$$\tilde{W}(\ldots, [p_i, p_{i+1}], \ldots) = \tilde{W}([p_i, p_{i+1}]) \tilde{W}(p_1, \ldots, \hat{p}_i, \hat{p}_{i+1}, \ldots, p_n) \ . \tag{5.25}$$

By the spectral property the function $W(p_1, \ldots, p_n)$ is different from zero only if $p_1 \in \bar{V}_+$. The relation

$$\tilde{W}(p_1, p_2, \ldots, p_n) - \tilde{W}(p_2, \ldots, p_n, p_1) \\ = \sum_{i=2}^n \tilde{W}(\ldots, [p_1, p_i], \ldots) = \sum_{i=2}^n \tilde{W}([p_1, p_i]) \tilde{W}(\hat{p}_1, \ldots, \hat{p}_i, \ldots) \tag{5.26}$$

can then be solved for $p_1 \neq 0$, to yield

$$\tilde{W}(p_1, \ldots, p_n) = \sum_{i=2}^n \tilde{W}(p_1, p_i) \tilde{W}(\hat{p}_1, \ldots, \hat{p}_i, \ldots) \ . \tag{5.27}$$

This determines $\tilde{W}$ up to terms containing a factor $\delta^4(p_1)$ or a derivative thereof. But the existence of such terms is again excluded by condition (4.44) and the one-point normalization (5.5). Induction with respect to $n$, starting from $n = 2$, leads at once to the cluster expansion (5.13).

The value of the normalization constant $b$ still needs to be fixed. It is instructive to consider this problem in the system of units of remark 2 made at the end of the previous chapter. In this system $\hbar$ is a dimensional constant but $c$ is still $= 1$, so that the units of length and time coincide. For reasons of consistency the parameter $m$ in (5.1) must have the dimension $[\text{length}]^{-1}$, *not* the dimension of a mass. If we assume as usual that the Lagrangian $L = \int dx\, \mathcal{L}(x)$ has the dimension of an action, then the field $\varphi$ has dimension $[\text{mass}]^{1/2}[\text{length}]^{-1/2}$. The dimension of the $W$-function

$$W(x,y) = (2\pi)^{-3} b \int dk\, e^{-ikx}\, \theta(k_0)\, \delta(k^2 - m^2) = -i\, b\, \Delta_+(x-y) \quad (5.28)$$

is then $[mass][length]^{-1}$, which gives $b$ the dimension of an action. Since $\hbar$ is the only available constant with that dimension, we find that $b$ must be of the form $b = r\hbar$, with $r$ a real number. This implies that the $W$-functions vanish in the classical limit, which makes sense: the $W$-functions describe quantum fluctuations of the field in the vacuum, and these vanish in the classical limit.

The choice $r = 0$ leads clearly to an extremely uninteresting theory. For $r < 0$ we find a theory in a state space with a negative-definite scalar product (see below), which is physically meaningless, so that we are left with positive $r$. There is no further restriction: we can choose $r$ to be any positive number we like. This statement needs some elaboration. We refer to the discussion of observables given in Chap. 4 in connection with the locality postulate. In a field theory we must demand that any observable $\mathcal{O}$, which is physically defined by a measurement prescription, be represented in the mathematical formulation of the theory by an operator, likewise called $\mathcal{O}$, which is a function of the fundamental fields, in the present case of $\varphi$. The correct choice of this function is in the last resort an empirical problem; it is not fixed by any a priori requirements. Let us now choose an arbitrary normalization constant $r > 0$, and construct the corresponding field $\varphi(x)$. Let the observable $\mathcal{O}$ be represented in this theory by the functional $\mathcal{O}(\varphi)$, Now let $r'$ be a different choice of normalization, $\varphi'$ the corresponding field. It follows immediately from our results that $\varphi'(x) = \sqrt{r'/r}\, \varphi(x)$, so that $\mathcal{O}'(\varphi') = \mathcal{O}\left(\sqrt{r/r'}\, \varphi'\right)$ represents the same observable in the primed theory, and that existing relations between different observables are conserved by this operation. This simple transformation between physically equivalent theories is called a "field renormalization". We use the consequent freedom of choice to demand $r = 1$ without restriction of generality, so that

$$b = \hbar\,. \quad (5.29)$$

This accords with the value of this constant in the canonical formalism (see below).

Finally we want to make a remark on the spin–statistics theorem. The function $\Delta_+(\xi)$ is invariant under orthochronous Lorentz transformations. Since for a spacelike $\xi$ one can always find such a transformation which transforms it into $-\xi$, we find $\Delta_+(-\xi) = \Delta_+(\xi)$ for $\xi^2 < 0$, hence

$$W([x,y]) = -i\Delta(x-y) = 0 \quad \text{for } (x-y)^2 < 0, \tag{5.30}$$

in agreement with the sign assignment in the locality postulate (4.45). The other sign would lead to an inconsistency, because

$$W(x,y) + W(y,x) = -ib\bigl(\Delta_+(x-y) + \Delta_+(y-x)\bigr)$$

does not vanish for $(x-y)^2 < 0$ unless $b = 0$. This is the simplest special case of the spin–statistics theorem. The exceptional case $b = 0$ is realized for the Faddeev–Popov ghost fields $c$, $\bar{c}$, of non-abelian gauge theories, which therefore can be treated as Fermi fields with impunity. In that case $b = 0$ does not imply the vanishing of $c$ and $\bar{c}$, as it would for our single field $\varphi$, because the mixed 2-point function $(\Omega, c\bar{c}\Omega)$ may still be non-zero.

## Relation to the Canonical Formalism

In this subsection we wish to show that our approach leads to the same results as the canonical approach used in most introductory texts of quantum field theory.

From Theorem (5.1) we find that

$$\begin{aligned} W(\ldots,[x,y],\ldots) &= W(\ldots,x,y,\ldots) - W(\ldots,y,x,\ldots) \\ &= W(\ldots)W([x,y]), \end{aligned} \tag{5.31}$$

which implies

$$[\varphi(x),\varphi(y)] = W([x,y]) = -i\hbar\Delta(x-y); \tag{5.32}$$

the commutator of two fields at arbitrary points is a c-number! This is a peculiarity of free fields. It is not true for interacting fields.

From the explicit expression (5.11) we find the *equal-time commutators*

$$\begin{gathered} [\varphi(t,\vec{x}),\varphi(t,\vec{y})] = 0 \quad , \quad [\dot{\varphi}(t,\vec{x}),\dot{\varphi}(t,\vec{y})] = 0, \\ [\dot{\varphi}(t,\vec{x}),\varphi(t,\vec{y})] = -i\hbar\,\delta^3(\vec{x}-\vec{y}). \end{gathered} \tag{5.33}$$

These are the *"canonical commutation relations"* (ccr's), which are one of the basic inputs in the canonical formulation of quantum field theory. In our approach they are a derived result. We prefer not to use ccr's as a basic ingredient, because they are objectionable on the ground that quantum fields, being distributions, cannot necessarily be defined at sharp times. Indeed, we will find in Chap. 10 that in interacting QED expressions of the type (5.33) do not exist. From now on we set again $\hbar = 1$. We can use the relations (5.33) to give an explicit construction of the field operators $\varphi(x)$ and the state space on which they act. Our description is somewhat cursory, because these results will not be used later on, and a similar explicit construction of interacting fields is impossible.

Because of the field equation $(p^2 - m^2)\tilde{\varphi}(p) = 0$ the Fourier transform $\tilde{\varphi}$ of $\varphi$ can be written as

$$\tilde{\varphi}(p) = \delta(p^2 - m^2)\left(a(\vec{p})\theta(p_0) + a^*(-\vec{p})\theta(-p_0)\right). \tag{5.34}$$

$a$ and its adjoint $a^*$ are operator-valued distributions. Hermiticity of $\varphi$ has been assumed. The ccr's (5.33) yield

$$[a(\vec{p}), a(\vec{q})] = 0, \quad [a(\vec{p}), a^*(\vec{q})] = 2\omega(\vec{p})\,\delta^3(\vec{p} - \vec{q}). \tag{5.35}$$

From the definition (4.25) of the momentum operators $P_\mu$ we obtain

$$[P_\mu, a(\vec{p})] = -p_\mu a(\vec{p}), \quad [P_\mu, a^*(\vec{p})] = p_\mu a^*(\vec{p}), \tag{5.36}$$

where for the energy component we must set $p_0 = \omega(\vec{p})$. This means that application of $a(\vec{p})$ to a given state lowers its 4-momentum by $(\omega, \vec{p})$, $a^*(\vec{p})$ raises it by that amount. The $a$ are called annihilation operators, the $a^*$ creation operators. As a consequence, the vector $a(\vec{p})\,\Omega$ has the negative energy $-\omega(\vec{p})$ and must therefore vanish:

$$a(\vec{p})\,\Omega = 0. \tag{5.37}$$

We define the kets

$$|\vec{p}_1, \ldots, \vec{p}_n\rangle = \prod_{i=1}^n a^*(\vec{p}_i)\,\Omega, \quad |\rangle = \Omega. \tag{5.38}$$

Their scalar products are

$$\langle \vec{q}_1, \ldots, \vec{q}_m | \vec{p}_1, \ldots, \vec{p}_n \rangle = \delta_{nm} \prod_{i=1}^n (2\omega(\vec{p}_i)) \sum \delta^3(\vec{p}_1 - \vec{q}_{i_1}) \cdots \delta^3(\vec{p}_n - \vec{q}_{i_n}), \tag{5.39}$$

the sum extending over all permutations of $\vec{q}_1, \ldots, \vec{q}_n$. The scalar product defined in this way is positive definite. The states (5.38), $n = 0, 1, 2, \ldots$, form an improper orthogonal basis of a Hilbert space $\mathcal{F}$ called *Fock space*. A proper basis of $\mathcal{F}$ can be found as follows. Let the functions $f_i(\vec{p}) \in \mathcal{S}$, $i = 1, 2, \ldots$, form an orthonormal basis of the $\mathcal{L}_2$-space over $R^3$ with respect to the measure $\frac{d^3p}{2\omega(\vec{p})}$. This means that the $f_i$ form a maximal set with

$$\int \frac{d^3p}{2\omega(\vec{p})}\,f_i^*(\vec{p})\,f_j(\vec{p}) = \delta_{ij}. \tag{5.40}$$

Define
$$a_i = \int \frac{d^3p}{2\omega(\vec{p})} f_i^*(\vec{p}) a(\vec{p}) \tag{5.41}$$

and $a_i^*$ its adjoint. Then the vectors

$$|n_1, \ldots\rangle = \prod_{i=1}^{\infty} \frac{(a_i^*)^{n_i}}{\sqrt{n_i!}} \Omega , \quad n_i \in \mathbb{Z}_+ , \quad \sum n_i < \infty , \tag{5.42}$$

form a basis of $\mathcal{F}$. The action of $a$ and $a^*$ on $\mathcal{F}$ is given by

$$a^*(\vec{q}) |\vec{p}_1, \ldots, \vec{p}_n\rangle = |\vec{q}, \vec{p}_1, \ldots, \vec{p}_n\rangle ,$$
$$a(\vec{q}) |\vec{p}_1, \ldots, \vec{p}_n\rangle = 2\omega(\vec{q}) \sum_{i=1}^{n} \delta^3(\vec{q} - \vec{p}_i) |\vec{p}_1, \ldots, \hat{\vec{p}}_i, \ldots, \vec{p}_n\rangle . \tag{5.43}$$

Notice that the condition $f_i \in \mathcal{S}$ in the choice of basis is not really necessary: we might have chosen $f_i \in \mathcal{L}_2$. The distributions $a(\vec{p})$ and $a^*(\vec{p})$ can be extended to this larger test space.

The states $a^*(f_i)\Omega$ are eigenstates of the *mass operator* $M^2 = P_0^2 - \sum P_i^2$ with eigenvalue $m$. In particle physics the possession of a sharp value of the mass is considered a characteristic property of a particle. The above states are therefore interpreted as states of a free particle of mass $m$ with the $p$-space wave function $f_i(\vec{p})$. Correspondingly we interpret $|\vec{p}_1, \ldots, \vec{p}_n\rangle$ as an $n$-particle state with particle momenta $\vec{p}_i$. In this way the notion of particles emerges in free field theory. It should be noted that this notion has at first sight nothing to do with the more intuitive definition of a particle as an object with a sharp localization in space at all times. Rather it corresponds to what has been called a "quantum" in early quantum theory. In QED we will find it convenient to use a different way of introducing particles, which is closer to the intuitive notion.

For the $P_\mu$ we obtain the explicit representations

$$P_0 = \frac{1}{2} \int d^3p \, a^*(\vec{p}) a(\vec{p}) ,$$
$$P_i = \frac{1}{2} \int d^3p \, \frac{p_i}{\omega(\vec{p})} a^*(\vec{p}) a(\vec{p}) , \tag{5.44}$$

which satisfy the defining relations (4.25). More exactly, this is true for our normalization $b = 1$. For a different choice of $b$ the commutation relations (5.35) would acquire an aditional factor $b$ on the right-hand side, and in (5.44) we would have to insert a factor $b^{-1}$. This is an example of the renormalization procedure discussed above. Clearly, $b = 1$ is the simplest choice.

## 5.2 Dirac Spinors

The free Dirac spinors $\psi$ and $\bar{\psi} = \psi^* \gamma_0$ satisfy the free Dirac equations

$$(i\slashed{\partial} - m)\psi(x) = 0 , \quad \bar{\psi}(i\overleftarrow{\slashed{\partial}} + m) = 0 . \tag{5.45}$$

## 5. Free Fields

We will not consider the most general case, but only the Dirac spinors as they occur in the gauge invariant QED. That is, we postulate the existence of a conserved current $j_\mu$ (see below for its expression in terms of $\psi, \bar{\psi}$) and a hermitian charge operator

$$Q = \int_{x^0=t} d^3x\, j^0(x) \tag{5.46}$$

satisfying

$$Q\Omega = 0, \quad [Q, \psi(x)] = -e\,\psi(x), \quad [Q, \bar{\psi}(x)] = e\,\bar{\psi}(x) \tag{5.47}$$

with $e$ the charge of the positron. Current conservation implies $t$-independence of the definition (5.46). $Q$ is the infinitesimal generator of a global gauge group.

The construction of the $W$-functions of this theory follows narrowly the procedure used in the scalar case. We introduce the following notation: variables of a $\psi$-field are denoted by Roman letters: $x, y, \ldots, p, q, \ldots$, those of $\bar{\psi}$-fields by barred Roman letters: $\bar{x}, \bar{y}, \ldots, \bar{p}, \bar{q}, \ldots$. The spinor indices are not shown explicitly, except where they are essential, in which case they are indicated by subscripts in front of the variables. For example, $_\rho x$ refers to the field $\psi_\rho(x)$. Thus we write

$$\left(\Omega,\, \psi_\rho(x)\, \bar{\psi}_\sigma(y)\, \Omega\right) = W(_\rho x, _\sigma\bar{y}), \tag{5.48}$$

where the indices $\rho$ and $\sigma$ might also be omitted. If a given variable, say $y$, occurs in an equation once as the argument of a $\bar{\psi}$-field, and once as an argument of a $\psi$-field or some other function, it will be barred in the first position, but not in the second one: the symbols $y$ and $\bar{y}$ occurring at different places in one and the same equation refer to the same variable; they are numerically identical. The general $W$-function with an arbitrary number of fields is written as $W(X, \bar{Y})$, where $X = \{x_1, \ldots, x_r\}$ is the set of $\psi$ variables, $\bar{Y} = \{\bar{y}_1, \ldots, \bar{y}_s\}$ the set of $\bar{\psi}$ variables. These variables may occur in an arbitrary ordering: our notation does not imply that all $\psi$-fields are adjacent. By $|X| = r$, $|\bar{Y}| = s$ we denote the number of variables in $X$ and $\bar{Y}$, respectively.

$W(X, \bar{Y})$ must satisfy the differential equations

$$(i\slashed{\partial}_j - m)\, W(X, \bar{Y}) = W(X, \bar{Y})\, (i\overleftarrow{\slashed{\partial}}_k + m) = 0 \tag{5.49}$$

for every $x_j \in X$ and every $\bar{y}_k \in \bar{Y}$. And $W$ must be translation invariant and transform under the Lorentz transformation $\Lambda$ as

$$W(\Lambda X, \Lambda \bar{Y}) = \prod_X S^{(j)}(\Lambda)\, W(X, \bar{Y}) \prod_Y \left(S^{(k)}(\Lambda)\right)^{-1}. \tag{5.50}$$

$S^{(j)}$ acts on the spinor index belonging to $x_j$, $S^{(k)}$ on that to $\bar{y}_k$. Condition (5.47) implies that the $W(X, \bar{Y})$ with $|X| \neq |\bar{Y}|$ vanish (charge conservation). This holds in particular for the one-point functions $W(x)$, $W(\bar{y})$.

## 5.2 Dirac Spinors

Because of
$$(i\partial + m)(i\partial - m) = -(\Box + m^2) \tag{5.51}$$
the non-vanishing 2-point functions $W(x,\bar{y})$ and $W(\bar{y},x)$ satisfy the Klein–Gordon equation in each variable. As in the scalar case their Fourier transform must have the form
$$\tilde{W}(_\rho p,_\sigma \bar{q}) = \delta^4(p+q)\, F_{\rho\sigma}(p)\, \delta_+(p) \tag{5.52}$$
and similarly for $\tilde{W}(\bar{q},p)$. The $4\times 4$-matrix $F(p)$, which is needed only at the positive mass shell, serves to satisfy the covariance requirement (5.50). It implies, written in matrix notation,
$$F(\Lambda p) = S(\Lambda)\, F(p)\, S^{-1}(\Lambda)\, . \tag{5.53}$$
The most general expression with this property that can be constructed out of the available quantities is
$$F(p) = a\,\mathbf{1} + b\, p_\mu \gamma^\mu\, , \tag{5.54}$$
with two free parameters $a, b$. The first Dirac equation (5.49) implies
$$(\slashed{p} - m)\, F(p) = 0$$
which is satisfied if and only if $b = a$. Inserting this result into (5.52) and transforming back into $x$-space we obtain
$$W(_\rho x,_\sigma \bar{y}) = -i\,a\,(i\partial_x + m)_{\rho\sigma} \Delta_+(x-y)\, . \tag{5.55}$$
We introduce the notation
$$S^\pm(\xi) = (i\partial + m)\, \Delta_\pm(\xi)\, , \tag{5.56}$$
with which (5.55) becomes
$$W(x,\bar{y}) = -i\,a\,S^+(x-y)\, . \tag{5.57}$$
The coefficient $a$ must be real on account of the hermiticity requirement (4.39), which reads in our case
$$\gamma_0\, W(x,\bar{y})^*\, \gamma_0 - W(y,\bar{x}) \tag{5.58}$$
in matrix notation. The actual value of $a$ will be fixed presently.

The other 2-point function $W(\bar{x}, y)$ is determined for $(x-y)^2 < 0$ by the locality requirement (4.45):
$$\begin{aligned} W(_\rho \bar{x},_\sigma y) &= -W(_\sigma y,_\rho \bar{x}) \\ &= i\,a\,(i\partial_y + m)_{\sigma\rho} \Delta_+(y-x) \\ &= i\,a\,(i\partial_y + m)_{\sigma\rho} \Delta_+(x-y)\, . \end{aligned}$$

In the last equality the Lorentz invariance of $\Delta_+$ has been used. The last form given has the correct analyticity properties of a $W$-function for all values of $x$ and $y$, not only for spacelike separations. Therefore it is the correct expression for $W(\bar{x}, y)$:

$$W(\bar{x}, y) = i\,a\,(i\partial_y + m)^T \Delta_+(x - y) = -i\,a\left(S^-(y-x)\right)^T . \tag{5.59}$$

To summarize: the 2-point functions of the free Dirac fields are

$$W(x, y) = W(\bar{x}, \bar{y}) = 0 ,$$
$$W({}_\rho x, {}_\sigma \bar{y}) = -i\,a\,S^+_{\rho\sigma}(x-y) , \quad W({}_\rho \bar{x}, {}_\sigma y) = -i\,a\,S^-_{\sigma\rho}(y-x) . \tag{5.60}$$

The non-vanishing higher functions $W(X, \bar{Y})$ with $|X| = |Y| > 1$ are given by a cluster expansion similar to (5.13). Remember that the notation $W(X, \bar{Y})$ does not indicate the ordering of fields inside the vacuum expectation value. $\psi$ and $\bar{\psi}$ variables may occur interlaced in an arbitrary order, which we call the original ordering.

**Theorem 5.2.** $W(X, \bar{Y})$ with $|X| = |Y| = n$ is given by

$$W(X, \bar{Y}) = \sum_{\{\pi_i\}} \pm \prod_{i=1}^{n} W(x_i, \bar{y}_{\pi_i}) . \tag{5.61}$$

*The sum extends over all permutations $\{\pi_1, \ldots, \pi_n\}$ of $\{1, \ldots, n\}$. The order of the variables in each 2-point factor is as in the original ordering, i.e. not necessarily as shown. The sign for a given pairing is positive (negative) if the ordering of the variables in the corresponding term of (5.61) is an even (odd) permutation of the original ordering.*

The proof of this theorem follows exactly the proof of Theorem 5.1, except that in (5.18) the commutator $[x_i, x_{i+1}]$ must be replaced by the anticommutator $\{x_i, x_{i+1}\}$. This necessitates some obvious sign changes in the rest of the proof.

In the same way as the commutation relations (5.33) we find the equal-time anticommutators

$$\{\psi(t,\vec{x}), \psi(t,\vec{y})\} = \{\bar{\psi}(t,\vec{x}), \bar{\psi}(t,\vec{y})\} = 0 ,$$
$$\{\psi_\rho(t,\vec{x}), \bar{\psi}_\sigma(t,\vec{y})\} = a\,\gamma^0_{\rho\sigma}\,\delta^3(\vec{x}-\vec{y}) . \tag{5.62}$$

The value of the normalization constant $a$ needs still to be fixed. For this purpose we do not now appeal to the Lagrangian formalism, as we did for the scalar field, but to a more detailed version of the QED requirements (5.46), (5.47). We start from the classical expression (3.16)

$$j^\mu_{cl} = e\,\bar{\psi}\,\gamma^\mu\,\psi \tag{5.63}$$

for the current density. This already fixes the dimension of $\psi$ and $\bar\psi$, which ought to be the same classically and quantum mechanically, to be $[\text{length}]^{-2}$, even if $\hbar$ has a non-trivial dimension. The "mass" parameter $m$ must have dimension $[\text{length}]^{-1}$, in order to make the Dirac equation dimensionally consistent. One finds then from (5.60) that $a$ must be a pure number: the canonical anticommutation relations (5.62) do not contain a factor $\hbar$, in contrast to the corresponding commutators (5.33) of the scalar field.

We wish the quantum expression of $j^\mu$ in terms of $\psi$ and $\bar\psi$ to be as close as possible to the classical form (5.63). This form cannot be taken over as it stands, however. Firstly, the product of the distributions $\bar\psi(x)$ and $\psi(x)$ is not defined. Secondly, even if we ignore this difficulty and calculate formally, we find that the vacuum expectation value of $j^\mu$ does not vanish and that therefore the charge $Q$ defined by (5.46) does not annihilate the vacuum. In fact, $Q$ is not even defined on $\Omega$, because $(\Omega, j^0 \Omega)$ does not vanish for $|\vec x| \to \infty$. But both difficulties can be solved at one stroke by defining

$$j^\mu(x) = e :\bar\psi(x)\,\gamma_\mu\,\psi(x): , \qquad (5.64)$$

where the "Wick product" $:\cdots:$ is defined by

$$:\bar\psi(x)\,\gamma^\mu\,\psi(x): = \bar\psi(x)\,\gamma^\mu\,\psi(x) - (\Omega, \bar\psi(x)\,\gamma^\mu\,\psi(x)\,\Omega) . \qquad (5.65)$$

This definition is at first sight meaningless because both terms on the right-hand side are divergent. We give the expression a meaning in the context of our free theory as follows. We first note that

$$L(x,y) = :\bar\psi(x)\,\gamma^\mu\,\psi(y): = \bar\psi(x)\,\gamma^\mu\,\psi(y) - (\Omega, \bar\psi(x)\,\gamma^\mu\,\psi(y)\,\Omega) \qquad (5.66)$$

is perfectly well defined if $x$ and $y$ are independent variables. Furthermore we note that that the operator $L(x,y)$ is completely determined by its matrix elements

$$(\Omega, F_1(x_1) \cdots F_n(x_n)\, L(x,y)\, F_{n+1}(x_{n+1}) \cdots F_m(x_m)\, \Omega) ,$$

where $F_i$ is either $\psi$ or $\bar\psi$. But these matrix elements can be calculated from the cluster expansion (5.61). The result is this: calculate the desired matrix elements by replacing $L$ simply by its unsubtracted form $\bar\psi(x)\,\gamma^\mu\,\psi(y)$, then drop all the terms in the expansion which contain a factor $W(x,\bar y)$. After this we can set $y=x$ without problems; and the result multiplied by $e$ defines $j_\mu(x)$.

Equations (5.46), (5.47) demand that

$$\left(\Omega, \bar\psi(y) \int_{x^0 = z^0} d^3x\, [:\bar\psi(x)\,\gamma^0\,\psi(x):, \psi(z)]\, \Omega \right) = -\left(\Omega, \bar\psi(y)\,\psi(z)\,\Omega\right) , \qquad (5.67)$$

and analogously for the $\psi(y)$–$\bar\psi(z)$ case. The left-hand side can be evaluated with the anticommutators (5.62) (notice that the subtraction term in (5.65) is a c-number, hence does not contribute to the commutator), and one finds that (5.67) is satisfied only for $a = 1$, if we discard the trivial possibility $a = 0$. Hence the natural choice (5.64) of $j_\mu$ determines $a$ uniquely: there is no freedom of renormalization. However, it will be seen in the next chapter that this is a special feature of the free theory. In interacting QED the situation is not nearly that clear-cut, and the admission of field renormalizations becomes unavoidable.

The subtraction term in the definition (5.64) of $j^\mu$ has the consequence that $j^0$ is no longer a positive quantity, as it is in the classical case. This becomes apparent from (5.47), which tells us that $\bar\psi$ applied to a state raises its charge by $e$, $\psi$ lowers it by $e$. And $\psi$ and $\bar\psi$ enter the formalism on an equal footing. This *charge symmetry* can be formalized as follows. We define an unitary operator $U_C$ on $\mathcal{V}$ by

$$U_C \Omega = \Omega ,$$
$$U_C \psi(x) U_C^* = \gamma^2 \gamma^0 \bar\psi^T(x) =: \psi'(x) ,$$
$$U_C \bar\psi(x) U_C^* = \psi^T(x) \gamma^0 \gamma^2 =: \bar\psi'(x) . \tag{5.68}$$

This transformation is called *charge conjugation*. The strange dependence on the Dirac matrices holds for our special choice (2.48). The superscript $T$ denotes transposition in spinor space. A lengthy but elementary calculation shows that $\psi'$ and $\bar\psi'$ satisfy again the Dirac equations (5.45). Moreover one finds that

$$j'^\mu = e :\bar\psi' \gamma^\mu \psi': = U_C j^\mu U_C^* = -j^\mu ; \tag{5.69}$$

the current changes its sign. This is best seen by writing $:\bar\psi(x) \gamma^\mu \psi(x):$ in a different way. We start again by splitting points and write [3]

$$\bar\psi(x) \gamma^\mu \psi(y) = \frac{1}{2} \gamma^\mu_{\rho\sigma} \left( [\bar\psi_\rho(x), \psi_\sigma(y)] + \{\bar\psi_\rho(x), \psi_\sigma(y)\} \right) . \tag{5.70}$$

The anticommutator in this expression is a c-number and thus coincides with its vacuum expectation value. Therefore it disappears after the Wick subtraction has been performed. The Wick subtraction in the commutator term is

$$\frac{1}{2} \left( W(_\rho \bar x, _\sigma y) - W(_\sigma y, _\rho \bar x) \right)_{y=x} = \frac{i}{2} \left( S^+_{\sigma\rho}(y-x) - S^-_{\sigma\rho}(y-x) \right)_{y=x}$$
$$= \frac{i}{2} (i\partial\!\!\!/_y + m) \left[ \Delta_+(y-x) + \Delta_+(x-y) \right]_{y=x} .$$

Formally we can argue like this: since the square bracket is an even function of $x - y$, its first derivatives, hence the $\partial\!\!\!/_y$-term, vanish at $y = x$. The surviving $m$-term occurs in $j^\mu$ multiplied with $\mathrm{tr}\,\gamma^\mu = 0$. As a result we find

$$j^\mu(x) = \frac{e}{2} [\bar\psi(x), \gamma^\mu \psi(x)] . \tag{5.71}$$

---

[3] For spinor indices we use the summation convention without regard to their position, upstairs or downstairs.

This formal result becomes rigorous if we *define* the commutator in it as

$$[\bar{\psi}_\rho(x), \psi_\sigma(x)] = \lim_{n\to\infty} \int d\xi\, \delta_n(\xi)\, [\bar{\psi}_\rho(x), \psi_\sigma(x+\xi)]\,, \qquad (5.72)$$

where $\{\delta_n\}$ is a sequence of test functions $\delta_n \in \mathcal{D}$ with compact support, satisfying $\delta_n(-\xi) = \delta_n(\xi)$, which converges in $\mathcal{D}'$, i.e. as distributions, to $\delta^4(\xi)$. From the form (5.71) the claim (5.69) can be easily verified.

The result is this: we can build up our state space from the vacuum either by acting on it with polynomials in $\psi, \bar{\psi}$ or with polynomials in $\psi', \bar{\psi}'$. Both procedures look exactly alike. But $\psi$ carries the charge $-e$, $\psi'$ the charge $e$, and the reverse holds for $\bar{\psi}, \bar{\psi}'$. Hence one formulation becomes the other one simply through replacing $e$ by $-e$: both signs of the charge have equal rights.

From the definition (5.65) it is obvious that $(\Omega, j^\mu \Omega) = 0$. Moreover, with the Wick definition of $j^0$ the charge operator $Q = \int d^3x\, j^0$ is well defined, is invariant under translations, and changes sign under charge conjugation. The state $Q\Omega$ is then also translation invariant, hence a multiple of $\Omega$, because of the uniqueness of the vacuum. This proportionality constant, the eigenvalue of Q, is zero because of the invariance of $\Omega$ under charge conjugation.

The non-trivial equal-time relation (5.62) becomes

$$\{\psi(t, \vec{x}), \bar{\psi}(t, \vec{y})\} = \gamma^0\, \delta^3(\vec{x} - \vec{y})\,, \qquad (5.73)$$

without a factor $\hbar$. As in the scalar case this result is an *assumption* in the canonical formalism. It must be said, though, that in this case the word "canonical" has to be taken with a grain of salt. The replacement of commutators by anticommutators is not really in the canonical spirit, but was introduced more or less ad hoc as a means of getting around the problems which lie at the bottom of the spin-statistics theorem. Contrary to the scalar case, quantization with the "wrong" statistics, i.e. with commutators instead of anticommutators, would now be perfectly possible but would lead to a state space with an indefinite scalar product. This is inacceptable because in the noninteracting limit of QED the free $\psi$ and $\bar{\psi}$ create *physical* electron–positron states; hence they must generate a Hilbert space.

An explicit representation of $\psi, \bar{\psi}$ as operators on a Fock space can be derived in close analogy to the scalar case. This will not be done here because Fock spaces will not be used later on. The construction is described in most textbooks on QED or advanced quantum mechanics. Let us only note that the particles obtained in this way carry spin $1/2$.

## 5.3 The Electromagnetic Field

For constructing the theory of the free electromagnetic field we can in principle work directly with the observable field strengths $F^{\mu\nu}$, which satisfy the

## 5. Free Fields

Maxwell equations

$$\partial_\mu F^{\mu\nu} = 0, \qquad \varepsilon_{\alpha\beta\mu\nu} \partial^\beta F^{\mu\nu} = 0 . \qquad (5.74)$$

These equations imply the wave equation

$$\Box F^{\mu\nu} = 0 . \qquad (5.75)$$

Hence the $F^{\mu\nu}$ are hermitian boson fields satisfying the Klein–Gordon equation with $m = 0$. The methods used for the scalar field yield the 2-point function

$$\tilde{W}\left({}^{\alpha\beta}p_1, {}^{\gamma\delta}p_2\right) = \delta^4(p_1 + p_2)\, b_{\alpha\beta\gamma\delta}(p_1)\, \delta_+(p_1) , \qquad (5.76)$$

where the coefficients $b$ must be chosen such that the Maxwell equations are satisfied in both variables, and that the correct Lorentz behaviour is present. The result is, in general units,

$$\begin{aligned}&\left(\Omega, F^{\alpha\beta}(x)\, F^{\gamma\delta}(y)\, \Omega\right) \\ &= i\hbar \left(g^{\alpha\gamma}\partial^\beta\partial^\delta - g^{\alpha\delta}\partial^\beta\partial^\gamma - g^{\beta\gamma}\partial^\alpha\partial^\delta + g^{\beta\delta}\partial^\alpha\partial^\gamma\right) D_+(x - y) .\end{aligned} \qquad (5.77)$$

$D_+$ is the function $\Delta_+$ for the special case $m = 0$. The normalization is chosen such that the energy operator $P_0$ has the Wick-corrected classical form

$$P_0 = \frac{1}{2} \int d^3x\, :|\vec{E}(t,\vec{x})|^2 + |\vec{B}(t,\vec{x})|^2: , \qquad (5.78)$$

where the Wick product is defined as in (5.65). The higher $W$-functions are obtained from this 2-point function by the cluster expansion given in Theorem 5.1.

It might be objected that the normalization of the 2-point functions (5.77) is not free, because the fields $F^{\alpha\beta}$ are observables. The justification for our choice can be derived from Remark 2 at the end of Chap. 4, in which the connection between the translation equations (4.53) and the value of Planck's constant is discussed. In fact, our choice of normalization merely amounts to defining the value of $\hbar$. Remember that Planck's constant first entered physics in connection with black-body radiation, i.e. in a context of free electromagnetic fields.

But interacting QED cannot be formulated in a tractable way, especially in a way fitting into our framework, without the use of 4-potentials (see Chap. 3). Hence we must also formulate the free theory in these terms. In order that the methods used in the previous cases can also be used here, we insist on the potentials $A^\mu$ being covariant and local fields, even though these requirements are not physically necessary, the $A^\mu$ having no direct physical meaning. This insistence on a nice mathematical setup leads to difficulties with the physical interpretation of the theory, which difficulties are solved by the *Gupta–Bleuler formalism* [Gu 50, Bl 50].

## 5.3 The Electromagnetic Field

The hermitian fields $A^\mu$ must satisfy the free version of the equations (3.20) and (3.22), which read in momentum space

$$p_\mu \tilde{A}^\mu = 0 , \qquad (5.79)$$
$$p^2 \tilde{A}^\mu - \lambda p^\mu p_\nu \tilde{A}^\nu = 0 , \qquad (5.80)$$

with $\lambda \neq 1$ an arbitrary real number. The case $\lambda = 1$ (Landau gauge) needs a separate, more complicated treatment. We do not consider it, because it will not be used later on.

At first we ignore the Lorentz condition (5.79), which is at the heart of the promised trouble, and consider only the equation of motion (5.80). The parameter $\lambda$ is then not redundant. Contracting (5.80) with $p_\mu$ we find $(1 - \lambda) p^2 p_\nu \tilde{A}^\nu = 0$, i.e.

$$p^2 \left( p_\nu \tilde{A}^\nu \right) = 0 . \qquad (5.81)$$

This implies

$$(p^2)^2 \tilde{A}^\mu(p) = 0 . \qquad (5.82)$$

The 1-point function $W(^\mu x) = (\Omega, A^\mu(x) \Omega)$ must be a constant transforming as a 4-vector. Such constants are not available in the theory, hence $W(^\mu x) = 0$. For the 2-point function we find from (5.80), (5.82), and the covariance, spectral, and hermiticity requirements, the form

$$\tilde{W}(^\mu p,^\nu q) =$$
$$\delta^4(p+q)\, \theta(p^0) \left\{ a \left[ g^{\mu\nu} \delta(p^2) - \frac{\lambda}{1-\lambda} p^\mu p^\nu \delta'(p^2) \right] + \rho\, p^\mu p^\nu \delta(p^2) \right\} , \qquad (5.83)$$

with $a$, $\rho$, real constants. In order to achieve agreement with the $F$-normalization (5.77) we must choose $a = -\hbar$. The $\lambda$-term and the $\rho$-term do not contribute to the $F^{\mu\nu}$-functions, hence $\lambda$ and $\rho$ are still free.

The distributions $\theta(p^0)\, \delta(p^2)$ and $\theta(p^0)\, p^\mu p^\nu \delta'(p^2)$ are defined as

$$\int dp\, \theta(p^0)\, \delta(p^2)\, f(p^0, \vec{p}) = \int \frac{d^3 p}{2|\vec{p}|}\, f(|\vec{p}|, \vec{p}) ,$$
$$\int dp\, \theta(p^0)\, p^\mu p^\nu \delta'(p^2)\, f(p^0, \vec{p}) = -\int d^3 p \left[ \frac{\partial}{\partial p^0} \left( f(p) \frac{p^\mu p^\nu}{4 p^0 |\vec{p}|} \right) \right]_{p^0 = |\vec{p}|} \qquad (5.84)$$

for $f(p) \in \mathcal{S}$. Both of these integrals exist as ordinary Riemann integrals. Hence they can be approximated arbitrarily well by their values taken for test functions which vanish in an open neighbourhood of $\{p^0 = 0\}$. This fact will be helpful in the sequel.

The higher $W$-functions are given by a cluster expansion in complete analogy to (5.13).

56    5. Free Fields

The 2-point function (5.83) and the ensuing higher functions define a field theory on a state space $\mathcal{V}$ with an indefinite scalar product, a phenomenon that we encounter here for the first time. This follows from the indefiniteness of the expression (5.83). In the case $\lambda \neq 0$ it is a simple consequence of the indefiniteness of the distribution $p^\mu p^\nu \delta'(p^2)$, which holds even for $\mu = \nu$: it is easy to find a positive test function $f(p) > 0$, $f \in \mathcal{S}$ such that $\int dp\, p^{02}\, \delta'(p^2)\, f(p) < 0$. For $\lambda = 0$ the $g^{\mu\nu}$ term dominates the $\rho$ term for sufficiently small $p$. But $g^{\mu\nu}$ is negative for $\mu = \nu = 0$, positive for $\mu = \nu = i$.

A short comparison of our results with the results of the canonical approach is in order. For simplicity we consider only the case $\lambda = 0$ (Feynman gauge). In that case the field equation (5.80) becomes in $x$-space the wave equation $\Box A^\mu = 0$. It can be derived from the Lagrangian density

$$\mathcal{L}(x) = -\frac{1}{4} F_{\mu\nu}(x)\, F^{\mu\nu}(x) - \frac{1}{2}\, (\partial_\rho A^\rho(x))^2 \, , \tag{5.85}$$

where $F_{\mu\nu}$ must be expressed in terms of the potentials $A_\mu$. The canonical prescriptions yield the equal-time commutators (see [IZ], p.128).

$$[A_\mu(t,\vec{x}), A_\nu(t,\vec{y})] = \left[\dot{A}_\mu(t,\vec{x}), \dot{A}_\nu(t,\vec{y})\right] = 0 \, ,$$

$$\left[\dot{A}_\mu(t,\vec{x}), A_\nu(t,\vec{y})\right] = i\, g_{\mu\nu}\, \delta^3(\vec{x}-\vec{y}) \, . \tag{5.86}$$

From the expression (5.83) we obtain in the same way as (5.32) (we set again $\hbar = 1$)

$$[A_\mu(x), A_\nu(y)] = i\, \{g_{\mu\nu} - \rho\, \partial_\mu \partial_\nu\}\, D(x-y) \, , \tag{5.87}$$

$D$ being defined by (5.11) with $m = 0$. Using $\Delta(\xi)|_{\xi^0=0} = 0$, $\partial_0 \Delta(\xi)|_{\xi^0=0} = \delta^3(\vec{x}-\vec{y})$ we can also calculate equal-time commutators. They agree with (5.86) only if $\rho = 0$: our expressions are more general than the canonical ones. This might at first appear as a flaw of our procedure. However, as Källén [Kä 52] has pointed out, and as will be seen in Part III, such an additional term is actually present in the asymptotic free fields of interacting QED. It is important for discussing scattering in our manifestly covariant framework and must therefore not be overlooked. As a starting point of the developments of Part II we will however work with $\rho = 0$, because only this choice leads to a renormalizable theory. Since the $\rho$-terms drop out in going over from the unobservable $A^\mu$ to the observable $F^{\mu\nu}$, we have that freedom.

The Lorentz condition $\partial_\mu A^\mu = 0$ has as yet been ignored. It remains to be integrated into the formalism. Taken as an operator identity it is only consistent with the 2-point function (5.83) if $a = 0$, which is not a value leading to a meaningful theory: we need $a = -1$. This means that the Lorentz condition is certainly not satisfied on all states of the space $\mathcal{V}$ that we have just constructed. And this means that the free Maxwell equation $\partial_\mu F^{\mu\nu} = 0$

does not hold on $\mathcal{V}$. But it must hold on physical states, i.e. on states that can in principle be produced in a laboratory. We already know, however, that not all vectors of $\mathcal{V}$ can represent physical states, because of the indefiniteness of the scalar product. On the other hand, $\mathcal{V}$ certainly does contain physical states, namely those produced by applying polynomials in the observable fields $F^{\mu\nu}$ to the vacuum.

We need then a simple characterization of the physical states of the theory. They must form a vector space $\mathcal{V}_{\mathrm{ph}}$ equipped with a positive-definite scalar product, which contains the $F$-generated states mentioned above. $\mathcal{V}_{\mathrm{ph}}$ cannot be defined as the kernel of

$$B(x) = \partial_\mu A^\mu(x) \tag{5.88}$$

in $\mathcal{V}$, because the vacuum $\Omega$ is physical, but $B(x)\Omega \neq 0$, as is again seen from (5.83).

But we may be satisfied with less. Suppose we can find a subspace $\mathcal{V}'$ of $\mathcal{V}$ such that (a) $\mathcal{V}'$ contains the vacuum $\Omega$, (b) $B$ and $F^{\mu\nu}$ map $\mathcal{V}'$ into itself, (c) the matrix elements $(\Phi, B\Psi)$ vanish for all $\Phi, \Psi \in \mathcal{V}'$, and (d) the scalar product $(\cdot, \cdot)$ is positive on $\mathcal{V}'$, though not necessarily positive-definite. We then restrict our attention to $\mathcal{V}'$, discarding the rest of the original space $\mathcal{V}$. The assumed positivity of the scalar product implies the Schwarz inequality. Hence, if $\Phi_0 \in \mathcal{V}'$ has a vanishing norm $(\Phi_0, \Phi_0) = 0$, then $|(\Phi_0, \Psi)| \leq \sqrt{(\Phi_0, \Phi_0)} \sqrt{(\Psi, \Psi)} = 0$ for all $\Psi \in \mathcal{V}'$: $\Phi_0$ is orthogonal to all of $\mathcal{V}'$. These $\Phi_0$ form a linear subspace $\mathcal{V}_0$ of $\mathcal{V}'$. Identifying vectors whose difference lies in $\mathcal{V}_0$, we obtain the quotient space $\mathcal{Q} = \mathcal{V}'/\mathcal{V}_0$,[4] on which the given scalar product induces a positive-definite scalar product. $\mathcal{Q}$ is thus a pre-Hilbert space, and the physical state space is defined as its closure

$$\mathcal{H}_{\mathrm{ph}} = \overline{\mathcal{Q}} \tag{5.89}$$

in the Hilbert norm topology. It is easily seen that $F^{\mu\nu}$ and $U(\Lambda, a)$ map equivalence classes into equivalence classes, hence extend to operators defined on $\mathcal{Q}$. The bounded $U(\Lambda, a)$ can then be extended to $\mathcal{H}_{\mathrm{ph}}$ by continuity.

The space $\mathcal{H}_{\mathrm{ph}}$ satisfies the requirements of positivity (by construction) and of the validity of the Maxwell equations: let $\Phi \in \mathcal{V}'$, then by assumption $B\Phi \in \mathcal{V}'$ and $\langle B\Phi | B\Phi \rangle = \langle \Phi | B | B\Phi \rangle = 0$, hence $B\Phi = 0$ as an element of $\mathcal{H}_{\mathrm{ph}}$. This is equivalent to the validity of the Maxwell equations.

It remains to prove the existence of $\mathcal{V}'$, which is done by explicit construction. From the 2-point function (5.83) we obtain the important formulae

---

[4] See the analogous construction used in the proof of the Wightman reconstruction theorem, Theorem 4.1. Remember that such an identification has already been effected in $\mathcal{V}$. But since $\mathcal{V}'$ is smaller than $\mathcal{V}$, there may exist elements $\Phi_0$ of $\mathcal{V}'$ of the type indicated, which are $\neq 0$ as elements of $\mathcal{V}$.

58    5. Free Fields

$$\left(\Omega, \tilde{B}(p)\, \tilde{A}^\mu(q)\, \Omega\right) = -\left(\Omega, \tilde{A}^\mu(p)\, \tilde{B}(q)\, \Omega\right)$$
$$= -\frac{i}{1-\lambda} p^\mu\, \delta^4(p+q)\, \delta_+(p)\,, \tag{5.90}$$
$$\left(\Omega, \tilde{A}^\mu(p)\, \tilde{F}^{\rho\sigma}(q)\, \Omega\right) = -\left(\Omega, \tilde{F}^{\rho\sigma}(p)\, \tilde{A}^\mu(q)\, \Omega\right)$$
$$= i(g^{\mu\sigma}\, p^\rho - g^{\mu\rho}\, p^\sigma)\, \delta^4(p+q)\, \delta_+(p)\,, \tag{5.91}$$
$$\left(\Omega, \tilde{B}(p)\, \tilde{F}^{\rho\sigma}(q)\, \Omega\right) = \left(\Omega, \tilde{F}^{\rho\sigma}(p)\, \tilde{B}(q)\, \Omega\right) = 0\,, \tag{5.92}$$
$$\left(\Omega, \tilde{B}(p)\, \tilde{B}(q)\, \Omega\right) = 0\,. \tag{5.93}$$

We split $\tilde{B}$ into an annihilation part $\tilde{B}_+$ and a creation part $\tilde{B}_-$: $\tilde{B} = \tilde{B}_+ + \tilde{B}_-$—analogously to the more explicit decomposition (5.34) of the free scalar field—by defining

$$\left(\Omega, \tilde{A}^\mu(p)\, \tilde{B}_+(q)\, \Omega\right) = 0\,, \quad \left(\Omega, \tilde{B}_+(p)\, \tilde{A}^\mu(q)\, \Omega\right) = \left(\Omega, \tilde{B}(p)\, \tilde{A}^\mu(q)\, \Omega\right),$$
$$\left(\Omega, \tilde{A}^\mu\, \tilde{B}_-\, \Omega\right) = \left(\Omega, \tilde{A}^\mu\, \tilde{B}\, \Omega\right)\,, \quad \left(\Omega, \tilde{B}_-\, \tilde{A}^\mu\, \Omega\right) = 0\,. \tag{5.94}$$

This determines all matrix elements $\left(\Omega, \prod_1^n \tilde{A}^{\mu_i}(p_i)\, \tilde{B}_\pm(p)\, \prod_1^m \tilde{A}^{\nu_j}(q_j)\, \Omega\right)$ via their cluster expansion, and thus defines $\tilde{B}_\pm$ as operators on $\mathcal{V}$. The hermiticity of $B(x)$ yields

$$\tilde{B}_+^*(p) = \tilde{B}_-(-p)\,. \tag{5.95}$$

We claim that the kernel $\mathcal{V}'$ of $B_+$,

$$\mathcal{V}' = \{\Phi \in \mathcal{V}:\ \tilde{B}_+(p)\, \Phi \equiv 0\}\,, \tag{5.96}$$

satisfies the requirements (a)–(d) introduced above. The requirement (a) is a simple consequence of the first equation of (5.94), which implies at once that $B_+\Omega = 0$. The commutators $[B_+, B]$ and $[B_+, F^{\mu\nu}]$ are c-numbers, as is seen exactly like in the proof of (5.32) in the scalar field theory. From (5.92) and (5.93) these commutators are then found to vanish identically. This yields $B_+\, F^{\mu\nu}\, \Phi = f^{\mu\nu}\, B_+\, \Phi = 0$ if $\Phi \in \mathcal{V}'$, i.e. $F^{\mu\nu}\, \Phi$ is also in $\mathcal{V}'$, and similarly for $B\,\Phi$. Hence condition (b) is verified. Condition (c) follows from

$$\left(\Phi, \tilde{B}(p)\, \Psi\right) = \left(\Phi, \tilde{B}_+(p)\, \Psi\right) + \left(\tilde{B}_+(-p)\, \Phi, \Psi\right) = 0$$

if $\Phi, \Psi \in \mathcal{V}'$.

The remaining condition (d) stating the positivity of the scalar product on $\mathcal{V}'$ is most easily verified by a more explicit analysis of $\mathcal{V}'$, which is also of considerable interest in itself. We have already seen that states obtained

## 5.3 The Electromagnetic Field

by applying any number of $F$-operators to $\Omega$ are in $\mathcal{V}'$. We claim that in fact these states span $\mathcal{V}'$.

**Lemma 5.3.** *The kernel $\mathcal{V}'$ of the field $B_+$ defined on $\mathcal{V}$ consists exactly of the states which are obtained by applying a polynomial in the fields $F^{0j}$ to the vacuum $\Omega$.*

**Proof.** By definition $\mathcal{V}$ consists of the states $\mathcal{P}(A^\mu)\Omega$ with $\mathcal{P}$ any polynomial in the fields $A^\mu$ smeared over arbitrary test functions. It suffices in fact to admit only the fields $\tilde{A}^\mu(p)$ at points $p$ with $p^0 \leq 0$. Because, if a $\tilde{A}$ with $p^0 > 0$ is contained in a term of $\mathcal{P}$, we can commute it through to the right until it acts directly on the vacuum, which it annihilates. The commutator $[A^\mu, A^\nu]$ being a c-number, the resulting expression no longer contains $\tilde{A}^\mu(p)$ as an operator. Next we note that $\mathcal{V}$ is also spanned by the vectors $\mathcal{P}(A^0, F^{0j})\Omega$, again only formed with fields at negative energies. This is so because

$$\tilde{A}^j(p) = \frac{1}{p^0}\left(i\tilde{F}^{0j}(p) + p^j\tilde{A}^0(p)\right).$$

The singularity of this expression at $p^0 = 0$ is harmless because of the remark made after (5.84). For the same reason we need only consider fields $\tilde{A}^0(p)$ in $\mathcal{P}$ which are integrated over test functions $f(p)$ with support in $\{p^0 < 0\}$. But such an $f$ we can write as $f(p^0, \vec{p}) = g(p^2, \vec{p})$, where $g$ is again a test function. We split

$$\int_{p^0 \leq 0} dp\, \tilde{A}^0(p)\, g(p^2, \vec{p}) = \int_{p^0 \leq 0} dp\, \tilde{A}^0(p)\, g(0, \vec{p})$$
$$+ \int_{p^0 \leq 0} dp\, p^2\, \tilde{A}^0(p)\, \frac{g(p^2, \vec{p}) - g(0, \vec{p})}{p^2}.$$

Because of the $\{p^2 = 0\}$-support of $\tilde{A}^0$ these expressions exist despite the non-decrease of $g(0, \vec{p})$ for $p^0 \to -\infty$. The product $p^2 \tilde{A}^0(p)$ in the last term is found to be

$$p^2\, \tilde{A}^0(p) = i\frac{\lambda}{1-\lambda} p_j\, \tilde{F}^{0j}(p)$$

with the help of the field equation (5.80). Thus this term is expressible through $F^{0j}$, and $A^0$ enters genuinely only through the term $\int dp\, \tilde{A}^0(p)\, g(0, \vec{p})$, which is non-vanishing only if $g(0, \vec{p}) \neq 0$.

We call a state $\Phi = \mathcal{P}(A^\mu)\Omega$ depending only on $A^\mu$ at momenta with $p^0 < 0$ an "$n$-field state" if $\mathcal{P}$ is homogeneous of degree $n$. Non-vanishing states with different $n$ are linearly independent because such states are mutually orthogonal, and to a non-vanishing state $\Phi_n$ with field number $n$ one can always find another such state $\Psi_n$ with $(\Psi_n, \Phi_n) \neq 0$. Hence the lemma is proved if we can prove that

$$\tilde{B}_+(p)\, \mathcal{P}_n(\tilde{A}_0, \tilde{F}^{0j})\, \Omega \neq 0 \tag{5.97}$$

whenever $\mathcal{P}_n \Omega \neq 0$ and $\mathcal{P}_n$ is a form of degree $n$ containing at least one field $\tilde{A}_0(q)$ at a value $q$ on the mass shell $q^2 = 0$. Such a state can be expanded as

$$\Phi_n = \mathcal{P}_n \Omega = \sum_{m=0}^{n} \mathcal{F}_m(\tilde{A}_0)\, \mathcal{F}'_{n-m}(\tilde{F}^{0j})\, \Omega\, ,$$

with $\mathcal{F}$ and $\mathcal{F}'$ forms of degree $m$ and $n - m$ respectively, and with at least one non-vanishing term with $m > 0$. Let $M$ be the largest such $m$. Acting with $B_+{}^M := \prod_{k=1}^M \tilde{B}_+(q_k)$ on $\Phi_n$ annihilates the terms with $m < M$. Hence

$$B_+{}^M \Phi_n = B_+{}^M \mathcal{F}_M \mathcal{F}'_{n-M} \Omega = [B_+{}^M, \mathcal{F}_M]\, \mathcal{F}'_{n-M}\, \Omega\, .$$

The commutator $[B_+{}^M, \mathcal{F}_M]$ can be evaluated and can be shown to be non-zero. Hence $B_+{}^M \Phi_n \neq 0$ for a suitable choice of the $q_k$, hence there exist $q$'s for which $B_+(q)\, \Phi_n \neq 0$. This concludes the proof of Lemma 5.3.

The scalar product on $\mathcal{V}'$ is determined by the 2-point function

$$\left(\Omega,\, \tilde{F}^{\alpha\beta}(p)\, \tilde{F}^{\gamma\delta}(q)\, \Omega\right) =$$
$$-\delta^4(p+q)\, \delta_+(p)\, (g^{\alpha\gamma}\, p^\beta\, p^\delta - g^{\alpha\delta}\, p^\beta\, p^\gamma - g^{\beta\gamma}\, p^\alpha\, p^\delta + g^{\beta\delta}\, p^\alpha\, p^\gamma)\, , \tag{5.98}$$

obtained from (5.91). This expression agrees with (5.77), which was derived in a different, more direct way.

The positivity of the scalar product is now easy to show. The $n$-field subspace $\mathcal{V}'_n$ of $\mathcal{V}'$ is spanned by the improper vectors $\prod_{r=1}^n \tilde{F}^{0i_r}(p_r)\, \Omega$ with $p_r^0 < 0$. The ordering of the factors is immaterial because the negative-energy fields commute. The $\mathcal{V}'_n$ are mutually orthogonal. $\mathcal{V}'_n$ consists of the states

$$|\ell\rangle = (n!)^{-\frac{1}{2}} \int \prod_1^n d^3 p_r\, \ell_{i_1 \cdots i_n}(\vec{p}_1, \ldots, \vec{p}_n) \prod_r F^{0i_r}(p_r)\, \Omega\, , \tag{5.99}$$

where the functions $\ell_{\cdots}$ are restrictions of totally symmetric test functions $L_{\cdots i_r \cdots}(\ldots, p_r, \ldots)$ to the negative mass shell $p_r^0 = -|\vec{p}_r| =: -\omega_r$. From (5.98) the scalar product is found to be

$$\langle \ell | \ell' \rangle = \frac{1}{2} \int \prod_r (\omega_r d^3 p_r)\, \overline{\ell_{\cdots i_r \cdots}(\ldots, \vec{p}_r, \ldots)}$$
$$\times \prod_r \left( \delta^{i_r j_r} - \frac{p_r^{i_r} p_r^{j_r}}{\omega_r^2} \right) \ell_{\cdots j_r \cdots}(\ldots, \vec{p}_r, \ldots)\, . \tag{5.100}$$

This product is positive but not positive-definite: If $\ell_{\cdots i_r \cdots}(\cdots) = p_{i_r}\, \hat{\ell}_{\cdots \hat{i}_r \cdots}(\cdots)$ and $\hat{\ell} \neq 0$, then $|\ell\rangle \neq 0$ as a vector in $\mathcal{V}$ but $\langle \ell | \ell \rangle = 0$. Hence $\mathcal{V}' = \oplus_n \mathcal{V}'_n$ satisfies all the requirements postulated in our construction of the physical state space $\mathcal{H}_{\text{ph}}$.

As has been remarked before, the expression (5.98) which lies at the heart of the construction of $\mathcal{H}_{\mathrm{ph}}$ could also have been obtained without a detour through the theory of the potentials $A^\mu$ with its unappealing indefinite scalar product. However, such a shortcut is not possible for interacting QED. There the use of potentials is quite unavoidable, and moreover the relation of $\mathcal{V}'$ to $\mathcal{V}$ will turn out to be more indirect than in the free case.

We conclude the section with a remark on the need of using an indefinite metric for the $A^\mu$-theory, and the possibility of avoiding this necessity by relaxing our requirements. After all, the $A^\mu$ have no direct observable meaning, hence we might allow them some unphysical traits like non-covariance or non-locality. We first note that in our construction of the theory the locality assumption has never been used. Like in the case of the free scalar field, locality is a consequence of the other assumptions and can therefore not be abolished if we want to retain the other properties. It has been shown by Strocchi [Str 70] that an indefinite product also cannot be avoided if one is willing to sacrifice manifest Lorentz covariance but wants to retain locality. A theory with gauge potentials acting on a physical state space with a positive scalar product ("physical gauge") is only possible if one is ready to give up both locality and covariance. Such an approach has many practical disadvantages, especially for the extension to the interacting theory, and will therefore not be pursued in this book.

## 5.4 The Functions $\Delta_+, D_+,$ and Related Functions

We wish to discuss the properties of the function $\Delta_+(\xi)$ defined by (5.10) and of some related functions—collectively known as "invariant functions"—which will play an important role in Parts II and III.

We note first that $\delta_+(p) = \theta(p_0)\,\delta(p^2 - m^2)$ is invariant under orthochronous Lorentz transformations, and so is the differential $dp := d^4p$, hence $\Delta(\xi)$ is invariant,

$$\Delta_+(\Lambda\,\xi) = \Delta_+(\xi) \tag{5.101}$$

for $\Lambda \in L$.

Secondly, the definition (5.10) can be generalized to complex arguments. Let $\xi, \eta$, be two real 4-vectors and define $\zeta = (\xi + i\eta) \in \mathcal{C}^4$. We define

$$\begin{aligned}\Delta_+(\zeta) &= \frac{i}{(2\pi)^3} \int dp\, \delta_+(p)\, e^{-ip\zeta} \\ &= \frac{i}{(2\pi)^3} \int dp\, \delta_+(p)\, e^{-ip\xi}\, e^{p\eta}\ .\end{aligned} \tag{5.102}$$

The support of $\delta_+$ is the "mass shell" $\{p^2 = m^2,\ p^0 \geq 0\}$. If we choose $\eta$ in the open backward cone $V_-$, then $e^{-ip\zeta}$ decreases rapidly for $p \to \infty$ inside this support: $e^{-ip\zeta}$ is an admissible test function for the distribution $\delta_+$. Hence $\Delta_+(\zeta)$ exists pointwise and can easily be shown to depend continuously on $\zeta$. The same is true for its complex derivatives,

$$\frac{\partial}{\partial \zeta^\mu} \Delta_+(\zeta) = (2\pi)^{-3} \int dp\, \delta_+(p)\, p_\mu\, e^{-ip\zeta}\ .$$

This means that the function $\Delta_+(\zeta)$ is analytic[5] in the "backward tube"

$$\mathcal{T}_- := \{\zeta \in \mathcal{C}^4 \ :\ \Im\zeta \in V_-\}\ . \tag{5.103}$$

Moreover it can be shown (see e.g. [StW]) that the distribution $\Delta_+(\xi)$ is a "boundary value" of the analytic function $\Delta_+(\zeta)$ in the following sense: the value of the distribution $\Delta_+$ on the test function $f(\xi) \in \mathcal{S}$ is given by

$$\Delta_+(f) = \lim_{\eta \to 0} \int d\xi\, f(\xi)\, \Delta_+(\xi + i\eta)\ , \tag{5.104}$$

the limit $\eta = 0$ being approached from inside $V_-$. This statement will be proved and generalized at the end of the section.

For a more explicit discussion let us first consider the special case $m = 0$, i.e. the distribution

$$D_+(\xi) = \frac{i}{(2\pi)^3} \int dp\, \theta(p_0)\, \delta(p^2)\, e^{-ip\xi}\ . \tag{5.105}$$

$D_+$ has the scaling property

$$D_+(\lambda \xi) = \lambda^{-2} D_+(\xi) \tag{5.106}$$

for any positive real $\lambda$. There are only two linearly independent Lorentz invariant distributions with this scaling behaviour: $[(\xi^0 + i\varepsilon)^2 - (\vec{\xi})^2]^{-1}$ and $[(\xi^0 - i\varepsilon)^2 - (\vec{\xi})^2]^{-1}$. Only the second one of them is the boundary value of an analytic function in $\mathcal{T}_-$, namely $(\zeta^2)^{-1}$. Hence we find

$$D_+(\xi) = \frac{c}{(\xi^0 - i\varepsilon)^2 - \vec{\xi}^2}\ . \tag{5.107}$$

The coefficient $c$ is found by evaluating $D_+$ at $\xi^0 = 0$,

$$D_+(0, \vec{\xi}) = \frac{i}{(2\pi)^3} \int \frac{d^3p}{2|\vec{p}|}\, e^{i(\vec{p},\vec{\xi})} = \frac{i}{(2\pi)^2} \frac{1}{|\vec{\xi}|^2}\ ,$$

a well-known result from potential theory. This gives $c = -i\,(2\pi)^{-2}$.

In the case $m > 0$ we use a more elaborate strategy to arrive at an explicit expression for $\Delta_+$. In $\mathcal{T}_-$ we can replace the independent variables

---

[5] A function of $n$ complex variables $z_1, \ldots, z_n$ is called analytic in the domain $\mathcal{D} \subset \mathcal{C}^n$ if it is continuous on $\mathcal{D}$ and is there analytic as a function of each $z_i$ separately.

5.4 The Invariant Functions

$\zeta^\mu$ by $z = (\zeta, \zeta) = g_{\mu\nu} \zeta^\mu \zeta^\nu$ and $\zeta^i$. This substitution is regular. Hence we can write $\Delta_+(\zeta) = F(z, \zeta^i)$ where $F$ is again an analytic function. But $\Delta_+(\zeta)$ is invariant under the real Lorentz transformations $\zeta \to \Lambda\zeta = \Lambda\xi + i\Lambda\eta$. The variable $z$ is invariant under these transformations, but the $\zeta^i$ are not and also cannot be combined into invariant expressions. Thus $F$ cannot depend on $\zeta^i$, and we find

$$\Delta_+(\zeta) = F(z) = \Delta_+ \left( \zeta^0 = (z - \zeta_i \zeta^i)^{\frac{1}{2}}, \zeta^i \right) . \quad (5.108)$$

The sign of the square root is chosen such that its imaginary part is negative. For any $z$ in the complex $z$-plane cut along the real axis we can find a $\zeta \in \mathcal{T}_-$, e.g. $\zeta = (\sqrt{z}, \vec{0})$. Hence $F(z)$ is analytic in this cut plane. Note that this domain contains the real negative $z$ so that $\Delta_+(\zeta)$ can be continued analytically to real spacelike arguments $\xi$. For lightlike and timelike arguments $\Delta_+(\xi)$ is a boundary value of $F$,

$$\Delta_+(\xi) = \lim_{\varepsilon \downarrow 0} F\left( (\xi^0 - i\varepsilon)^2 - (\vec{\xi}\,)^2 \right) , \quad (5.109)$$

the limit being taken in the sense of distributions like in (5.104).

For the explicit determination of $F$ we use the special $\zeta$'s mentioned above and find

$$F(z) = \frac{i}{(2\pi)^3} \int \frac{d^3p}{2\omega} e^{-i\omega\sqrt{z}} , \quad (5.110)$$

which integral converges for $\Im\sqrt{z} < 0$. Remember $\omega = (\vec{p}^{\,2} + m^2)^{1/2}$. Expression (5.110) can be transformed into

$$\begin{aligned} F(z) &= \frac{i}{(2\pi)^2} \int_m^\infty d\omega \sqrt{\omega^2 - m^2}\, e^{-i\omega\sqrt{z}} \\ &= -\frac{m^2}{8\pi} \frac{1}{m\sqrt{z}} H_1^{(2)}(m\sqrt{z}) , \end{aligned} \quad (5.111)$$

with $H_1^{(2)}$ a Hankel function of the second kind.[6] This is the desired explicit form of $\Delta_+$.

From the known properties of the Hankel functions we can derive some important properties of $\Delta_+(\xi)$. $H_1^{(2)}(w)$ has a branch point at $w = 0$ but can elsewhere be analytically continued from the lower half-plane to the real axis. This implies that $\Delta_+(\xi)$ is $C^\infty$ everywhere except at the light-cone $\xi^2 = 0$. The behaviour near the light-cone follows from the known expansion of $H_1^{(2)}(w)$ at small $w = m\sqrt{z}$. We find for small $\xi^2$

$$\Delta_+(\xi) = -\frac{i}{(2\pi)^2} \frac{1}{(\xi^0 - i\varepsilon)^2 - \vec{\xi}^{\,2}} + \frac{i\,m^2}{8\pi^2} \log\left( m |\sqrt{\xi^2}| \right) + \text{bounded} . \quad (5.112)$$

---
[6] For the definition and properties of Bessel and Hankel functions see vol.2, chapter VII, of the Bateman project [EMOT].

Note that the strongest singularity is the same as in the massless case (5.107). But now there is also a logarithmic singularity present.

The behaviour for large $\zeta$ or $\xi$ will also be important. If the real argument $\xi$ tends to infinity in a spacelike direction, $\Delta_+(\xi)$ decreases exponentially like $(-\xi^2)^{-3/4} \exp[-m(-\xi^2)^{1/2}]$. For $\xi \to \infty$ in a timelike direction $\Delta_+(\xi)$ behaves like $(\xi^2)^{-3/4} \exp[\pm im(\xi^2)^{1/2}]$. The sign in the oscillatory factor depends on the sign of $\xi^0$. Constant coefficients have been suppressed in both cases. For $\zeta \to \infty$ in a complex direction the asymptotic behaviour of $\Delta(\zeta)$ is given by $(\zeta^2)^{-3/4} \exp[-im(\zeta^2)^{1/2}]$. Remember that we define $\sqrt{\zeta^2}$ with a negative imaginary part and that $\eta = \Im\zeta$ lies in $V_-$. Hence $\Delta_+(\zeta)$ decreases exponentially in complex directions inside $\mathcal{T}_-$.

The functions $-D_-(\xi)$ and $-\Delta_-(\xi)$ are boundary values of the same analytic functions as $D_+$ and $\Delta_+$, the boundary being approached from a different direction: the $i\varepsilon$-terms in (5.107) and (5.109) reverse their sign. The real $\xi$ are approached from $\mathcal{T}_+$, defined as in (5.103) but with $\eta \in V_+$. $\mathcal{T}_+$ is contained in the domain of analyticity of $F(\zeta^2)$. These considerations apply also to the case $m = 0$ with $F(z) = -i(2\pi)^{-2}z^{-1}$. For spacelike $\xi$, the real value $z = \xi^2 < 0$ lies in the domain of analyticity of $F(z)$. Therefore the sign of the $\varepsilon$-prescription is irrelevant there, and we have

$$\Delta_-(\xi) = -\Delta_+(\xi) \quad \text{if} \quad \xi^2 < 0, \tag{5.113}$$

so that

$$\Delta(\xi) = \Delta_+(\xi) + \Delta_-(\xi) = 0 \quad \text{if} \quad \xi^2 < 0, \tag{5.114}$$

in accordance with the locality of the free scalar field, in connection with which the functions $\Delta_\pm, \Delta$ have first been introduced.

Another related function that will become important is the "Feynman propagator"

$$\Delta_F(\xi) = \theta(\xi^0)\,\Delta_+(\xi) - \theta(-\xi^0)\,\Delta_-(\xi), \tag{5.115}$$

and similarly $D_F$. It is the boundary value of $F(\zeta^2)$ from $\mathcal{T}_-$ if $\xi^0 > 0$, from $\mathcal{T}_+$ if $\xi^0 < 0$. For $\xi^0 = 0$, $\vec{\xi} \neq 0$, we are in the analyticity domain of $F$, so that the two prescriptions coincide. Only at $\xi = 0$ is this analytic definition not possible. This problem is not serious as long as we are only interested in $\Delta_F$ by itself, because limits of the form (5.104) exist separately for the two terms in (5.115) despite the discontinuous $\theta$-factors. In fact one finds that $\Delta_F$ can be expressed as

$$\Delta_F(\xi) = \lim_{\varepsilon \downarrow 0} F(\xi^2 - i\varepsilon), \tag{5.116}$$

where $F(z)$ is the analytic function occurring in (5.109). This means that

$$\int d\xi\, \Delta_F(\xi)\, f(\xi) = \lim_{\varepsilon \downarrow 0} \int d\xi\, f(\xi)\, F(\xi^2 - i\varepsilon) \tag{5.117}$$

exists for all $f \in \mathcal{S}$ and defines a tempered distribution. Nevertheless, the point $\xi = 0$ remains singular. This manifests itself in particular in the fact that

$$\lim_{\varepsilon \downarrow 0} \int d\xi\, f(\xi) \left[ F(\xi^2 - i\varepsilon) \right]^2$$

does not exist if $f(0) \neq 0$. This problem will become serious later on: it lies at the heart of the notorious UV-difficulties.

The fact that $\Delta_+(\xi)$ is a boundary value of the analytic function $\Delta(\zeta)$, and generalizations of it, will play a vital role later on. We will therefore provide an exact statement and a proof of the relevant theorem. This is desirable because such a proof is hard to find in the literature, even though the theorem itself is widely known.

**Theorem 5.4.** *Let $z_j = x_j + i y_j$, $j = 1, \ldots, n$, be complex 4-vectors. Define $V_-^r = \{q : |\vec{q}| < -r q_0\}$ with $0 < r < 1$. Let the function $F(z_1, \ldots, z_n)$ be analytic in*

$$\mathcal{A} = \left\{ \{z_j\} : y_j \in V_+^r,\, \sum y_{j0} \leq K \right\} \tag{5.118}$$

*for a positive constant $K$. Assume that there exist non-negative real numbers $c_1, \ldots, c_n$, $a$, $b$, and a positive integer $N$ such that*

$$\prod_j y_{j0}^{c_j} |F(z_1, \ldots, z_n)| \leq a + b \left( \sum_j |x_j| \right)^N \tag{5.119}$$

*in $\mathcal{A}$. Then the boundary value*

$$\int \prod dx_j\, F(x_1, \ldots, x_n)\, f(x_1, \ldots, x_n)$$
$$= \lim_{y_j \to 0} \int \prod dx_j\, F(x_1 + i y_1, \ldots)\, f(x_1, \ldots) \tag{5.120}$$

*exists for every test function $f \in \mathcal{S}$ and defines a tempered distribution, if the limit $y_j = 0$ is approached from within $V_+^r$.*

$|x|$ is the Euclidean length of the 4-vector $x$. The theorem remains of course true if the backward cone $V_-^r$ is replaced by the forward cone $V_+^r = \{q \in R^4 : r q_0 > |\vec{q}|\}$.

**Proof of the theorem.** Consider first the case $n = 1$. Choose $\bar{z}_0 = K i$. Then $\bar{z} = (\bar{z}_0, \vec{z})$ is in $\mathcal{A}$ if $|\vec{y}| < rK$, $\vec{y} = \Im \vec{z}$. For $z \in \mathcal{A}$ define

$$F_1(z) = \int_{\bar{z}_0}^{z_0} d\zeta_0\, F(\zeta_0, \vec{z})\,,$$

the integral being taken along a path inside $\mathcal{A}$. $F_1(z)$ is again analytic in $\mathcal{A}$, and $F = (\partial/\partial z_0)\, F_1$. Choosing a path that first runs in a real direction from $iK$ to $x_0 + iK$, then in an imaginary direction from there to $z_0$, and inserting the estimate (5.119) in the integrand, we find that $|F_1|$ satisfies an estimate of the form (5.119) with new constants $a_1, b_1, N_1$, and $c_1 = c - 1$ if $c > 1$, $c_1$ positive but arbitrarily small if $c = 1$, $c_1 = 0$ if $c < 1$. Iterating this procedure we find that there exists a non-negative integer $M$ and a function $F_M$ analytic in $\mathcal{A}$ such that

$$F(z) = \frac{\partial^M}{\partial z_0{}^M} F_M(z) \tag{5.121}$$

and

$$|F_M(z)| \leq a_M + b_M |x|^M \tag{5.122}$$

in $\mathcal{A}$. $F_M(z)$ is bounded on $\mathcal{A}$. The next iteration $F_{M+1}(z)$ has even a bounded $y^0$-derivative. Thus we find for $f(x) \in \mathcal{S}$

$$\int dx\, f(x)\, F(x + iy) = (-1)^M \int dx\, \frac{\partial^{M+1}}{\partial x_0{}^{M+1}} f(x)\, F_{M+1}(x + iy),$$

and $\lim_{y \to 0}$ of this integral exists and depends continuously on $f$. In the case $n > 1$ we repeat this procedure $n$ times.

The function $\Delta_+$ satisfies the assumptions of the theorem. We have already seen that $\Delta_+(\zeta)$ is analytic in $\mathcal{A}$. That the estimate (5.119) is satisfied is seen by noting that $\delta_+(p)$ is a positive measure, so that for $\zeta = \xi + i\eta$, $\eta \in V_-^r$ we have

$$\left| \int dp\, \delta_+(p)\, e^{-ip\xi}\, e^{p\eta} \right| \leq \int dp\, \delta_+(p)\, e^{p\eta}$$

$$\leq \int dp\, \theta(p_0)\, \delta(p^2 - m^2)\, e^{(1-r)p_0\eta_0}$$

$$= \frac{1}{(1-r)^2 \eta_0{}^2} \int dq\, \theta(q_0)\, \delta\big(q^2 - m^2(1-r)^2 \eta_0{}^2\big)\, e^{-q_0}.$$

The $q$-integral remains bounded for $\eta_0 \to 0$, hence the condition (5.119) is satisfied with $b = 0$, $c = 2$, $a$ suitable.

# Chapter 6

# An Outline of Interacting QED

This chapter contains a heuristic decription of the basic ideas underlying the formulation of interacting QED, and of the various problems with which this program is beset. The claims of mathematical rigour are set aside in this provisional account, except as far as they need be considered to locate the major problems.

## 6.1 General Description

QED is a quantized form of classical electrodynamics as outlined in Chap. 3. This means that it is a quantum field theory as discussed in Chap. 4, whose basic fields are quantum versions of the classical fields of Chap. 3 satisfying quantum versions of the field equations introduced there.

The field strengths $F^{\mu\nu}(x)$ and the current densities $j^\mu(x)$ represent observables and must therefore be present in the theory. Moreover, they should satisfy the Maxwell equations (3.6) and (3.7),

$$\partial_\nu F^{\nu\mu} = j^\mu \,, \tag{6.1}$$

$$\varepsilon_{\mu\nu\rho\sigma} \partial^\nu F^{\rho\sigma} = 0 \,. \tag{6.2}$$

The current $j^\mu$ must satisfy the continuity equation

$$\partial_\mu j^\mu = 0 \,, \tag{6.3}$$

so that the charge

$$Q = \int_{x^0=t} d^3x \, j^0(x) \tag{6.4}$$

is conserved. All this is exactly as in the classical case, except that now $F^{\mu\nu}$, $j^\mu$, and $Q$ are operators on an infinite dimensional vector space $\mathcal{V}$.

## 6. Interacting QED

The observable fields $F^{\mu\nu}$ and $j^\mu$ must be local for the reasons explained in Postulate 6 of Chap. 4, and they must transform as a tensor field of rank 2 and a vector field, respectively, in order to ensure the relativistic invariance of the theory.

The field $\varphi(x)$ is said to carry charge $q$ if

$$[Q, \varphi(x)] = q\,\varphi(x) \ . \tag{6.5}$$

This implies that $\varphi(x)\,v$ is an eigenstate of $Q$ with eigenvalue $q' + q$ if $v$ is an eigenstate with eigenvalue $q'$: application of $\varphi$ to a state raises its charge by $q$. Note that $Q$ is empirically known to possess a purely discrete spectrum, hence its eigenstates should span the whole state space. By taking the adjoint of (6.5) and noting that $Q$, as an observable, must be hermitian, we find

$$[Q, \varphi^*(x)] = -q\,\varphi^*(x) \ ; \tag{6.6}$$

the adjoint $\varphi^*$ of the field $\varphi$ carries the reverse charge of $\varphi$. This implies that hermitian fields carry charge zero. In particular this is true for $F^{\mu\nu}$ and $j^\mu$ which, being observable, must be hermitian: the electromagnetic field carries no charge, and so does, somewhat paradoxically, the current $j^\mu$ despite of its describing the movement of charges.

In order to have charges present we must therefore introduce additional, charged, fields. Since it is our aim to describe the interaction of the electromagnetic field with electrons and positrons, we choose for this purpose spin 1/2 fields, i.e. a Dirac spinor $\psi(x)$ and its twin $\bar\psi(x) = \psi^*(x)\,\gamma^0$. $\psi$ shall carry the charge $-e$, $\bar\psi$ the charge $e$, the charge of the positron. Since $\psi$ and $\bar\psi$ are not hermitian, they cannot be observables. Nevertheless we desire them to be local and covariant for the reasons explained in chapter 4.

But here we already meet the first serious problem: the requirements enunciated as yet are incompatible, as has been shown in [FPS 74]. This is seen as follows (a rigorous version of the argument will be given in the next chapter). From (6.1) and (6.4) we find, remembering that $F^{00} = 0$,

$$Q = \int_{x^0=t} d^3x\, j^0(x) = \int_{x^0=t} d^3x\, \partial_i F^{i0}(x) \tag{6.7}$$

independently of $t$, hence

$$[Q, \psi(y)] = \int_{x^0=y^0} d^3x\, \frac{\partial}{\partial x^i}\left[F^{i0}(x), \psi(y)\right] \ . \tag{6.8}$$

But because of locality the commutator $\left[F^{i0}(y^0, \vec x), \psi(y^0, \vec y)\right]$ vanishes at $\vec x \neq \vec y$, in particular for $|\vec x| \to \infty$, so that the expression (6.8) vanishes by virtue of Gauß' theorem, in contradiction to the assumption that $\psi$ carries a non-vanishing charge. We met such a quandary already in developing the theory of the free 4-potential $A^\mu$. There the problem could have been solved by giving up locality and Lorentz covariance for unphysical fields. But we preferred to relinquish instead the universal validity of the Maxwell equations. The same alternative exists here. If $\psi$ is non-local, then $\left[F^{i0}(x), \psi(y)\right]$ may not tend to zero for $|\vec{x}| \to \infty$ sufficiently fast to make the theorem of Gauß applicable. But again, we prefer to sacrifice the universal validity of the Maxwell equation (6.1). If this equation is not satisfied, the equation (6.7) is no longer correct and the contradiction disappears. The price to pay is again the appearance of unphysical states in the formalism.

In fact the two problems are closely related. Before we can see this we must consider the nature of the interaction between the spinor fields and the electromagnetic field. This interaction is modelled after the classical description given in Chap. 3. Since the current $j$ describes the movement of the charges generated by $\psi$ and $\bar{\psi}$, it must be expressible as a function of these fields. For the form of this dependence we use the charge-symmetrical ansatz (5.71),

$$j^\mu(x) = \frac{e}{2}\left[\bar{\psi}(x),\, \gamma^\mu\, \psi(x)\right] . \tag{6.9}$$

As equations of motion for the spinors we introduce the Dirac equations (3.15):

$$(i\slashed{\partial} - e\slashed{A} - m)\psi = 0 \quad,\quad \bar{\psi}(i\overleftarrow{\slashed{\partial}} + e\slashed{A} + m) = 0 , \tag{6.10}$$

in which the electromagnetic field enters through the potentials $A_\mu$ defined by

$$F_{\mu\nu} = \partial_\mu A_\nu - \partial_\nu A_\mu . \tag{6.11}$$

It is not possible to write down local field equations for $\psi$, $\bar{\psi}$ in terms of $\psi$ and $F$ only. We therefore use $A_\mu$, $\psi$, $\bar{\psi}$ as the fundamental fields of the formalism, treating $F$ as a function of $A$ given by (6.11). The Maxwell equation (6.1) becomes

$$\Box A^\mu - \partial^\mu \partial_\nu A^\nu = j^\mu . \tag{6.12}$$

As has been pointed out in Sect. 5.2, the commutator in (6.9) is not defined as it stands, since quantum fields are distributions, not functions. The same is true for the product $\slashed{A}(x)\,\psi(x)$ in the Dirac equation. But for the moment we ignore this problem.

An important formal property of the field equations (6.10) and (6.12) is their *gauge invariance*, as has already been discussed in the classical case. Let $G(x)$ be a real function or distribution on $R^4$, possibly operator-valued. Let the fields $A^\mu(x)$, $\psi(x)$, $\bar{\psi}(x)$ solve (6.10) and (6.12). Then the gauge-transformed fields

$$A_G^\mu(x) = A^\mu(x) + \partial^\mu G(x) ,$$
$$\psi_G(x) = e^{-ieG(x)}\,\psi(x) \quad,\quad \bar{\psi}_G(x) = \bar{\psi}(x)\,e^{ieG(x)} , \tag{6.13}$$

are again a solution. The observable fields $F$ and $j$ are invariant under gauge transformations, while the unobservable $A$ and $\psi$ are not. More generally, only gauge invariant operators can represent observable quantities. And since the observable content of the theory consists in relations between observables, the transformed fields and the original fields should define physically equivalent theories. This is what is known as the gauge invariance of QED. A mathematically rigorous analysis of this notion is however highly complicated, much more so in QED than in classical electrodynamics, and will not be attempted in this book. A discussion of the problem can be found in [StrW 74]; an even approximately complete solution does not yet exist to the best of my knowledge. Let me only mention that the crux of the problem lies in the fact that in general gauge transformations are not implementable within a single state space. They change not only the field operators but also the spaces on which they act. And this change of space is hard to control.

Like in the classical case we can choose to work in mathematically convenient gauges singled out by gauge conditions, without restriction of physical generality. And this brings us back to our problem: that, if we insist on the Maxwell equations being satisfied as an operator equation, then $\psi$ cannot be a local field and $A^\mu$ can be neither local nor covariant. Hence we must relax some of the assumptions hitherto discussed as being desirable. Because of the gauge freedom we have some choice in the matter. Most attractive from a physical point of view are the "physical gauges" defined as gauges in which the equations of motion (6.10) and (6.12) hold without restriction, and whose state spaces are physical in the sense that they are spanned by experimentally realizable "physical" states. But then $A$ and $\psi$ are by necessity neither local nor Lorentz covariant. The best-known example of such a gauge is the "Coulomb" or "radiation" gauge characterized by the gauge condition $\partial_i A^i(x) = 0$, which is to be satisfied in addition to the field equations. Unfortunately these physical gauges are mathematically awkward. Locality and manifest covariance are powerful tools in developing a detailed formalism, especially for establishing a consistent perturbation theory. And the unavailability of these tools in physical gauges makes itself sorely felt. We will therefore use an extension to the interacting case of the Gupta–Bleuler method presented in the preceding chapter for the free electromagnetic field. In this method the basic fields $A$, $\psi$, $\bar\psi$ are local and covariant. The equations of motion are the modified Maxwell equation (3.22),

$$\Box A^\mu - \lambda\, \partial^\mu \partial_\nu\, A^\nu = j^\mu\,, \tag{6.14}$$

with $\lambda \neq 1$ a real number and $j^\mu$ given by (6.9), and the Dirac equations (6.10). We call (6.14) the p-Maxwell (for pseudo-Maxwell) equation. In addition we demand the Lorentz condition

$$\partial_\mu A^\mu = 0 \tag{6.15}$$

as a gauge condition. Equation (6.15) together with (6.14) implies the true Maxwell equation (6.1). However, we have seen in the preceding chapter that the stated requirements are incompatible. The way out of this difficulty is the same as that followed in the free case. In a first step a field theory is developed based on the field equations (6.10) and (6.14) and the Wightman properties of Chap. 5, but disregarding the Lorentz condition. The resulting theory is not yet the desired theory, because its state space $\mathcal{V}$ is not a Hilbert space and the Maxwell equations are not satisfied. But starting from this unphysical theory a new state space $\mathcal{H}_{\mathrm{ph}}$ can be constructed, which *is* a Hilbert space and on which the Lorentz condition is satisfied in addition to the field equations, but on which the explicit Lorentz covariance is lost.

The traditional method of achieving this, which is not the method we will use, runs in outline as follows.[1] The state space $\mathcal{V}$ of the first step consists of the states

$$\mathcal{P}(A_\mu, \psi, \bar{\psi}) \Omega , \qquad (6.16)$$

where $\mathcal{P}$ is any polynomial in the fields $A_\mu$, $\psi$, $\bar{\psi}$ smeared over test functions. On this space an invariant but indefinite scalar product is defined. As has been indicated above, this space does not contain any charged physical states, because the charge (6.7) vanishes identically on $\mathcal{V}$. But charged physical states may be obtained as limits of sequences of $\mathcal{V}$-states embedded in a larger space. The standard procedure is to define on $\mathcal{V}$ a second, noncovariant but positive scalar product $(\cdot,\cdot)_H$, thereby turning $\mathcal{V}$ into a pre-Hilbert space. This space can then be completed to a Hilbert space $\mathcal{H}$ as usual: $\mathcal{H}$ consists of the Cauchy sequences of elements of $\mathcal{V}$ with respect to the norm topology induced by $(\cdot,\cdot)_H$. This new space may then be sufficiently large to contain a full set of physical states. But $\mathcal{H}$ is not yet the desired space because the Maxwell equations are still not satisfied. The physical space $\mathcal{H}_{\mathrm{ph}}$ is then obtained as a subspace of $\mathcal{H}$, more exactly a quotient of two subspaces, precisely as explained in Chap. 5 for the free electromagnetic field. Like there the divergence

$$B(x) = \partial_\mu A^\mu(x) \qquad (6.17)$$

is introduced as an auxiliary field. From the p-Maxwell equation and current conservation it is found that $B$ satisfies the free wave equation

$$\Box B(x) = 0 . \qquad (6.18)$$

The results of Sect. 5.1 are applicable. In particular, the representation (5.34),

$$\tilde{B}(p) = \delta(p^2 - m^2) \left( b(\vec{p}) \, \theta(p_0) + b^*(-\vec{p}) \, \theta(-p_0) \right) , \qquad (6.19)$$

holds. Here $b^*$ is a creation operator, $b$ an annihilation operator. The annihilation part of (6.19) is called $B_+$, the creation part $B_-$, and the subspace $\mathcal{H}'$ of $\mathcal{H}$ is defined as

$$\mathcal{H}' = \{\Phi \in \mathcal{H} \; : \; B_+(x) \, \Phi \equiv 0\} , \qquad (6.20)$$

which is the generalization of (5.96) to interacting QED. On $\mathcal{H}'$ the two scalar products $(\cdot,\cdot)$ and $(\cdot,\cdot)_H$ coincide. $\mathcal{H}_{\mathrm{ph}}$ is the quotient of $\mathcal{H}'$ by its

---
[1] For a careful and detailed discussion of this method see [MSt 83].

subspace $\mathcal{H}^0$ of elements with vanishing norm. For $\Phi$, $\Psi \in \mathcal{H}_{\text{ph}}$ one finds $(\Phi, B(x)\Psi) \equiv 0$: in this sense the Maxwell equations are satisfied.

But this procedure is not really satisfying. For one thing, the positive scalar product $(\cdot, \cdot)_H$ is introduced ad hoc, without a recognizable physical justification. More serious is the following objection which will be discussed in more detail in the next chapter. We can distinguish three different definitions of the charge operator:

(a) The "current charge" $Q_C$ is defined by (6.4).
(b) The "Maxwell charge" $Q_M$ is defined by the last expression in (6.7).
(c) The "gauge charge" $Q_G$ is defined by (6.5), or in more detail by

$$[Q_G, A^\mu(x)] = 0 , \quad [Q_G, \psi(x)] = -e\,\psi(x) , \quad [Q_G, \bar{\psi}(x)] = e\,\bar{\psi}(x) ,$$
$$Q_G \Omega = 0 . \tag{6.21}$$

The transformation

$$A^\mu(x) \rightarrow A^\mu(x) ,$$
$$\psi(x) \rightarrow e^{i\alpha Q_G}\,\psi(x)\,e^{-i\alpha Q_G} = e^{-i\alpha e}\,\psi(x) ,$$
$$\bar{\psi}(x) \rightarrow e^{i\alpha Q_G}\,\bar{\psi}(x)\,e^{-i\alpha Q_G} = e^{i\alpha e}\,\bar{\psi}(x) , \tag{6.22}$$

for $\alpha \in \mathbb{R}$ transforms solutions of the field equations again into solutions. This transformation is called a "global gauge transformation" or gauge transformation of the first kind, whilst the more general transformations (6.13) are called "local" or of second kind. The global transformations form a Lie group of which $Q_G$ is the infinitesimal generator.

The Maxwell charge $Q_M$ vanishes identically on $\mathcal{V}$, as has been indicated above, while $Q_G = Q_C \neq 0$. That $Q_G \neq 0$ is easily seen because the state

$$\prod_{i=1}^{\alpha} A^\mu(x_i) \prod_{j=1}^{\beta} \psi(y_j) \prod_{k=1}^{\gamma} \bar{\psi}(z_k) \Omega$$

is an eigenstate of $Q_G$ with eigenvalue $e(\gamma - \beta) \neq 0$ if $\beta \neq \gamma$. For the equality of $Q_G$ and $Q_C$ we refer to Chap. 7. On the other hand, on $\mathcal{H}_{\text{ph}}$ the three definitions must coincide. Now, according to the traditional wisdom a physical state $\Phi$ is obtained as a limit of a sequence $\Phi_1, \Phi_2, \ldots$ of states in $\mathcal{V}$, the limit being taken in the Hilbert completion $\mathcal{H}$ of $\mathcal{V}$. Assume that $\Phi$ is a charge eigenstate with a non-vanishing eigenvalue. These states are total in $\mathcal{H}_{\text{ph}}$ because $Q$ is known to possess a purely discrete spectrum. We have then $\Phi = \lim_{i \to \infty} \Phi_i$ but $Q_M \Phi \neq \lim_{i \to \infty} Q_M \Phi_i$: $Q_M$ cannot be a closed operator. Put otherwise: there is an operator $Q_M$ on $\mathcal{V}$ and another operator $Q_M$ on $\mathcal{H}_{\text{ph}}$, and the two have little or nothing to do with each other. And this suggests that it is perhaps not such a good idea to insist on considering $\mathcal{V}$ and $\mathcal{H}_{\text{ph}}$ as subspaces of one and the same vector space $\mathcal{H}$.

We will therefore employ a different, more indirect method of constructing $\mathcal{H}_{\text{ph}}$ from $\mathcal{V}$. Individual physical states are not defined as limits of sequences of $\mathcal{V}$-states, but only the space $\mathcal{H}_{ph}$ as a whole is constructed from $\mathcal{V}$ as a limit of sorts via a detour through the Wightman functions, using the Wightman reconstruction theorem. Let $\varphi(x)$ be a generic symbol for the basic fields $\psi$, $\bar{\psi}$, $A$, $F$. We will introduce a family of gauge transformations $\varphi(x) \to \varphi_\xi(x)$ of the form (6.13) with operator-valued gauge functions $G_\xi$ which are parametrized by a real parameter $\xi > 0$. The fields $\varphi_\xi$ are defined on $\mathcal{V}$, but the limits $\varphi_0 = \lim_{\xi \to 0} \varphi_\xi$ do not exist as operators on $\mathcal{V}$. However, the limiting Wightman functions

$$W(\ldots, x_i, \ldots) = \left(\Omega, \prod_1^n \varphi_0(x_i) \Omega\right) = \lim_{\xi \to 0} \left(\Omega, \prod \varphi_\xi(x_i) \Omega\right) \quad (6.23)$$

do exist and satisfy the assumptions of the reconstruction theorem, except locality and Lorentz covariance. The limit $\xi \to 0$ must be taken in all the factors of (6.23) at the same time. The reconstruction theorem yields then a field theory with non-local and non-covariant fields but with a positive scalar product and the correct equations of motion. In other words: we end up in a physical gauge. But this we do only after the relevant quantities have been fully constructed in the Gupta–Bleuler gauge, making full use of locality and covariance, so that the eventual loss of these helpful properties is no longer bothersome. By the same token we can also eliminate at this late stage the unphysical potentials $A^\mu$ in favour of the observable field strengths $F^{\mu\nu}$. The explicit form of the transformation will be described in Chap. 12 in the context of perturbation theory. A rigorous discussion of the procedure outside of perturbation theory is still lacking.

## 6.2 Solving the Field Equations

There remains the task of solving the field equations (6.10) and (6.14). In trying to do this we are at once confronted with two serious problems.

(a) Until now we have talked of fields as though they were functions. But in fact they are distributions. And the product of two distributions is a priori undefined. Hence the interaction terms in the field equations are at first meaningless. This problem is known as the ultraviolet (UV) problem. That name is due to the fact that the formal Fourier transform of the distribution product $D_1(x) D_2(x)$ is the convolution $\int dk\, \tilde{D}_1(k)\, \tilde{D}_2(p-k)$ which is also undefined in general. But here the problem lies not in the local singularities of $\tilde{D}_1$ and $\tilde{D}_2$, but in their possibly insufficient decrease at infinity. If the $D_i$ are free fields, then the support of $\tilde{D}_i(k)$ is the hyperboloid $k^2 = m^2$, so that $k \to \infty$ in the support of $\tilde{D}_i$ implies that the energy $k_0$ becomes large, hence "UV-problem".

(b) The field equations are partial differential equations which must be expected to possess a vast number of different solutions, if they possess solutions at all. We must therefore formulate subsidiary conditions which single out the physically relevant solution or solutions.

## 6. Interacting QED

Let us first discuss the second problem, (b). The standard way of selecting the correct solution of a set of equations of motion is that of specifying "canonical commutation relations", in analogy to what is known from systems with a finite number of degrees of freedom. We will not use this method, therefore it need not be described in detail. Suffice it to say that for each of the fundamental fields $\varphi_i$ a canonically conjugate field $\pi_i$ is defined, and (anti-)commutation relations

$$[\varphi_i(\vec{x},t)\,,\,\varphi_j(\vec{y},t)]_\pm = [\pi_i(\vec{x},t)\,,\,\pi_j(\vec{y},t)]_\pm = 0\,,$$

$$[\varphi_i(\vec{x},t)\,,\,\pi_j(\vec{y},t)]_\pm = i\,\delta_{ij}\,\delta^3(\vec{x}-\vec{y}) \tag{6.24}$$

are postulated. The relations (5.62) and (5.86) are examples of this. It is essential that the two factors are taken at equal times, which is only possible if the fields $\varphi_i$ and $\pi_i$ can be restricted to sharp times despite their distributionality. This is true for free fields but in general not for interacting fields. The 2-point function $(\Omega,\varphi_i^*\,\varphi_i\,\Omega)$ can be written as a Fourier integral

$$\left(\Omega,\varphi_i^*(\vec{x},x^0)\,\varphi_i(\vec{y},y^0)\,\Omega\right) = \int dp\,\rho_i(p)\,e^{i(p,x-y)},$$

so that at equal times $x^0 = y^0 = t$ we find the value

$$\int d^3p\,e^{-i(\vec{p},\vec{x}-\vec{y})} \int dp_0\,\rho_i(\vec{p},p_0)\,.$$

One finds in perturbation theory that for all the fields of QED as well as other 4-dimensional field theories $\rho_i$ does not decrease sufficiently fast for $|p_0| \to \infty$ to make this integral converge, even after integration over a test function in $\vec{x}-\vec{y}$. This might be different in an exact theory. But little is known about the properties of exact QED, or even about its very existence, and it would be unwise to rely on such hopes. It might also happen that the equal-time commutators exist despite the non-existence of the fields at sharp times, thanks to cancellations between the two terms of the commutator. This is also not the case in perturbation theory.

Moreover, $\pi_i$ is formally defined as a derivative of the Lagrangian density which, being a local polynomial in the fields, suffers from the deficiencies mentioned in point (a) even to a higher degree than the equations of motion. Therefore $\pi_i$ is a priori ill defined.

A remark on a popular variant of the above procedure, namely the use of the interaction picture instead of the Heisenberg picture, is perhaps apposite at this point. The interaction picture is a formulation of quantum dynamics which is intermediate between the Heisenberg and Schrödinger pictures. Operators evolve according to the free part of the Hamiltonian, states according to the interaction part. Apart from destroying the manifest Lorentz invariance, this method is invalidated in relativistic field theory by Haag's theorem [StW, BLOT] stating that in such theories the interaction picture is mathematically unsound, unless the fields in question are free.

Another possibility of selecting the correct solution consists in specifying initial conditions at $t = -\infty$. This can be done if asymptotic conditions hold for the fundamental fields. Crudely speaking, these conditions state that the interacting fields $\varphi_i$ converge for $t \to -\infty$ to free fields $\varphi_i^{\text{in}}$ defined on the same state space. There are various formulations of such conditions, using different definitions of convergence, which we need not discuss at the moment (see below and Chap. 13). Since the theory of free fields is fully understood, the requirement that asymptotic conditions should hold can be used to single out the relevant solution, provided the requirement *can* be satisfied. Asymptotic conditions can be proved in the Wightman framework under some additional assumptions on the spectrum of the mass operator, i.e. on the particle content of the theory, which are unfortunately not satisfied in QED: one finds that the charged fields $\psi, \bar{\psi}$ do not satisfy asymptotic conditions which are simple enough to be useful.

Luckily we will be able to avoid these problems by singling out the desired physical solution not with the help of initial conditions, but by using the mathematically well defined Wightman postulates as subsidiary conditions to be satisfied in addition to the equations of motion. More concretely, we will make use of the reconstruction theorem, Theorem 4.1. The field equations imply a system of differential equations for the Wightman functions, and this system can be shown in perturbation theory to possess a unique solution satisfying the W-properties.

## 6.3 Renormalization and the UV Problem

We return to the UV problem (a), the problem of defining the distribution products occurring in the field equations. The solution of this problem is found by noticing that it is intimately tied up with yet another problem, that of renormalization. This problem emerges in the endeavour to bring QED into contact with experiment. For this we need to identify the measurable quantities of QED, and to assign numerical values to the as yet undetermined constants $e$ and $m$. This *renormalization problem* we must now discuss. We do this at first independently of the UV problem, proceeding as though the fields were functions instead of distributions.

We were confronted with the two problems and their connection already in the study of the free fields $\psi, \bar{\psi}$. There we tried to find the correct expression in terms of $\psi, \bar{\psi}$ of the current $j^\mu$, and we found that the naive ansatz $j^\mu = \bar{\psi}\gamma^\mu\psi$ is undefined because $\psi$ and $\bar{\psi}$ are distributions. This is the UV problem. Second, even ignoring this mathematical difficulty, the vacuum expectation value of the naive $j^\mu$ is not zero, and therefore the charge operator $Q$ does not annihilate the vacuum. This is part of the renormalization problem. We found that both problems can be solved at one stroke by replacing the ordinary product $\bar{\psi}\gamma^\mu \psi$ by the Wick product $:\bar{\psi}\gamma^\mu \psi:$ defined in (5.65). But the proof of this result relied heavily on the cluster expansion (5.61), which is no longer true for interacting fields. For them the higher $W$-functions cannot be expressed in terms of the 2-point functions. Also, the 2-point functions of the interacting fields do not have the simple form (5.60). These two facts together also invalidate the derivation of the commutation relations (5.62), and therefore the chosen normalization of the fields $\psi$, $\bar{\psi}$. Similar problems occur for the electromagnetic field $A^\mu$ and its normalization using the Hamiltonian (5.78). Here we have the additional problem that the interacting Hamiltonian does not even formally have the simple form (5.78). But it has already been remarked in Sect. 5.1 that the normalization chosen there for the scalar field $\varphi$ is not enforced by any physical principles, but is conventional. We are free to change that field into $c\varphi$ for any positive constant $c$, without altering the physical content of the theory, provided we change the definitions of the observables accordingly. The same considerations apply to $\psi$, $\bar{\psi}$ and $A$. We leave therefore the field normalization open for the moment and return to it later.

The as yet unknown exact expression of $j^\mu$ in terms of $\psi$ and $\bar{\psi}$, and the unavailability of the commutation relations (5.62), precludes the identification of the *coupling constant e* occurring in the field equations (6.10) and (6.12) with the *elementary charge e* occurring in the relation (6.21). From now on we will therefore provisionally denote the coupling constant by $e_0$. $e_0$ is also called the "unrenormalized" or "bare" coupling constant, $e$ the "renormalized" or "physical" coupling constant. We cannot expect these two constants to coincide. But we will find in perturbation theory that the charge condition (6.21) can be satisfied with a charge parameter $e$ which is proportional to $e_0$: there exists a "renormalization constant" $C_e$ with

$$e_0 = C_e\, e \ . \tag{6.25}$$

$C_e$ is a function of $e$ and depends on the chosen normalization of the fields $A$ and $\psi$. In the limit $e_0 \to 0$ we find $C_e = 1$, if the free fields are normalized like in Chap. 5. This indicates that the smallness of $e_0$ implies the smallness of $e$, and vice versa. This relation between the two constants implies that the elementary charge $e$ can be used as a measure of the coupling strengths instead of $e_0$. It is convenient to do so, because $e$ is directly measurable, which $e_0$ is not. This introduction of $e$ as "physical coupling constant" into the formalism is effected by replacing $e_0$ in the field equations by the expression (6.25). $C_e$ is determined by requiring that the condition (6.21) shall hold. That this is possible is a non-trivial result of perturbation theory. The current density entering the $A$-equation (6.12) takes then the provisional form

$$j'^\mu(x) = \frac{e}{2} C_e \left[\bar{\psi}(x), \gamma^\mu \psi(x)\right] \ , \tag{6.26}$$

where the prime serves as a reminder that other renormalizations are yet to follow. The substitution (6.25) is of course also made in the Dirac equations

## 6.3 Renormalization and the UV Problem

(6.10). This transformation of the field equations is called "*charge renormalization*" or more properly "*renormalization of the coupling constant*".

We turn to the other free parameter, the mass parameter $m$ in the Dirac equations. From our experience with the coupling constant we should be prepared to expect that also this parameter has no direct physical meaning, in particular that it does not coincide with the measured mass of the electron. This premonition is not unfounded, and we therefore rebaptize that constant $m_0$. $m_0$ is called the "unrenormalized electron mass".

In the free theory $m_0$ *does* stand for the electron mass, when electrons and positrons are defined as particles by the Fock-space method explained for the scalar theory in Sect. 5.1. In our discussion of the interacting theory there has as yet been no mention of particles. But as a rule the experimentalist observes particles, not fields. Therefore a way must be found of decribing particles and their interactions in a field theory like QED. Let us at first assume that asymptotic conditions hold: the interacting fields $\varphi_i$ converge for $t \to \pm\infty$ in a sense to be concretized presently to free fields $\varphi_i^{\text{ex}}$ where ex stands for *out* or *in* in the case $t \to \infty$ or $t \to -\infty$, respectively. The description of scattering events is then based on these two observations:

(*i*) In a scattering experiment particles are observed only long before or after the scattering takes place. And these observed particles behave within experimental accuracy like free particles, because they are too far apart from one another to experience a perceptible interaction.

(*ii*) The relation between free particles and free fields is well understood (see Sect. 5.1).

Starting from these two facts one proceeds as follows. The initial and final states $\Phi_{\text{in}}$, $\Phi_{\text{out}}$, which are present long before and after the interaction takes place, are free particle states. They are described as states in the Fock spaces of the free fields $\varphi_i^{\text{ex}}$, which are then written as limits of the interacting $\varphi_i$. This yields an expression for the transition amplitude $(\Phi_{\text{out}}, \Phi_{\text{in}})$ which can be calculated if the field theory has been solved. The free fields $A^{\text{ex}}$, $\psi^{\text{ex}}$, $\bar\psi^{\text{ex}}$, satisfy the field equations of QED with vanishing coupling constant, the mass parameter $m$ in the Dirac equation being the observed electron mass. The gauge parameter $\lambda$ has no observational meaning and is for the moment left open. For making the connection with the interacting fields we must specify the asymptotic conditions. We use the best-known formulation of such a condition, the LSZ-condition named after its inventors Lehmann, Symanzik, and Zimmermann. We give here only a formal, non-rigorous version of the condition. Let $f(x)$ be a smooth positive-frequency solution of the Klein–Gordon equation

$$\left(\Box + m_i^2\right) f(x) = 0 , \qquad (6.27)$$

with $m_i = m$ or $0$ as the case may be. This means

$$f(x) = (2\pi)^{-\frac{3}{2}} \int \frac{d^3p}{2\omega} \, \hat{f}(\vec{p}) \, e^{i\vec{p}\cdot\vec{x}} \, e^{-i\omega x^0} \qquad (6.28)$$

# 6. Interacting QED

with $\omega = \omega(\vec{p}) = \sqrt{\vec{p}^2 + m_i^2}$ and $\hat{f}$ a sufficiently smooth function on $R^3$, e.g. a tempered test function. We define for the field $\varphi_i$

$$\varphi_i(f, t) := -i \int_{x^0 = t} d^3x\, f(x)\, \overset{\leftrightarrow}{\partial}_0 \varphi_i(x) ,\tag{6.29}$$

with

$$a(x) \overset{\leftrightarrow}{\partial}_0 b(x) := a(x)\, \partial_0 b(x) - \bigl(\partial_0 a(x)\bigr)\, b(x) ,\tag{6.30}$$

a notation that will also be used later on. $\varphi_i(f, t)$ becomes in $p$-space

$$\varphi_i(f, t) = -\int dp\, \frac{p_0 + \omega(\vec{p})}{2\,\omega(\vec{p})}\, e^{it(\omega(\vec{p}) - p_0)}\, \hat{f}(\vec{p})\, \tilde{\varphi}_i(p) .\tag{6.31}$$

We assume that the free field $\varphi_i^{\text{ex}}$ also satisfies the Klein–Gordon equation (6.27). This is true for the spinors $\psi^{\text{ex}}, \bar{\psi}^{\text{ex}}$. For $A_\mu^{\text{ex}}$ it is true only for the choice $\lambda = 0$ of the gauge parameter $\lambda$. But the value of $\lambda$ is physically irrelevant, so for now we make that choice and return to the point later on. Under this assumption $\tilde{\varphi}_i^{\text{ex}}(p)$ contains a factor $\delta(p^2 - m_i^2)$, and as a result the expression $\varphi_i^{\text{ex}}(f, t)$ formed with an asymptotic field is $t$-independent. We can therefore formulate the *asymptotic condition*

$$\lim_{t \to \pm\infty} \varphi_i^{(*)}(f, t) = \varphi_i^{\text{ex}(*)}(f) .\tag{6.32}$$

The bracketed star means that the relation holds both for $\varphi_i$ and for its adjoint $\varphi_i^*$. As an illustration of what this condition means in terms of $W$-functions, let us consider the 2-point function

$$(\Omega,\, \tilde{\varphi}_i(p)\, \tilde{\varphi}_j(q)\, \Omega) = \delta^4(p + q)\, w_{ij}(p) .$$

Let $g(q) \in \mathcal{S}$ and define $\varphi_j(g) = \int dq\, g(q)\, \tilde{\varphi}_j(q)$. We find

$$(\Omega,\, \varphi_i(f, t)\, \varphi_j(g)\, \Omega) = \int dp\, w_{ij}(p)\, e^{it(\omega - p_0)}\, \hat{f}^*(\vec{p})\, g(-p) .$$

The desired limit $(\Omega,\, \varphi_i^{\text{out}}(f)\, \varphi_j(g)\, \Omega)$ for $t \to \infty$ is obtained if

$$w_{ij}(p) = A_{ij}(\vec{p})\, \delta_+(p, m_i) + \sigma_{ij}(p) ,\tag{6.33}$$

where the first term is the free form of $w_{ij}$, and $\sigma_{ij}$ is a sufficiently smooth, e.g. locally integrable, function of $p$. In QED, $A_{ij}$ is obtained by Fourier transform from (5.60) in the case of the spinor functions, from (5.83) in the $A$-case (remember $\lambda = 0$), and vanishes for the mixed functions. The contribution of the $\sigma$-term to the $T$-limit vanishes by the lemma of Riemann–Lebesgue, and the $\delta_+$-contribution is $t$-independent and has the desired limit form. Indeed, it is the presence of such "one-particle singularities" in the 2-point functions which is responsible for the phenomenological appearance of particles at large

## 6.3 Renormalization and the UV Problem

times. This will be discussed in Part III for perturbative QED. The observed mass $m$ of the emerging particle is fixed by the position of the $\delta_+$ singularity in $w$. In a free theory, e.g. of a Dirac spinor, this mass coincides with the mass parameter $m_0$ in the Dirac equation. In interacting QED this is no longer the case, as will be seen in Part II. The 2-point function $(\Omega, \bar{\psi}\psi\Omega)$ does contain a 1-particle singularity, which is however not a $\delta$-function (see below), and which occurs at $p^2 = m^2$ with $m \neq m_0$. But as in the case of the coupling constant we prefer to use the directly measurable $m$ instead of the inaccessible $m_0$ as a free parameter of the theory. This we achieve by rewriting the Dirac equation in the form

$$(i\not{\partial} - m)\psi = C_e\, e\, \not{A}\,\psi - (m - m_0)\,\psi\,, \tag{6.34}$$

considering the mass shift term $-\delta m\,\psi$, $\delta m = m - m_0$, as part of the interaction term. This makes sense because $\delta m = 0$ if $e = 0$, so that the $\delta m$-term is present only in the interacting case. In perturbation theory $\delta m$ is a formal power series in $e$ starting with a term of second order, so that a factor $e^2$ can be drawn in front of it. $\delta m$ must be determined such that the 1-particle singularity of the spinorial 2-point function lies at the physical position $p^2 = m^2$. The replacement of $m_0$ by $m$ effected in (6.34) is called a *mass renormalization*. $m_0$ is the "unrenormalized" or "bare", $m$ the "renormalized" or "physical" mass.

The photon field $A^\mu$ satisfies the LSZ condition if its 2-point function contains as its most singular part a term of the free form (5.83) without a $\delta'$-term:

$$\tilde{W}(^\mu p,^\nu q) = -\delta^4(p+q)\left[(g^{\mu\nu} - \rho\, p^\mu p^\nu)\,\theta(p^0)\,\delta(p^2) + \text{smooth terms}\right]\,. \tag{6.35}$$

We will find in perturbation theory that a term of the form (6.35) is indeed present without a need for the renormalization of the photon mass. However, because the normalization of the $A$-field has not yet been fixed, the term is multiplied with a constant factor $C_A^{-2} \neq 1$, if we start with the original photon field called $A_0^\mu$. This defect can be cured by replacing $A_0^\mu$ by the "renormalized field" $A^\mu$ defined by

$$A_0^\mu(x) = C_A\, A^\mu(x)\,. \tag{6.36}$$

The renormalization constant $C_A$ must be determined such that the $\delta$-term in (6.35) gets the proper coefficient 1. $C_A$ enters then the field equations.

But there is yet another problem with the 2-point function. If we work with a constant gauge parameter $\lambda$, even with $\lambda = 0$, there appears in $\tilde{W}(p,q)$ a non-vanishing term of the form $c\,p^\mu p^\nu\,\delta'(p^2)$, which plays havoc with the LSZ condition. This can be remedied by using a rather more complicated formulation of the asymptotic condition (see [Na 74]; or [JR], Suppl. S1). However, working with this condition is very cumbersome. We will therefore use a different method. It turns out that the gauge parameter $\lambda$ can be made

$e$-dependent in such a way that the unwanted $\delta'$-term disappears and the 2-point function really takes the form (6.35), with $\rho$ a real function of $e$. We call the gauge defined in this way the "Källén gauge" since it was used, without explicit discussion, in a seminal paper by Källén [Kä 52].

The normalization of $\psi$ is a more serious problem. We will find that in this case the assumption (6.33) is unwarranted: the 2-point function of the interacting electron field does not contain a $\delta_+$ contribution. Rather, we find in finite orders of perturbation theory, after mass renormalization, more severe singularities of the type $(p^2 - m^2)^{-1} \left[ \log(p^2 - m^2) \right]^n$, which can however be summed over all orders to yield a branch point singularity $(p^2 - m^2)^{-r}$ with $0 < r < 1$, which is less severe than $\delta(p^2 - m^2)$. This summation happens as in

$$\sum_{\nu=0}^{\infty} \frac{(ce^2)^\nu}{\nu!} \left[\log(p^2 - m^2)\right]^\nu = \exp\left(ce^2 \log(p^2 - m^2)\right)$$
$$= (p^2 - m^2)^{ce^2} \ .$$

In both cases the LSZ condition does not hold. The phenomenon is due to the vanishing photon mass. This allows the existence of photons with arbitrarily small energies, which fact makes it impossible to distinguish a lonely electron experimentally from an electron accompanied by one or several photons of sufficiently small energies. These states with varying numbers of "soft photons" are also intricately mixed dynamically, which leads to a washing out of the sharp mass shell of the electron. This problem is called the *infrared* (IR) *problem* of QED, because it is produced by low-frequency photons. It will be shown in Part III that electrons exist nevertheless as asymptotic particles in the observational sense of the word, though their theoretical description is necessarily more complicated than that provided by the LSZ or similar asymptotic conditions.

The weakness of the $(p^2 - m^2)$-singularity is in accordance with the requirement of positivity that must be satisfied in $\mathcal{H}_{\text{ph}}$. This requirement implies that the 2-point functions of the physical fields must be measures in momentum space, so that their mass-shell singularities cannot be stronger than $\delta$-functions. And we will find that in going over from the unphysical space $\mathcal{V}$ to $\mathcal{H}_{\text{ph}}$ the strengths of this singularity in the electron 2-point function is not changed. Hence it must not be too strong already in the Gupta–Bleuler formulation. This will be an important consideration later on for fixing the exact form of the renormalization conditions.

But at the moment our main concern is that the complicated mass shell singularity prevents us from normalizing $\psi$ in the way used for $A$. We therefore leave this normalization open. Nevertheless we distinguish the original "*unrenormalized field*" $\psi_0$ from the "*renormalized field*" $\psi$, which are related by

$$\psi_0(x) = C_\psi \, \psi(x) \ , \tag{6.37}$$

## 6.3 Renormalization and the UV Problem

because $\psi_0$ will turn out to be divergent. The renormalization constant $C_\psi$ must be chosen such that $\psi(x)$ exists, which condition determines it up to a finite factor. In the perturbative treatment of Part II it will be fixed implicitly by the construction of the $W$-functions given there, which contains an arbitrariness concerning this point.

After these renormalizations the field equations read

$$C_A^2 \Box A^\mu = C_e\, C_A\, C_\psi^2\, \frac{e}{2}\, [\bar{\psi}, \gamma^\mu \psi] + C_A^2\, \lambda\, \partial_\mu \partial_\nu A^\nu ,$$

$$C_\psi^2\, (i\slashed{\partial} - m)\, \psi = C_e\, C_A\, C_\psi^2\, e\, \slashed{A} \psi - C_\psi^2\, \delta m\, \psi ,$$

$$C_\psi^2\, \bar{\psi}\, (i\overleftarrow{\slashed{\partial}} + m) = -C_e\, C_A\, C_\psi^2\, e\, \bar{\psi}\, \slashed{A} + C_\psi^2\, \delta m\, \bar{\psi} . \qquad (6.38)$$

The first equation has been multiplied with $C_A$, the second and third with $C_\psi$, so that the interaction term on the right-hand side has in all cases the same coefficient. In this form the equations are formally the Euler–Lagrange equations of the Lagrangian (see Chap. 2)

$$\mathcal{L}(x) = -\frac{1}{4}\, C_A^2\, F_{\mu\nu}\, F^{\mu\nu} + C_\psi^2\, \bar{\psi}\, (i\slashed{\partial} - m)\, \psi + C_\psi^2\, \delta m\, \bar{\psi}\, \psi$$
$$- C_e\, C_A\, C_\psi^2\, \frac{e}{2}\, \slashed{A}\, [\bar{\psi}, \psi] - \frac{1-\lambda}{2}\, C_A^2\, (\partial_\nu A^\nu)^2 .$$

This will become useful later on. The coupling constant $e$ is the charge of the positron, $m$ its mass. The renormalization "constants" $C_e$, $C_A$, $C_\psi$, $\lambda$, and $m$ are functions of the coupling constant $e$. They must be determined such that the normalization conditions (6.21) with $Q_G = Q_C$ and (6.35) are satisfied and that the dominant singularity of the 2-point function $(\Omega, \bar{\psi}(p)\,\psi(q)\,\Omega)$ lies at the physical value $p^2 = m^2$. For the free case $e = 0$ this is satisfied for the choice $C_e(0) = C_A(0) = C_\psi(0) = 1$, $\lambda(0) = \delta m(0) = 0$. Therefore the right-hand sides of (6.38) vanish in the free limit and may be considered to describe the effects of the interaction between the fields. Our notation concerning the renormalization constants is unconventional. The $C_a$ are related to the constants $Z_i$ which are current in the literature by

$$C_A = \sqrt{Z_3}, \quad C_\psi = \sqrt{Z_2}, \quad C_e = \frac{Z_1}{Z_2 \sqrt{Z_3}} .$$

As yet these considerations have been purely formal, because we have disregarded the UV problem, i.e. the missing definition of the products $[\bar{\psi}, \gamma^\mu \psi]$ and $\slashed{A}\psi$. However, to our delight it turns out that renormalization *solves* this problem, at least according to the evidence of perturbation theory. Crudely speaking, what happens is this: if we split the left-hand side of the first equation of (6.38) as $\Box A^\mu + (C_A^2 - 1)\,\Box A^\mu$, move the second term to the right-hand side, proceed analogously for the second and third equation, and calculate the renormalization constants by the prescriptions given above, then we find that the UV divergences of the various terms in the right-hand sides cancel one another, so that the equations become meaningful. This is meant by the statement that QED is renormalizable. Obviously this statement needs some elaboration since expressions of the type $0\cdot\infty$ or $\infty-\infty$ are not defined a priori. The standard way of making sense out of these undefined expressions is that of working with a *regularization*. Both the field products and the renormalization constants are defined as limits of modified ("regularized") but well-defined quantities in such a way that the limit of the composite object (product or difference) exists, though it does not exist for the constituents separately. A typical example is the method of point splitting, in which e.g. $j_\mu$ is written as

$$j_\mu(x) = e \lim_{y \to x} a(y-x) [\bar{\psi}(y), \gamma^\mu \psi(x)] . \tag{6.39}$$

Before taking the limit, $x$ and $y$ are treated as independent variables, which makes a product like $\bar{\psi}(y) \gamma^\mu \psi(x)$ well defined. $a(\xi)$ is a smooth function which is $\neq 0$ for $\xi \neq 0$ and which vanishes at $\xi = 0$ in such a way that the limit (6.39) exists if $y$ approaches $x$ from a spacelike direction, and such that this limit has the properties required of a current density. Clearly this construction is difficult to control, since the regularized version of $j^\mu$ violates locality, possibly also Lorentz invariance, and current conservation cannot even be formulated for it (see [Br 69] and [Br 70]). The other popular regularizations are beset with similar difficulties. We will therefore use a different method, the BPHZ-method (the letters stand for Bogolubov–Parasiuk–Hepp–Zimmermann), which needs no regularization. The strategy is the following. In perturbation theory we will not calculate the field operators, but the $W$- and related functions. They satisfy equations of the form (6.38) in each variable. This system of differential equations is solved inductively. The right-hand sides are determined by the $\sigma^{\text{th}}$ step of induction; the $(\sigma+1)^{\text{th}}$ step consists then in solving for the unknown functions on the left-hand sides. Each term on the right-hand side taken by itself is divergent. But it is possible to combine these terms in a natural way into a single expression which is then found to converge: the original divergences are due to an inappropriate splitting of the interaction terms into parts.

This method will be explained and discussed in detail in Chap. 10. Unfortunately the method relies heavily on the special structures of perturbation theory. It cannot be used in its present formulation for defining renormalization in a possibly existing exact version of QED. This is a drawback of the method which it shares with some of the known methods of regularization, but not with all of them. For example, point splitting has at least a fighting chance of working also outside perturbation theory. And we would certainly like to have a rigorous definition of what is meant by "exact QED", even without having any information on the possible existence of a theory satisfying this definition. Such a formulation could also be helpful for developing approximation schemes other than perturbation theory. But whether a certain proposed definition has a chance of possessing a solution is almost impossible to assess at the present stage of our knowledge. The problem is rather ill defined and we will not consider it further, but will be content with being able to at least develop perturbation theory with the maximal possible rigour.

Another remark concerning the use of perturbation theory is in order. Perturbation theory consists in expanding the quantities of interest in power series in the coupling constant $e$. This necessitates considering the theory at unphysical values of $e$, namely at all values which are smaller than the measured charge of the positron. We have no information on the electron mass in such unphysical theories. But this mass is a free parameter of the theory. Hence we are justified in fixing $m$ at the observed physical value even for unphysical $e$, because there is no other assignment imposing itself in an obvious way.

Concerning the relation between renormalization and the removal of UV divergences it must be stressed that these are at first hand quite different problems. Renormalization, i.e. writing the field equations in the complicated form (6.38), is necessary independent of the occurrence of UV divergences, if we want to describe the theory in terms of directly measurable parameters. Remember that in the derivation of (6.38) the UV problem was never mentioned. That renormalization also solves the UV problem is just a beneficial side effect. This important point is often overlooked in the existing literature. Textbooks on QED usually start from the unrenormalized equations (6.10) and (6.14), point out that the resulting expressions are UV divergent, and then explain how these divergences can be removed by an ingenious procedure called renormalization. This way of presenting things is apt to look to the unwary reader like a dirty trick, by which the undesirable divergences are merely swept under the rug. It is even sometimes declared to be such by the authors. But nothing could be farther from the truth.

Finally it should be noted that identifying the coupling constant with an easily observable constant of motion, the charge of the positron, is possible in QED but not in non-abelian gauge theories. As a consequence, the introduction of a "running coupling constant" which is a function of the energy is rather less natural in QED than in Yang–Mills theories, even though it can be convenient for certain practical purposes.

# Chapter 7

# The Electric Charges

In the preceding chapter the charge operators $Q_G$, $Q_C$, $Q_M$, have been defined and their relations discussed in a non-rigorous way. We wish to rigorize these important matters as far as is possible without an explicit solution of the theory. We assume the existence of a rigorously defined QED as a theory of quantum fields $A^\mu$, $\psi$, $\bar\psi$ satisfying the p-Maxwell equations (6.14) and the Dirac equations (6.10). We need not know the exact definition of the singular products $\bar\psi \gamma^\mu \psi$ and $\not{A}\psi$, as long as the current $j^\mu$ thus defined is conserved. The theory must satisfy the Postulates 1–6 or, equivalently, the Properties W1–W5 described in Chap. 4.

We need rigorous definitions of the three charges on the state space $\mathcal{V}$ of this theory. The definition (6.21) of $Q_G$ is already rigorous and not in need of improvement. Define the *"field monomial"* $\mathcal{M}$ by

$$\mathcal{M} = \int \prod_{h=1}^{n} dx_h\, f(x_1, \ldots, x_n) \prod_{h=1}^{n} \varphi_h(x_h)\,, \qquad (7.1)$$

where $\varphi_h$ is a component of any of the fundamental fields and $f \in \mathcal{S}$. The "monomial states" $\mathcal{M}\Omega$ span $\mathcal{V}$. If $\mathcal{M}$ contains $n_+$ $\psi$-fields and $n_-$ $\bar\psi$-fields, then the state $\mathcal{M}\Omega$ is an eigenstate of $Q_G$ with eigenvalue

$$q_M = e(n_+ - n_-)\,, \qquad (7.2)$$

and

$$[Q_G, \mathcal{M}] = q_M\, \mathcal{M}\,. \qquad (7.3)$$

$q_M$ is called the charge of $\mathcal{M}$. The exponential $U(\alpha) = \exp\{i\,\alpha\, Q_G\}$, $\alpha \in R$, is defined by

$$U(\alpha)\,\mathcal{M}\,\Omega = e^{i\alpha q_M}\,\mathcal{M}\,\Omega\,. \qquad (7.4)$$

This defines $U$ on $\mathcal{V}$ because $\mathcal{V}$ is finitely spanned by the vectors $\mathcal{M}\Omega$. If $\mathcal{M}$ and $\mathcal{N}$ are monomials with charges $q_M$ and $q_N$, respectively, we find from

(6.21) and the hermiticity of $Q_G$ that

$$(\mathcal{N}\Omega, \mathcal{M}\Omega) = 0 \quad \text{if} \quad q_M \neq q_N . \tag{7.5}$$

A simple calculation shows that $U(\alpha)$ is unitary. Furthermore we find $U(\alpha+\beta) = U(\alpha) U(\beta)$: the $U(\alpha)$ form a unitary representation of the global gauge group whose elements are the transformations

$$\begin{aligned} A^\mu(x) &\to A^\mu(x) , \\ \psi(x) &\to e^{-i\alpha e} \psi(x), \quad \bar\psi(x) \to e^{i\alpha e} \bar\psi(x). \end{aligned} \tag{7.6}$$

Finally, $U(\alpha)$ satisfies the differential equation

$$\frac{d}{d\alpha} U(\alpha) = i Q_G U(\alpha), \tag{7.7}$$

which justifies the exponential notation $U(\alpha) = \exp[i\alpha Q_G]$. In particular we find

$$Q_G = -i \left. \frac{d}{d\alpha} U(\alpha) \right|_{\alpha=0} ; \tag{7.8}$$

$Q_G$ is the "infinitesimal generator"[1] of the representation $U(\alpha)$.

In Chap. 6 we have defined $Q_C$ and $Q_M$ as integrals over charge densities:

$$Q = \int_{x^0=\text{const}} d^3x \, j^0(x^0, \vec{x}) , \tag{7.9}$$

with $j^0$ the zero-component of a conserved vector current. In the case of $Q_C$ this current is given by the p-Maxwell equation as

$$j^\mu_C = \Box A^\mu - \lambda \partial^\mu \partial_\nu A^\nu , \quad \lambda \neq 1 , \tag{7.10}$$

for $Q_M$ by the true Maxwell equation as

$$j^\mu_M = \Box A^\mu - \partial^\mu \partial_\nu A^\nu . \tag{7.11}$$

This use of the Maxwell equations for defining $j^\mu$ saves us from the embarrassment of not possessing an exact definition of $j^\mu$ in terms of $\psi$ and $\bar\psi$. Conservation of $j^\mu_C$ does not follow from the definition (7.10) and must be assumed. The distinction between $\lambda = 1$ and $\lambda \neq 1$ is irrelevant for our present purposes, so that we can treat the two cases together.

The problem with the definition (7.9) is that it does not make sense for distribution-valued fields for two reasons. First, the restriction of the distribution $j^0(x)$ to the sharp time $x^0 = \text{const}$ is not defined as a distribution in the remaining variables $\vec{x}$. Second, even if it were so defined, the existence

---
[1] More exactly this is the definition of an infinitesimal generator current among physicists. Mathematicians prefer to define this notion without the factor $-i$.

of its integral over $\mathbf{R}^3$ would not be assured, the constant 1 not being a test function. There exists an extensive literature on how to handle these problems in local field theories with Hilbert spaces as state spaces (see e.g. [V]). We can take over the solutions offered there to our somewhat different case, but not large parts of the proofs, which therefore must be developed anew.

We start from the formal expression (7.9), where $Q$ may be either $Q_C$ or $Q_M$. Because of current conservation $Q$ is formally independent of $x^0$. Hence, if $\alpha(x^0)$ is a test function with compact support (a function in $\mathcal{D}$ in the notation of Schwartz) satisfying

$$\int dx^0\, \alpha(x^0) = 1 , \tag{7.12}$$

we can also write

$$Q = \int dx\, \alpha(x^0)\, j^0(x) , \tag{7.13}$$

thus obviating the necessity of restricting distributions to lower-dimensional manifolds. The remaining difficulty of defining the integral $\int d^3x$ is handled as usual by approximating it by a sequence of cut-off integrals. The following method of doing this has turned out to be most convenient. Choose $\chi(\vec{x}) \in \mathcal{S}$ with $\chi(0) = 1$. Then $\chi(R^{-1}\vec{x})$ tends for $R \to \infty$ locally uniformly to 1. Hence we make the ansatz

$$Q = \lim_{R\to\infty} \int dx\, \alpha(x^0)\, \chi(R^{-1}\vec{x})\, j^0(x) . \tag{7.14}$$

We would like to prove that this limit exists as a hermitian operator on $\mathcal{V}$ and that it does not depend on the special choice of the auxiliary functions $\alpha$ and $\chi$. This we are not able to do on the basis of the stated assumptions. We can only prove the weaker result:

**Theorem 7.1.** *Let $Q_R$ be the integral (7.14) and $\mathcal{M}$ a field monomial. Assume that*

$$\lim_{R\to\infty} Q_R \Omega = 0 \tag{7.15}$$

*for any choice of $\alpha$, $\chi$. Then the limit*

$$\lim_{R\to\infty} Q_R \mathcal{M}\Omega = Q\mathcal{M}\Omega \tag{7.16}$$

*exists and is independent of the choice of $\alpha$ and $\chi$.*

Remember that the monomial states span $\mathcal{V}$. Hence the theorem states that $Q$ exists on all of $\mathcal{V}$ if it exists on the vacuum. In Chap. 11 $Q\Omega$ will be shown to vanish in perturbation theory. Using the definition of convergence in $\mathcal{V}$ we find that Theorem 7.1 is a simple consequence of the following fact.

**Lemma 7.2.** *Let* $\mathcal{M}$, $\mathcal{N}$ *be field monomials. Then*

$$\lim_{R \to \infty} (\Omega,\, \mathcal{N}\,[Q_R,\, \mathcal{M}]\,\Omega)$$

*exists and does not depend on the choice of* $\alpha$ *and* $\chi$.

**Proof.** Let $Y = \{y_1, \ldots, y_n\}$, $Z = \{z_1, \ldots, z_m\}$, be the arguments of the fields in $\mathcal{N}$ and $\mathcal{M}$ respectively, $f(Y)$, $g(Z)$, the corresponding test functions, and $n(Y)$, $m(Z)$, the corresponding unintegrated field products. We must study the expression

$$E_R = \int dx\, dY\, dZ\, f(Y)\, g(Z) \\ \times \alpha(x^0)\, \chi(R^{-1}\vec{x})\, (\Omega,\, n(Y)\,[j^0(x),\, m(Z)]\,\Omega)\,, \qquad (7.17)$$

where $dx$, $dY$, $dZ$ are products of the differentials of all the components of the variables $x$, $y_i$, $z_j$. By locality the support of the vacuum expectation value in this expression is contained in the set

$$S = \{(x - z_i)^2 \geq 0 \ \forall\ z_i \in Z\}\,. \qquad (7.18)$$

Assume that $\alpha(x^0) \equiv 0$ for $|x^0| \leq C < \infty$. Then the product of the vacuum expectation value with $\alpha$ vanishes if

$$|\vec{x}| > |\vec{z}_i| + |z_i^0| + C \quad \forall \quad z_i \in Z\,, \qquad (7.19)$$

because in that case we have $|\vec{x}| > |\vec{z}_i|$, and thus for $|x^0| \leq C$:

$$|\vec{x} - \vec{z}_i| \geq |\vec{x}| - |\vec{z}_i| > |z_i^0| + C \geq |z_i^0 - x^0|\,,$$

so that $(x - z_i)^2 < 0$. But $\left(|\vec{z}_i| + |z_i^0| + C\right)^2 \leq 3\left(|\vec{z}_i|^2 + (z_i^0)^2 + C^2\right)$, hence the condition (7.19) is satisfied if

$$(\vec{x}, \vec{x}) > 3\sum_Z \left((\vec{z}_i, \vec{z}_i) + (z_i^0)^2 + C^2\right) =: A(Z)\,. \qquad (7.20)$$

$A$ is a polynomial. Let $\beta(u)$ be a $C^\infty$ function with

$$\beta(u) = \begin{cases} 1 & \text{for } u \leq 0 \\ 0 & \text{for } u \geq C\,. \end{cases} \qquad (7.21)$$

Then $\beta(\vec{x}, Z) := \beta((\vec{x}, \vec{x}) - A(Z))$ is a $C^\infty$ function which is $\equiv 1$ on $S$. Hence we can multiply the integrand of $E_R$ with this function without changing the integral. But $F = f(Y)\, g(Z)\, \alpha(x^0)\, \beta(\vec{x}, Z)$ is a tempered test function, as is its product with $\chi(R^{-1}\vec{x})$, and this product converges for $R \to \infty$ towards $F \in \mathcal{S}$ in the topology of $\mathcal{S}$. This proves the existence of $\lim_{R \to \infty} E_R$. The result is simply the value of the vacuum expectation value in (7.17) integrated over the test function $F$, hence it is independent of $\chi$. The $\alpha$-independence is seen as follows. Let $\alpha_1(x^0), \alpha_2(x^0)$ be test functions with support in $\{|x^0| \leq C\}$ and $\int dx^0 \, \alpha_i(x^0) = 1$. Then their difference $\Delta = \alpha_1 - \alpha_2$ has the same or a smaller support and its integral vanishes. This implies that $\Delta$ is the derivative of another test function: $\Delta(x^0) = \frac{d}{dx^0}\Gamma(x^0)$, $\Gamma \in \mathcal{D}$. The difference of the two limits formed with these $\alpha_i$ is

$$\int dx \, dY \, dZ \, f(Y) \, g(Z) \, \beta(\vec{x}, Z) \frac{\partial}{\partial x^0} \Gamma(x^0) \left( \Omega, \, n(Y)[j^0(x), m(Z)] \, \Omega \right) . \quad (7.22)$$

Integration by parts over $x^0$ yields the factor $[\partial_0 j^0(x), m] = -[\partial_i j^i(x), m]$ by current conservation. Another integration by parts transfers the $\partial_i$-derivatives to the factor $\beta(\vec{x}, Z)$. But $\partial_i \beta$ vanishes in the support of the integrand, hence the difference (7.22) vanishes: Lemma 7.2 is proved.

Next we show that under our assumptions $Q_M$ vanishes identically on $\mathcal{V}$, as was already indicated in Chap. 5. $Q_M$ is defined by (7.14) with

$$j^\mu(x) = \partial_\nu F^{\nu\mu}(x) , \quad (7.23)$$

hence $j^0 = \partial_i F^{i0}$ is a 3-divergence. We assume that $Q_M \Omega = 0$. For proving $Q_M \equiv 0$ it suffices to prove the vanishing of

$$(\Omega, \mathcal{N}[Q_M, \mathcal{M}] \, \Omega) = \int dx \, dY \, dZ \, f(Y) \, g(Z)$$
$$\times \alpha(x^0) \, \beta(\vec{x}, Z) \left( \Omega, \, n(Y) \, [\partial_i F^{i0}(x), m(Z)] \, \Omega \right) . \quad (7.24)$$

But the derivatives can be transferred to the factor $\beta$ through integration by parts. The result vanishes because $\partial_i \beta$ vanishes on the support of the integrand of (7.24).

The next point to be discussed is the identity of $Q_G$ and $Q_C$ on $\mathcal{V}$. This identity can easily be proved for free fields. Free fields *can* be restricted to sharp times, hence we can write, using the time independence of $Q_C$:

$$[Q_C, \psi(x)] = \lim_{R \to \infty} \int_{y^0 = x^0} d^3y \, \chi(R^{-1}\vec{y}) \, [j^0(y), \psi(x)] . \quad (7.25)$$

Plugging in the form (5.64) of $j^0$ and using the anticommutators (5.62) we obtain at once the desired result $-e \, \psi(x)$. The same method works for $\bar\psi$, and trivially also for $A^\mu$ if we use that the free $\psi$, $\bar\psi$ commute with the free $A^\mu$ at all points because they are completely independent from each other. But neither simple commutation relations nor an exact definition of $j^\mu$ in terms of $\psi$, $\bar\psi$, are available in the interacting case. We can therefore not prove that $Q_C = Q_G$, except that the special case $[Q_C, A^\mu] = 0$ follows from the hermiticity of $Q_C$ and $A^\mu$ (see (6.6)). The "*charge identity*"

$$Q_C = Q_G , \quad (7.26)$$

which must hold in QED, must be introduced as a new condition to be satisfied by the correct solution of the equations of motion of QED.

This identity is usually discussed in the literature in connection with the Noether theorem. For classical field theories this theorem in its simplest form states the following. Let $\mathcal{L}$ be the Lagrangian density of a local field theory with fields $\varphi_1, \ldots, \varphi_f$. Let $\mathcal{L}$ be invariant under a family of transformations $T(\alpha)$:

$$T(\alpha) \quad : \quad \varphi_i(x) \to \varphi_i^\alpha(x) \tag{7.27}$$

depending differentiably on the real parameter $\alpha$. Then

$$j^\mu(x) = \sum_{i=1}^{f} \frac{\partial \mathcal{L}}{\partial(\partial_\mu \varphi_i)} \left. \frac{\partial \varphi_i^\alpha}{\partial \alpha} \right|_{\alpha=0} \tag{7.28}$$

defines a conserved vector current, and $Q = \int d^3x \, j^0(t, \vec{x})$, if this integral exists, is a conserved charge. In the case of electrodynamics the classical Lagrangian (3.12) and (3.13) is invariant under the gauge transformation (7.6). This remains true if a "gauge fixing term" depending only on $A$ is added to $\mathcal{L}$. The corresponding conserved Noether current is precisely the electromagnetic current (3.16) and the Noether charge is $Q_C$. This establishes a relation between $Q_C$ and the gauge group (of the first kind) (7.6). This theorem can be taken over *formally* into quantum field theory and can then be expanded to show $Q_C = Q_G$. However, $\mathcal{L}$ and even more so the expression (7.28) are ill defined in quantum field theory. Therefore the rigorous status of the Noether theorem is uncertain, and we will refrain from using it, except possibly as a heuristic guide.

Concerning the physical state space $\mathcal{H}_{\text{ph}}$ we are on less certain ground. Even though $\mathcal{H}_{\text{ph}}$ must be a Hilbert space, the standard results mentioned earlier cannot be applied because of the unavoidable nonlocality of the basic fields. In the present chapter dealing with rigorous results we must therefore be content with proving a negative result: a no-go theorem.

The standard textbook approach to the problem of constructing $\mathcal{H}_{\text{ph}}$ from $\mathcal{V}$ is to follow as closely as possible the procedure that succeeded with the free electromagnetic field. The vanishing of $Q_M$ on $\mathcal{V}$ indicates however that some slight generalization of the procedure will be necessary. Remember that on $\mathcal{H}_{\text{ph}}$ the three charge definitions ought to coincide, hence $Q_M$ must be non-trivial. We can try to use the following idea. Suppose that we can find a vector space $\bar{\mathcal{V}}$ which is an extension of $\mathcal{V}$, and a hermitian nondegenerate scalar product on $\bar{\mathcal{V}}$ which coincides on $\mathcal{V}$ with the already known product, and such that $\mathcal{V}$ is dense in $\bar{\mathcal{V}}$ with respect to the weak topology induced by this scalar product. The fields $A$, $\psi$, $\bar{\psi}$ should be defined on a dense subset of $\bar{\mathcal{V}}$ including $\mathcal{V}$ and satisfy the given field equations. Suppose further that we can find a subspace $\mathcal{V}' \in \bar{\mathcal{V}}$ on which the true Maxwell equations hold in the sense of sesquilinear forms (like in the free case), which is mapped into itself by gauge invariant operators like $F^{\mu\nu}$ and $j^\mu$, and on which the scalar product $(\cdot,\cdot)$ is positive. From there on we proceed as in the free case: we find the subspace $\mathcal{V}_0 \in \mathcal{V}'$ of the states with vanishing square norm, form the quotient $\mathcal{Q} = \mathcal{V}'/\mathcal{V}_0$, and complete it to the Hilbert space $\mathcal{H}_{\text{ph}}$. We can hope to find in this way a sufficiently large physical space. In the conventional formulation of this procedure (see [MSt 80] for a careful exposition) $\bar{\mathcal{V}}$ is a Hilbert space with a positive-definite scalar product $\langle\cdot,\cdot\rangle$. In this space exists a bounded self-adjoint operator $\eta$ with $\eta^2 = 1$, such that $(\Phi, \Psi) = \langle\Phi, \eta\,\Psi\rangle$, and such that $\mathcal{V}$ is dense in $\bar{\mathcal{V}}$ in the Hilbert sense. It is easy to see that then $\mathcal{V}$ is also weakly dense in $\bar{\mathcal{V}}$ in our sense. That is, our assumptions are satisfied in the conventional approach, but they are more general, because we do not postulate the existence on $\mathcal{V}$ of the new, positive but non-invariant scalar product $\langle\cdot,\cdot\rangle$. In this framework we are confronted with the following problem.

**Theorem 7.3.** *Let $\mathcal{V}$, $\bar{\mathcal{V}}$, $\mathcal{V}'$, $\mathcal{V}_0$, $\mathcal{Q}$, $\mathcal{H}_{\text{ph}}$, be defined as above. Assume that $Q_M$ as defined by (7.9) and (7.23) exists as a densely defined, hermitian, gauge invariant, operator on all these spaces. Then $Q_M \equiv 0$.*

**Proof.** We know that $Q_M = 0$ on $\mathcal{V}$. Choose $\Phi \in \bar{\mathcal{V}}$ in the domain of definition of $Q_M$, and $\Psi \in \mathcal{V}$. Then we have $(Q_M\,\Phi, \Psi) = (\Phi, Q_M\,\Psi) = 0$: $Q_M\,\Phi$ is orthogonal to $\mathcal{V}$, hence to $\bar{\mathcal{V}}$, because $\mathcal{V}$ is dense in $\bar{\mathcal{V}}$. Therefore $Q \equiv 0$ on $\bar{\mathcal{V}}$, hence also on its subspace $\mathcal{V}'$. Because of its gauge invariance it defines an operator, still called $Q_M$, on $\mathcal{Q}$, which is also the operator zero. We repeat the same argument in $\mathcal{H}_{\text{ph}}$. Let $\Phi \in \mathcal{H}_{\text{ph}}$, $\Psi \in \mathcal{Q}$. Then $(Q_M\,\Phi, \Psi) = (\Phi, Q_M\,\Psi) = 0$ for all $\Psi$ from the dense subspace $\mathcal{Q}$ of $\mathcal{H}_{\text{ph}}$, hence $Q_M\,\Phi = 0$, which proves the theorem.

Hence we cannot construct charged physical states by this method. We need to find a more indirect way leading from $\mathcal{V}$ to $\mathcal{H}_{\text{ph}}$. This way will be described in Chap. 12 in the framework of perturbation theory.

# Part II

# Perturbation Theory

In this part the equations of QED will be solved in perturbation theory, at first in an unphysical Gupta–Bleuler gauge. The general framework is outlined in Chap. 8: the equations to be solved and the subsidiary conditions to be respected are formulated and explained. This program is then carried out in several stages in the following three chapters. For the sake of clarity it will not be tried to solve all the problems at once. The construction starts at a purely formal level, firmly disregarding all problems of existence, which are then dealt with at the later stages of the procedure. The final Chap. 12 is devoted to the construction of the $W$-functions of a physical gauge out of the unphysical functions found before.

The contents of Chaps. 8–11 can be generalized without serious difficulties to non-abelian gauge theories. The same is not true for Chap. 12. Non-abelian theories will, however, not be considered in this book.

# Chapter 8

# The Program of Perturbation Theory

The $W$-functions of QED fully determine the theory, as has been explained in Chap. 4. QED is therefore known if its $W$-functions have been found. This is the problem that we now intend to attack using the Gupta–Bleuler formalism. To this end we need to enumerate the conditions which the $W$-functions must satisfy. There are three sets of conditions: (a) the equations of motion, i.e. partial differential equations which the $W$ must satisfy, (b) the W-properties explained in Chap. 4, and (c) the renormalization conditions indicated in Chap. 6.

## 8.1 The Problem

### (a) Equations of Motion

For the reasons explained in Chap. 6 we work in a special Gupta–Bleuler gauge, which we call the Källén gauge. Its field equations we write in the form

$$\begin{aligned}
\Box A^\mu(x) &= e\, N_1\big(\bar\psi(x)\,\gamma^\mu\,\psi(x)\big) =: j^\mu(x)\ ,\\
(i\slashed\partial - m)\,\psi(x) &= e\, N_2\big(\slashed A(x)\,\psi(x)\big)\ ,\\
\bar\psi(x)(i\overleftarrow{\slashed\partial} + m) &= -e\, N_3\big(\bar\psi(x)\,\slashed A(x)\big)\ ,
\end{aligned} \qquad (8.1)$$

so that the interaction terms stand on the right-hand side. $e$ is the observed charge of the positron, $m$ its observed mass. $N_i$ stands for "*normal product*" and signifies the renormalization procedure that is necessary to turn the undefined local products $\bar\psi_\rho(x)\,\psi_\sigma(x)$ etc. into mathematically well-defined fields. No rigorous definition of these normal products which can be shown to work outside of perturbation theory is known at present, so we simply assume the existence of such a definition. Formally the $N_i$ are defined by (6.38). We rewrite them in a form adapted to perturbation theory:

$$N_1^\mu(x) := N_1(\bar\psi \gamma^\mu \psi)$$
$$= \frac{1}{2}[\bar\psi, \gamma^\mu \psi] + \frac{R_1}{2}[\bar\psi, \gamma^\mu \psi] + R_2 \Box A^\mu + C_A{}^2 \lambda \partial_\mu \partial_\nu A^\nu ,$$
$$N_2(x) := N_2(\slashed{A}\psi)$$
$$= \slashed{A}\psi + R_1 \slashed{A}\psi + R_3(i\slashed{\partial} - m)\psi - C_\psi{}^2 \delta m\, \psi ,$$
$$N_3(x) := N_3(\bar\psi \slashed{A})$$
$$= \bar\psi \slashed{A} + R_1 \bar\psi \slashed{A} + R_3 \bar\psi(i\overleftarrow{\slashed{\partial}} + m) + C_\psi{}^2 \delta m\, \bar\psi . \tag{8.2}$$

All fields are taken at the point $x$. The *renormalization constants* $R_1 = C_e C_A C_\psi{}^2$, $R_2 = 1 - C_A{}^2$, $R_3 = 1 - C_\psi{}^2$, $\lambda$, $\delta m$ are functions of the coupling constant $e$ which vanish at $e = 0$. They are fixed by current conservation, the renormalization conditions to be specified below, and the requirement that the $N_i$ must be local and covariant fields. The individual terms in the expressions (8.2) diverge, so that these definitions do not make sense as they stand. We will return to this problem later on.

The Lorentz condition $\partial_\mu A^\mu = 0$ is omitted in the list (8.1) of field equations. In this and the next few chapters we develop the field theory defined by (8.1), disregarding the Lorentz condition. Only in Chap. 12 will it be shown how this condition can be incorporated into the formalism.

The field equations (8.1) imply differential equations for the $W$-functions. Before writing them down we introduce some streamlined notation. The Dirac operators will be represented by the symbols

$$\mathcal{D} = (i\slashed{\partial} - m) , \quad \overleftarrow{\mathcal{D}} = (i\overleftarrow{\slashed{\partial}} + m) . \tag{8.3}$$

The variables of fields $\psi$ or $\tilde\psi$ occurring in a $W$-function are denoted by lower-case Roman letters: $x, \ldots, p, \ldots$, variables of $\bar\psi$ or $\tilde{\bar\psi}$ fields by barred Roman letters: $\bar x, \ldots, \bar p, \ldots$, those of $A$ or $\tilde A$ fields by sans serif characters: $\mathsf{x}, \ldots, \mathsf{p}, \ldots$ . The vector and spinor indices of the fields are in general not shown. If this is desirable, they appear as subscripts or superscripts in front of the variable. E.g. $_\rho x$ is the argument of a factor $\psi_\rho(x)$, $^\mu\mathsf{p}$ that of the field $\tilde A^\mu(\mathsf{p})$. One and the same letter occurring in an equation in different fonts refers each time to the same 4-vector: $x = \bar x = \mathsf{x}$ numerically if these symbols occur in the same equation. The set of all $\psi$ variables $x_i$ in a given $W$-function is abbreviated by the corresponding capital letter $X = \{x_1, \ldots, x_n\}$, and analogously for the other fields. The symbol $|X| = n$ denotes the number of variables in $X$. Hence $W(X, \bar Y, \mathsf{Z})$ stands for the vacuum expectation value of a product of $|X|$ $\psi$-fields, $|\bar Y|$ $\bar\psi$-fields, and $|\mathsf{Z}|$ $A$-fields. The information on the ordering of the fields within the product is suppressed in this notation. Where it is important, it is indicated explicitly. The ordering of variables in various $W$-functions in an equation is the same in all terms, except if noted otherwise.

Applying the field equations (8.1) to the fields in $W(X, \bar{Y}, \mathsf{Z})$ yields the following system of $(|X|+|Y|+|Z|)$ differential equations, with $x$, $\bar{y}$, $\mathsf{z}$, generic elements of $X$, $\bar{Y}$, $\mathsf{Z}$:

$$\Box_z W(\ldots,{}^\mu\mathsf{z},\ldots) = e\ (\Omega, \cdots N_1^\mu(z) \cdots \Omega)\ ,$$
$$\mathcal{D}_x W(\ldots, x, \ldots) = e\ (\Omega \cdots N_2(x) \cdots \Omega)\ ,$$
$$W(\ldots, \bar{y}, \ldots) \overleftarrow{\mathcal{D}}_y = -e\ (\Omega, \cdots N_3(y) \cdots \Omega)\ . \tag{8.4}$$

Inserting for the normal products the formal expressions (8.2) we see that the right-hand sides of these equations can also be written in terms of $W$-functions. We assume that this remains true for the exact definitions of $N_i$. A *consistency condition* for the $N_i$ is obtained by writing $\mathcal{D}_x \Box_z W(\ldots, x, \ldots, \mathsf{z}, \ldots)$ in two different ways, obtaining

$$\mathcal{D}_x\ (\Omega, \cdots \psi(x) \cdots N_1(\mathsf{z}) \cdots \Omega) = \Box_z\ (\Omega, \cdots N_2(x) \cdots A(\mathsf{z}) \cdots \Omega)\ . \tag{8.5}$$

Similar relations are derived from the other pairings of (8.1).

Equations (8.4) are the basic equations of our formulation of QED. Solving them is our main goal.

## (b) W-Properties

The W-properties W1–W5 explained in Chap. 4 must be satisfied by $W(X, \bar{Y}, \mathsf{Z})$ with the following qualifications.

The hermiticity condition (4.39) assumes the form

$$W(X, \bar{Y}, \mathsf{Z})^* = \prod_Y \gamma_j^0\ W^\leftarrow(\bar{X}, Y, \mathsf{Z}) \prod_X \gamma_i^0\ . \tag{8.6}$$

Here the matrix $\gamma_i^0$ is the Dirac matrix $\gamma^0$ acting on the spinor indices of the field $\bar{\psi}(x_i)$, $\gamma_j^0$ acts on $\psi(y_j)$. The notation $W^\leftarrow$ signifies that the ordering of the variables is the reverse of that in $W$. A special example of this relation has been given in (5.58).

The Lorentz representations $R^{(i)}$ in the covariance relations (4.40) are $R(\Lambda) = \Lambda$ for $A$-fields, $R(\Lambda) = S(\Lambda)$ for $\psi$-fields, and $R(\Lambda) = S^{-1^T}(\Lambda)$ for $\bar{\psi}$-fields. The superscript $T$ denotes transposition.

In the locality condition (4.45) $\psi$ and $\bar{\psi}$ enter as Fermi fields, $A^\mu$ as Bose fields. The spectral property and the weak cluster condition (4.44) need no comment.

## (c) Renormalization Conditions

The reason for demanding renormalization conditions has been explained in Chap. 6. They are needed in particular for fixing the definitions of the normal products $N_i$. Remember that their formulation is inspired by the desire to find a particle interpretation of states at large times. The exact conditions under which such a description emerges are not known. We must therefore rely on intuition and the experience gained from perturbation theory and other approximations. The conditions to be enumerated are formulated in a way which is known to lead to a working theory, at least perturbatively. The given formulations are not claimed to be optimal; they cannot be derived from first principles.

With respect to the *coupling constant e* we are still on safe ground. The normalization condition (6.21) is imposed on the $W$-functions in the form of the charge identity (7.26). Implicit in this condition is the requirement that the current $j^\mu$ in the field equations (8.1) is conserved.

Concerning the *renormalization of masses and fields* we formulate for the sake of simplicity rather strong conditions, which are however not so strong as to prevent the development of perturbation theory. For the *electron field* we demand that the function $w(p)$ in its 2-point function $\tilde{W}(p,\bar{q}) = \delta^4(p+q)\,w(p)$ (see (4.43)) be continuous outside the mass shell $p^2 = m^2$, but divergent at the mass shell. The singularity at the mass shell must be compatible with the desired particle behaviour at large times. As will be seen later, this requirement is met if we demand that the singularity at $p^2 = m^2$ is the weakest that is compatible with the other conditions. We return to this problem in Chap. 10 in connection with the determination of the normal-order prescriptions $N_i$. For the *photon field* we postulate the more precise condition that its 2-point function be of the form (6.35):

$$\tilde{W}(^\mu p, ^\nu q) = -\delta^4(p+q)\left[(g^{\mu\nu} - \rho\, p^\mu p^\nu)\,\delta_+(p) + \Xi(p)\right], \qquad (8.7)$$

where $\delta_+(p) = \theta(p^0)\,\delta(p^2)$, and $\Xi(p)$ is a continuous function. $\rho$ is a function of $e$ whose form is left open for the moment. The $\rho$-term is physically irrelevant because it drops out in going over from the unphysical fields $A^\mu$ to the physical $F^{\mu\nu}$. The ensuing freedom of choice of this function $\rho(e)$ is essential for the success of perturbation theory (see later). The condition (8.7) selects a special gauge, the Källén gauge. In general Gupta–Bleuler gauges a term of the form $p^\mu p^\nu \theta(p^0)\,\delta'(p^2)$ would also be present. It would have a considerable nuisance value in our formalism, though it would not invalidate it. The special choice of gauge is made in the interest of simplicity. It must be remarked, however, that retaining a free gauge parameter $\lambda$ would have a practical advantage: since measurable quantities must be gauge independent, an explicit $\lambda$-dependence in a calculated expression for such a quantity would indicate that an error has been made. We sacrifice this useful method of checking calculations in favour of conciseness.

## 8.2 Perturbation Theory

No exact solution of the problem just outlined is known. We will solve it however in an approximate way, using *perturbation theory*. In perturbation theory the quantities of interest are expanded into power series in the coupling constant $e$ and the expansion coefficients calculated. In particular we expand the $W$-functions as

$$W(X, \bar{Y}, \mathsf{Z}) = \sum_{\sigma=0}^{\infty} e^{\sigma} W_{\sigma}(X, \bar{Y}, \mathsf{Z}) \,, \tag{8.8}$$

where the $W_{\sigma}$ do not depend on $e$. It is not known whether these expansions converge, but it is the prevalent feeling that they do not. The series (8.8), as well as similar expansions for other quantities of interest, must therefore be considered as a formal power series. Because of the smallness of the coupling constant $e$, excellent approximations may and do result if the perturbation series are broken off after a finite number of terms. Indeed, this method yields numerical results which agree with experiment with an astounding accuracy (see e.g. [Ki 1990]). But in the present work we are not so much concerned with numerical results as with structural studies. We aim at proving results which are correct to all orders of perturbation theory and may therefore be indicative of rigorous properties. In such a context the size of the coupling constant is of little importance.

Inserting the expansion (8.8) into (8.4) and equating the terms of order $e^{\sigma}$ on both sides we find

$$\begin{aligned}
\Box_z W_{\sigma}(\ldots,{}^{\mu}\mathsf{z},\ldots) &= (\Omega, \cdots N_1^{\mu}(z) \cdots \Omega)_{\sigma-1} \,, \\
\mathcal{D}_x W_{\sigma}(\ldots, x, \ldots) &= (\Omega, \cdots N_2(x) \cdots \Omega)_{\sigma-1} \,, \\
W_{\sigma}(\ldots, \bar{y}, \ldots) \overleftarrow{\mathcal{D}}_y &= -(\Omega, \cdots N_3(y) \cdots \Omega)_{\sigma-1} \,.
\end{aligned} \tag{8.9}$$

We will find that in perturbation theory the divergent individual terms in the $N$-definitions (8.2) can be combined in a natural way into convergent expressions, thus giving a well-defined meaning to the right-hand sides of (8.9), expressing them in terms of the $W$-functions of order $\sigma - 1$. This remarkable fact is called the "renormalizability of QED". The structure of (8.9) immediately suggests solving them by iteration. For $\sigma = 0$ the right-hand sides are zero; for $\sigma > 0$ they are known if the problem has been solved up to order $\sigma - 1$. $W_{\sigma}$ can then be determined as a solution of a system of linear partial differential equations.

The consistency condition (8.5) becomes

$$\mathcal{D}_x \big(\Omega, \cdots \psi(x) \cdots N_1(z) \cdots \Omega\big)_{\sigma-1} = \Box_z \big(\Omega, \cdots N_2(x) \cdots A(z) \cdots \Omega\big)_{\sigma-1} \,. \tag{8.10}$$

Similar relations hold for the other two pairings of equations of (8.9). These relations are formally satisfied if the field equations are derivable from a Lagrangian, which is the case for (8.1). But they also admit more general theories.

The solution of (8.9), assuming there is one, is not unique. The relevant solution is selected by using the W-properties and the renormalization conditions as generalized "boundary conditions". All the W-properties except the

100    8. The Program

weak cluster property are linear in the $W$-functions. Hence they must be satisfied for each coefficient $W_\sigma$ separately if they are satisfied for $W$ in the sense of formal power series. If we write the cluster property (4.44) symbolically as $W^{IJ} \to W^I W^J$, then its perturbative version reads

$$W^{IJ}_\sigma \to \sum_{\rho=0}^{\sigma} W^I_\rho W^J_{\sigma-\rho} \, . \tag{8.11}$$

It must be said that this condition does not follow rigorously from (4.44) without further assumptions. The condition (4.44) involves taking the limit $\lambda \to \infty$. $W_\sigma$ is defined as $\frac{d^\sigma}{de^\sigma} W$ at $e = 0$. And there is no a priori reason why this $e$-derivative should commute with the $\lambda$-limit. But since we know nothing even about the existence of $W$, let alone its precise analytic properties, we take the liberty of ignoring these scruples.

The renormalization conditions also play a role in the uniqueness problem, but mostly they are needed for fixing the $N$-definitions. We must therefore also discuss what becomes of them in perturbation theory. Here we are confronted with problems similar to the one just encountered. The definition of $Q_C$ and $Q_M$ involves limits which do not necessarily commute with differentiation with respect to $e$. And the fact that the term $\Xi(p)$ in (8.7) is a continuous function does not imply that the same is true for its perturbative coefficients. This holds also for the continuity of the 2-point $\psi$-function. Hence we are forced to a pragmatic procedure, if we want to get anywhere. We assume that the charge identity holds separately in each order of PT, that the function $\tilde{W}_\sigma(\mathsf{p},\mathsf{q})$ has the form (8.7) for every $\sigma$, and that the only discontinuity of $\tilde{W}_\sigma(p,\bar{q})$ lies at $p^2 = m^2$ and is the weakest one compatible with the other requirements. Since the $g_{\mu\nu}$-term in (8.7) is $e$-independent it is of course only present in the lowest order $\sigma = 0$. The $\rho$-term is determined by a new requirement that we need to introduce at this point: the *requirement of renormalizability*. It states that the various divergent terms in the $N_i$-definitions (8.2) can indeed be combined into convergent expressions, as has been promised above. What this exactly means will become clear in Chap. 10.

This non-rigorous way of motivating the renormalization conditions in PT is justified a posteriori by its success in creating a mathematically consistent and physically meaningful formalism.

## 8.3  Uniqueness

Before starting the construction of a solution of the problem just outlined we want to settle the *problem of uniqueness*: we want to make sure that the conditions as enunciated determine the solution completely. We find that there is at most one solution, up to an immaterial ambiguity in the normalization of $\psi$. Concerning this ambiguity we refer to the discussion in Sect. 5.1 and to the actual handling of the problem in the physical context studied in Part III.

The uniqueness of the normal products $N_i$ can only be discussed together with their definition in Chap. 10. At present we assume that unique $N_i$-definitions are given.

The proof of uniqueness proceeds by induction with respect to $\sigma$. The induction starts at $\sigma = 0$. In this case the right-hand sides of the equations (8.9) vanish. The problem reduces then to the solution of the free-field equations, a problem that has been solved in Chap. 5. Of course, we must now consider the combined $A$–$\psi$–$\bar{\psi}$-theory, but this is a trivial complication. As in Chap. 5 we find that the 1-point functions vanish (this is true in every order $\sigma$) and that the higher functions are fully determined by the 2-point functions through cluster expansions of the type (5.13) and (5.61). The mixed 2-point functions $W_0(x, \mathsf{z})$ and $W_0(\bar{y}, \mathsf{z})$ vanish because of charge conservation. As a result the joint cluster expansion of $W_0(X, \bar{Y}, \mathsf{Z})$ is simply the product $W_0(X, \bar{Y}) W_0(\mathsf{Z})$ of the separate expansions (5.61) and (5.13), the latter adapted to the case of a vector field. The spinorial 2-point functions are given by (5.57) and (5.59):

$$W_0(x, \bar{y}) = -i\, a\, S^+(x-y)\,, \qquad W_0(\bar{x}, y) = -i\, a\, \left(S_-(y-x)\right)^T. \qquad (8.12)$$

In contrast to Chap. 5 we leave the value of the normalization constant $a > 0$ open, in accordance with the ambiguity of the $\psi$-normalization. The photonic 2-point function is given by (5.83) which reads in the Källén gauge (8.7)

$$\tilde{W}_0(^\mu\mathsf{p},{}^\nu\mathsf{q}) = -\delta^4(p+q)\left(g^{\mu\nu} - \rho_0 p^\mu p^\nu\right) \delta_+(p)\,.$$

The presence of the $\rho$-term would destroy renormalizability (see Chap. 10), hence we set $\rho_0 = 0$, obtaining the final result

$$\tilde{W}_0(^\mu\mathsf{p},{}^\nu\mathsf{q}) = -\delta^4(p+q)\, g^{\mu\nu}\, \delta_+(p)\,. \qquad (8.13)$$

Transforming this expression into $x$-space we find

$$W_0(^\mu\mathsf{x},{}^\nu\mathsf{y}) = i\, g^{\mu\nu}\, D_+(x-y)\,. \qquad (8.14)$$

$D_+(\xi)$ is the $m = 0$ case of the function $\Delta_+(\xi)$ defined in (5.10).

In the case $\sigma > 0$ we assume that solutions $W_\tau$ of (8.9) have been found for all $\tau < \sigma$. We wish to prove that $W_\sigma$, if it exists at all, is determined uniquely up to the $\psi$-ambiguity mentioned above. Let $W^1_\sigma$, $W^2_\sigma$ be two solutions of the problem. Define

$$h_\sigma(X, \bar{Y}, \mathsf{Z}) = W^1_\sigma(X, \bar{Y}, \mathsf{Z}) - W^2_\sigma(X, \bar{Y}, \mathsf{Z})\,. \qquad (8.15)$$

$h_\sigma$ satisfies the homogeneous equations

$$\Box_z h_\sigma(\ldots, z, \ldots) = 0 ,$$
$$\mathcal{D}_x h_\sigma(\ldots, x, \ldots) = h_\sigma(\ldots, \bar{y}, \ldots) \overleftarrow{\mathcal{D}}_y = 0 \tag{8.16}$$

and the W-properties, the weak cluster condition (8.11) in the form

$$h_\sigma^{IJ} \to W_0^I h_\sigma^J + h_\sigma^I W_0^J . \tag{8.17}$$

(We assume $h_\tau = 0$ for $\tau < \sigma$.) This is again the problem solved in Chap. 5, except for the slightly changed cluster condition. As there, we find that the 1-point functions and the 2-point functions $h_\sigma(x, z)$, $h_\sigma(\bar{y}, z)$, $h_\sigma(x_1, x_2)$, $h_\sigma(\bar{y}_1, \bar{y}_2)$ vanish. The determination of the remaining 2-point functions does not involve the cluster property. Hence we find like in order $\sigma = 0$

$$\tilde{h}_\sigma(p, \bar{q}) = r_\sigma^h \tilde{W}_0(p, \bar{q}) ,$$
$$\tilde{h}_\sigma(^\mu\mathsf{p}, ^\nu\mathsf{q}) = \delta^4(p+q) \rho_\sigma^h p^\mu p^\nu \delta_+(p) , \tag{8.18}$$

with real coefficients $r_\sigma^h$, $\rho_\sigma^h$. But if $W^1$ is part of a renormalizable theory, then the addition of the $\rho$-term destroys renormalizability, so that the $W^2$-theory is not renormalizable (see Chap. 10), except if $\rho_\sigma^h = 0$. The coefficient $r_\sigma^h$ need not vanish, however. The $h_\sigma$ with more than two variables can also be determined by the procedure of Chap. 5. The equivalent of (5.27) takes the more complicated form

$$\tilde{h}_\sigma(p_1, \ldots, p_n) = \sum_{i=2}^n \pm \big[\tilde{h}_\sigma(p_1, p_i) \tilde{W}_0(\hat{p}_1, \ldots, \hat{p}_i, \ldots)$$
$$+ \tilde{W}_0(p_1, p_i) \tilde{h}_\sigma(\hat{p}_1, \ldots, \hat{p}_i, \ldots)\big] . \tag{8.19}$$

Here the $p_i$ are generic variables of arbitrary type, not necessarily $\psi$-variables. The sign in front of the square bracket is the usual fermionic sign. Equation (8.19) allows the determination of all $h_\sigma$ by induction with respect to the number $n$ of variables, starting from the known cases $n = 1, 2$. The result is (see (5.13) and (5.61))

$$\tilde{h}_\sigma(p_1, \ldots, p_n) = r_\sigma^h f \sum_{\text{pairings}} \pm \tilde{W}_0(p_{i_\alpha}, p_{j_\alpha}) , \tag{8.20}$$

where $f$ is the number of $\psi$-variables, which is the same as a the number of $\bar{\psi}$-variables. If we define the two solutions $W^1$, $W^2$ up to order $\sigma$ by

$$W^i = \sum_{\tau=0}^{\sigma-1} W_\tau + W_\sigma^i , \tag{8.21}$$

we find that we have, up to this order,

$$W^1(X, \bar{Y}, Z) = \left(1 + r_\sigma^h\right)^{|X|} W^2(X, \bar{Y}, Z) , \tag{8.22}$$

the claimed uncertainty in the $\psi$-normalization: $\psi$ and $\bar\psi$ are rescaled by the factor $(1+r_\sigma^h)^{1/2}$. If such a factor was already present at lower orders,

$$W^1_\tau = \sum_{\kappa=0}^{\tau} \left(r^{|X|}\right)_\kappa W^2_{\tau-\kappa} \quad \text{for} \quad \tau < \sigma\,,$$

then we include it by replacing $W^2$ by $W'^2 = r^f W^2$ in our proof, and obtain the result that there exists an $r(e) = \sum r_\sigma e^\sigma$ such that

$$W^1(X,\bar Y,\mathsf{Z}) = r^{|X|}\, W^2(X,\bar Y,\mathsf{Z}) \tag{8.23}$$

up to the arbitrary order $\sigma$. This completes the proof of essential uniqueness.

The actual construction of an iterative solution of (8.9) will be carried out in the next three chapters. We will not try to solve all the problems at one stroke, but proceed in several stages of increasing mathematical rigour. In Chap. 9 a formal solution of the problem, the "unrenormalized $W$-functions", will be derived, firmly ignoring all questions of existence. The renormalization conditions will also be partially ignored at this stage. The result is expressed as a sum over multi-dimensional integrals, which can be symbolized by generalized Feynman graphs. These integrals suffer from two different types of divergences, UV and IR. It will be shown in Chap. 10 that renormalization removes the UV divergences. In Chap. 11 the apparent IR divergences of the $W_\sigma$ will be shown to be spurious. It must be said, however, that the IR problem encountered at this stage is quite mild. A more serious problem will appear in Part III in connection with the introduction of particles into the theory.

The result is obtained not by a constructive procedure, but by a guess based on simple examples: low-order cases and the well-known Feynman rules for time-ordered functions (see following section). In order to validate this guess, it will be shown at each stage of the procedure that the proposed solution indeed satisfies the necessary requirements at the actual level of rigour. In many cases the "proofs" of one stage carry over essentially unchanged to the next stage. In these obvious cases the proof will not be stated anew. Only the points requiring more precision will be noted. The final result is stated in Sect. 11.3 and shown there to solve the problem rigorously, again not repeating those parts of the proofs that can be simply taken over from the earlier stages.

No regularization, either UV or IR, will ever be used, except possibly for heuristic purposes.

## 8.4 Time-Ordered Products

Our construction also yields expressions for a generalization of the $W$-functions: the *completely or partially time-ordered functions*, which are important for many applications of the theory, and also for the comparison of our results with those of the more conventional approaches.

The *time-ordered product* $T^+(x_1, \ldots, x_n)$ of the fields $\varphi^{(1)}(x_1), \ldots, \varphi^{(n)}(x_n)$ is formally defined by

$$T^+(x_1, \ldots, x_n) = \sum_\pi \sigma_\pi \, \theta(x^0_{\pi_1} - x^0_{\pi_2}) \cdots \theta(x^0_{\pi_{n-1}} - x^0_{\pi_n}) \, \varphi^{(\pi_1)}(x_{\pi_1}) \cdots \varphi^{(\pi_n)}(x_{\pi_n}) \, . \quad (8.24)$$

The sum extends over all permutations $\pi : (1, \ldots, n) \to (\pi_1, \ldots, \pi_n)$ of the index set $I = (1, \ldots, n)$. $\pi$ induces a permutation $\pi_f$ of the subset $I_f \subset I$ consisting of the indices numbering Fermi fields. The coefficient $\sigma_\pi$ is $+1$ if $\pi_f$ is an even permutation, $-1$ if $\pi_f$ is odd. This choice of $\sigma_\pi$ and local commutativity guarantee that $T^+$ transforms covariantly under the orthochronous Poincaré group despite the non-invariance of the function $\theta(\xi^0)$:

$$T^+(\ldots, \Lambda x_\alpha + a, \ldots) = U(\Lambda, a) \prod_\alpha R^{(\alpha)}(\Lambda) \, T^+(\ldots, x_\alpha, \ldots) \, U^*(\Lambda, a) \, . \quad (8.25)$$

As an example we find, using the notations introduced for the $W$-functions,

$$T^+(x, \bar{y}) = \theta(x^0 - y^0) \, \psi(x) \, \bar{\psi}(y) - \theta(y^0 - x^0) \, \bar{\psi}(y) \, \psi(x)$$

and

$$T^+(\Lambda x, \Lambda \bar{y}) = \theta\big((\Lambda(x-y))^0\big) \, \psi(\Lambda x) \, \bar{\psi}(\Lambda y) - \theta\big((\Lambda(y-x))^0\big) \, \bar{\psi}(\Lambda y) \, \psi(\Lambda x) \, .$$

If $\xi = x - y$ is timelike or lightlike then $\theta(\pm(\Lambda \xi)^0) = \theta(\pm \xi^0)$ for orthochronous $\Lambda$, so that (8.25) is satisfied. For $\xi^2 < 0$ $\psi(x)$ and $\bar{\psi}(y)$ anticommute and we obtain

$$T^+(x, \bar{y}) = \big[\theta(x^0 - y^0) + \theta(y^0 - x^0)\big] \, \psi(x) \, \bar{\psi}(y) = \psi(x) \, \bar{\psi}(y) \, ,$$

which is clearly covariant.

The *anti-time-ordered product* $T^-(x_1, \ldots, x_n)$ is defined in the same way, replacing the factors $\theta(x^0_{\pi_\alpha} - x^0_{\pi_{\alpha+1}})$ in (8.24) by $\theta(x^0_{\pi_{\alpha+1}} - x^0_{\pi_\alpha})$.

The expression (8.24) has no rigorous meaning because the fields $\varphi^{(\alpha)}(x_\alpha)$ are distributions which cannot be multiplied with the discontinuous $\theta$-functions. We need therefore a better definition of time ordering. This definition should satisfy the following requirements.

(a) $T^\pm(X)$, $X = \{x_1, \ldots, x_n\}$, is covariant, i.e. it satisfies (8.25).
(b) $T^\pm(X)$ satisfies the symmetry property

$$T^\pm(X_\pi) = \sigma_\pi \, T^\pm(X) \, , \quad (8.26)$$

where $X_\pi = \{x_{\pi_1}, \ldots, x_{\pi_n}\}$ is a permutation of $X$ and $\sigma_\pi$ is defined as in (8.24).

(c) Let $X$, $Y$ be any two non-overlapping sets of field variables. Then the *ordering relations*

$$T^+(X,Y) = T^+(X)\,T^+(Y) \quad \text{if} \quad x_\alpha^0 > x_\beta^0 \,\forall\, x_\alpha \in X,\ y_\beta \in Y\ ,$$
$$T_-(X,Y) = T_-(X)\,T_-(Y) \quad \text{if} \quad x_\alpha^0 < y_\beta^0 \,\forall\, x_\alpha \in X,\ y_\beta \in Y\ , \quad (8.27)$$

hold. These relations capture the essence of time ordering.

(d) The reality condition

$$\bigl(T^+(x_1,\ldots,x_n)\bigr)^* = \pm T^-({}^*x_n,\ldots,{}^*x_1) \qquad (8.28)$$

holds. The sign is positive (negative) if the reversal of the ordering of Fermi variables in $X$ is an even (odd) permutation. ${}^*x_i$ is the argument of the field $\varphi^*(x_i)$. In applying this relation we must remember that $\bar\psi \neq \psi^*$, so that in writing (8.28) for $T^+(X,\bar Y,\mathsf{Z})$ $\gamma^0$-factors must be introduced like in (8.6).

These conditions are satisfied by the formal definition (8.24). The ordering relation determines $T^\pm(X)$ uniquely at the points where no two zero-components $x_\alpha^0$ and $x_\beta^0$ coincide. For such a point there is a unique permutation $\pi$ such that $x_{\pi_1}^0 > x_{\pi_2}^0 > \cdots > x_{\pi_n}^0$, and we obtain

$$T^+(X) = \sigma_\pi\, \varphi^{(\pi_1)}(x_{\pi_1})\cdots\varphi^{(\pi_n)}(x_{\pi_n})\ , \qquad (8.29)$$

and analogously for $T^-$. If two or more $x_\alpha^0$ are equal but the corresponding space parts $\vec x_\alpha$ are different, we can find a Lorentz frame in which all $x_\alpha^0$ are different and (8.29) is applicable. We then find $T^+$ in the original frame by using covariance. There remains an ambiguity with support at the points where two or more $x_\alpha$ coincide. These ambiguities in $T$-products with different numbers of variables are related through the ordering relations (8.27), and they are further restricted by the covariance and symmetry requirements.

A final natural restriction of ambiguity can be introduced if we define the $T$-products in the spirit of our approach by specifying their matrix elements

$$\Bigl(\Omega,\ \varphi^{(\alpha_1)}(x_1)\cdots\varphi^{(\alpha_n)}(x_n)\,T^\pm(Y)\,\varphi^{(\beta_1)}(z_1)\cdots\varphi^{(\beta_m)}(z_m)\,\Omega\Bigr) \qquad (8.30)$$

between a total set of $\mathcal{V}$-states. As usual $\varphi^{(\alpha)}$ denotes the generic field: it is either $\psi$ or $\bar\psi$ or $A$. Let us consider the more general functions

$$\mathcal{W}(X_1,s_1|X_2,s_2|\cdots|X_N,s_N) := \bigl(\Omega,\ T^{s_1}(X_1)\,T^{s_2}(X_2)\cdots T^{s_N}(X_N)\,\Omega\bigr)\ , \qquad (8.31)$$

where the $s_i$ are signs and the $X_i$ are non-overlapping finite sets of field variables. Field types and component indices have been suppressed. (They may be considered to be included in the $x_i$.) Special cases of $\mathcal{W}$-functions are the following.

(i) If all $|X_i| = 1$ we find

$$\mathcal{W}(x_1,s_1|\cdots|x_n,s_n) = W(x_1,\ldots,x_n)\ , \qquad (8.32)$$

irrespective of the choice of the signs $s_i$. Here we have used the definition

$$T^+(x) = T^-(x) = \varphi(x) \tag{8.33}$$

of a time-ordered product of a single field.

(ii) If $|X_i| = 1$ for all but one $i$ we obtain the matrix elements (8.30). Again, the choice of the $s_i$ other than the exceptional one is irrelevant.

(iii) If $N = 1$ we obtain the *time-ordered* and *anti-time-ordered functions* $\tau^\pm(x_1, \ldots, x_n)$. The $\tau^+$ are also known as the Green's functions of the theory. They are the basic quantities of the traditional formulations of QED and other quantum field theories, because they are easier to calculate than the $\mathcal{W}$-functions if canonical methods or path integrals are used.[1] Deriving expressions for these functions is therefore essential for establishing the equivalence of our results with the familiar ones.

(iv) From the hermiticity condition (8.28) we find

$$\left(\Omega, T^{+\,*}(x_1, \ldots, x_n) T^+(y_1, \ldots, y_m) \Omega\right)$$
$$= \pm \mathcal{W}(\{x_1, \ldots, x_n\}, -|\{y_1, \ldots, y_m\}, +) \, . \tag{8.34}$$

These functions will play an important role in Part III, since inclusive cross sections can be written in terms of them.

The $\mathcal{W}$ must satisfy the conditions that obviously follow from the conditions (a)–(d) explained above. They allow the determination of the general $\mathcal{W}$-functions from the more special $\mathcal{W}$-functions, up to terms with support at coinciding points within a set $X_i$. These ambiguities are restricted by the conditions already noted. Moreover we introduce the requirement of maximal scaling degree as an additional restriction. We must first define this notion. Let $T(u_1, \ldots, u_\ell)$ be a tempered distribution on $\mathbf{R}^\ell$. We say that $T$ has the *scaling degree* $d$ if

$$\lim_{\lambda \to 0} \lambda^\beta T(\lambda u_1, \ldots, \lambda u_\ell) = 0 \tag{8.35}$$

for every real $\beta > d$, but not for any $\beta < d$. The $\lambda$-limit must be taken in the sense of distributions. This means that for $\beta > d$ and for every test function $f \in \mathcal{S}$ we have

$$\lim_{\lambda \to 0} \lambda^\beta \int \prod du_i \, T(\ldots, \lambda u_i, \ldots) f(\ldots, u_i, \ldots)$$
$$= \lim_{\lambda \to 0} \lambda^{\beta - \ell} \int \prod dv_i \, T(\ldots, v_i, \ldots) f(\ldots, \lambda^{-1} v_i, \ldots) = 0 \, , \tag{8.36}$$

while for $\beta < d$ there is at least one $f \in \mathcal{S}$ for which this limit does not exist. A unique real constant $d$ with this property exists for any distribution $T$. It may have the value $+\infty$ but not $-\infty$. The value of $d$ gives information on the short-distance behaviour of $T$, as can be seen from the following examples.

---

[1] This holds also for such less familiar methods as e.g. the causal Bogoliubov–Epstein–Glaser approach as applied to the Green's functions of QED in [BS 75].

(i) The form
$$T(\ldots, u_i, \ldots) = \sum_{\sum \alpha_i = A} c_{\alpha_i} u_i^{\alpha_i}$$

of degree $A$ has the scaling degree $A$. The scaling degree of a finite sum of forms, of a polynomial in other words, is the lowest degree of the forms contained in it. This result is not changed if the form is multiplied by logarithmic factors, e.g. by $\log|u_i|$: logarithmic singularities are not distinguished from bounded non-vanishing functions as far as the scaling degree is concerned.

(ii) The derivative $D\delta^\ell(u)$ of the $\delta$-function at the origin, with $D$ a derivative of order $|D|$, has the scaling degree $-\ell - |D|$.

(iii) If the support of $T$ does not contain the origin, then its scaling degree is $d = \infty$.

With this definition we can formulate the *condition of maximal scaling degree* for the $\mathcal{W}$-functions. We demand that $\mathcal{W}$ have the maximal possible scaling degree that is compatible with the other requirements. In other words: we demand that $\mathcal{W}$ be locally as smooth as possible, in particular that it contains no high derivatives of $\delta$-functions in difference variables. Notice that due to translation invariance the scaling degree of $\mathcal{W}$ does not depend on the choice of the origin in $x$-space. The maximality condition is introduced for convenience, in order to reduce as far as possible the ambiguity inherent in the definition of time ordering. The physical content of the theory is not prejudiced by the condition. In perturbative QED the maximality condition together with the other requirements determines the $\mathcal{W}$ uniquely in terms of the simpler $W$-functions,[2] and the scaling degree is the same for all $\mathcal{W}$-functions with the same arguments.

As an example, consider the time-ordered 2-point functions of the free fields. We define the *scalar Feynman propagator* by

$$\Delta_F(\xi) = \theta(\xi^0)\,\Delta_+(\xi) + \theta(-\xi^0)\,\Delta_+(-\xi) \;. \tag{8.37}$$

In the case of a vanishing mass $m = 0$ this function is called $D_F(\xi)$. The singularity of $\Delta_+$ at $\xi = 0$ is weak enough to allow multiplication with $\theta(\xi^0)$. This is seen by writing the terms in (8.37) as Fourier integrals, yielding

$$\Delta_F(\xi) = -(2\pi)^{-4} \int dp\, (p^2 - m^2 + i\varepsilon)^{-1}\, e^{-ip\xi} \;, \tag{8.38}$$

which is a perfectly meaningful expression if interpreted as a Fourier transform in the sense of distributions. Remember that the value of the distribution

---

[2] No such result is known in the purely axiomatic framework. In fact it is not even known whether the mere existence of the $\mathcal{W}$-functions follows from the W-properties, assuming that the $W$-functions exist. The existence of the $\mathcal{W}$ has been rigorously established for some non-trivial models in 2- and 3-dimensional space-time [EEF 76, EE 79] but not for any interacting theories in 4 dimensions.

$T = (u \pm i\varepsilon)^{-1}$ for the test function $f(u)$ is defined as

$$T(f) = \lim_{\varepsilon \downarrow 0} \int du \, \frac{f(u)}{u \pm i\varepsilon} \,. \tag{8.39}$$

$\Delta_F$ is clearly Lorentz invariant. Its scaling degree is determined from the condition (8.36). The integral $\int d\xi \, \Delta_F(\lambda\xi) f(\xi)$, $f \in \mathcal{S}$, becomes in $p$-space $-(2\pi)^{-4}\lambda^{-2} \int dp \, (p^2 - \lambda m^2 + i\varepsilon)^{-1} \tilde{f}(p)$. The $p$-integral converges for $\lambda \to 0$ to $\int dp \, (p^2 + i\varepsilon)^{-1} \tilde{f}(p)$, which exists in the sense of distributions and does not vanish identically on $\mathcal{S}$. Hence the scaling degree of $\Delta_F$ is $-2$.

Starting from (8.14) we find

$$\tau_0^+ (^\mu x, ^\nu y) = i \, g^{\mu\nu} \, D_F(x-y) \,. \tag{8.40}$$

This function is covariant and satisfies the ordering relation $\tau_0^+(x,y) = W_0(x-y)$ if $x^0 > y^0$. Its scaling degree is that of $D_F$, i.e. $-2$. A possible ambiguity of $\tau_0^+$ is concentrated at $x = y$, i.e. it is a derivative of $\delta^4(x-y)$ and thus has a scaling degree $\leq -4$. The addition of such a term would lower the scaling degree of $\tau_0^+$ and thus violate the condition of maximal scaling degree.

Similar calculations yield the time-ordered fermion functions

$$\tau_0^+(x,\bar{y}) = -i \, a \, S_F(x-y) \,, \qquad \tau_0^+(\bar{x},y) = i \, a \, S_F(y-x) \,, \tag{8.41}$$

with

$$S_F(\xi) = (i\slashed{\partial}_\xi + m)\Delta_F(\xi) \,, \tag{8.42}$$

and the anti-time-ordered functions (using (8.28))

$$\tau_0^- (^\mu x, ^\nu y) = -i \, g^{\mu\nu} \, D_F^*(x-y) \,, \tag{8.43}$$

$$\tau_0^-(x,\bar{y}) = i \, a \, \gamma^0 S_F^*(y-x) \gamma^0 \,, \qquad \tau_0^-(\bar{x},y) = -i \, a \, \gamma^0 S_F^*(x-y) \gamma^0 \,. \tag{8.44}$$

$S_F$ is called the *spinorial Feynman propagator*.

That logarithmic singularities of $\mathcal{W}$ do not affect its scaling degree is of importance in assessing the relations between the perturbative results and the properties of a possibly existing exact theory. We will find that in perturbation theory the scaling degree of a given $\mathcal{W}$ is the same in all orders in which this $\mathcal{W}$ is different from zero. But this common degree is not necessarily also the degree of the corresponding exact $\mathcal{W}$-function. Due to the presence of increasingly high powers of logarithms in increasing orders the perturbative series may sum to an expression with a different degree, thus giving rise to the phenomenon of "anomalous dimensions". For example, the function $|u|^{ce}$ with $c$ a real number, has the scaling degree $ce$, but all the terms in its power series $\sum_n (n!)^{-1} c^n e^n (\log|u|)^n$ have scaling degree 0.

# Chapter 9

# Unrenormalized Solution

## 9.1 Generalities

In this chapter the general ideas employed for solving the equations of motion (8.9) are outlined by deriving a formal solution of the unrenormalized form of (8.9), which satisfies the W-properties but not the renormalization conditions. The unrenormalized equations of motion are defined by setting the renormalization constants $R_i$, $\lambda$, $\delta m$, in the $N$-definition (8.2) equal to zero. We give solutions both for the $x$-space functions $W_\sigma$ and their Fourier transforms $\tilde{W}_\sigma$. In both cases the solution is represented as a sum over integrals. That the solution is formal means the following. The integrands of the integrals just mentioned are products of distributions whose arguments are linear combinations of the external variables $x_i$ of $W_\sigma$ or $p_i$ of $\tilde{W}_\sigma$ and of a number of internal variables of integration. The arguments of the various factors are not necessarily independent variables, hence the product is in general not a direct product of distributions, which would be well defined. In the present chapter we disregard this problem, treating the integrand as though it were a direct product. Moreover, we also proceed as though the integrand decreased sufficiently fast to ensure convergence of the integral at infinity. This is also not true in general. The "proof" that the proposed expressions solve our problem consists in demonstrating properties of the integrands which would imply the desired properties of the integrals, if they existed and could be handled by the standard procedures of the integral calculus, including the use of regular substitutions of variables. $\tilde{\mathcal{W}}_\sigma$ is the Fourier transform of $W_\sigma$ also in this formal sense only.

In the $x$-space version of the rules this can be formulated more precisely. There the integrands in question are products of invariant functions as discussed in Sect. 5.4, with arguments which are differences of two variables of integration. This product is a $C^\infty$ function almost everywhere: it is singular only where the argument of a factor lies on the light cone. And the way in which one has to integrate over these singularities is uniquely specified by the rules of Sect. 5.4. This specification may or may not define a distribution. We say that the integrand is well defined as a "singular function", but not necessarily as a distribution. Our "formal proofs" of the necessary properties of $W_\sigma$ consist in showing that the integrands, considered as singular functions, possess properties which would guarantee the correct properties of the integrals, if these integrals existed, i.e. if the singular functions in question defined distributions which are sufficiently well behaved at infinity.

We will not derive the solution constructively, except in a cursory manner. We simply write it down and then show that it indeed satisfies the required conditions in the formal sense just indicated. An exception is the cluster property which depends essentially on the long-distance behaviour of the $x$-space integrand. For the sake of completeness we will nevertheless indicate the main ideas underlying its proof, as far as this is possible at the present formal level.

It is convenient to give these expressions directly for the general $\mathcal{W}$-functions instead of restricting ourselves to the more special Wightman functions $W$. This must not obscure the fact that the $W$-functions are the primary objects of the theory, which are constructed first by solving the problem outlined in Sect. 8.2. The more general $\mathcal{W}_\sigma$ are determined afterwards from $W_\sigma$, using the definitions given in Sect. 8.4. In particular, no differential equations of the type (8.9) are postulated for the $\mathcal{W}_\sigma$. The expressions will first be given in $x$-space for $\mathcal{W}_\sigma(X_1, s_1 | \cdots | X_N, s_N)$. The equations of motion for the Wightman functions and the W-properties referring to $x$-space are then verified. For verifying the spectral condition we need the rules for the Fourier transforms $\tilde{\mathcal{W}}_\sigma$. These will also be needed later on for implementing renormalization. Moreover, the $p$-space form is more suitable for most applications, in particular for the scattering formalism to be established in Part III.

We will exemplify our strategy by treating the lowest orders rather explicitly in Sect. 9.2. In Sect. 9.3 we write down the general unrenormalized expressions of $\mathcal{W}_\sigma$ both in $x$-space and in $p$-space. In Sect. 9.4 we will verify, in the formal sense explained above, that these expressions satisfy the required conditions apart from the renormalization conditions.

## 9.2 Special Examples in Low Orders

In order to get a feeling for the situation, let us first study simple cases in some detail: the few-point functions in the lowest orders of perturbation theory. The *zero-order* terms have already been determined in Chap. 8. The non-vanishing 2-point functions are

$$W_0(^\mu\mathsf{x},{}^\nu\mathsf{y}) = i\, g^{\mu\nu}\, D_+(x-y)\,,$$
$$W_0(x,\bar{y}) = -i\, S^+(x-y)\,, \qquad W_0(\bar{x},y) = -i\left(S^-(y-x)\right)^T, \qquad (9.1)$$

$$\tau_0^+ (^\mu \mathsf{x},^\nu \mathsf{y}) = i\, g^{\mu\nu}\, D_F(x-y)\,,$$
$$\tau_0^+ (x,\bar{y}) = -\tau_0^+ (\bar{y},x) = -i\, S_F(x-y)\,, \tag{9.2}$$

$$\tau_0^- (^\mu \mathsf{x},^\nu \mathsf{y}) = -i\, g^{\mu\nu}\, D_F^*(x-y)\,,$$
$$\tau_0^- (x,\bar{y}) = -\tau_0^- (\bar{y},x) = i\, S_F^*(x-y)\,. \tag{9.3}$$

We have set the indeterminate fermion normalization constant $a$ equal to 1, which we are free to do. At the moment we are in any case not concerned with finding a unique solution, since for this we need to take the renormalization conditions into account. The $\mathcal{W}_0$'s with more than two variables are given by a cluster expansion in terms of the 2-point functions, which we need not write down in detail. For the $W_0$ it has been described in Chap. 8.

In *first order* we meet for the first time inhomogeneous differential equations. As a typical example we determine the function $W_1(x,\bar{y},\mathsf{z})$ with the variables ordered as shown. It must satisfy the equations

$$\mathcal{D}_x W_1(x,\bar{y},{}^\mu \mathsf{z}) = W_0({}_\nu \mathsf{x},{}^\mu \mathsf{z})\,\gamma_x^\nu\, W_0(x,\bar{y})\,,$$
$$W_1(x,\bar{y},{}^\mu \mathsf{z})\, \overleftarrow{\mathcal{D}}_y = -\, W_0(x,\bar{y})\,\gamma_y^\nu\, W_0({}_\nu \mathsf{y},{}^\mu \mathsf{z})\,,$$
$$\Box_z W_1(x,\bar{y},{}^\mu \mathsf{z}) = -\, W_0(x,\bar{z})\,\gamma_z^\mu\, W_0(\bar{y},z)$$
$$+ \frac{1}{2} W_0(x,\bar{y}) \sum_{\rho,\sigma} \gamma_{\rho\sigma}^\mu \left( W_0({}_\rho \bar{z},{}_\sigma z) - W_0({}_\sigma z,{}_\rho \bar{z}) \right)\,. \tag{9.4}$$

The $W_0$ are given by (9.1)–(9.3). The Dirac matrix $\gamma_x^\nu$ acts on the spinor indices of $\psi(x)$, and similarly for $\gamma_y^\nu, \gamma_z^\mu$. The $\sum_{\rho,\sigma}$-term in the last equation can be shown to vanish by an argument used earlier after (5.70).

From the definitions (8.37) and (8.42) we find

$$\Box D_F(\xi) = \delta^4(\xi)\,, \quad \mathcal{D} S_F(\xi) = -\delta^4(\xi)\,, \quad S_F(-\xi)\,\overleftarrow{\mathcal{D}}_\xi = \delta^4(\xi)\,. \tag{9.5}$$

This means that $D_F(\xi), -S_F(\xi)$, and $S_F(-\xi)$, respectively, are *Green's functions*, or *elementary solutions* in the language of distributions, of the differential operators in (9.4). Note that in the context of quantum field theory the name "Green's functions" is also often applied to the time-ordered functions $\tau^+$. This is a generalized use of the expression inspired by the relations (8.40) and (8.41). We find that

$$A_1(x,y,{}^\mu \mathsf{z}) = -\int du\, S_F(x-u)\,\gamma_u^\mu\, S_+(u-y)\, D_+(u-z)\,,$$
$$A_2(x,\bar{y},{}^\mu \mathsf{z}) = -\int du\, S_+(x-u)\,\gamma_u^\mu\, S_F(u-y)\, D_+(u-z)\,,$$
$$A_3(x,\bar{y},{}^\mu \mathsf{z}) = \int du\, S_+(x-u)\,\gamma_u^\mu\, S_-^T(u-y)\, D_F(u-z) \tag{9.6}$$

solve, respectively, the first, second, and third equation (9.4). Moreover, each of these solutions is annihilated by the differential operators in the other two equations, e.g. $A_1 \overleftarrow{\mathcal{D}}_y = \square_z A_1 = 0$. Hence

$$W_1(x, \bar{y}, z) = \sum_{i=1}^{3} A_i(x, \bar{y}, z) \tag{9.7}$$

solves the system (9.4). It is the correct solution if it satisfies the W-properties. Before discussing this problem let us briefly study the existence of the $u$-integrals in (9.6), even though we have decided not yet to worry about existence in this chapter. The three $A_i$ can be treated similarly. Therefore let us concentrate on $A_1$. All the factors in the integrand are true distributions, not locally or even globally integrable functions. Hence the integral is at best defined in the sense of distributions. A Fourier transform yields an expression of the form

$$\tilde{A}(p, \bar{q}, r) = \delta^4(p + q + r)\, \delta_-(q)\, \delta_-(r) \left((q + r)^2 - m^2 + i\varepsilon\right)^{-1} \mathcal{P}(q, r), \tag{9.8}$$

with $\mathcal{P}$ a polynomial. This defines a tempered distribution if the product of the three $q$–$r$-dependent factors (not including the $\delta^4$) is a distribution in $q$ and $r$. The integral of this product over the test function $f(q, r)$ is

$$\int \frac{d^3q\, d^3r}{\omega(\vec{q})\, |\vec{r}|} \left(2\omega(\vec{q})\, |\vec{r}| - (\vec{q}, \vec{r})\right)^{-1} f\!\left(\omega(\vec{q}), \vec{q}, |\vec{r}|, \vec{r}\right). \tag{9.9}$$

But $\omega(\vec{q}) > |\vec{q}|$, hence the factor $(\cdots)^{-1}$ is regular everywhere except at $\vec{r} = 0$, where it is singular of first order. Together with the factor $|\vec{r}|^{-1}$ this gives a singularity of second order which is still integrable. Hence the integral exists.

Now to the W-properties. Translation invariance of the $A_i$ is obvious. Lorentz invariance is easy to check. So is the spectral condition from the form (9.8) and analogous expressions for $A_{2,3}$. The weak cluster property is satisfied if it is satisfied for each $A_i$ separately. For $A_1$ it states that if we insert in the integrand $I(\vec{q}, \vec{r})$ of (9.9) an additional factor $\gamma(\lambda \vec{r})$ or $\gamma(\lambda(\vec{q} + \vec{r}))$, $\gamma \in \mathcal{S}$, the limit $\lambda \to \infty$ vanishes. But this follows from the integrability of $I$ and the fact that

$$\lim_{\lambda \to \infty} \gamma(\lambda \vec{r})\, I(\vec{q}, \vec{r}) = \lim_{\lambda \to \infty} \gamma\!\left(\lambda(\vec{q} + \vec{r})\right) I(\vec{q}, \vec{r}) = 0$$

in $\mathcal{L}_1$. For checking the hermiticity and locality conditions we would need expressions for the other 3-point functions with different orderings of the variables, which we have not written down. We refrain from doing this here and refer to the proof of the W-properties for the general case to be given later. It is interesting to note that the conditions that we did prove suffice already to fix $W_1$ uniquely. Any solution of the homogeneous part of (9.4) which satisfies the covariance and spectral conditions is in $p$-space of the form $\delta^4(p + q + r)\, \delta_+(p)\, \delta(q^2 - m^2)\, \delta_-(r)\, T(\vec{q}, \vec{r})$ with $T$ a sufficiently regular distribution. The mass in the $\delta_+$-factors is $m$, in the $\delta_-$-factor zero. But the support of this product is contained in the manifold $\{r = 0\}$, and a term with this support is excluded by the weak cluster condition. Hence neither the remaining two W-properties nor the renormalization conditions are needed to fix the solution in this simple case. This is no longer so in higher orders.

As a second example let us look at the 2-photon function in second order, which shows already most of the essential complications of the general situation. This function must satisfy

$$\Box_x W_2({}^\mu \mathsf{x}, {}^\nu \mathsf{y}) = \frac{1}{2} \gamma^\mu_{\rho\sigma} \left( W_1(\rho \bar{x}, \sigma x, {}^\nu \mathsf{y}) - W_1(\sigma x, \rho \bar{x}, {}^\nu \mathsf{y}) \right)$$
$$=: R({}^\mu x, {}^\nu y) \tag{9.10}$$

and a similar equation with respect to $y$. For the evaluation of the right-hand side we need to know $W_1(\bar{y}, x, z)$. It looks very similar to $W_1(x, \bar{y}, z)$. In (9.6) we need only change $S_+$ to $S_-^T$ in $A_1$ and in $A_2$, and change the sign of $A_3$. A solution of (9.10) is again given by $\int du\, D_F(x-u)\, R({}^\mu u, {}^\nu y)$. But in contrast to the situation found in first order, $W_2$ is not this expression plus a corresponding solution of the $y$-equation. Consider first the $A_3$-contribution to $R$. It gives rise in the mentioned solution of (9.10) to a term

$$-\int du\, dv\, D_F(x-u)\, \gamma^\mu_{\rho\sigma} S^+_{\sigma\tau}(u-v)\, \gamma^\nu_{\tau\omega} S^-_{\omega\rho}(v-u)\, D_F(v-z) \,. \tag{9.11}$$

This term occurs also in the corresponding solution of the $y$-equation. Summing the two solutions would thus lead to a double counting which would violate (9.10): the term (9.11) occurs only once in the correct expression for $W_2$.

The $A_1$- and $A_2$-contributions to $R(x,y)$ read

$$\frac{1}{2} \gamma^\mu_{\rho\sigma} \gamma^\nu_{\tau\omega} \int du \left[ S^F_{\sigma\tau}(x-u)\, S^+_{\omega\rho}(u-x) - S^F_{\sigma\tau}(x-u)\, S^-_{\omega\rho}(u-x) \right.$$
$$\left. + S^+_{\sigma\tau}(x-u)\, S^F_{\omega\rho}(u-x) - S^-_{\sigma\tau}(x-u)\, S^F_{\omega\rho}(u-x) \right] D_+(u-y). \tag{9.12}$$

Its contribution to the solution of (9.10) is obtained by convolution with $D_F(x)$. No double counting problem is present here because the expression (9.12) is annihilated by $\Box_y$. A corresponding $A_{1,2}$-solution of the $y$-equation must be added in $W_2$.

The complicated form (9.12) is not suitable for generalization to higher orders. For that we need to rewrite its integrand in a simpler way: we will show that the square bracket can be replaced by $2\, S^F_{\sigma\tau}(x-u)\, S^F_{\omega\rho}(u-x)$ without changing the value of the full expression (9.12). From the definitions (5.56) and (8.42) we find that

$$S_F(\xi) = \begin{cases} S_+(\xi) & \text{for } \xi^0 > 0 \\ -S_-(\xi) & \text{for } \xi^0 < 0 \,. \end{cases} \tag{9.13}$$

Split the integral (9.12) into a contribution for $u^0 > x^0$ and one for $u^0 < x^0$. At the present level of rigour we need not worry about what happens at $u^0 = x^0$. Set $\xi = x - u$. For both signs of $\xi^0$ the square bracket is equal to

$$2 S^F_{\sigma\tau}(\xi) S^F_{\omega\rho}(-\xi) + S^-_{\sigma\tau}(\xi) S^-_{\omega\rho}(-\xi) + S^+_{\sigma\tau}(\xi) S^+_{\omega\rho}(-\xi) , \qquad (9.14)$$

which expression is then valid for all $u$. The $S^F$-$S^F$-terms sum to the claimed result

$$\int du \, \mathrm{Tr}\{\gamma^\mu S_F(x-u) \gamma^\nu S_F(u-x)\} D_+(u-y) \qquad (9.15)$$

as $A_{1,2}$-contribution to $R$. The curly bracket is a sum over products of $\gamma$-matrices and "Tr" denotes taking the trace of the ensuing $4 \times 4$ matrix.

Here we encounter for the first time a serious existence problem: the product $S^F(x-u) S^F(u-x)$ is not defined as a distribution, as can be seen from the singularities of $S^F$ as described after (5.117). This is part of the UV problem which will be studied in Chap. 10, but is ignored at present. But the offending singularity is confined to the lower-dimensional set $(x-u)^2 = 0$, outside of which $S^F(\pm(x-u))$ is regular. The integrand in (9.15) is thus defined as a singular function in the sense explained in the previous section. The $u$-integral (9.15), however, does not exist as it stands.

The contribution of the $S^+$-$S^+$-term in (9.14) to the integral (9.12) we define as

$$\int du \, S^+(x_1 - u) S^+(u - x_2) D^+(u-y) \Big|_{x_1=x_2=x} . \qquad (9.16)$$

The $\gamma$-matrices and spinor indices have been dropped as irrelevant to the following discussion. As long as $x_1$ and $x_2$ are treated as independent, this integral is of a similar structure as the expressions (9.6) and can be shown to exist as a distribution by the methods employed there. We claim that this distribution vanishes. Analogously to (9.9) its Fourier transform integrated over a test function is of the form

$$\int dp \, dq \, \delta^m_+(p) \, \delta^m_-(q) \, \delta_+(p+q) \, f(p,q)$$
$$= \int d^3p \, d^3q \, \delta\big((\omega_p - \omega_q)^2 - (\vec{p}+\vec{q})^2\big) \, \hat{f}(\vec{p},\vec{q}) , \qquad (9.17)$$

with $\hat{f}$ a bounded function of strong decrease at infinity, which also contains the polynomial factors of $\tilde{S}^+$. The argument $-A(\vec{p},\vec{q})$ of the $\delta$-factor is negative except at $\vec{q} = -\vec{p}$ where it vanishes. We write $\delta(-A) = \frac{d}{d\tau}\theta(\tau - A)\big|_{\tau=0}$, so that (9.17) becomes

$$\frac{d}{d\tau} \int d^3p \, d^3q \, \hat{f}(\vec{p},\vec{q}) \, \theta(\tau - A(\vec{p},\vec{q})) \Big|_{\tau=0} . \qquad (9.18)$$

The support of the $\theta$-factor extends to infinity in the directions in which $\vec{q} = -\vec{p}$. But even in these directions the integral is cut off by the strongly decreasing factor $\hat{f}$. In all other directions the $\theta$-support has a diameter of order $\tau^{1/2}$ for $\tau \to 0$. This leads to a $\tau^{3/2}$-behaviour of the integral in (9.18) and to a $\tau^{1/2}$-behaviour of its derivative, which therefore vanishes at $\tau = 0$. That is, the expression (9.16) vanishes before identifying $x_1$ and $x_2$. But then this identification is clearly unproblematic and leads again to a vanishing result: the $S^+$–$S^+$-term in (9.14) does not contribute to (9.12), and the same holds for the $S^-$–$S^-$-term. Hence our contention that the expressions (9.12) and (9.15) are identical is proved.

Needless to say, these considerations are not offered as a proof of anything, but merely as a plausibility argument showing that the replacement of the formal expression (9.12) by the equally formal but simpler expression (9.15) is meaningful, and that therefore (9.15) is acceptable as a basis for the generalizations to higher orders to be described in the next section. The correctness of the new expression is established a posteriori as a special case of the general verifications given in Sect. 9.4.

Finally we remark that the resulting formal 2-point function $W_2(\mathsf{x},\mathsf{y})$ does not satisfy the renormalization condition (8.7). But this problem, like the UV problem, will be postponed to the next chapter.

A somewhat related problem occurs for the 4-point functions in second order, e.g. $W_2(x_1, x_2, \bar{y}_1, \bar{y}_2)$. They cannot be determined in the simple way used until now, solving the individual equations of motion by using Feynman propagators as Green's functions and adding these solutions with appropriate combinatorial factors. The result of such a procedure would violate locality. In order to cure this defect a suitable solution of the homogeneous system corresponding to (8.9) must be added. This is however not easily seen, so we will not discuss this point here, but refer to the general case treated in the following two sections.

## 9.3 The General Solution

The constructive method of solving the equations of motion that has been used in the preceding section for low orders can in principle be extended to higher orders. But it becomes rapidly intractably involved. We use therefore a different method: that of making an educated guess as to the form of the general solution and then verifying that this expression indeed satisfies the necessary requirements.

### Configuration Space

We write down at once the rules for the general function $\mathcal{W}_\sigma(X_1, s_1| \cdots | X_N, s_N)$ in $x$-space, where for the moment the $X_\alpha$ are sets of field variables of arbitrary type, not only of $\psi$-variables. The individual variables are consecutively numbered: $x_1, \ldots, x_n$, with $n = \sum_\alpha |X_\alpha|$. The rules to be stated have first been obtained by Ostendorf [Os 84] for the special case $s_\alpha = +$ for all $\alpha$, in [Ste 93] for the general case.

$\mathcal{W}_\sigma$ is a sum of terms of the general form already encountered in low orders, e.g. (9.6) and (9.11). Each term is an integral over a product of "propagators" $D_+$, $D_F$, $S_+$, $S_F$, and of certain constant factors. The propagators depend on differences between variables from a set comprising the $n$ external variables $x_i$ and $\sigma$ internal variables of integration $u_1, \ldots, u_\sigma$. The exact form of these terms can best be explained with the help of a graphical representation leading to "generalized Feynman graphs" or "sector graphs". Draw first an ordinary Feynman graph, which we call the "scaffolding" of the sector graph. It consists of $n$ "external points" and $\sigma$ "internal points" or "vertices", which are connected by lines. There are two kinds of lines: directed electron (or fermion) lines and undirected photon (or boson) lines. The fermion lines are drawn as unbroken lines with an arrow indicating the direction, the boson lines as dashed lines. With each external point we associate one of the external variables $x_i$, with each vertex an integration variable $u_j$, a 4-vector. At each external point $x_i$ one line ends. It is a boson line if $x_i$ is a photon variable, a fermion line directed away from the point if $x_i$ is a $\psi$ variable, a fermion line directed into the point if $x_i$ is a $\bar{\psi}$ variable. At each vertex a boson line meets two fermion lines, one of them directed into the vertex, one out of it. The scaffolding need not be connected, but it must not contain any connected components with less than two external points. It must also not contain any subgraphs ("tadpoles") without external points, which are connected to the rest of the graph by exactly one line. The connected scaffoldings of orders $\sigma = 1, 2, 3$ are shown in Fig. 9.1, where the convention of characterizing the types of external variables by different characters is again used. The convention does not apply to the internal points which are all of the same kind. An important feature of these rules is this: the fermion lines in a scaffolding are arranged in directed linear chains, which we call "trajectories". A trajectory either forms a closed loop, a "loop" for short, or it starts at an external $\psi$ point and ends at an external $\bar{\psi}$ point.

A sector graph is generated from a scaffolding by partitioning it into nonoverlapping subgraphs called its "sectors", such that the external points of a given set $X_\alpha$ belong to the same sector, but variables from different $X_\alpha$'s to different sectors. These $X_\alpha$-sectors are called "external". In general there exist also "internal sectors" containing only vertices but no external points. The lines connecting points of the same sector are parts of that sector. We call such lines "sector lines", those crossing sector boundaries "cross lines". The sectors are either of type $T^+$ or $T^-$. For external sectors this sign is given by $s_\alpha$. To each sector $S$ we assign a number $\nu(s)$ according to the following rules.

(i) $\nu(S) = \alpha$ for the external $X_\alpha$-sector.

(ii) If $s_\alpha = s_{\alpha+1}$ there may exist an internal sector $S$ numbered $\nu(S) = \alpha + \frac{1}{2}$. Its type $T^{s_{\alpha+\frac{1}{2}}}$ is the reverse of that of the adjacent sectors:

$s_{\alpha+\frac{1}{2}} = -s_\alpha$. No such intermediate internal sectors exist if $s_\alpha \neq s_{\alpha+1}$, nor are there internal sectors with numbers $< 1$ or $> N$.

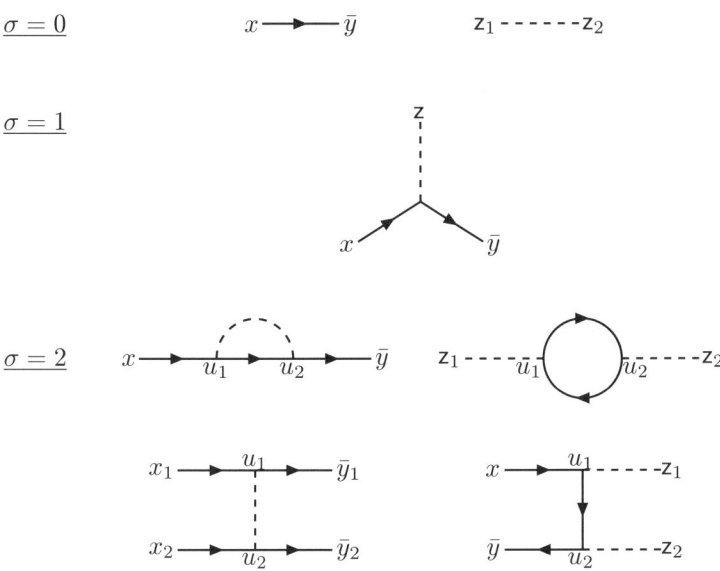

**Fig. 9.1.** Connected scaffoldings of low orders

Non-connected sectors are admitted even for connected scaffoldings. The sector graphs contributing to the explicit low-order functions given in Sect. 9.2. are shown in Fig. 9.2. Sectors are surrounded by thin unbroken lines. The $\sigma = 1$ graphs are those contributing to $W_1(x_1, \bar{x}_2, \mathsf{x}_3) = \mathcal{W}_1(x_1, +|\bar{x}_2, +|\mathsf{x}_3, +)$. The numbering of the variables is the same in all graphs. The first three graphs correspond to the terms in (9.6). The graphs with an internal 1-point sector give a vanishing contribution in this simple case. $S_\nu^s$ denotes a sector of type $s$ with number $\nu$. The $\sigma = 2$ graphs are those for $W_2(\mathsf{x}_1, \mathsf{x}_2)$. The first graph represents the expression (9.11), the second one the contribution obtained from (9.15) by convolution with $D_F(x_1 - u_1)$, the third one an analogous term with $x_1$ and $x_2$ interchanged. The fourth graph is formally a solution of the homogeneous part of the relevant system of differential equations. It is needed to satisfy locality. Graphs with 1-point internal sectors or disconnected sectors give vanishing contributions and have been omitted.

To the sector graph $G$ with variables $x_1, \ldots, x_n, u_1, \ldots, u_\sigma$, we associate as integrand a singular function of its variables, which we rename for the moment $z_1, \ldots, z_{n+\sigma}$. This integrand is constructed as follows.

$\sigma = 1$

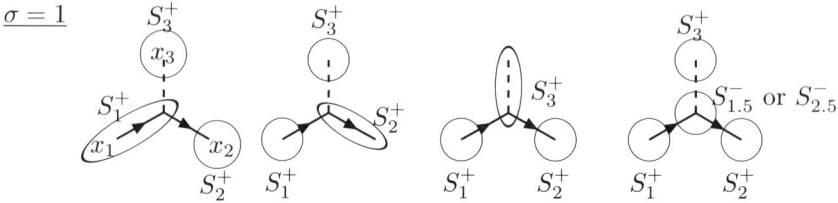

$\sigma = 2$

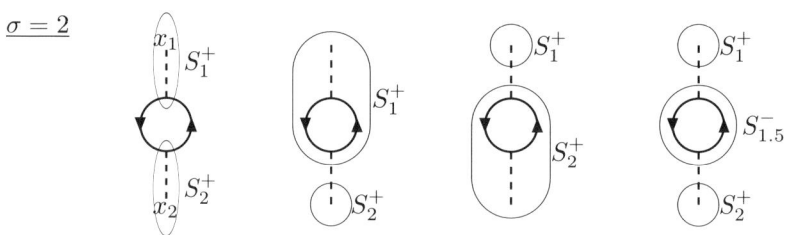

**Fig. 9.2.** Simple sector graphs

(i) Each *vertex* carries a factor $\mp i\gamma^\mu$ if it belongs to a $T^\pm$-sector.

(ii) A photon line connecting the points with variables $z_i$ and $z_j$ carries a *photon propagator*, which is $i g_{\mu\nu} D_F(z_i - z_j)$ inside a $T^+$-sector, $-i g_{\mu\nu} D_F^*(z_i - z_j)$ in a $T^-$-sector, $i g_{\mu\nu} D_+(z_i - z_j)$ if $z_i$ and $z_j$ are belonging to different sectors with $z_i$ lying in the lower-numbered sector of the two. The indices $\mu, \nu$, accord with those of the $\gamma$-matrices at the end points of the line if these are vertices, with the Minkowski index in $A_\mu(z)$ at external points.

(iii) Similar rules apply to an electron line. If its end points $z_i, z_j$ lie in the same $T^+$-sector, the line being directed from $z_i$ to $z_j$, it carries the *electron propagator* $-i S_F(z_i - z_j)$, in a $T^-$-sector the propagator is $i \bar{S}_F(z_j - z_i)$ with

$$\bar{S}_F(\xi) = (i\partial_\xi + m)\Delta_F^*(\xi) = \gamma^0 S_F^*(-\xi)\gamma^0 \ . \tag{9.19}$$

If the $z_i \to z_j$ line leads from a lower-numbered to a higher-numbered sector its propagator is $-i S_+(z_i - z_j)$, in the reverse case $-i S_-(z_i - z_j)$. The lines and vertices of a given trajectory carry propagators and vertex factors which are spinor matrices. They must be arranged in the order indicated by the trajectory and multiplied. The spinor indices of the resulting $4 \times 4$ matrix correspond with those of the initial and final point of the trajectory. For a closed loop the procedure starts at an arbitrary point and the trace of the resulting matrix product is formed. The result is independent of the choice of the starting point because of the invariance of the trace under cyclic permutations of the factors.

(iv) Rearrange the external fermion variables such that the two endpoints of each open trajectory stand together in the order indicated by the direction of the trajectory. If this ordering is obtained from the original ordering in $x_1 \ldots, x_n$ by an odd permutation, the integrand acquires a factor $-1$. Each closed trajectory contributes a factor $(-1)^{L_C+1}$ with $L_C$ the number of cross lines in the trajectory which lead from a higher to a lower sector.

The contribution of $G$ to $\mathcal{W}_\sigma(X_1, s_1| \cdots)$ is then obtained by integrating its integrand over the internal variables $u_j$. $\mathcal{W}_\sigma$ is the sum over all sector graphs $G$ that can be formed with the variables at hand according to the prescriptions just explained.

In the special case of the time-ordered function $\tau^+(x_1,\ldots,x_n) = \left(\Omega, T^+(x_1,\ldots,x_n)\Omega\right)$ these rules coincide with the familiar Feynman rules of the standard approach. This requirement was one of the decisive inputs for finding the rules [Os 84].

The rules applying to $T^-$-sectors can be stated in an alternative way which will be helpful later on at various occasions. The vertex factor $i\gamma^\mu$ can be written as $\gamma^0(-i\gamma^\mu)^*\gamma^0$, the fermion propagator as $i\gamma^0 S_F^*(z_i - z_j)\gamma_0$. At a vertex the $\gamma^0$'s of the vertex factor multiply $\gamma^0$'s of the incident fermion propagator to give $\gamma^{0\,2} = 1$. This effect removes all the $\gamma^0$'s except those at end points of cross lines and at external points. Hence we can state the $T^-$-rules as follows. The vertex factors and the sector propagators are the adjoints of their $T^+$ counterparts. Each fermion line leaving or entering the sector and each external fermion point in the sector carries a factor $\gamma^0$. The directions of the fermion lines inside the sector are reversed with respect to the original situation.

As to the formal treatment of this integrand as a singular function we refer to Sect. 9.1. Note that if two such integrands with the same singularity sets agree outside the singularities, then their integrals also agree if they exist. This is the situation met with in those proofs of the following section which are carried out in $x$-space. The question of the local existence of the integrands as distributions is the UV problem which will be solved in Chap. 10. The existence of the integrals at large distances is the IR problem and will be treated in Chap. 11.

The following result will be very useful in the sequel.

**Theorem 9.1: Furry's theorem.** *Graphs containing a closed trajectory with an odd number of lines do not contribute to $\mathcal{W}_\sigma$.*

**Proof.** In a graph $G$ consider a closed trajectory with vertex variables $z_1, \ldots, z_n$, $n$ odd. There exists another graph $G'$ differing from $G$ only by the direction in which this trajectory is traversed. Each line of the trajectory carries a factor $(i\slashed{\partial}_i + m)\Delta_\alpha(z_i - z_j)$ with $\Delta_\alpha$ an invariant function. From the graph rules explained above, including especially the sign factors of rule (iv), we find that the reversal of direction changes in each line the sign of the $\slashed{\partial}$-factors, everything else remaining unchanged. Calculating the corresponding integrands involves taking traces of products of $\gamma$-matrices along the trajectory. This trace does not depend on the direction of traversal. And the trace of the product of an odd number of $\gamma$'s vanishes. But there are in the trajectory an odd number of vertex-$\gamma$'s. Hence only terms with an odd number of $\slashed{\partial}$ survive. But these have different signs in the two cases, hence the contributions of $G$ and $G'$ cancel each other.

As a corollary of this result it is easy to see that the $\mathcal{W}_\sigma$ with no fermion variables and an odd number of photon variables vanish.

## Momentum Space

The momentum space version of the above graph rules is found by Fourier transform. We define

$$\tilde{\mathcal{W}}_\sigma(P_1, s_1| \cdots |P_N, s_N)$$
$$= (2\pi)^{-\frac{5}{2}n} \int \prod_{j=1}^{n} dx_j \, e^{i\sum p_j x_j} \, \mathcal{W}_\sigma(X_1, s_1| \cdots |X_N, s_N) \, . \qquad (9.20)$$

The set $P_\alpha$ of 4-vectors $p_i$ consists of the conjugate momenta of the $x$-space variables $X_\alpha$. The variables in $p$-space are consecutively numbered: $p_1, \ldots, p_n$, like in $x$-space, and comprise variables of all three field types. The contribution of a given sector graph can be computed from its $x$-space form by inserting the definitions (5.10), (5.56), (8.38), (8.42), and (9.19), of the invariant functions. The known properties of Fourier transforms, e.g. that they turn products into convolutions and vice versa, are used formally without regard to existence problems.

The result is the following. $\tilde{\mathcal{W}}_\sigma$ is a sum over integrals associated with the same graphs as in the case of $\mathcal{W}_\sigma$, except for two changes. Firstly, the photon lines are now also given a direction. For cross lines we choose this direction as pointing from the lower to the higher sector. For sector lines the direction is chosen arbitrarily. The result of the final integration does not depend on this choice. Secondly, variables are now assigned to the lines of the graph, not to its points. An "external" line, i.e. a line beginning or ending at the external point with number $i$, carries the variable $p_i$ if it begins at the point, the variable $-p_i$ if it ends at it. For a line connecting the two external points

$i$ and $j$ either assignment may be used because the integrand yet to be specified contains a factor $\delta^4(p_i + p_j)$. Each "internal line", i.e. a line connecting two vertices, carries a 4-vector $k_j$ as integration variable, with $j = 1, \ldots, I$, $I$ being the number of internal lines of the graph. Two simple examples are shown in Fig. 9.3.

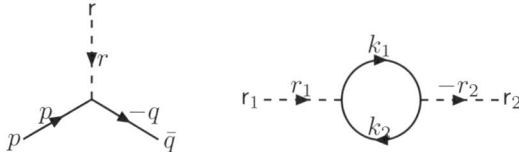

**Fig. 9.3.** Momentum assignment to graph lines.
Shown are the scaffoldings of $\tilde{W}_1(p, \bar{q}, r)$ and $\tilde{W}_2(r_1, r_2)$. The momentum assignment to the lines does not depend on the sectoring

To a sector graph in momentum space we assign an integrand by the following rules.

(i) The vertex $V$ with incident variables $\ell_1, \ell_2, \ell_3$, carries the factor $\mp i(2\pi)^{-1/2}\gamma^\mu \delta^4\left(\sum_1^3(\pm \ell_i)\right)$. The overall sign in front is negative (positive) in $T^+$ ($T^-$) sectors. The sign of $\ell_i$ is positive if its line points into $V$, negative otherwise.

E.g. the $\delta^4$-factor at the vertex of the first graph in Fig. 9.3. has the argument $p - (-q) + r = p + q + r$, the $\delta$-arguments of the vertices of the second graph are $r_1 - k_1 + k_2$ and $k_1 - k_2 + r_2$.

(ii) The propagator of a photon line with momentum $\ell$ is $-g_{\mu\nu}\,\theta(\ell_0)\,\delta(\ell^2)$ for cross lines, $\mp i(2\pi)^{-1}g_{\mu\nu}\,(\ell^2 \pm i\varepsilon)^{-1}$ for sector lines in a $T^\pm$ sector.

(iii) Electron cross lines carry propagators $\pm(\slashed{\ell} + m)\,\delta_\pm(\ell)$ if they point in the direction of $\{\begin{smallmatrix}\text{increasing}\\\text{decreasing}\end{smallmatrix}\}$ sector numbers, electron sector lines the propagators $\pm i(2\pi)^{-1}(\slashed{\ell} + m)\,(\ell^2 - m^2 \pm i\varepsilon)^{-1}$ in $T^\pm$ sectors. The rules for arranging factors along trajectories are as in $x$-space.

A notational simplification for the propagators inside sectors arises if we notice that $(\slashed{\ell} + m)(\slashed{\ell} - m) = \ell^2 - m^2$. We can then write $(\slashed{\ell} + m)\,(\ell^2 - m^2 \pm i\varepsilon)^{-1} = (\slashed{\ell} - m \pm i\varepsilon\ell_0)^{-1}$. The singularity at $\ell^2 = m^2$ manifests itself in the fact that at these points the matrix $(\slashed{\ell} - m)$ is not invertible. The $i\varepsilon$-term is usually not written down explicitly. It must be understood.

(iv) The trajectory signs of the $x$-space version are also present in $p$-space.

(v) Lines connecting two external points $i$ and $j$ acquire an extra factor $\delta^4(p_i + p_j)$.

The derivation of these rules from the $x$-space rules is straightforward and will not be given here, except for one detail which is useful also in other contexts. Let $n$ be the number of external points of a graph, $\sigma$ the number of its vertices, and $L$ the number of its lines. Then the number of line endings is $2L$. On the other hand, since one line ends at each external point, three at each vertex, the same number is also given by

$$2L = n + 3\sigma \ . \tag{9.21}$$

The $p$-space rules can be simplified by noting that $4\sigma$ of the totally $4L$ integrations – including those over test functions in the external variables – can be carried out trivially with the $\delta^4$-factors of the vertices. By this operation the rules are changed as follows. There are no more $\delta^4$-factors present. But the 4-momenta assigned to the various lines are no longer independent. Instead we assign independent variables $\ell_1, \ldots, \ell_{L-\sigma}$, to suitably chosen lines, and to the remaining lines we assign linear combinations of the $\ell_i$ such that momentum is conserved at each vertex. A simple example is shown in Fig. 9.4. We call these rules the $\delta$-reduced rules. Of course, the basic momenta must be selected such that they are not linearly dependent as a result of the $\delta^4$. For example, we cannot include in this set the momenta of all three lines meeting at a vertex. But for $\sigma > 0$ there are always various possible choices of such a basis. This arbitrariness is occasionally embarrassing in proving statements involving algebraic relations between different graphs, so that for such purposes it is more prudent to retain the $\delta^4$. In applications, e.g. numerical calculations, use of the $\delta$-reduced rules is clearly indicated. If we are interested in $\tilde{\mathcal{W}}_\sigma(p_1, \ldots)$ as a distribution not yet integrated over a test function, we integrate only over the internal variables of a given graph. In this case there remains a $\delta^4$- factor in the sum of the external variables of each connected component of the graph, which $\delta^4$'s indicate momentum conservation.

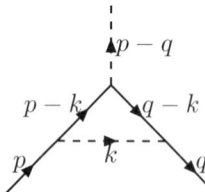

Fig. 9.4. Momentum assignment with conservation at the vertices

Concerning the formal treatment of the product of distributions forming the integrand of a graph we are in a less favorable situation than in $x$-space. Inside a sector and using the $\delta$-reduced rules the situation is indeed the same: the Feynman propagators occurring there are $C^\infty$ everywhere except at the mass shell, and are defined as distributions at these exceptional points by an

$i\varepsilon$-prescription. But the $\delta_\pm$-factors of the cross lines are not boundary values of analytic functions. Hence for "proofs" proceeding in $p$-space we use the somewhat crude strategy explained in Sect. 9.1.

## 9.4 Verifications

We want to verify that the Wightman functions $W_\sigma$ as given in the preceding section indeed satisfy the correct equations of motion and the $W$-properties in the weak sense explained in Sect. 9.1, and that the $\mathcal{W}_\sigma$ satisfy in addition the conditions explained in Chap. 8. An additional difficulty that must be clarified is this: a single field $\varphi(x)$ may be considered either a $T^+$ or a $T^-$ product with one factor only. Hence we must have

$$\mathcal{W}_\sigma(\cdots|x,+|\cdots) = \mathcal{W}_\sigma(\cdots|x,-|\cdots) . \qquad (9.22)$$

But the individual sector graphs of these two expressions differ, because in one case the external variable $x$ lies in a $T^+$ sector, in the other case in a $T^-$ sector. Hence we must show that nevertheless the sum of all graphs of order $\sigma$ coincide for the two expressions.

The proofs proceed partly by induction with respect to $\sigma$. We already know the correctness of our ansatz in zeroth order. In the proof of its correctness in order $\sigma$ we assume that it has all the required properties in order $\sigma - 1$.

We start by demonstrating that the *equations of motion* (8.9) are satisfied. Consider the first of these equations, with the photon variable $^\mu z$ as the distinguished variable. Insert our ansatz for $W_\sigma(\ldots,{^\mu z},\ldots)$ into its left-hand side, i.e. apply the wave operator $\Box_z$ to this ansatz. The wave operator annihilates all graphs in which the variable $z$ occurs in a $D_+$-propagator, because $\Box_\xi D_+(\pm\xi) = 0$. There remain the graphs in which the external $z$-point is connected by its line to an internal point in the same sector. Assume that we have defined $A^\mu(z) = T^+({^\mu z})$, so that this sector is a $T^+$ sector. Summing over all the surviving graphs we find, using (9.5),

$$\Box_z W_\sigma(\ldots,{^\mu z},\ldots) = \gamma^\mu_{\rho\tau}\left(\Omega,\cdots T^+({_\rho \bar{z}},{_\tau z})\cdots\Omega\right)_{\sigma-1} . \qquad (9.23)$$

But this is apparently not the desired equation (8.9), where instead of $T^+({_\rho\bar{z}},{_\tau z})$ we need to have $\frac{1}{2}[{_\rho\bar{z}},{_\tau z}] := \frac{1}{2}[\bar{\psi}_\rho(z), \psi_\tau(z)]$ in the vacuum expectation value on the right-hand side. We must show that these two expressions coincide in the sense that the corresponding graph integrands coincide as singular functions. This is not true for individual graphs. But it is true for the sum over all graphs with the same scaffolding. In this class the arguments of the various integrands can be trivially identified between graphs, so that the sum of integrands is well defined, provided the integrands themselves are well defined. This is not the case for scaffoldings in which the two $z$-points are directly connected by a line, because this line would carry a factor $S_\alpha(0)$

which diverges. But both $z$-points being external, this critical line is disconnected from the rest of the graph. It represents a free 2-point function as discussed in Sect. 5.2 and can be handled by the methods used there. In the case of $\gamma^\mu[\bar{z}, z]$ we are dealing with the vacuum expectation value of the free current (5.71), and this vanishes if defined by (5.72). The same definition yields $S_F(z - z) = 0$, so that the dangerous graphs also do not contribute to the right-hand side of (9.23). In both cases these definitions anticipate the renormalization prescriptions to be explained in the next chapter. For the remaining graphs without a direct line between the two $z$-points the integrands are well defined as singular functions despite the coinciding arguments. Moreover, at their regular points they can be obtained by point splitting, e.g. from $\gamma^\mu \left(\Omega, \cdots T^+(\bar{z}_1, z_2) \cdots \Omega\right)_{\sigma-1}$ as the limit $z_1 \to z_2 = z$. More exactly this means that the integrand is continuous in $z_1, z_2$ in an open neighbourhood of $z_1 = z_2 = z$, as long as none of the other variables $u$ directly connected to a $z$-point lies on the manifold $(z - u)^2 = 0$. We are then free to approach the coinciding points $(z, z)$ from the region $z_2 = z$, $z_1^0 > z^0$, $(z_1 - z)^2 < 0$. Since $T$-ordering and locality are satisfied in order $\sigma - 1$ we find that in $(\Omega, \cdots \Omega)_{\sigma-1}$ we can replace $T^+(\bar{z}_1, z)$ by $\bar{\psi}(z_1)\psi(z)$, and also $\frac{1}{2}[\bar{\psi}(z_1), \psi(z)]$ by $\bar{\psi}(z_1)\psi(z)$, which proves the claimed equality. Notice that in the same way it can also be shown that $T^+(\bar{z}, z)$ can be replaced by $T^-(\bar{z}, z)$ in the considered expression without changing it. This will become important later on. The other two, spinorial, equations (8.9) are verified in the same way.

The proofs of the remaining points (W-properties, $T^\pm$-properties, $s$-independence for $T^s(x)$-factors) occasionally assume that some of the other points have already been established. We are therefore forced to give these proofs in a seemingly haphazard order. We start with *covariance*. If we shift all the external variables by the same 4-vector $a$: $x_i \to x_i + a$, $i = 1, \ldots, n$, and at the same time substitute $u'_j = u_j + a$ for $u_j$ as variables of integration, we find that the integrand of any given graph is not changed, and that therefore $\mathcal{W}_\sigma(x_1 + a, \ldots, x_n + a) = \mathcal{W}_\sigma(x_1, \ldots, x_n)$. The singular functions $\Delta_\pm$ and $\Delta_F$ are clearly Lorentz invariant. From the transformation property (2.49) of the Dirac matrices we find that the $S_\alpha$ transform as

$$S_\alpha(\Lambda\xi) = S(\Lambda)\, S_\alpha(\xi)\, S(\Lambda)^{-1} \ , \tag{9.24}$$

with $S(\Lambda)$ the spinor representation of $L$. Using the identity (2.55) the same property is also derived for $\bar{S}_F$. And we know that

$$D^\alpha_{\mu\nu}(\Lambda\xi) = \Lambda_\mu{}^\kappa \Lambda_\nu{}^\lambda D^\alpha_{\kappa\lambda}(\xi) \ , \tag{9.25}$$

when $D^\alpha_{\mu\nu} = g_{\mu\nu} D^\alpha$ and $D^\alpha$ is any of the functions $D^\pm, D^F$. Consider now the contribution $\mathcal{W}^G(x_1, \ldots, x_n)$ of graph $G$ to $\mathcal{W}_\sigma(x_1, \ldots, x_n)$, and its Lorentz transformed $\mathcal{W}^G(\Lambda x_1, \ldots, \Lambda x_n)$. For the variables of integration $u_j$ we substitute $u'_j = \Lambda^{-1} u_j$. The corresponding Jacobian is 1. Take a given vertex $V$ in the transformed graph. From the electron line ending at $V$ we obtain a factor $S(\Lambda)^{-1}$, from the electron line starting from $V$ a factor $S(\Lambda)$, hence we have the matrix product $S^{-1}\gamma^\mu S = \Lambda^\mu{}_\nu \gamma^\nu$. But from the photon line at $V$ we have a factor $\Lambda_\mu{}^\kappa$, and the two $\Lambda$'s multiply to $\delta^\kappa_\nu$ by (2.24). As a result, the vertex factors retain their original untransformed form. The only matrices $S$, $S^{-1}$, $\Lambda$ that survive are those connected with external points, and they give the desired relation

$$\mathcal{W}^G(\ldots, \Lambda x_i, \ldots) = \prod_i R^{(i)}(\Lambda)\, \mathcal{W}^G(\ldots, x_i, \ldots)\,, \tag{9.26}$$

which holds for all graphs contributing to $\mathcal{W}_\sigma$, hence to their sum.

Next we address the *spectral property*. This is again satisfied for each graph separately. Let $\tilde{\mathcal{W}}^G(p_1, \ldots, p_n)$ be the contribution of graph $G$ to $\tilde{\mathcal{W}}(P_1, s_1|\cdots|P_N, s_N)$. Define $\bar{p}_\alpha = \sum_{P_\alpha} p_i$ and $\bar{p}_A = \sum_{\alpha \leq A} \bar{p}_\alpha$, $1 \leq A < N$. We want to show that $\tilde{\mathcal{W}}^G$ vanishes unless $\bar{p}_A \in \bar{V}_+$ $\forall A$. Let $\Gamma$ be a subgraph of $G$ consisting of a subset $G_\Gamma$ of points of $G$ and the lines connecting them. Call $k_1, \ldots, k_\Gamma$, the momenta of the lines entering or leaving $\Gamma$. Then, using the identity $\delta^4(p-r)\,\delta^4(q+r) = \delta^4(p-r)\,\delta^4(p+q)$ repeatedly, we find from the vertex-$\delta$'s of $\Gamma$ that the $G$-integrand contains a factor $\delta^4(\sum_1^\Gamma \pm k_i)$, where the upper sign applies to lines entering $\Gamma$, the lower one to lines leaving $\Gamma$. Choose $G_\Gamma$ to consist of the points in $\bigcup_{\nu(s_\alpha) \leq A} S_\alpha$. The lines entering or leaving the corresponding subgraph $\Gamma_A$ are the external lines with momenta $p_i \in P_A = \bigcup_{\alpha \leq A} P_\alpha$ and the cross lines with momenta $\ell_j, \cdots$ leading to sectors with numbers $\nu(S) > A$ (see Fig. 9.5). These cross lines contain factors $\delta_+(\pm \ell_j)$ depending on whether they leave or enter $\Gamma_A$. Thus we find that all $\pm \ell_j$ in the factor $\delta^4$ associated with $\Gamma_A$ lie in $\bar{V}_-$, hence $\sum \pm \ell_j \in \bar{V}_-$, hence this $\delta^4$ is different from zero only if $\bar{p}_A \in \bar{V}_+$, which is what we claimed.

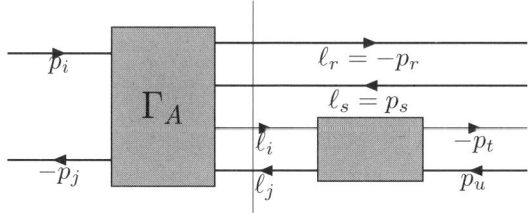

**Fig. 9.5.** Graph partition for proof of spectral property.
The grey boxes are subgraphs. The thin vertical line denotes a sector boundary. All possibilities as to the direction of lines entering or leaving $\Gamma_A$ are shown. The type of line (boson or fermion) is immaterial. The external variables to the left of $\Gamma_A$ belong to $P_A$, those to the right of $\Gamma_A$ to $P \backslash P_A$

The *permutation symmetry* inside $T^\pm$ factors is trivial to see. The graph rules depend on the ordering of the external variables in a sector only through the trajectory signs of point (iv) of the rules given in Sect. 9.3, and this observation yields at once the desired symmetry.

We turn to the *ordering relations* (8.27), which we prove for the $T^+$-case. The $T^-$-case is handled in the same way. The proof given here is due to Ostendorf [Os 84]. Let $X'$ and $Y$ be two non-empty sets of field variables. We want to show that

$$I_1 := \mathcal{W}_\sigma(\cdots|X',Y,+|\cdots) = \mathcal{W}_\sigma(\cdots|X',+|Y,+|\cdots) =: I_2 \qquad (9.27)$$

if $x_i^0 > y_j^0$ for all $x_i \in X'$, $y_j \in Y$. We select an arbitrary variable from $X'$. Its type is irrelevant for our argument, but for fixing the ideas we assume it to be a photon variable $^\mu\mathsf{x}$. We write $X' = \{\mathsf{x}, X\}$ with $X = X'\backslash\mathsf{x}$. $X$ may be empty. From the graph rules we find

$$\begin{aligned}
I_1 = \int &du\, \gamma^\mu_{\rho\tau} \{D_F(x-u)\,\mathcal{W}_{\sigma-1}(\cdots|_\rho \bar{u},_\tau u, X, Y, +|\cdots) \\
&+ D_+(x-u)\sum_R \pm \left[\mathcal{W}_{\sigma-1}(\cdots|X,Y,+|\cdots|_\rho\bar{u},_\tau u,\pm|\cdots)\right. \\
&\left. \pm \mathcal{W}_{\sigma-1}(\cdots|X,Y,+|\cdots|_\rho\bar{u},_\tau u, X_\alpha, \pm|\cdots)\right] \\
&+ D_+(u-x)\sum_L \pm \left[\mathcal{W}_{\sigma-1}(\cdots|_\rho\bar{u},_\tau u,\pm|\cdots|X,Y,+|\cdots)\right. \\
&\left. \pm \mathcal{W}_{\sigma-1}(\cdots|_\rho\bar{u},_\tau u, X_\alpha, \pm|\cdots|X,Y,+|\cdots)\right]\},
\end{aligned}$$

$$\begin{aligned}
I_2 = \int &du\, \gamma^\mu_{\rho\tau} \{D_F(x-u)\,\mathcal{W}_{\sigma-1}(\cdots|_\rho\bar{u},_\tau u, X, +|Y,+|\cdots) \\
&+ D_+(x-u)\left[-\mathcal{W}_{\sigma-1}(\cdots|X,+|_\rho\bar{u},_\tau u, -|Y,+|\cdots)\right. \\
&\left. + \mathcal{W}_{\sigma-1}(\cdots|X,+|_\rho\bar{u},_\tau u, Y, +|\cdots)\right. \\
&\left. + \sum_R (\pm\mathcal{W}_{\sigma-1}(\cdots|X,+|Y,+|\cdots|_\rho\bar{u},_\tau u,\pm|\cdots)\right. \\
&\left. \pm \mathcal{W}_{\sigma-1}(\cdots|X,+|Y,+|\cdots|_\rho\bar{u},_\tau u, X_\alpha, \pm|\cdots))\right] \\
&+ D_+(u-x)\sum_L \left[\pm \mathcal{W}_{\sigma-1}(\cdots|_\rho\bar{u},_\tau u,\pm|\cdots|X,+|Y,+|\cdots)\right. \\
&\left. \pm \mathcal{W}_{\sigma-1}(\cdots|_\rho\bar{u},_\tau u, X_\alpha, \pm|\cdots|X,+|Y,+|\cdots)\right]\}. \qquad (9.28)
\end{aligned}$$

The $R$-sums ($L$-sums) extend over graphs in which the $x$-line ends in a sector whose number is higher (lower) than that of the $X$–$Y$-sector or those of the $X$- and $Y$-sectors, respectively. We assume that the ordering relations hold in order $\sigma - 1$. Then the $L$- and $R$-sums agree in $I_1$ and $I_2$ if all variables in $X'$ are later than those in $Y$. Under this condition we can find a real number $r$ such that $x^0 > r$, $x_i^0 > r > y_j^0$ for all $x_i \in X$, $y_j \in Y$. Consider the integrands in (9.28) first at points $u$ with $u^0 \geq r$. There the $D_F$-term in $I_1$ coincides with the $D_F$-term in $I_2$ because in the former we have $|\bar{u}, u, X, Y, +| = |\bar{u}, u, X, +|Y, +|$ from the ordering relations in order $\sigma - 1$. Likewise we can set $|\bar{u}, u, Y, +| = |\bar{u}, u, +|Y, +|$ in the third term of $I_2$, whereupon it cancels the second term, because $|\bar{u}, u, -| = |\bar{u}, u, +|$, as has been shown in the verification of the equations of motion. For $u^0 \leq r$ we have $x^0 \geq u^0$ and therefore $D_F(x-u) = D_+(x-u)$. Also, in the first term of $I_1$ we have $|\bar{u}, u, X, Y, +| = |X, +|\bar{u}, u, Y, +|$, so that this term is equal to the third term in $I_2$. And the first and second term in $I_2$ cancel because of $|\bar{u}, u, X, +| = |X, +|\bar{u}, u, +| = |X, +|\bar{u}, u, -|$.

In this proof we have assumed that the $x$-line ends at a vertex. But if it ends at another external point, then this line is disconnected from the rest of the graph, and the ordering relation follows from its validity in $0^{th}$ order.

We come to the *sign independence* (9.22), which is slightly more difficult to establish than the points considered up to now. We first prove a lemma which will also be used later on for other purposes.

**Lemma 9.2.** *Let $\Gamma$ be a subgraph with $N$ points, including $E$ external ones, of a scaffolding $G$ with variables $Z = \{z_1, \ldots, z_n\}$ of arbitrary type. Let $\Gamma_L \cup \Gamma_R = \Gamma$ be any partition of $\Gamma$ into a $T^+$ sector $\Gamma_L$ and a $T^-$ sector $\Gamma_R$ with $\nu(\Gamma_L) < \nu(\Gamma_R)$. The corresponding subsets of $Z$ are called $Z_L$ and $Z_R$. One of them may be empty. Let $I_{LR}$ be the integrand of this sector subgraph. Then*

$$\sum_{\Gamma_L} (-1)^{E_L} I_{LR}(Z) = 0 \tag{9.29}$$

*if $E_L$ is the number of external points in $\Gamma_L$.*

Under a subgraph $\Gamma$ of $G$ we understand a non-empty subset of the points of $G$ and of lines connecting them, but not including any lines with only one endpoint in $\Gamma$. The lemma can be proved in the same way under the reverse ordering assumption $\nu(\Gamma_L) > \nu(\Gamma_R)$. In this case the factor $(-1)^{E_L}$ is replaced by $(-1)^{E_R}$.

**Proof of the Lemma.** We need to introduce some notation. Let $z_i, z_j$, be any two variables of $\Gamma$, $\xi = z_i - z_j$. If their points are connected by a line we define

$$P_+(\xi) = \begin{cases} i\,g_{\mu\nu}\,D_+(\xi) & \text{for a photon line} \\ -i\,(i\slashed{\partial}+m)\Delta_+(\xi) & \text{for an electron line from } z_i \text{ to } z_j \\ -i\,(-i\slashed{\partial}+m)^T\Delta_+(\xi) & \text{for an electron line from } z_j \text{ to } z_i \,, \end{cases} \tag{9.30}$$

and

$$P_F(\xi) = \begin{cases} i\,g_{\mu\nu}\,D_F(\xi) \\ -i(i\slashed{\partial}+m)\,\Delta_F(\xi) \\ -i(-i\slashed{\partial}+m)^T\Delta_F(\xi) \,, \end{cases} \qquad \bar{P}_F(\xi) = \begin{cases} -i\,g_{\mu\nu}\,D_F^*(\xi) \\ i(i\slashed{\partial}+m)\Delta_F^*(\xi) \\ i(-i\slashed{\partial}+m)^T\Delta_F^*(\xi) \,, \end{cases} \tag{9.31}$$

with the same case distinction. One sees immediately that

$$P_F(\xi) = P_+(\xi) \quad \text{if} \quad \xi^0 > 0\,, \qquad \bar{P}_F(\xi) = P_+(\xi) \quad \text{if} \quad \xi^0 < 0\,. \tag{9.32}$$

Consider a partition $\Gamma_{L,R}$ with $L$ points in $\Gamma_L$. Splitting the range of variables according to their time ordering, we write its contribution to (9.29) as

$$i^{N-E}(-1)^{L-E_L}\sum_{\pi_L,\pi_R}\theta(z_1^{\pi_L}-z_2^{\pi_L})\cdots\theta(z_{L-1}^{\pi_L}-z_L^{\pi_L})$$
$$\times\left[\theta(z_L^{\pi_L}-z_{L+1}^{\pi_L})+\theta(z_{L+1}^{\pi_R}-z_L^{\pi_L})\right]\theta(z_{L+2}^{\pi_R}-z_{L+1}^{\pi_R})\cdots\theta(z_N^{\pi_R}-z_{N-1}^{\pi_R})$$
$$\times\left\{z_1^{\pi_L},\ldots,z_L^{\pi_L},z_{L+1}^{\pi_R},\ldots,z_N^{\pi_R}\right\}\ .$$
(9.33)

We have defined $\theta(\xi)=\theta(\xi^0)$. The sum extends over all permutations $(z_1^{\pi_L},..,z_L^{\pi_L})$ of $Z_L=(z_1,\ldots,z_L)$ and $(z_{L+1}^{\pi_R},\ldots,z_N^{\pi_R})$ of $Z_R=(z_{L+1},\ldots,z_N)$. $\{z_1^{\pi_L},\ldots,z_N^{\pi_R}\}$ is the product over the propagators $P_+(z_i-z_j)$ of all lines of $\Gamma$, with $z_i$ standing to the left of $z_j$ in the ordering $\{z_1^{\pi_L},\ldots\}$, and over the vertex-$\gamma$'s which are the same in all terms. Take the contribution of the second term in the square bracket for a particular pair $(\pi_L,\pi_R)$:

$$i^{N-E}(-1)^L\cdots\theta(z_{L-1}^{\pi_L}-z_L^{\pi_L})\theta(z_{L+1}^{\pi_R}-z_L^{\pi_L})\theta(z_{L+2}^{\pi_R}-z_{L+1}^{\pi_R})\cdots\{z_1^{\pi_L},\ldots,z_N^{\pi_R}\}\ .$$

This is cancelled by the contribution from the first term in $[\cdots]$ and the same ordering $\{\cdots\}$ in the partition $Z_L'=Z_L\cup z_{L+1}^{\pi_R}$, $Z_R'=Z_R\setminus z_{L+1}^R$. If $\Gamma_L=\Gamma$ or $\emptyset$, then there is no square bracket, but cancellation occurs nevertheless against a term with $\Gamma_R=z_N^{\pi_L}$ or $\Gamma_L=z_1^{\pi_R}$, respectively. This completes the proof of the lemma.

For proving sign independence we need the special cases $E=0,1$ of the lemma. In the second case, let $x$ be the single external variable in $\Gamma$. We use the following notation: $S_x^s$ denotes the integrand of a $T^s$ sector containing $x$, $S^s$ that of an internal $T^s$ sector. We define e.g. the product $S_x^+S^-$ as

$$S_x^+S^-=\sum S^+(x,U_1)\,S^-(U_2)\ ,\qquad(9.34)$$

the sum extending over all partitions of the internal points $U$ of $\Gamma$ into two subsets $U_1,U_2$. $U_1$ may be empty but $U_2$ not. The factors in the sum are the integrands of the corresponding sectors. The cross lines connecting the two factors are included. The generalizations of this definition are obvious. The relevant cases of Lemma 9.2 and its order-reversed form read in this notation

$$S_x^- - S_x^+ - S_x^+S^- + S^+S_x^- = 0\ ,\quad S_x^- - S_x^+ + S_x^-S^+ - S^-S_x^+ = 0\ ,$$
$$S^- + S^+ + S^+S^- = 0\ ,\quad S^- + S^+ + S^-S^+ = 0\ .$$
(9.35)

Consider now the field variable $|x,+|$ or $|x,-|$ in (9.22). Assume at first that both neighbouring factors are of the same sign, for instance, both $T^+$, so that we want to prove, in obvious notation, that $|Y,+|x,+|Z,+|=|Y,+|x,-|Z,+|$. We consider a scaffolding and fix all sectors except those lying between the $Y$ and $Z$ sectors. These are the $x$ sector and possibly one or two internal sectors. What we want to prove reads then in our notation

$$S_x^+ + S^-S_x^+ + S_x^+S^- + S^-S_x^+S^- = S_x^-\ .$$

But the difference between the two sides is equal to

$$(S_x^+ + S_x^+ S^- - S_x^- - S^+ S_x^-) + S^- (S_x^+ + S_x^+ S^- - S_x^- - S^+ S_x^-)$$
$$+ (S^+ + S^- + S^- S^+) S_x^-,$$

which vanishes because all the brackets vanish by (9.35). For neighbours of different sign we have to prove e.g. that $|Y, +|x, +|Z, -| = |Y, +|x, -|Z, -|$, which reduces in our graph notation to $S_x^+ + S^- S_x^+ = S_x^- + S_x^- S^+$, which is one of the relations (9.35). If $x$ is an extremal variable of $\mathcal{W}_\sigma$, e.g. $x = x_1$, so that no $T^s$ stands to its left but for instance a $T^+$ to its right, the relation to be proved is $S_x^+ + S_x^+ S^- = S_x^-$, which is one of the relations (9.35) if we note that $S^+ S_x^-$ vanishes in this case by an argument like that used in the proof of the spectral property: if there is no external sector to the left of $S^+$, all the momenta of the cross lines attached to $S^+$ lie in the forward cone and cannot sum to zero, as they should because of momentum conservation.

We can now attack the remaining W-properties. To establish *locality*, consider the function $\mathcal{W}_\sigma(\cdots |x, +|y, +|\cdots)$ with two adjacent one-field factors at spacelike separation $(x-y)^2 < 0$. Because of the sign independence that has just been proved we can asume these factors to be $T^+$. We can find a frame of reference in which $x^0 > y^0$, so that by the ordering relation we have $\mathcal{W}_\sigma(\cdots |x, +|y, +|\cdots) = \mathcal{W}_\sigma(\cdots |x, y, +|\cdots)$, which is equal to $\pm \mathcal{W}_\sigma(\cdots |y, x, +|\cdots)$ by the symmetry of $T^+(x, y)$. By Lorentz invariance this is also equal to $\pm R(\Lambda^{-1}) \mathcal{W}_\sigma(\cdots |\Lambda y, \Lambda x, +|\cdots)$ for $\Lambda \in L$ with $R$ a finite-dimensional product representation of $L$. We can find a $\Lambda$ such that $(\Lambda y)^0 > (\Lambda x)^0$ and obtain, again applying the ordering relation: $\pm R(\Lambda^{-1}) \mathcal{W}_\sigma(\cdots |\Lambda y, +|\Lambda x, +|\cdots) = \pm \mathcal{W}_\sigma(\cdots |y, +|x, +|\cdots)$ as claimed. The sign in front is negative if both fields concerned are Fermi fields.

For demonstrating *hermiticity* we need expressions for the graph integrands contributing to $\mathcal{W}_\sigma(X_1, s_1| \cdots |X_N, s_N)^*$. It is convenient to use the alternative version of the rules in $T^-$ sectors given after the listing of the $x$-space rules in Sect. 9.3: the integrand of a $T^-$ sector is the adjoint of the corresponding $T^+$ integrand multiplied with a $\gamma^0$ at each external point and each cross line. Consider a graph $G$ contributing to $\mathcal{W}_\sigma(\cdots)$ and its integrand. Taking its adjoint reverses the ordering of the sectors and exchanges $\psi$- and $\bar\psi$-variables. And apart from the $\gamma^0$'s just mentioned it turns a $T^-$ sector into a $T^+$ sector and vice versa. The propagator $i\,g_{\mu\nu} D_+(x-y)$ of a photon cross line becomes $-i\,g_{\mu\nu} D_+^*(x-y) = i\,g_{\mu\nu} D_+(y-x)$, in accordance with the reversal of sector ordering. The spinorial cross propagators $-i\,S^\pm(x-y)$ become $-i(i\overleftarrow{\partial}_y^* + m)\Delta_\pm(y-x)$. Using $\gamma^0 \partial^* = \partial \gamma^0$ it is then easily seen that the new cross propagators conform with the adjoint graph rules. But the $\gamma^0$'s at the external points must be adapted to the new situation. We obtain

$$\left(\mathcal{W}_\sigma(X_1, s_1| \cdots |X_N, s_N)\right)^* = \prod_\psi \gamma_j^0 \mathcal{W}_\sigma(\overleftarrow{X}_N, -s_N| \cdots | \overleftarrow{X}_1, -s_1) \prod_{\bar\psi} \gamma_i^0 \,.$$

(9.36)

$\overleftarrow{X}_\alpha$ is $X_\alpha$ in reversed order with $\psi$- and $\bar\psi$-variables interchanged. A special case is

$$W_\sigma(X,\bar Y,Z)^* = \prod_Y \gamma_j^0 \, W_\sigma^\leftarrow(\bar X, Y, Z) \prod_X \gamma_i^0 \,, \qquad (9.37)$$

the perturbative version of the hermiticity condition (8.6).

The *cluster property* is more problematic to treat at the present non-rigorous level, because it cannot be reduced even formally to a local property of graph integrands. Nevertheless, some of the pertinent information is already at hand, and for the sake of completeness we may as well describe this now, in preparation for the final proof which can only be given in Chap. 11. This proof will proceed in $x$-space, hence we argue also now in this space. The $x$-space form of the cluster property is obtained from (4.44) and (8.11) by Fourier transform. Let $X$ and $Y$ be two sets of field variables, $f(X)$ and $g(Y)$ two test functions, and $\hat\gamma(\vec w)$ a test function with $\int d^3w\, \hat\gamma(\vec w) = 1$. Define

$$f_\lambda(x) = \lambda^{-3} \int d^3w\, f(\ldots, x_i^0, \vec x_i - \vec w, \ldots)\, \hat\gamma(\lambda^{-1}\vec w)\,. \qquad (9.38)$$

Then the weak cluster property states that

$$\lim_{\lambda\to\infty} \int dX\, dY\, f_\lambda(X)\, g(Y)\, W_\sigma(X,Y)$$
$$= \sum_{\rho\le\sigma} \int dX\, f(X)\, W_\rho(X) \int dY\, g(Y)\, W_{\sigma-\rho}(Y)\,. \qquad (9.39)$$

For $\lambda\to\infty$ the function $\lambda^{-3}\hat\gamma(\lambda^{-1}\vec w)$ converges to zero uniformly on every compact subset of $R^3$. On the other hand it gets spread out more and more over $R^3$, i.e. larger values of $|\vec w|$ become more important. The same is true for $f_\lambda$: $f_\lambda(X)$ converges locally to zero for $\lambda\to\infty$, but its decrease at large $\vec x_i$ becomes slower, so that large $\vec x$-values become more important and in fact determine the asymptotic behaviour of the left-hand side of (9.39). But no such effect is present for $g$: $g(Y)$ contributes substantially only for $Y$'s from some bounded set, e.g. from the compact support of $g$ if we choose $g$ from $\mathcal D$ for simplicity. The same is true for the $x_i^0$-components. This means that the integral on the left-hand side of (9.39) is for large $\lambda$ dominated by the integrands at points where all $\vec x_i$ are far apart from all $\vec y_j$. Consider a graph $G$ contributing to this expression. Choose a sequence of lines in $G$ which connects an $X$-variable $x_i$ to a $Y$-variable $y_j$. Let $x_i, u_1, \ldots, u_\ell, y_j$ be the variables of the points on this sequence listed sequentially. Call $u_0 = x_i$, $u_{\ell+1} = y_j$. Then these lines contribute the product $\prod_{\alpha=0}^\ell P_\alpha(u_\alpha - u_{\alpha+1})$ of propagators to the integrand. But if $|\vec x_i - \vec y_j| \to \infty$, then at least one of the distances $|\vec u_\alpha - \vec u_{\alpha+1}|$ tends to infinity. For fixed 0-components this means that the argument of $P_\alpha$ tends to infinity in a spacelike direction. But according to Sect. 5.4, $P_\alpha$ decreases in such a direction exponentially if it is an electron propagator, of second order if it is a photon propagator. This decrease leads to the vanishing of the integral for $\lambda \to \infty$ as long as we do not worry about the uniformity of this estimate in the $u_\alpha^0$. The discussion of the influence of large $u_\alpha^0$ to the argument is beyond our present possibilities, hence our consideration does not claim to be even a formal proof. This problem will be solved in Chap. 11.

The argument as given clearly does not work if no sequence of lines of the specified type exists in $G$. This is the case if $G$ consists of two disconnected subgraphs $G_X$ and $G_Y$ containing the external points of $X$ and $Y$, respectively. $G_Y$ contributes a factor $\int dY\, g(Y)\, W_{\sigma-\rho}(Y)$ if $\sigma - \rho$ is the number of its vertices. The $\lambda$-dependence is confined to the $G_X$-contribution. Let $u_1, \ldots, u_\rho$, be the vertex variables of $G_X$. Introducing $\vec{z}_i = \vec{x}_i - \vec{w}$ and $\vec{v}_h = \vec{u}_h - \vec{w}$ as new variables and using the translation invariance of $G_X$ we find that the $G_X$-part does in fact not depend on $\lambda$ but is equal to $\int dX\, f(X)\, W_\rho(X)$. These disconnected graphs sum to the right-hand side of (9.39), which equation is thus verified.

## 9.5 Current Conservation and the Charge Identity

Another topic that can profitably be considered at the present formal level is that of current conservation and the charge identity (7.26).

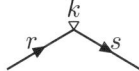

**Fig. 9.6.** The current vertex

Consider the $W$-function $\tilde{W}_\sigma(\cdots|^\mu k|\cdots)$ with an arbitrary number of basic fields and one current component $\tilde{j}^\mu(k)$. The graphs of this function have the standard structure except for the presence of an additional external vertex of the form shown in Fig. 9.6, to which the current variable $k$ is attached. It carries the vertex factor $(2\pi)^{-\frac{3}{2}}\gamma^\mu \delta^4(r+k-s)$. We wish to prove that

$$k_\mu \tilde{\mathcal{W}}_\sigma(\cdots|^\mu k|\cdots) = 0\,. \tag{9.40}$$

Consider a graph $G$ contributing to $\tilde{\mathcal{W}}_\sigma$, and in it the trajectory containing the $k$-vertex (see Fig. 9.7) and the corresponding product of vertex factors and propagators in the non-$\delta$-reduced formulation. In (9.40) the factor $\gamma^\mu$ of the $k$-vertex is contracted with $k_\mu$, to yield $\slashed{k} = (\slashed{s}-m) - (\slashed{r}-m)$, where momentum conservation at the vertex has been used. Take the first term $(\slashed{s} - m)$ and multiply it into the $s$-propagator which contains a factor $(\slashed{s}+m)$. This yields the product $(s^2-m^2)$. If the $s$-line is a cross line, then this term annihilates its factor $\delta_+(s)$. If the $s$-line is a sector line e.g. in a $T^+$ sector, its denominator is cancelled and we are left with the constant $i/2\pi$. An $s$-dependence remains only in the $\delta$-factors $\delta^4(k+r-s)\,\delta^4(s+\ell_i-s') = \delta^4(k+r-s)\,\delta^4(r+\ell_i+k-s')$. The integration over $s$ removes the first of these factors. Up to a constant factor we obtain a trajectory from which the $k$-vertex and the $s$-line have disappeared, and where the $\ell_i$-vertex carries the $\delta$-factor $\delta^4(r-s'+\ell_i+k)$. The $(\not{r}-m)$-term in $\not{k}$ leads in the same way to a trajectory without the $r$-line and with the $\ell_{i-1}$ vertex factor $-\delta^4(r'+\ell_{i-1}+k-s)$. But the position of the $k$-vertex on the trajectory is arbitrary. Taking the case in which the $k$- and the $\ell_i$-vertex are interchanged, all else remaining unchanged, we find that the $s$-term of the original ordering and the $r$-term of the new one cancel. Summing over all possible positions of the $k$-vertex we obtain in this way a complete cancellation, irrespective of whether the trajectory is open or closed. This proves the current conservation (9.40).

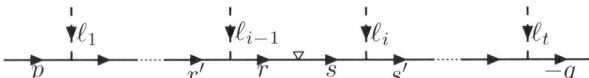

**Fig. 9.7.** Trajectory of the $k$-vertex

The same kind of argument will repeatedly play an important role later on. For instance it can be used to prove the *"Ward–Takahashi identities"* (WT identities). They state the following. Let $\tilde{\tau}_\sigma^\pm(P,\bar{Q},\mathsf{R}|^\mu k)$ be a time-ordered or anti-time-ordered function with one current factor $\tilde{j}^\mu(k)$. Then we have, for $\sigma \geq 1$,

$$k_\mu\,\tilde{\tau}_\sigma^\pm(P,\bar{Q},\mathsf{R}|^\mu k) = \pm i(2\pi)^{-\frac{5}{2}}\Bigg[-\sum_{p_i\in P}\tilde{\tau}_{\sigma-1}^\pm(p_1,\ldots,p_i+k,\ldots,p_f),\bar{Q},\mathsf{R})$$
$$+\sum_{q_j\in Q}\tilde{\tau}_{\sigma-1}^\pm(P,q_1,\ldots,q_j+k,\ldots,q_f,\mathsf{R})\Bigg]. \tag{9.41}$$

We turn to the charge identity (7.26). The proof that $Q_C\,\Omega = 0$ needs some sophisticated methods for which the ground has not yet been laid. So we simply assume that relation. Then, according to Theorem 7.1, $(\Omega,\mathcal{N}\,Q_C\,\mathcal{M}\,\Omega)$ exists for all field monomials $\mathcal{N}$ and $\mathcal{M}$ and is independent of the special choice of the auxiliary functions $\alpha$ and $\chi$ in the definition (7.14) of $Q_C$. We must establish the identity

$$(\Omega,\mathcal{N}\,Q_C\,\mathcal{M}\,\Omega) = (\Omega,\mathcal{N}\,Q_G\,\mathcal{M}\,\Omega) = q_M\,(\Omega,\mathcal{N}\,\mathcal{M}\,\Omega)\,, \tag{9.42}$$

with $q_M$ the charge of $\mathcal{M}$ given by (7.2). We can use the $\alpha$-independence of $Q_C$ by considering the family

$$\alpha_\lambda(x^0) = \lambda\,\alpha(\lambda\,x^0)\,, \qquad \lambda > 0\,, \tag{9.43}$$

of such functions, with $\alpha(u) \in \mathcal{D}$ and $\int du\, \alpha(u) = 1$. The first term in (9.42) does not depend on $\lambda$, hence can be evaluated as the $\lim_{\lambda \to 0}$ of the corresponding expression. Consider a graph $G$ contributing to

$$(\Omega, \mathcal{N}\tilde{j}^{\mu}(k)\mathcal{M}\Omega)_{\sigma} =: F^{\mu}(k) \,. \tag{9.44}$$

Its contribution to the first term in (9.42) is

$$C_G = \int dk_0\, \beta_\lambda(k_0)\, F^{\mu}(k)\Big|_{\vec{k}=0} \,. \tag{9.45}$$

Here $\beta_\lambda(k_0) = \beta(k_0/\lambda)$ with $\beta$ the Fourier transform of $\alpha$. It satisfies $\beta(0) = 1$. The restriction to $\vec{k} = 0$ is the $p$-space equivalent of taking the limit $R \to \infty$ in $x$-space, which limit we know to exist. Assume that the $k$-sector is $T^+$. The $k$-dependence of the $G$-integrand is contained in the vertex of Fig. 9.6. Assume at first that this vertex is internal to the $k$-sector, so that both the $r$- and the $s$-line are sector lines carrying Feynman propagators. After integrating out the $\delta^4$-factor of this vertex we are confronted with the integral (up to uninteresting constant factors)

$$\int dk_0\, \beta_\lambda(k_0)\, S_F(r)\, \gamma^0 S_F(r_0 + k_0, \vec{r}) \,. \tag{9.46}$$

The integrand contains the singular factors $(r^2 - m^2 + i\varepsilon)^{-1}\big((r_0 + k_0)^2 - \vec{r}^2 - m^2 + i\varepsilon)^{-1}\big)$. This product is well defined as a distribution in $r$ even for $k_0 = 0$: considered as a function of $k_0$ (i.e. after integration over a test function in $r$) it is continuous. But $\beta_\lambda(k_0) \to 0$ pointwise in $k_0 \neq 0$, while remaining bounded by a strongly decreasing function; hence the expression (9.46) converges to zero for $\lambda \to 0$: the corresponding contribution to $C_G$ can be neglected.

Assume now that the $s$-line is a cross line e.g. pointing from the $k$-sector to a higher sector. Then we have to evaluate instead of (9.46) the expression

$$C'_G = \frac{\slashed{r}+m}{r^2-m^2} \int dk_0\, \beta_\lambda(k_0)\, \delta(r_0 - s_0 - k_0)\, \delta^3(\vec{r}-\vec{s})\, \gamma^0(\slashed{s}+m)\, \delta_+(s) \,, \tag{9.47}$$

again omitting powers of $2\pi$ and $i$, which of course combine to give the correct result in the end. We find, defining $u = r_0 - s_0$ and $\omega = \omega(\vec{r}) = \omega(\vec{s})$,

$$C'_G = \beta_\lambda(u)\left(2u(\omega+u) + i\varepsilon\right)^{-1}(\slashed{s} + u\gamma^0 + m)\, \gamma^0(\slashed{s}+m)\, \delta_+(s)\, \delta^3(\vec{r}-\vec{s}) \,.$$

Since we intend to let $\lambda \to 0$ we can set $u = 0$ wherever the $u$-dependence is smooth: we may drop the $u\gamma^0$ term and replace $(\omega + u)$ by $\omega$ in the denominator. Using $(\slashed{s}+m)\gamma^0 = 2s_0 - \gamma^0(\slashed{s}-m)$ we obtain finally

$$C'_G = \frac{\beta_\lambda(r_0-s_0)}{r_0-s_0+i\varepsilon}\,(\slashed{s}+m)\,\delta_+(s) \,. \tag{9.48}$$

## 9. Unrenormalized Solution

There is yet an $r$-dependence present in the factor $\delta^4(P-r)$ of the starting point of the $r$-line, with $P$ the sum of the other momenta incident at that vertex. But again the factor $\delta(P_0 - r_0)$ is continuous in $r_0$ if considered a distribution in $P_0$, hence we can replace $r_0$ by $s_0$ in the limit $\lambda \to 0$. Carrying out the $r$-integration we obtain thus from $C'_G$ the contribution

$$\int du \, \frac{\beta_\lambda(u)}{u+i\varepsilon} (\slashed{s} + m) \delta_+(s) . \tag{9.49}$$

Inserting this into the full graph integrand $I_G$ we obtain the integrand of a graph contributing to $(\Omega, \mathcal{N}\mathcal{M}, \Omega)_{\sigma-1}$ multiplied with the factor $\frac{i}{2\pi} \int du \, \frac{\beta_\lambda(u)}{u+i\varepsilon}$, if we reinsert the omitted constants. In more detail: assume without restriction of generality that the neighbouring sectors of the $k$-sector are external $T^-$ sectors. Then the disappearance of the $k$-vertex as explained above turns the original $k$-sector into an internal $T^+$ sector $S$, as is appropriate between two $T^-$ sectors. One of the cross lines of this internal sector originated from the $k$-bearing cross line of the $k$-sector. But this may have been any of the cross lines. We must sum over all the possibilities. Then we get from each of the $V_H$ fermion lines leaving $S$ to a higher sector the factor $\frac{i}{2\pi} \int du \, \frac{\beta_\lambda(u)}{u+i\varepsilon}$ and the negative of this factor for each of the $U_H$ lines entering $S$ from a higher sector, and the factor $\pm \frac{i}{2\pi} \int du \, \frac{\beta_\lambda(u)}{u-i\varepsilon}$ for each of the $\left\{ \begin{matrix} V_L \\ U_L \end{matrix} \right\}$ lines $\left\{ \begin{matrix} \text{leaving for} \\ \text{entering from} \end{matrix} \right\}$ a lower sector. These numbers satisfy the identity $U_L + U_H = V_L + V_H$. Summing over all cross lines of $S$ we obtain

$$\frac{i}{2\pi}(V_H - U_H) \int du \, \beta_\lambda(u) \left( \frac{1}{u+i\varepsilon} - \frac{1}{u-i\varepsilon} \right) = (V_H - U_H) \beta_\lambda(0) = V_H - U_H . \tag{9.50}$$

Using $(V_H - U_H)e = q_M$ we then obtain the desired relation (9.42).

There may be lines crossing $S$ without meeting any vertices of $S$. They emerge from trajectories in the original $k$-graph in which both the adjacent $r$- and $s$-lines are cross lines. In that case one can easily calculate the integrals involved with the help of the two $\delta_+$-functions, finding that the above result is not changed.

# Chapter 10

# Renormalization and the UV Problem

In this chapter the UV divergences are discussed and shown to be removed by renormalization. We proceed as in Chap. 9. In Sect. 10.1 the problem and its solution will be explained in detail for simple but typical examples in low orders. Then the general formulation of the renormalization procedure will be stated in Sect. 10.2 and shown to satisfy the renormalization conditions. In these two first sections we work with the $p$-space form of the graph rules, because the renormalization conditions have been formulated in Sects. 8.1 and 8.2 for that space, and because the rules are easier to describe and motivate there. But the proof that these rules indeed remove the UV divergences is simpler in $x$-space. Also, the proof that renormalization does not destroy the W-properties needs to be carried out partly in $x$-space. We therefore describe the $x$-space form of renormalization in Sect. 10.3 and give the necessary proofs, as far as feasible at this stage. The final complete verifications that all is as it should be can only be given in the next chapter, after the IR problem for Wightman functions has been solved.

## 10.1 Low Orders

### The Photon 2-Point Function

An UV problem occurs for the first time in second order for the 2-point functions $\tilde{W}_2(p, \bar{q})$ and $\tilde{W}_2(\mathsf{p}, \mathsf{q})$. Let us first consider the photon function $\tilde{W}_2(\mathsf{p}, \mathsf{q}) = \delta^4(p+q)\, w_2(\mathsf{p})$. We use the $\delta$-reduced graph rules for $w_2$. The sector graphs contributing to $w_2$ are shown in Fig. 10.1. We associate to each of these graphs an integral according to the rules of Chap. 9. Concerning these integrals we are confronted with three types of problems.

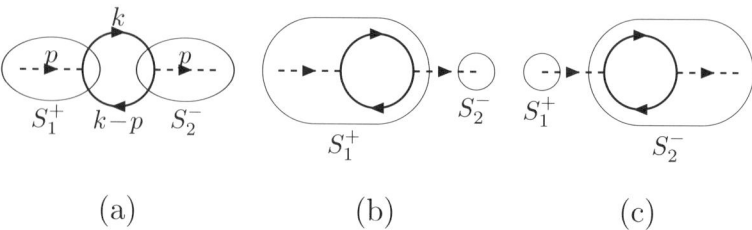

**Fig. 10.1a–c.** Sector graphs contributing to $w_2(p)$.
The momentum assignment is the same in all graphs shown. $S_j^s$ is a sector of type $T^s$

**Problem A.** In the graphs (b) and (c) the external lines contribute to the integrand a product $(p^2 + i\varepsilon)^{-1} \delta_+(p)$ or $\delta_+(p)(p^2 - i\varepsilon)^{-1}$, respectively, which is not defined as a distribution. A similar problem seems also to be present in graph (a), where the external lines contribute the divergent product $(p^2 + i\varepsilon)^{-1} (p^2 - i\varepsilon)^{-1}$. However, this appearance is deceptive because the two internal lines contribute the product $\delta_+(k) \delta_+(p-k)$ which vanishes for $p^2 < 4m^2$, so that the dangerous set $p^2 = 0$ lies outside the support of the integrand.

This problem is concerned with the local singularities of the $p$-space integrands: it is part of the extended IR problem for $W$-functions, which will be generally solved only in Chap. 11 and will be essentially ignored in the present chapter. However, it is perhaps as well already to describe its solution for our particular example, in order to give credibility to the contention that the problem is not serious and can be savely ignored at first. We first note that according to Lemma 9.2 the loop integrals of the three graphs, i.e. the $k$-integrals without the propagators of the external lines, add to zero. (We still disregard the UV divergences of these integrals.) But the loop integral of graph (a) vanishes in a neighbourhood of $\{p^2 = 0\}$, so that the loop integrals $L_b(p)$ and $L_c(p)$ of graphs (b) and (c) add to zero for $p^2 \sim 0$, i.e. at the support of $\delta_+(p)$. Moreover, $L(p) := L_b(p) = -L_c(p)$ is analytic at $p^2 = 0$. We need only consider points with $p^0 \geq 0$. There we have $2\pi \delta_+(p) = i\left[(p^2 + i\varepsilon)^{-1} - (p^2 - i\varepsilon)^{-1}\right]$. Inserting this into the graphs (b) and (c) we get for their sum the expression

$$-\frac{1}{(2\pi)^2} L(p) \left[(p^2 + i\varepsilon)^{-2} - (p^2 - i\varepsilon)^{-2}\right] = -\frac{i}{2\pi} L(p) \delta'(p^2) , \qquad (10.1)$$

which *is* a perfectly well defined distribution. As a result we find that the problem-A singularities cancel in the sum over the graphs with the same scaffolding. This will be found in Chap. 11 to be generally true.

**Problem B.** The loop integral in graph (b) is

$$L_b(p) = (2\pi)^{-3} \int dk \, \frac{1}{k^2 - m^2 + i\varepsilon} \, \frac{1}{(p-k)^2 - m^2 + i\varepsilon} \\ \times \mathrm{Tr}\{\gamma^\mu(\slashed{k}+m)\gamma^\nu(\slashed{k}-\slashed{p}+m)\} \, . \quad (10.2)$$

The integrand decreases only like $|k|^{-2}$ for $k \to \infty$, if $|k|$ is the Euclidean length of the 4-vector $k$. This is not sufficient to make the 4-dimensional $k$-integral converge at infinity, even if the local singularities are of an integrable type. This is the UV problem. It is also present in the same way in graph (c), but not in graph (a), where the corresponding integral reads

$$L_a(p) = -(2\pi)^{-1} \int dk \, \delta_+(k)\, \delta_+(p-k) \\ \times \mathrm{Tr}\{\gamma^\mu(\slashed{k}+m)\gamma^\nu(\slashed{k}-\slashed{p}+m)\} \, . \quad (10.3)$$

The $k$-supports of the two factors $\delta_+$ have a compact intersection for fixed $p$, hence there is no convergence problem at infinity.

**Problem C.** Even disregarding existence problems, the sum of the graphs (a)–(c) does not satisfy the renormalization condition (8.7).

We address Problem C first, before Problem B, still working formally like in Chap. 9, ignoring the UV divergences. We need to know how the condition (8.7) is violated. This condition regulates the behaviour of $w_2(\mathsf{p})$ near the positive light cone $p^2 = 0$, $p_0 \geq 0$. It has already been seen that graph (a) vanishes in that region and that the graphs (b) and (c) sum to the expression (10.1) with $L(p)$, now written more exactly as $L^{\mu\nu}(p)$, given by equation (10.2). $\gamma^\mu$ and $\gamma^\nu$ are the two vertex factors of the loop. We note that $L^{\mu\nu}$ is $C^\infty$ for $p^2 < 4m^2$. To see this we first integrate over $k_0$ in the integral (10.2). The integrand is meromorphic in $k_0$ with four real poles (for $\varepsilon = 0$) at $u_1 = \omega(\vec{k})$, $u_2 = p_0 + \omega(\vec{p}-\vec{k})$, $u_3 = -\omega(\vec{k})$, $u_4 = p_0 - \omega(\vec{p}-\vec{k})$. The $i\varepsilon$-prescription tells us to integrate around these poles along the path $\mathcal{C}_1$ shown in Fig. 10.2. The poles at $u_1$ and $u_2$ can coalesce for particular values of $p$, $\vec{k}$, and so can $u_3$ and $u_4$. But no coalescence is possible between $u_{1,2}$ and $u_{3,4}$. For example, $u_1 = u_4 = u$ would mean that the 4-vectors $\ell_1 = (u, \vec{k})$ and $\ell_2 = (p_0 - u, \vec{p} - \vec{k})$ lie on the positive mass shell $\{\ell^2 = m^2, \ell_0 > 0\}$, hence $p^2 = (\ell_1 + \ell_2)^2 \geq 4m^2$ in contradiction to our assumption. As a result the contour $\mathcal{C}_1$ can be deformed into the complex contour $\mathcal{C}_2$ of Fig. 10.2, which never comes close to a singularity. The integrand along $\mathcal{C}_2$ is then an analytic function of $k$ and $p$, and thus the integral is analytic in $p$, if the UV problem is ignored.

Furthermore $L^{\mu\nu}$ transforms under the Lorentz group as a tensor and must therefore be of the form

$$L^{\mu\nu}(p) = g^{\mu\nu} A(p^2) + p^\mu p^\nu B(p^2) \, , \quad (10.4)$$

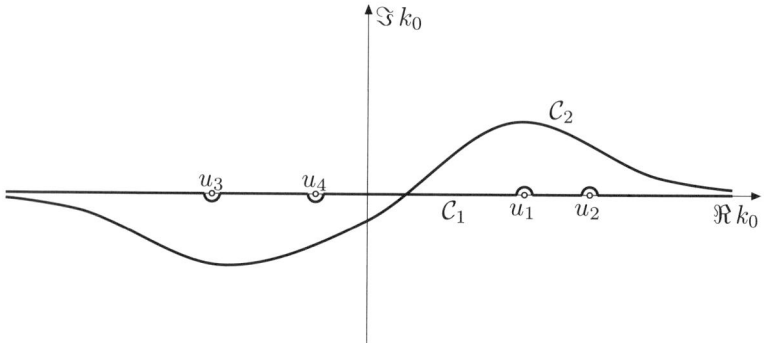

**Fig. 10.2.** Integration paths

this being the most general tensor function of a single 4-vector $p$. The WT identities (9.41) tell us that $p_\mu L^{\mu\nu} = p^\nu(A + p^2 B) = 0$, so that $A = -p^2 B$. Thus we find

$$L^{\mu\nu}(p) = (p^2 g^{\mu\nu} - p^\mu p^\nu) B(p^2) \, . \tag{10.5}$$

$B(u)$ is analytic for $u < 4m^2$ because of the analyticity of $L^{\mu\nu}$. Inserting the Taylor expansion $B(p^2) = b_0 + b_1 p^2 + \cdots$ we find from (10.1) and (10.5) that the unrenormalized $w_2$, now called $w_{2,\mathrm{nr}}$, is

$$w_{2,\mathrm{nr}}^{\mu\nu}(p) = \frac{i}{2\pi} \theta(p_0) \left( b_0 g^{\mu\nu} \delta(p^2) + b_0 p^\mu p^\nu \delta'(p^2) - b_1 p^\mu p^\nu \delta(p^2) \right) \tag{10.6}$$

for $p^2 < 4m^2$. The third term in this expression is the contribution of second order to the $\rho$-term in condition (8.7). But, as explained in Sect. 8.2, the first two terms must not be present. The second of them can be removed by activating the hitherto ignored $\lambda$-term in the normal-product definition (8.2). In the spirit of perturbation theory we expand $\lambda(e)$ in a power series

$$\lambda(e) = \sum_{\sigma=2}^{\infty} \lambda_\sigma \, e^{\sigma - 1} \, . \tag{10.7}$$

The $e^0$ term in the expansion of $C_A$ is 1, hence in second order we can set $C^A = 1$ in the product $C_A{}^2 \lambda$. In the recursive expression for $p^2 w_2^{\mu\nu}(\mathbf{p})$ we have then the additional term $\lambda_2 p^\mu p_\rho w_0^{\rho\nu} = -\lambda_2 p^\mu p^\nu \delta_+(p)$ yielding a term $\lambda_2 p^\mu p^\nu \theta(p_0) \, \delta'(p^2)$ in $w_2$, which must be added to (10.6). The choice

$$\lambda_2 = -\frac{i}{2\pi} b_0 \tag{10.8}$$

cancels the undesirable $\delta'$ in (10.6). $b_0$ is purely imaginary, hence $\lambda_2$ is real. The first unwanted term $T(p) = \mathrm{const} \cdot \delta_+(p)$ can be dropped because it solves

## 10.1 Low Orders

the homogeneous equation $p^2 T(p) = 0$.[1] Remember that $w_{2,\text{nr}}^{\mu\nu}$ was obtained as a solution of $p^2 w_2^{\mu\nu} = known$, which solution contains an ambiguity of the form $T$. After the two eliminations we find the *renormalized* 2-point function

$$w_{2,\text{r}}^{\mu\nu}(\mathsf{p}) = w_{2,\text{nr}}^{\mu\nu}(\mathsf{p}) + \frac{ib_0}{2\pi}\,\theta(p_0)\left(p^2 g^{\mu\nu} - p^\mu p^\nu\right)\delta'(p^2)\,. \qquad (10.9)$$

It solves, still formally, the $p$-space form of (8.9) for $\tilde{W}_2(\mathsf{p},\mathsf{q})$ with the full renormalized normal product $N_1$ defined in (8.2). The $R_1$-term does not yet contribute in second order. It is easily checked that the proofs of the W-properties in Chap. 9 still apply. In addition, the renormalization condition (8.7) is also satisfied.

**Fig. 10.3.** Renormalization graphs for $w_2(p)$. The new vertex is marked by a circlet

It remains to be seen that renormalization, i.e. replacing $w_{2,\text{nr}}$ by $w_{2,\text{r}}$, also solves the UV problem B. For this we must include the renormalization terms in (10.9) in the graph rules. This is done by introducing a new two-line vertex as shown in Fig. 10.3. This new vertex carries the vertex factor

$$\pm 2\pi i\,\lambda_2\left(p^2 g^{\mu\nu} - p^\mu p^\nu\right) \qquad (10.10)$$

with $\lambda_2$ given by (10.8). The overall sign accords with the sector type. The indeterminacy of the products $\delta(p^2)\left(p^2 g^{\mu\nu} - p^\mu p^\nu\right)(p^2 \pm i\varepsilon)^{-1}$ in the two graphs is handled as in (10.1): the two graphs of Fig. 10.3 sum to

$$\lambda_2\left(p^2 g^{\mu\nu} - p^\mu p^\nu\right)\theta(p_0)\,\delta'(p^2)\,, \qquad (10.11)$$

a well-defined distribution, if $\lambda_2$ were finite. The $\lambda_2$-vertex is to be counted as of *second* order, in contrast to the interaction vertex of Chap. 9, which is of first order. This means that the perturbative order of a graph is the sum of the orders of its vertices. For instance, the graphs of Fig. 10.3 are contributions to the second-order function $w_2$ despite of their containing only one vertex.

We can now at last approach the UV problem. The graph (a) of Fig. 10.1 has been found to be UV convergent. The graphs (b) and (c) and the renormalization graphs sum to

$$\frac{i}{2\pi}\,\theta(p_0)\,\delta'(p^2)\left[\left(p^2 g^{\mu\nu} - p^\mu p^\nu\right)\left(B(p^2) - b_0\right)\right]\,. \qquad (10.12)$$

---

[1] The $R_2$-term in (8.2) does not contribute to the right-hand side of the $w_2$ equation because it reads $-R_2\, p^2 \tilde{w}_0(p) = 0$.

Since $L^{\mu\nu}(p)$ is (formally) analytic in a neighbourhood of $p = 0$, it can be developed into a Taylor series. The first two terms vanish because of the factor $(p^2 g^{\mu\nu} - p^\mu p^\nu)$, the term of second order is $b_0 (p^2 g^{\mu\nu} - p^\mu p^\nu)$. Hence the square bracket in (10.12) is obtained by expanding $L^{\mu\nu}$ into a power series around $p = 0$ and dropping the first three terms, two of which are not present anyway.

$L^{\mu\nu}(p)$ is defined by (10.2). The $k$-integrand in this expression is of the form

$$I = \frac{N(k,p)}{D(k, k-p)}. \qquad (10.13)$$

The numerator $N$ is a polynomial in $k, p$ of degree 2, the denominator $D$ a homogeneous polynomial of degree 4, hence for fixed $p$ a polynomial in $k$ of degree 4. Let $|k|$ be the Euclidean length of the 4-vector $k$. Ignoring for the time being the singularities of $I$ at the zeros of $D$, we find a decrease of order $|k|^{-2}$ for $|k| \to \infty$. This does not suffice to ensure existence of the integral at infinity: the integral $J(p)$ is UV divergent. A Taylor expansion of $J(p)$ is therefore at first meaningless. Let us nevertheless write down this expansion in a purely formal way, blithely exchanging $k$-integration and $p$-differentiation, i.e. expanding the integral by first expanding the integrand, which *can* be done meaningfully, and integrating afterwards. In order to see what happens, let us first look at a general expression $Q(k,p)$ of the form (10.13) with $N$ a polynomial of degree $n$, $D$ a form of degree $d$, so that the quotient is decreasing at least like $|k|^{n-d}$ for $|k| \to \infty$, but in general not faster.[2] Define $Q_0(k) = Q(k,0)$. A trivial calculation yields

$$Q(k,p) = Q_0(k) + \frac{N(k, k-p) D(k,k) - N(k,k) D(k, k-p)}{D(k, k-p) D(k,k)}.$$

The numerator of the quotient in this expression vanishes at $p = 0$, so that this quotient is of the form $p_\mu \frac{N^\mu(k,p)}{D'(k, k-p)}$ with $N^\mu$ a polynomial of degree $n + d - 1$, $D'$ a form of degree $2d$. Hence $N^\mu/D'$ decreases for $|k| \to \infty$ at least like $|k|^{n-d-1}$, faster by one order than the original $N/D$. Repeating this procedure on $N^\mu/D'$, and then once again, we obtain

$$Q(k,p) = Q_0(k) + p_\mu Q_1^\mu(k) + p_\mu p_\nu Q_2^{\mu\nu}(k) + R(k,p), \qquad (10.14)$$

where the $Q_i$ and $R$ are rational functions. $R$ vanishes at $p = 0$ of third order and decreases for $|k| \to \infty$ at least like $|k|^{n-d-3}$. The formula (10.14) clearly gives the Taylor expansion of $Q$ to second order in $p$. Applying the formula to the special case $Q = I$, we find that the first three terms of this expansion are the terms that are removed by renormalization. After renormalization we

---

[2] The following calculations can be immediately extended to graphs of higher order, where there will be several $p$'s and $k$'s. We will not discuss this generalization explicitly later on.

obtain then the integral

$$J_{\text{ren}}(p) = \int dk\, R(k,p) , \qquad (10.15)$$

which is convergent at infinity. Thus the contention that renormalization removes the UV divergences of $w_2(\mathsf{p})$ is proved. That this result is not invalidated by the existence of local singularities in $R$ will be shown later.

We still seem to have arrived at this finite result in a rather formal way, as a difference of divergent expressions. But we can remove this objection by reformulating our procedure as follows. We need to define the normal products (8.2) in a rigorous way. This we do not do by directly defining the $N_i$ as operators, but by specifying their matrix elements $(\Omega, \mathcal{M}\,\tilde{N}_i(p)\,\mathcal{N}\,\Omega)$ between monomial states, as was done in Chap. 7 for the current $j^\mu$. In the present instance we need to know

$$(\Omega, \tilde{N}_1^\mu(p)\,\tilde{A}^\nu(q)\,\Omega)_{\sigma=1} =: \delta^4(p+q)\,\hat{n}_{1,1}(p) . \qquad (10.16)$$

This we *define* as

$$\hat{n}_{1,1}(p) := -\frac{1}{(2\pi)^2}\,L_a^{\mu\nu}(p)\,\frac{1}{p^2} - \frac{i}{2\pi}\,L_{\text{ren}}^{\mu\nu}(p)\,\delta_+(p) , \qquad (10.17)$$

where

$$L_{\text{ren}}^{\mu\nu}(p) = \frac{1}{(2\pi)^3}\,J_{\text{ren}}^{\mu\nu}(p) \qquad (10.18)$$

and $L_a^{\mu\nu}$ is the UV convergent loop integral (10.3). Apart from the subscript "ren", the expression (10.17) is what we would have found by calculating the Fourier transform of $(\Omega, \bar{\psi}(x)\,\gamma^\mu\,\psi(x)\,A^\nu(y)\,\Omega)_1$ with the rules of Chap. 9. But the renormalized version (10.17) is not obtained from the unrenormalized one by adding the new terms in $N_1$ as explained above and a subsequent juggling with divergent quantities. Rather, the finite $L_{\text{ren}}^{\mu\nu}$ is introduced directly as part of an exact *definition* of $N_1$. The function $(\Omega, \tilde{N}_1^\mu(p)\,\tilde{A}^\nu(q)\,\Omega)_1$ defined in this way has all the required properties. This is easily seen for those properties that can be proved in $p$-space, like covariance, spectral property, current conservation, and hermiticity. For the latter we need also $(\Omega, A^\nu\,N_1^\mu\,\Omega)_1$, but this is defined in a completely analogous way. The proof of the other properties will have to await the formulation of renormalization in $x$-space, which will be given in Sect. 10.3.

The pertinent equation (8.9) reads in $p$-space

$$-p^2\,\tilde{W}_2(\mathsf{p},\mathsf{q}) = (\Omega, \tilde{N}_1(p)\,\tilde{A}^\nu(q)\,\Omega)_1 \qquad (10.19)$$

and is solved by the renormalized $\tilde{W}_2^{\text{ren}}$ that has been constructed above. It is given as a sum over the graphs of Fig. 10.1, where however now the loop integrals in (b) and (c) take their finite subtracted forms $L_b^{\text{ren}}$ and $L_c^{\text{ren}}$ as explained before (i.e. the subtractions are carried out in the integrands). The

142    10. Renormalization

(b) and (c) terms must of course be combined into a single one like in (10.1), in order to solve Problem A. This $\tilde{W}_2$ solves also the $q$-equation (8.9), which looks similar to (10.19). And it enjoys again all the required properties, but now rigorously, not merely formally. Apart from the more obvious properties indicated above, including the renormalization condition (8.7), a proof of this statement must wait until the general formalism has been developed.

We close this discussion of $W_2$ with a remark on the status of the canonical commutation relations. Notice that only graph (a) contributes to $w_2(\mathsf{p})$ for large values of $p_0$ if $\vec{p}$ is kept fixed or is integrated over a test function with compact support. It is then easily seen that $w_2$ decreases like $p_0^{-2}$ for $p_0 \to \infty$. This implies that $W_2(x,y)$ is still defined as a distribution in $\vec{x}-\vec{y}$ at $x^0 = y^0$, but $\partial W_2/\partial x^0$ is not. Therefore the canonical equal-time commutation relations make no sense in this order of perturbation theory, as was announced in Chap. 5.

## The Electron 2-Point Function

We turn to the fermion 2-point function $\tilde{W}_2(p,\bar{q}) = \delta^4(p+q)\,w_2(p)$ in second order. Its unrenormalized form is given by the graphs shown in Fig. 10.4, in close analogy to the photon case of Fig. 10.1. Not surprisingly, the problems A–C encountered there show up here as well.

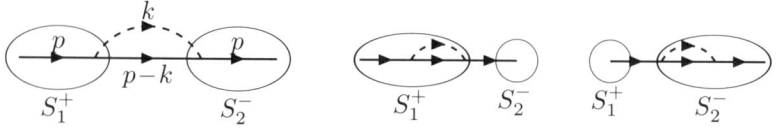

**Fig. 10.4.** Sector graphs contributing to $w_2(\mathsf{p})$.
The conventions of Fig. 10.1 apply

Problem A concerns the local singularities of the graphs, in particular the undefined products of the two external propagators of each graph. This problem occurs now also for the first graph. Nevertheless it can be solved similarly to the photon case, by summing the three graphs. This time we apply Lemma 9.2 to the full scaffolding instead of only to its loop part. The result is

$$w_2(p) = \theta(p_0)\left[t_2^+(p) + t_+^-(p)\right], \qquad (10.20)$$

where $t^\pm$ is defined by $\tilde{\tau}^\pm(p,\bar{q}) = \delta^4(p+q)\,t^\pm(p)$. $t_2^\pm(p)$ is given by a graph with the scaffolding of Fig. 10.4, but consisting of a single $T^\pm$ sector. For example, we find

$$t_2^+(p) = -(2\pi)^{-2}\frac{\not{p}+m}{p^2-m^2+i\varepsilon}L_b(p)\frac{\not{p}+m}{p^2-m^2+i\varepsilon}. \tag{10.21}$$

In $t_2^-$ the sign of the $i\varepsilon$-terms is reversed and $L_b$ replaced by $L_c$. $L_{b,c}$ are the loop integrals of the graphs (b) and (c) respectively. We must prove that this product is defined as a distribution. All its factors can be shown to be analytic functions of the complex 4-vector $p = p' + ip''$ in the domain $\{p_0' > -m,\ p'' \in V_+\}$. This is clear for the propagators $(p^2 - m^2 + i\varepsilon)^{-1}$ which are singular at $p^2 = m^2$ for $\varepsilon = 0$. But $(p' + ip'')^2 = m^2$ implies $p'^2 - p''^2 = m^2$, hence $p'^2 > m^2$, and $(p', p'') = 0$, which is impossible for $p'' \in V_+$. Going to the real limit $p'' \to 0$ from inside $V_+$ corresponds to the $i\varepsilon$-prescription in (10.21). The claimed analyticity of $L_b$—disregarding at the moment its UV divergence—is also not hard to see. Apart from a constant factor the integrand $I$ of $L_b$ is

$$\frac{\gamma^\mu(\not{p}-\not{k}+m)\gamma_\mu}{(k^2+i\varepsilon)[(p-k)^2-m^2+i\varepsilon]}. \tag{10.22}$$

As in the photon case we integrate first over $k_0$ along the complex contour $\mathcal{C}_2$ shown in Fig. 10.2, with pole positions $u_1 = |\vec{k}|$, $u_2 = p_0+\omega(\vec{p}-\vec{k})$, $u_3 = -|\vec{k}|$, $u_4 = p_0-\omega(\vec{p}-\vec{k})$. Because of the Lorentz covariance of the integral it suffices to prove analyticity for $\Im p_0 > 0$, $\vec{p}$ real. Giving $p_0$ a positive imaginary part moves $u_2$ and $u_4$ into the upper half plane. But $\mathcal{C}_2$ can be chosen such that $u_2$ still remains below, $u_4$ above it. And it is easy to see that in this situation no pinching of the contour between poles on different sides can occur for $p_0' > -m$. Hence the integrand is analytic everywhere on $\mathcal{C}_2$; therefore the integral is analytic in $p$. Thus $t_2^+$ is for real $p$, $p_0 \geq 0$, a boundary value of an analytic function in the "forward tube" $\{\Im p \in V_+\}$. This fact together with some easily proved estimates on the possible strength of divergence (not stronger than a negative power of $p_0''$) of the limit $p'' \to 0$, guarantees the existence of $t_2^+$ as a distribution (see Theorem 5.4). In the same way $t_2^-$ exists as boundary value of a function analytic in $\{\Im p \subset V_-\}$, and therefore $w_2(p)$ exists, at least as far as Problem A is concerned. At $p_0 = 0$ the functions $t_2^\pm$ are continuous, so that the factor $\theta(p_0)$ does not create any problems.

Problem B is the UV problem. The loop integral of graph (a) is UV convergent for the same reason as in the photon case. It is also locally convergent for all $p$ and defines a continuous function of $p$. The integrands of the loop integrals $L_b$ and $L_c$ are rational functions of $p$ and $k$ which decrease for $k \to \infty$ like $|k|^{-3}$. This is one degree better than in the photon case, but not sufficient to ensure existence of the integral at infinity. We will again find that this problem is automatically solved by the solution of Problem C.

Problem C is the renormalization problem. The renormalization condition for $w(p)$ is less clear-cut than that for $w(\mathsf{p})$. As formulated in Chap. 8 it states that $w(p)$ have a singularity at $p^2 = m^2$, which is the weakest one possible that is compatible with the other requirements. We must then determine the strength of the singularity of $w_2$ at $p^2 = m^2$, $p_0 \geq m$. This can be done by an extension of the arguments used in solving Problem A. From (10.20) and (10.21) we find that $w_2(p)$ has a singularity at the desired position $p^2 = m^2$. For determining its strength we must discuss the behaviour of $L_b(p)$ near the mass shell. Because of the known support $\{p^2 \geq m^2,\ p_0 > 0\}$ of $w_2$ and the Lorentz covariance of $L_b$ it suffices to consider the case $p_0 \geq m$, $\vec{p} = 0$. As in Problem A we integrate first over $k_0$ along the contour $\mathcal{C}_2$, but now for real $p$. Singularities occur where this path is pinched between two poles. We close the contour $\mathcal{C}_2$ by an infinite semicircle in the lower half plane, which gives a vanishing contribution on account of the strong decrease of the integrand (10.22) for $|k_0| \to \infty$. The integral can then be evaluated with the residue theorem. For $p_0 > 0$ the $u_2$-pole cannot coalesce with any of the other poles, hence its residue is a regular function of $p$ and $\vec{k}$, and is thus irrelevant for our present purpose. The $u_1$-residue is apart from a regular factor

$$\frac{1}{|\vec{k}| - i\varepsilon} \frac{1}{|\vec{k}| + \omega(\vec{k}) - p_0 - i\varepsilon} . \tag{10.23}$$

This expression must yet be integrated over $\vec{k}$. Both factors are well-defined distributions in $\vec{k}$. This is also true for arbitrarily high $p_0$-derivatives of the second factor, as long as $p_0 \neq m$. For $p_0 = m$ there occur for high $p_0$-derivatives strong singularities at $\vec{k} = 0$ which are not regularized by the $i\varepsilon$ prescription. The critical point is then $\vec{k} = 0$. Near this point the second denominator behaves like $(|\vec{k}| - (p_0 - m) + \mathcal{O}(|\vec{k}|^2))$. The $\mathcal{O}$-term is irrelevant for the $p_0$-singularity and can be neglected. Also, we need only integrate over a bounded neighbourhood of the critical point $\vec{k} = 0$, e.g. the ball $\{|\vec{k}| \leq R\}$, thus avoiding the UV problem. In the suppressed regular factor we may set $\vec{k} = 0$ and draw it in front of the integral. We obtain then an integral of the form, with $k = |\vec{k}|$,

$$\int_0^R dk\, \frac{k}{k - (p_0 - m) - i\varepsilon}$$
$$= R + (p_0 - m) \log\left(R - (p_0 - m + i\varepsilon)\right) - (p_0 - m) \log(p_0 - m + i\varepsilon) - i\pi.$$

The first and the last term are regular. The logarithmic singularity of the second term is spurious. It is due to the sharp cutoff at $|\vec{k}| = R$ and can be avoided by using a smooth cutoff instead. The relevant singularity is that of the third term. Going back to a general frame of reference ($\vec{p} \neq 0$) we can summarize this result as follows: $L_b(p)$ is for $p_0 > 0$ of the form

$$L_b(p) = A^+(p) + B^+(p)\,(p^2 - m^2)\,\log(p^2 - m^2 + i\varepsilon) , \tag{10.24a}$$

and analogously

$$L_c(p) = A^-(p) + B^-(p)\,(p^2 - m^2)\,\log(p^2 - m^2 - i\varepsilon) . \tag{10.24b}$$

$A^\pm$ and $B^\pm$ are analytic. $\log u$ is defined in the complex plane cut along the positive real axis, such that its value at the upper bord of this cut is real. Inserting (10.24) into (10.20) and using the known support $p^2 \geq m^2$ of $w_2(p)$ we find that $-A^-(p) = A^+(p) =: A(p)$ and $-B^-(p) = B^+(p) =: B(p)$ for $p^2 < m^2$, and therefore everywhere by analytic continuation.

For comparison with the literature it must be mentioned that $L_b(p)$ is called a "self-energy part" and is usually denoted by $\Sigma_2(p)$. We will henceforth also use this notation.

We find

$$w_2(p) = -\frac{i}{2\pi}\theta(p_0)(\not{p}+m)$$
$$\times \left[A(p)\,\delta'(p^2-m^2) - i\pi\,\delta(p^2-m^2) - B(p)\,\text{P.f.}\frac{\theta(p^2-m^2)}{p^2-m^2}\right](\not{p}+m), \tag{10.25}$$

where the "pseudo-function" P.f.$\frac{\theta(p^2-m^2)}{p^2-m^2}$ is defined as a distribution by

$$\text{P.f.}\int_0^\infty du\,\frac{f(u)}{u} = \lim_{\varepsilon\downarrow 0}\left[\int_\varepsilon^\infty du\,\frac{f(u)}{u} + f(0)\log\varepsilon\right] \tag{10.26}$$

for all $f \in \mathcal{S}$.

The singularities of this expression are at the correct position $p^2 = m^2$ with $m$ the physical mass of the electron. But the $\delta'$-singularity is decidedly too strong to be acceptable. It would play havoc already with the construction of a physical state space in Chap. 12, and even more so with the scattering formalism of Part III. But it can be removed by mass renormalization. This amounts to interpreting it as the second-order term in the perturbative expansion of $\delta(p^2 - m^2) = \delta(p^2 - (m_0 + \delta m)^2)$. That is, the singularities in all orders of perturbation theory are found to sum formally to a singularity of the right strength, but at the wrong position $p^2 = m_0^2$, and this must be shifted to the right position by mass renormalization. The P.f. singularity is also uncomfortably strong, namely stronger than the corresponding free singularity $\delta(p^2 - m^2)$.[3] But it is not of a form that can be removed by renormalization: it is the weakest possible singularity demanded in our renormalization condition. Luckily this turns out not to be a serious problem. It does not prevent us from establishing a scattering formalism, though it will have to be rather more elaborate than in purely massive theories, where this kind of problem does not occur. Actually, the strength of this singularity is an artifact of perturbation theory. As was indicated in Sect. 6.3 these mass shell singularities taken in all orders sum up to a singularity which is weaker than the free $\delta$-singularity.

---

[3] The singularity of a distribution is measured by the smoothness required for admissible test functions: $\delta'(u)$ is defined on $C^1$ functions, P.f.$(1/u)$ on Hölder continuous functions, $\delta(u)$ on continuous functions.

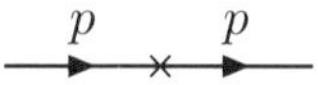

**Fig. 10.5.** Mass renormalization vertex

Mass renormalization is effected by introducing the $\delta m$-term in the $N_2$ definition (8.2) into our calculations. Analogously to the $\lambda$-terms in the photon function this introduces in second order the 2-line vertex shown in Fig. 10.5. It carries the vertex factor $\pm 2\pi i\,\delta m_2$, the sign conforming with the type of the sector in which the vertex occurs. The vertex is of second order in the sense explained after (10.11). Adding this term to $t_2^+$ we find in it the contribution

$$-\frac{1}{(2\pi)^2}\frac{1}{\not{p}-m}\left(A(p)+2\pi i\,\delta m_2\right)\frac{1}{\not{p}-m}\ . \tag{10.27}$$

$A(p)$ is a spin matrix. It transforms under the Lorentz group in the same way as the full 2-point function: $A(\Lambda p) = S(\Lambda)\,A(p)\,S^{-1}(\Lambda)$. With the available quantities $p_\mu$ and $\gamma^\mu$ we can only form two independent expressions with this behaviour: $\not{p}$ and the unit matrix $\mathbf{1}$. Suitably combining these expressions we obtain

$$A(p)=a(p^2)\,(\not{p}-m)+b(p^2)\ , \tag{10.28}$$

where $a$ and $b$ are analytic functions. The factor $\not{p}-m$ in the $a$-term cancels one of the denominators in (10.27): the undesirable pole of second order is entirely due to the scalar $b$-term. It is removed by choosing

$$\delta m_2 = \frac{i}{2\pi}\,b(m^2)\ . \tag{10.29}$$

We can now address the UV problem B. Mass renormalization does not yet render $w_2$ finite. The $R_3$-term in (8.2) does not help because it vanishes in second order for the same reason as the $R_2$-term in the photon case. But we must remember that the normalization of $\psi$ has not been fixed. We are therefore free to multiply $\tilde{W}(p,\bar{q})$, and thus $w(\tilde{p})$ and $\tau^\pm(p)$, with a positive constant $C := C_\psi{}^2 = 1 + C_2 e^2 + \cdots$. This introduces another second order vertex of the form shown in Fig. 10.5, this time with a vertex factor $\mp\frac{i}{2\pi}C_2(\not{p}-m)$. In $w_2$ this produces a term of the same form as the $\delta$-term in (10.25) but with an undetermined coefficient.

We proceed now as in the photon case. The self-energy part $\Sigma_2(p)$ is formally expanded into a power series in $p$ up to first order,

$$\Sigma_2(p)=\alpha(\not{p}-m)+\beta+\Sigma_2^{\rm f}(p)\ , \tag{10.30}$$

by first expanding its integrand $I(k,p)$ and integrating afterwards. The "finite part" $\Sigma_2^{\rm f} = \mathcal{O}(|p|^2)$ is UV convergent, the coefficients $\alpha$ and $\beta$ are divergent. This is shown exactly like the corresponding statement for the photon propagator. Comparing (10.30) with (10.28) we find $\alpha = a(0)$, $\beta = b(0)$. The renormalized form of $\Sigma_2$ is given by

$$\Sigma_2^{\text{ren}}(p) = \Sigma_2^{\text{f}}(p) + \left(\alpha - \frac{i}{2\pi}C_2\right)(\not{p} - m) + (\beta + 2\pi i\,\delta m_2)\,. \tag{10.31}$$

For the undetermined normalization constant $C_2$ we choose $C_2 = -2\pi i\,a(0)$, for $\delta m_2$ we insert the value (10.29), and obtain

$$\Sigma_2^{\text{ren}}(p) = \Sigma_2^{\text{f}}(p) + b(0) - b(m^2)\,. \tag{10.32}$$

But $\Sigma_2^{\text{f}} = [a(p^2) - a(0)](\not{p} - m) + [b(p^2) - b(0)]$ is UV convergent, hence the two square brackets are convergent for all $p$, hence

$$\delta m_2^{\text{f}} = \frac{i}{2\pi}[b(m^2) - b(0)] \tag{10.33}$$

is convergent. We can again give a rigorous UV convergent definition of $t_2^\pm$, $w_2$, by *defining* the self-energy insertion in graph (b) of Fig. 10.4 to be the finite part $\Sigma_2^{\text{f}}$ of the loop integral $L_b$, and adding a mass renormalization vertex of the form of Fig. 10.5 with the *finite* vertex factor $\pm 2\pi i\,\delta m_2^{\text{f}}$ to the graph rules. The superscript f will henceforth be dropped: unless noted otherwise, $\Sigma_2$ will from now on denote the finite part of the loop integral $L_b$, and $\delta m_2$ a finite constant which is chosen such that $t_2^\pm$ contains no pole of second order at $p^2 = m^2$.

The discussion of the renormalized equations of motion for $\tilde{W}_2(p,\bar{q})$, in particular the rigorous definition of $N_2$ and $N_3$, follows the example of the photon case and need not be given here.

The asymptotic behaviour of $\Sigma_2(p)$ for $p \to \infty$ will be of interest later on. It can be found by considering $\Sigma_2(\lambda p, m)$ for $\lambda \to \infty$, $p$ fixed. The dependence on $m$ is essential and is therefore displayed explicitly. By introducing $\ell = \lambda k$ as a new integration variable in the loop, we find $\Sigma_2(\lambda p, m) = \lambda \Sigma_2(p, \frac{m}{\lambda})$. The first-order subtraction of the factor $(\not{p} - \not{\ell} - \frac{m}{\lambda})^{-1}$ of the integrand contains the denominator $(\ell + \frac{m}{\lambda})^2$. For $\lambda \to \infty$ it engenders a singularity of second order at the origin $\ell = 0$ which together with the singularity $(\ell^2)^{-1}$ of the photon propagator causes a logarithmic divergence. Therefore $\Sigma_2(\lambda p)$ behaves for $\lambda \to \infty$ like $\lambda \log \lambda$, and thus $t_2^+(\lambda p)$ like $\lambda^{-1} \log \lambda$. Apart from the logarithmic factor this agrees with the behaviour of $t_0^+$, in accordance with the requirement of maximal scaling degree enunciated in Sect. 8.4: maximal smoothness at small distances in $x$-space is equivalent to a minimal increase for large $p$ in momentum space. The same result holds also for the 2-photon function.

## The Vertex Function

As the last fundamental low-order example we consider the 3-point function $\tilde{W}_3(p,\mathsf{k},-\bar{q})$ in third order. The functions with permuted variables can be treated in the same way, but the ordering shown here is the most instructive one. The scaffoldings contributing to this function are shown in Fig. 10.6. The dots inside the loops of the graphs (a) indicate that the finite parts of those loop integrals must be taken, the crosses designate the finite mass renormalization vertex introduced in the previous subsection. With these conventions the graphs (a) are individually UV convergent. The fat vertex in graph (c) corresponds to the $R_1$-terms in the normal products (8.2). Its vertex factor is $\mp \frac{i}{\sqrt{2\pi}} r_1$ with $r_1$ being a coefficient in the expansion $R_1 = \sum_{\rho=1}^{\infty} r_\rho e^{2\rho+1}$. That this single vertex can represent terms in all three equations is due to the fact that the same constant $R_1$ occurs in the three equations. This is enforced by the consistency condition (8.10). This condition is satisfied for all graphs of Fig. 10.6.

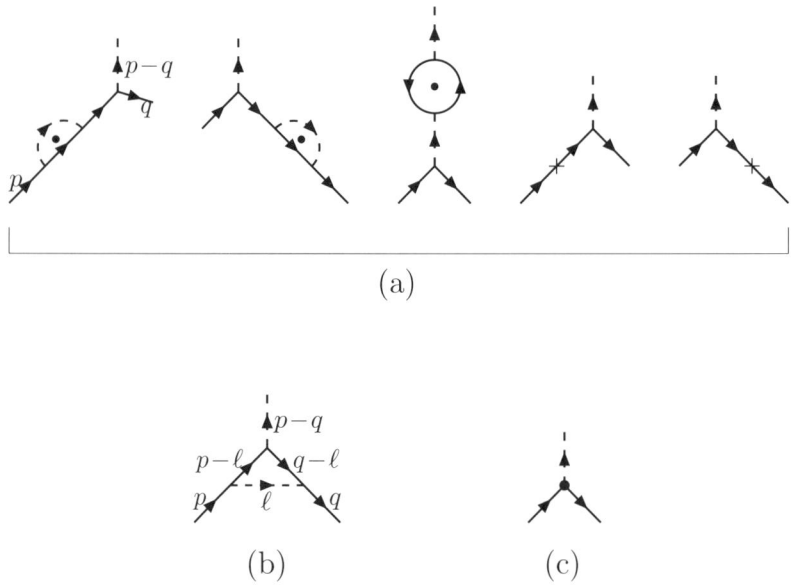

**Fig. 10.6a–c.** Scaffoldings of $\tilde{W}_3(p, k, -\bar{q})$.
The momentum assignment of the $\delta$-reduced rules is shown fully for graph (**b**) and partly for the first graph (**a**)

The possible sectorizations of these scaffoldings are clear and need not be shown explicitly. In order to avoid internal sectors we consider $\psi$ and $\bar\psi$ to be $T^-$ factors, $A^\mu$ a $T^+$ factor.

Problem A of the former cases (local singularities of the integrand) is also present here, albeit in a less severe form. We refrain from examining it, referring to its general solution in the following chapter. The loop integral of graph (b) is UV divergent if it occurs inside a sector: its integrand decreases for $\ell \to \infty$ like $|\ell|^{-4}$, which is just not sufficient for convergence. Following the 2-point example we define its finite part by subtracting its value at $p = q = 0$, which prescription is given a rigorous meaning by performing this subtraction in the integrand before integrating. The loop integral is a spinor matrix carrying also a vector index $\rho$. The only quantity with this behaviour available at $p = q = 0$ is $\gamma^\rho$. Hence the subtracted term is of the form $\pm i\alpha\,\gamma^\rho$, where the (divergent) coefficient $\alpha$ can be shown to be real. If we choose $r_1 = (2\pi)^{1/2}\alpha$ in graph (c), then graphs (b) and (c) sum to a modified graph (b) with a dot inside its loop, which signifies again taking the finite part of the loop integral.

The above choice of $r_1$ is enforced by the charge identity (7.26). A finite addition to $r_1$ would not destroy UV convergence, but it would violate the charge identity. That our choice does satisfy this identity is seen as follows. Denote the renormalized loop integral in a $T^+$ sector by $\Lambda^\rho(p,-\bar q)$. ($\Lambda$ is called the "vertex function" of third order.) Let $I_\Lambda^\rho(p,q,\ell)$ be the unsubtracted loop integrand and define $I'_\Lambda = (q-p)_\rho I_\Lambda^\rho(p,q,\ell)$. Then

$$(q-p)_\rho \Lambda_{\text{ren}}^\rho = \int d\ell\, (q-p)_\rho\, [I_\Lambda^\rho(p,q,\ell) - I_\Lambda^\rho(0,0,\ell)]$$
$$= \int d\ell\, [I'_\Lambda(p,q,\ell) - (q-p)_\rho I_\Lambda^\rho(0,0,\ell)] \,. \quad (10.34)$$

But the subtraction term in the second square bracket is simply the Taylor expansion to order 1 of $I'_\Lambda$ around $p = q = 0$. As in the proof of the WT identities (9.41) we find that

$$(q-p)_\rho I_\Lambda^\rho(p,q,\ell) = \frac{i}{2\pi}\left(I_\Sigma(p,\ell) - I_\Sigma(q,\ell)\right) , \quad (10.35)$$

where $I_\Sigma(p,\ell)$ is the unsubtracted integrand of the self-energy insertion $I_\Sigma(p)$. Expanding both sides of (10.35) into powers of $p$ and $q$ and dropping the terms of first order we find that (10.35) holds for the subtracted integrands, hence for the integrals,

$$(q-p)_\rho \Lambda^\rho(p,-\bar q) = \frac{i}{2\pi}\left(\Sigma(p) - \Sigma(q)\right) , \quad (10.36)$$

with $\Lambda$ and $\Sigma$ the renormalized vertex function and self-energy part, respectively. The proof of current conservation and of the charge identity is then a simple generalization of the proofs given in Sect. 9.5. The same holds for the WT identity (9.41) for the 3-point function in third order.

Before proceeding to renormalization in arbitrary orders, let us briefly mention a problem that occurs for the first time in $4^{th}$ order: the problem of *overlapping divergences*. Consider the 2-point electron function $t_4^+(p)$, in particular the contributions of the graphs shown in Fig. 10.7. In both cases the 8-dimensional $k$-$\ell$-integral without the external propagators is UV divergent,

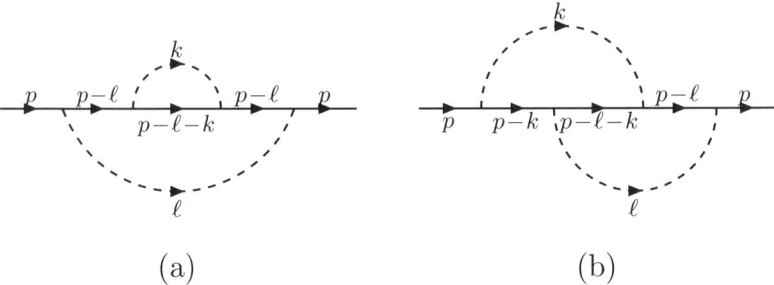

**Fig. 10.7a,b.** Examples of overlapping divergences.

because the integrands decrease for $(k, \ell) \to \infty$ only of order 7. In graph (a) the integral over the $k$-loop, for $\ell$ fixed, is already divergent in itself. The $\ell$-integral for fixed $k$ is convergent, but not the $\ell$-integral after the $k$-integration has been performed and renormalized. The correct procedure for defining the finite part is clear when we remember that our rules have been obtained by a recursive procedure. This suggests at once that the second-order $k$-loop be integrated over first and its finite part taken. It behaves for large $\ell$ like $(|\ell| \log |\ell|)$, hence the remaining $\ell$-integrand behaves as $(|\ell|^{-3} \log |\ell|)$. The finite part of the $\ell$-integral is then defined as before, by subtracting from its integrand the terms of $0^{th}$ and $1^{st}$ order in $p$. The logarithmic factor in the asymptotic expression does not invalidate this procedure.

For graph (b) both the $k$-loop and the $\ell$-loop integral are divergent by themselves. And neither of them is contained inside the other one; they occur at the same hierarchical level. This is the problem of "partially overlapping divergences". It cannot be solved as easily as that of graph (a). The correct procedure for defining the finite part of the subgraph formed by the two loops is the following. Let $I(p, k, \ell)$ be the integrand of this subgraph. Subtract first the $k$-loop part and the $\ell$-loop part independently. More explicitly: define $I_\ell(p, k, \ell)$ by setting $p = k = 0$ in the propagators of the $\ell$-loop, $I_k(p, k, \ell)$ analogously, and form $I'(p, k, \ell) = I(p, k, \ell) - I_k(p, k, \ell) - I_\ell(p, k, \ell)$. Then expand $I'$ in powers of $p$ and drop the first two terms. The result is the renormalized integrand $I_{\text{ren}}(p, k, \ell)$, and $\int dk \, d\ell \, I_{\text{ren}}(p, k, \ell)$ can be shown to be UV divergent, as are the subintegrals $\int dk \, I_{\text{ren}}$ and $\int d\ell \, I_{\text{ren}}$. We refrain from a fuller discussion of this point and of its connection with renormalization, since such a discussion would hardly be more enlightening than the corresponding considerations in the general case, which is studied in the following sections.

## 10.2 The General Case

The methods developed in the preceding section with the help of low-order examples must now be systematized, so as to become applicable to arbitrarily high orders. We use the BPHZ method as formulated by Zimmermann [Zi 67, Zi 69, Zi 71], based on previous work by Bogoliubov, Parasiuk, and Hepp. Our strategy is that followed in Chap. 9: we do not try to derive the correct prescriptions constructively, but simply write them down and verify them afterwards. This verification being easier in $x$-space than in $p$-space, it will be postponed to the next section. However, the general prescription is simpler to explain in $p$-space, hence we start with that space.

The order of proceeding is now the reverse of Sect. 10.1. We consider a graph $G$, define its UV finite part, and show that the result also satisfies the renormalization conditions. The graph $G$ is originally formed according to the $p$-space descriptions of Sect. 9.3, but also including finite mass renormalization vertices of the type introduced above (see Fig. 10.5), now with contributions $\delta m_\sigma$ of every even order $\sigma > 0$. The $\delta m_\sigma$ are chosen such that the 2-point electron function has the weakest possible singularity at the mass shell. Moreover, we admit the presence of external points representing the composite fields $\bar{\psi}\gamma^\mu\psi$ and $\slashed{A}\psi$. They are normal 3-prong vertices with an incident external line (photon or fermion, respectively) with momentum $p$, but with this external line having been "amputated" by multiplication with $-p^2$ or $(\slashed{p} - m)$, respectively.

In treating the UV problem we ignore the possible complications due to the local singularities of the integrand. The solution of this problem will be provided in Chap. 11.

As in the low-order examples it is easy to see that the UV problems are confined to the individual sectors. The integral over a loop containing cross lines is UV convergent, because it extends over a compact region of integration if the variables external to the loop are kept constant. We can therefore confine our attention to a single sector, which we choose to be $T^+$. In other words: we consider a one-sector graph $G$ as it appears in the graphical representation of a time-ordered function $\tilde{\tau}^+$, i.e. an ordinary Feynman graph or a subgraph thereof.

We need to introduce a number of definitions. A *"subgraph"* $\Gamma$ of $G$ consists of any number of points of $G$ and of lines connecting them. But it does not contain any lines with only one end in $\Gamma$. $\Gamma$ is called *"one-particle irreducible"*, abbreviated 1PI, if it is connected and cannot be split into two disconnected components by cutting a single line. We assume the full graph $G$ to be 1PI, because a reducible graph is UV convergent if all its 1PI subgraphs are. $\Gamma$ is called a *"proper subgraph"* if it is 1PI and contains at least two points. In order to determine the UV behaviour of a proper subgraph we need to study its integrand at large values of the integration variables. For the moment we use the non-$\delta$-reduced graph rules, with all lines car-

## 10. Renormalization

rying independent variables and the vertex factors containing $\delta^4$'s ensuring momentum conservation. Let $\Gamma$ be a proper subgraph with $V$ vertices, $L_\gamma$ photon lines, and $L_e$ electron lines, which is connected to the rest of $G$ by $A_\gamma$ photon lines and $A_e$ electron lines. A line $\notin \Gamma$ with both ends in $\Gamma$ gives a contribution 2 to $A$. The momenta $p_i$ of these external (to $\Gamma$) lines are kept fixed, the internal momenta are integrated over. The integration extends over a $4(L_\gamma + L_e)$-dimensional space. We need to know the behaviour of the integrand at large values of the internal variables. Consider first the vertex $\delta$'s. Using $\delta^4(a-b)\,\delta^4(b-c) = \delta^4(a-c)\,\delta^4(b-c)$ repeatedly, we find that one of these factors can be replaced by $\delta^4\left(\sum \pm p_i\right)$, which does not contain any internal variables. The remaining $(V-1)$ $\delta^4$'s do necessarily contain internal variables $\ell_j$. Let $P$ be a partial sum over external variables, $L$ a nonempty partial sum of $\ell_j$'s. Then $\delta^4(P + \rho L) = \rho^{-4}\delta^4\left(\frac{P}{\rho} + L\right)$, which decreases for $\rho \to \infty$ like $\rho^{-4}$ unless $L = 0$. A photon propagator with momentum $\ell$ decreases like $|\ell|^{-2}$, an electron propagator like $|\ell|^{-1}$. Hence in a generic direction $\{\rho \ell_j\}$, $\rho \to \infty$, the integrand decreases like $\rho^{-4(V-1)-2L_\gamma - L_e}$.

The difference

$$d(\Gamma) = 4(L_\gamma + L_e) - 2L_\gamma - L_e - 4(V-1) = 2L_\gamma + 3L_e - 4V + 4 \quad (10.37)$$

between the number of integration variables and the degree of decrease is called the *"dimension of $\Gamma$"*. If $d(\Gamma) \geq 0$ the integral is divergent. But $d(\Gamma) < 0$ does not guarantee UV convergence, because on certain submanifolds of the region of integration the decrease of the integrand may be slower than indicated above: some subintegrals may diverge. More exactly, choose a set $\ell_1, \ldots, \ell_s$ of internal variables and consider the integral over them, the other variables being kept fixed. Because of momentum conservation at the vertices, the selected $\ell_i$ can vary freely only if their lines form closed loops. This implies that they form one or several proper subgraphs, and the integrals over these subgraphs may be UV divergent even if $d(\Gamma) < 0$. A proper subgraph $\Gamma$ with $d(\Gamma) \geq 0$ is called a *"renormalization part"*, abbreviated R-part. A $\Gamma$ with $d(\Gamma) < 0$ is called *"superficially convergent"*. A R-part is *"primitively divergent"* if all its subgraphs are superficially convergent.

The renormalizability of QED rests on the fact that an estimate for $d(\Gamma)$ can be given which depends exclusively on the numbers $A_\gamma$, $A_e$, of the external lines of $\Gamma$.

**Theorem 10.1.** *Let $\Gamma$ be a proper subgraph with $A_\gamma$ external photon lines and $A_e$ external electron lines. Then*

$$d(\Gamma) \leq 4 - A_\gamma - \frac{3}{2} A_e \ . \quad (10.38)$$

*The equality sign holds if $\Gamma$ contains no $\delta m$-vertices.*

The proof is similar to the proof of (9.21). Assume first that there are no $\delta m$-vertices present. Then, noticing that one photon line and two electron lines are incident at each vertex, we find

$$V = 2L_\gamma + A_\gamma = L_e + \frac{1}{2} A_e$$

and hence

$$d(\Gamma) = V - A_\gamma + 3V - \frac{3}{2} A_e - 4V + 4 ,$$

the desired result. A $\delta m$-vertex introduces four additional variables but also an additional $\delta^4$ and an additional propagator, which improves the estimate for $d(\Gamma)$ by 1.

The only values of $(A_\gamma, A_e)$ which lead to superficial divergence are (2,0), (0,2), (1,2), (4,0). The case (3,0) does not occur because of Theorem 9.1 (Furry's theorem). The last case, that of a 4-photon graph, is somewhat special. Let us go over to the $\delta$-reduced rules. Let $F(p_1,\ldots,p_4)$, defined on $\{p_1 + \cdots + p_4 = 0\}$, be the sum over all proper $T^+$ graphs of order $\sigma$ with the external photon momenta $p_1,\ldots,p_4$. We can write $F$ as

$$F(p_1,\ldots,p_4) = F(0,\ldots,0) + \int_0^1 d\rho \, \frac{\partial}{\partial \rho} F(\rho p_1,\ldots,\rho p_4) .$$

The derivation in the last term reduces the degree of the integrands by 1, so that this term becomes superficially convergent, or even convergent if its subdivergences have been properly removed by the procedure to be explained presently. The UV divergences of $F$ then reside in $F(0,\ldots,0)$. But the WT identities (9.41), which are needed for current conservation, tell us that $\rho p_1^\mu F(\rho_\mu p_1,\ldots) = 0$, hence $p_1^\mu F(\rho_\mu p_1,\ldots) = 0$, hence $p_1^\mu F(\mu 0,\ldots) = 0$. But $F(\mu_1 p_1,\ldots,\mu_4 p_4)$ transforms as a tensor, and the only covariant tensor available at $p = 0$ is $g_{\mu\nu}$, so that $F(0,\ldots,0)$ must be of the form

$$F(\mu_1 0,\ldots,\mu_4 0) = c \left( g_{\mu_1\mu_2} g_{\mu_3\mu_4} + \text{permuted} \right) .$$

This expression is not annihilated by contraction with $p_1^{\mu_1}$ unless $c = 0$: we can expect that $F(p_1,\ldots,p_4)$ is actually UV convergent. However, this result is a consequence of cancellations between different graphs (see the proof of the WT identities), so that it does not apply to individual graphs. We must therefore retain the case (4,0) on our list of R-parts.[4]

We can now amend the graph rules such that they yield UV finite results. As explained before we need only consider a single sector, which we assume to be $T^+$. We use the formal rules of Chap. 9 in the $\delta$-reduced form. Let $\Gamma$ be a R-part with $n + 1$ external variables, and let its integrand be $I_\Gamma(p_1,\ldots,p_n,\ell_1,\ldots,\ell_m)$, where the $p_i$ form a complete set of independent external variables, the $\ell_j$ a maximal set of independent internal variables of

---

[4] We note for later reference that this argument applies to any 1PI subgraph with only photonic external lines: the integrands of sums over such subgraphs vanish at vanishing external momenta, and this holds also for the integrated subgraphs after renormalization.

$\Gamma$. We define $t_\Gamma I_\Gamma$ as the Taylor expansion of $I_\Gamma$ in the external $p_i$ up to order $d(\Gamma)$. $I_\Gamma$ depends on the $p_i$ through factors of the form $\left((\sum p_i + \sum \ell_j)^2\right)^{-1}$ and $(\sum \not{p}_i + \sum \not{\ell}_j - m)^{-1}$, the sums running over non-empty subsets of $\{p_i\}$ and $\{\ell_j\}$, respectively. Differentiating such a factor with respect to a $p_i$ improves its decrease for $|\ell| = \sum_j |\ell_j| \to \infty$ by one degree. Thus, if $\Gamma$ is primitively divergent, the "finite part"

$$I_\Gamma^{\rm f}(p_i, \ell_j) = I_\Gamma - t_\Gamma I_\Gamma =: (1 - t_\Gamma) I_\Gamma \tag{10.39}$$

is UV integrable in the internal variables $\ell_j$, i.e.

$$J_\Gamma^{\rm f}(\ldots, p_i, \ldots) = \int \prod d\ell_j\, I_\Gamma^{\rm f}(p_i, \ell_j) \tag{10.40}$$

exists, in complete analogy to the low-order examples discussed in the preceding section.

Handling the general R-parts is more involved. A *"forest"* $U$ of the graph $G$ is defined as a set $U = \{\Gamma_1, \ldots, \Gamma_r\}$ of R-parts, such that no two of its elements are partially overlapping. This means that for $i \neq j$ either $\Gamma_i \cap \Gamma_j = \emptyset$, or $\Gamma_i \subset \Gamma_j$, or $\Gamma_j \subset \Gamma_i$. Let $I_G(p_1, \ldots, p_n, \ell_1, \ldots, \ell_m)$ be the integrand of $G$, where now the $p_i$ are a complete set of independent external variables of $G$, the $\ell_j$ a complete set of independent internal variables. We claim

**Theorem 10.2.** *The integrand*

$$I_G^{\rm f} = \sum_U \prod_{\Gamma_i \in U} (-t_{\Gamma_i}) I_G \tag{10.41}$$

*is UV integrable, i.e. the integral*

$$J_G(p_1, \ldots, p_n) = \int \prod d\ell_j\, I_G^{\rm f}(p_1, \ldots, \ell_m) \tag{10.42}$$

*is UV convergent.*

The $U$-sum in (10.41) extends over all possible forests of $G$, including the empty forest. In the product $\prod_U$ the Taylor operator $t_{\Gamma_i}$ is applied before $t_{\Gamma_j}$ if $\Gamma_i \subset \Gamma_j$, while the order is irrelevant if $\Gamma_i \cap \Gamma_j = \emptyset$. The operator $t_{\Gamma_i}$ acts only on the propagators in $\Gamma_i$, not on propagators external to $\Gamma_i$ even if they depend on the external momenta of $\Gamma_i$. $J_G$ exists as a distribution in $p_1, \ldots, p_n$. Equation (10.41), which is due to Zimmermann, is called the "forest formula".

The proof of this theorem is easier to give for its Fourier transformed version and is therefore postponed to the subsequent section. But it is convenient to clear up one detail which is easier to handle in $p$-space. Suppose we have a graph $G$ with several sectors, e.g. two, $S_1$ and $S_2$, for simplicity. Then we claim that the corresponding integrand $I_G$ is also rendered UV integrable by the forest formula. This follows if Theorem 10.2 has been proved for individual sectors. We first define R-parts of $G$ as above but without regard to the nature of their lines, so that they may include cross lines. Let $\Gamma$ be a R-part overlapping both $S_1$ and $S_2$. We show that $t_\Gamma = 0$ in this case. Since $\Gamma$ is 1PI it contains at least two cross lines. Let $c_1, \ldots, c_r$ be the momenta of its cross lines, defined such that they flow from $S_1$ to $S_2$. Then all $c_i$ lie on a positive mass shell. Let $Q$ be the sum of the external momenta of $\Gamma$ in $S_1$. Apply $t_\Gamma$ to the integrand $I_\Gamma$ of $\Gamma$. $t_\Gamma I_\Gamma$ is evaluated at $Q = 0$. By momentum conservation we must then have $\sum c_i = 0$. This is impossible if at least one of the cross lines is fermionic. Hence in that case we have $t_\Gamma I_\Gamma = 0$. If all cross lines are bosonic, then $\prod_{i=1}^{r} \delta_+(c_i)$ vanishes at $Q = \sum c_i = 0$ if $r > 2$. More exactly this means that

$$C_r(Q) = \int \prod_{i=1}^{r} [dc_i\, \delta_+(c_i)]\, \delta^4\left(Q - \sum_i c_i\right)$$

vanishes for $Q \to 0$ like $|Q|^{2(r-2)}$, as is seen by a simple scaling argument. For $r = 2$ we find

$$C_2(Q) = \text{const} \cdot \theta(Q_0)\,\theta(Q^2)\,,$$

which remains bounded for $Q \to 0$ but does not vanish. This case can only occur if $\Gamma$ is a 4-photon part with exactly two of its external lines in $S_1$. Then $\Gamma \cap S_1$ is also a R-part and as such gets subtracted at $Q = c_1 = c_2 = 0$. After this subtraction $I_\Gamma$ vanishes at $Q = 0$, so that $t_\Gamma I_\Gamma = 0$. If $\Gamma$ is a 2-photon part, then $t_\Gamma$ involves derivations up to second order. Then the $|Q|^2$ behaviour of $C_3$ is not sufficient for $t_\Gamma I_\Gamma$ to vanish. But we find again that $\Gamma \cap S_1$ is a R-part, and its subtraction ameliorates the situation sufficiently to ensure $t_\Gamma I_\Gamma = 0$. As a result we find that in applying the forest formula to $I_G$ the forests containing such hybrid subgraphs can be omitted. The remaining forest sum is of the form $\sum_U = \sum_{U_1} \cdot \sum_{U_2}$ where the two factors sum over all forests of $S_1$ and $S_2$, respectively. This removes the UV divergences in both sectors, and the cross lines have already been shown to be UV harmless. That this procedure insures the validity of the renormalization conditions is shown as in the examples of Sect. 10.1. In particular, current conservation still holds for the subtracted R-parts. This is obvious for 2-line parts. For the 4-photon case it has been shown before, and for the vertex function it can be demonstrated by a simple generalization of the proof given in 3$^{\text{rd}}$ order (see (10.34)–(10.36)). Generally, the point is that the renormalization conditions involve graphs with $d(\Gamma) \geq 0$ only at the origin, and are thus saturated by the zero-order terms, because the UV finite integrals in higher orders vanish at the origin. The only exception is the condition of minimal singularity for the 2-point fermion function. Self-energy insertions $\Sigma_\sigma(p)$ occur only for even $\sigma$. Their renormalized value vanishes at $p = 0$. But in order to prevent the adjacent external propagators from combining into a pole of second order, $\Sigma_\sigma$ must vanish at the mass shell $p^2 = m^2$. This can be achieved, as shown in the case $\sigma = 2$, by adding an appropriate *finite* mass renormalization term $\delta m_\sigma$ of order $\sigma$.

That the result of our prescription also satisfies the W-properties referring to $p$-space is shown by the same arguments as used in Chap. 9 for the unrenormalized expressions.

## 10.3 Renormalization in Configuration Space

In $x$-space the UV divergences are due to the local singularities of the graph integrands, while the IR problem is one of their behaviour at large distances. Consider again the 2-point graphs of Fig. 10.1, but this time as graphs in $x$-space. Let $u, v$, be the variables of the two internal vertices and define $\xi = u - v$. The loop propagators of graph (a) contribute a product

$$P(\xi) = \big(d_1 \Delta_+(\xi)\big) \big(d_2 \Delta_+(\xi)\big) \tag{10.43}$$

to the integrand, with $d_i$ being constants or first derivatives. But according to Sect. 5.4, $\Delta_+(\xi)$ is the boundary value of a function $F(\zeta)$ which is analytic in the backward tube $\mathcal{T}_-$ and which diverges at most of finite order if $\zeta$ approaches a real point $\xi$ from inside $\mathcal{T}_-$. The same is true for $P(\xi)$, and thus this product is defined as a distribution as a result of Theorem 5.4.

In the case of graph (b) the $\Delta_+$ factors of (10.43) are replaced by $\Delta_F(\xi)$. But for $\xi^0 > 0$ we have $\Delta_F(\xi) = \Delta_+(\xi)$, hence $P(\xi)$ is still well defined there. The same holds analogously for $\xi^0 < 0$ where $\Delta_F(\xi) = -\Delta_-(\xi)$. For $\xi^0 = 0$ but $\vec{\xi} \neq 0$ $\Delta_F(\xi)$ is real analytic, so that in this region there is also no problem defining the product $P$. But at $\xi = 0$ these arguments break down. There $\Delta_F(\xi)$ diverges of second order if approached from a regular—i.e. not lightlike—direction, $d_i \Delta_F$ even of third order if $d_i$ is a derivation. Hence we have in $P$ a singularity of $6^{\text{th}}$ order at the worst. It is not integrable in the ordinary sense of integration theory. But in contrast to the $\Delta_+ \cdot \Delta_+$ case we also have no obvious way of defining $P$ as a distribution in the neighbourhood of $\xi = 0$. This is the UV problem in $x$-space.

Notice, however, that at the coincidence of the two end points of a single sector line, e.g. of an external line in graph (a) of Fig. 10.1, we have only a singularity of at most third order, which is still integrable. Therefore we find that like in $p$-space UV problems are exclusively due to proper subgraphs.

To come to the general case, let $G$ be a sector graph and $\Gamma \subset G$ a proper subgraph. Assume that $\Gamma$ intersects several sectors, e.g. two: $S_L$ and $S_R$. Let $u_1, \ldots, u_L$ be the variables of $\Gamma \cap S_L$, $v_1, \ldots, v_R$ those of $\Gamma \cap S_R$. The integrand $I_\Gamma$ of $\Gamma$ is translation invariant and can be written as a function of $L + R - 1$ independent difference vectors. For these we choose $\{\xi_i = u_1 - u_i, i = 2, \ldots, L\}$, $\{\eta_j = v_j - v_R, j = 1, \ldots, R-1\}$, and $\zeta = u_1 - v_R$. The propagators of the cross lines depend on these variables via factors $\Delta_+(\zeta + \eta_j - \xi_i)$ or derivatives thereof for some pair $(i, j)$, admitting the choices $i = 0$ or $j = R$ with $\xi_0 = \eta_R = 0$. Considered as a function of $\zeta$ these propagators are boundary values of analytic functions in $\zeta \in \mathcal{T}_-$, hence this is also true for their product. Furthermore this product satisfies the boundedness conditions of Theorem 5.4, so that it is everywhere defined as a distribution in $\zeta$. Problems with the definition of $I_\Gamma$ as a distribution can therefore only occur concerning the variables $\xi_i$ and $\eta_j$, i.e. inside the sectors $S_L$ and $S_R$. Thus we need only consider the corresponding subgraphs of $\Gamma$. Or, speaking more generally, we need only consider proper subgraphs $\Gamma$ inside individual sectors of $G$. This agrees with the analogous conclusion reached in $p$-space.

Take now such a proper subgraph $\Gamma$ of a $T^+$ sector. Let $u_1, \ldots, u_n$ be its variables. Consider the integrand $I_\Gamma(u_1, \ldots, u_n)$ on a domain $D$ in which no two time variables $u_j^0$ coincide anywhere. We can then order the variables chronologically. In the simplest but typical case we have $u_1^0 > \cdots > u_n^0$. By the definition (5.115) of $\Delta_F$ all the propagators of $I_\Gamma$ contain factors $\Delta_+(u_i - u_j)$ with $i < j$ or derivatives of them. Written as a function of the variables $\xi_i = u_i - u_{i+1}$, $i = 1, \ldots, n-1$, this product is the boundary value of an analytic function in $\mathcal{T}_-^{\otimes(n-1)}$ satisfying the assumptions of Theorem 5.4. Hence $I_\Gamma$ is defined on $D$ as a distribution. This remains true where two or more $u_i^0$ coincide but their spatial parts $\vec{u}_i$ are different. Hence problems of existence occur only where several $u_i$ coincide as 4-vectors. Consider the case that $all$ the $u_i \in \Gamma$ coincide: $u_1 = u_2 = \cdots = u_n$, assuming for the moment that the problems due to coincident subsets have already been solved. Whether $I_\Gamma$ is still defined as a distribution at this critical manifold depends on the strength of the singularity found there. This singularity is of order $2L_\gamma + 3L_e$ if $\Gamma$ contains $L_\gamma$ photon lines and $L_e$ electron lines. The $\xi$-space is $4(n-1)$-dimensional. Hence the singularity is integrable if $d(\Gamma) = 2L_\gamma + 3L_e - 4(n-1) < 0$. Not surprisingly, this is the criterion that has already been found in $p$-space: $d(\Gamma)$ is the dimension of $\Gamma$ defined in (10.37).

This reasoning may be less than convincing because $I_\Gamma$ is singular not only at $u_1 = \cdots = u_n$ but wherever a propagator variable $u_i - u_j$ is lightlike. Our way of arguing seems to mix "existence as a distribution" and "existence as an integrable function" in an objectionable way. In order to clear this point up let us look at the special case $J = \int_{\xi^0 \geq 0} d\xi \, P(\xi) \, f(\xi)$, where e.g. $P(\xi) = \partial \Delta_F(\xi) \, \partial \Delta_F(\xi)$ with the $\partial$ being first derivatives, and $f \in \mathcal{S}$. $P$ is for $\xi^0 > 0$ a boundary value of an analytic function and has at $\xi = 0$ a singularity of 6$^{\text{th}}$ order. From the proof of Theorem 5.4. we find that $P(\xi) = \partial_0^6 F(\xi)$ where $F(\xi)$ is $C^\infty$ for $\xi^2 < 0$ and continuous elsewhere except possibly at $\xi = 0$, where it may have a logarithmic singularity. Repeated integration by parts yields

$$J = \int_{\xi^0 \geq 0} d\xi \, F(\xi) \, \partial_0^6 f(\xi) + \sum_{\alpha=0}^5 \pm \int d^3\xi \partial_0^\alpha f(\xi) \, \partial_0^{5-\alpha} F(\xi) \Big|_{\xi^0 = 0}.$$

The first term exists. $\partial_0^{5-\alpha} F(\xi)$ is for $\xi^0 = 0$ singular at $\vec{\xi} = 0$ of order $5 - \alpha$, hence the $\alpha$-integrals exist for $\alpha > 2$ but diverge for general $f$ if $\alpha \leq 2$. But $J$ exists if $f$ vanishes at $\xi = 0$ together with all its derivatives of first and second order. This agrees with the conclusion reached above in a seemingly dubious way. This consideration can be extended to the general case of more than two coinciding variables.

The general situation is analogous to that encountered in $p$-space. Consider a sector graph $G$ constructed according to the rules of Sect. 9.3, with the unrenormalized integrand $I_\Gamma$ defined there, except that finite mass renormalization vertices appear as a new element. They are of the form shown in Fig. 10.5 and carry the vertex factor $\pm i\,\delta m_\sigma$ in the $T^\pm$ sectors. $\delta m_\sigma$ has the value explained in the preceding section. It must be computed in $p$-space because the renormalization conditions are formulated in that space. The UV problem is the problem of defining a renormalized finite part $I_G^f$ of the integrand $I_G$, which exists as a distribution in the variables (external and internal) of $G$. $I_G$ is so defined except possibly where two or more vertex variables of the same sector coincide. Subgraphs, proper subgraphs, the dimension of a subgraph, R-parts, and forests, are defined like in $p$-space. For a $N$-product vertex we use the following convention: the amputated line joining it is still included in $G$ with a propagator $\pm i\,\delta^4(x-u)$, where $x$ is the argument of a field $N_i(x)$ and $u$ the variable of an ordinary 3-prong or $\delta m$ vertex. Such $N$-lines occur only inside sectors, not as cross lines, and they cannot lie in a R-part.

Let $\Gamma$ be a R-part with dimension $d(\Gamma)$ and $\Gamma' = G\backslash\Gamma$ its complement in $G$. Let $y_1,\ldots,y_r$, be the variables of the $\Gamma$-vertices which are directly connected by lines to $\Gamma'$, $z_1,\ldots,z_s$, those of the other $\Gamma$-vertices, $x_1,\ldots,x_t$, those of the $\Gamma'$-vertices, including the external variables of $G$. Let $I_\Gamma(y_i, z_j)$ be the integrand of $\Gamma$: the product of its propagators and vertex factors. And let $E_\Gamma(x_k, y_i)$ be the integrand of $\Gamma'$. Its $y$-dependence comes from the lines connecting $\Gamma$ to $\Gamma'$. Define $\bar{y} = \frac{1}{r}\sum y_j$, $\hat{y}_i = y_i - \bar{y}$, and $\nabla_i$ as the gradient with respect to $y_i$. Then we define the $x$-space version of the subtraction operator $t_\Gamma$ as acting on $E_\Gamma$ by

$$t_\Gamma E_\Gamma(x_k, y_i) = \sum_{\rho=0}^{d(\Gamma)} \frac{1}{\rho!} \sum_{i_1,\ldots,i_\rho=1}^{r} \hat{y}_{i_1}\cdots\hat{y}_{i_\rho} \nabla_{i_1}\cdots\nabla_{i_\rho} E_\Gamma(x_k, y_i)\bigg|_{y_i=\bar{y}}. \quad (10.44)$$

The scalar product of each pair $\hat{y}_{i_\alpha}$, $\nabla_{i_\alpha}$ must be taken. $t_\Gamma E_\Gamma$ is formally the Taylor expansion of $E_\Gamma$ in $\{y_j\}$ to order $d(\Gamma)$ around the point $y_1 = \cdots = y_r = \bar{y}$. Of course $E_\Gamma$ is not differentiable but a singular function. As such we must also interpret the right-hand side of (10.44).

That this $x$-space version of $t_\Gamma$ corresponds indeed to the earlier $p$-space version is seen by (formal) Fourier transform. Since this involves lengthy calculations we will only indicate the argument, calculating formally, for the $\rho = 1$ terms, which are typical. We are only interested in the variables $y_i$, suppressing the $x$'s and $z$'s. Consider then the $\rho = 1$ term $J_1$ in the integral

$$J = \int \prod_1^r dy_j \, \text{tr}_\Gamma E(y_1, \ldots, y_r) \, I(y_1, \ldots, y_r) \,. \tag{10.45}$$

The $z$-variables in $I$ are assumed to have been integrated over, after possible divergences due to smaller R-parts have been subtracted. We write $E$ and $I$ as Fourier integrals, omitting uninteresting factors $2\pi$:

$$E(y_1, \ldots) = \int \prod_i dp_i \, \exp\left(i \sum_1^r p_i y_i\right) \tilde{E}(p_1, \ldots, p_r) \,,$$

$$I(y_1, \ldots) = \int \prod_i dq_i \, \exp\left(i \sum_1^r q_i y_i\right) \delta^4\left(\sum_1^r q_i\right) \hat{I}(q_1, \ldots, q_{r-1}) \,. \tag{10.46}$$

The special form $\delta^4 \cdot \hat{I}$ of $\tilde{I}$ is due to the translation invariance of $I$. We obtain

$$J_1 = \sum_{\alpha=1}^r \int \prod_1^r dp_i \, \tilde{E}(p_1, \ldots, p_r) \, p_\alpha \int \prod_1^r dq_j \, \delta^4\left(\sum_1^r q_j\right) \hat{I}(q_1, \ldots, q_{r-1})$$

$$\times \left(\partial_\alpha - \frac{1}{r}\sum \partial_j\right) \prod_1^r \delta^4\left(q_j + \frac{1}{r}\sum_1^r p_i\right) \,,$$

with $\partial_\alpha$ the gradient with respect to $q_\alpha$. We move these $q$-derivatives over to $\delta^4 \hat{I}$ by partial integration. They act only on $\hat{I}$, because $\left(\partial_\alpha - \frac{1}{r}\sum \partial_j\right)\delta^4\left(\sum q_s\right) = 0$. Noticing that

$$\delta^4\left(\sum q_i\right) \prod_j \delta^4\left(q_j + \frac{1}{r}\sum p_i\right) = \delta^4\left(\sum p_i\right) \prod \delta^4(q_j)$$

and that $\sum_1^r p_\alpha \left(\partial_\alpha - \frac{1}{r}\sum_1^r \partial_j\right) = \sum_1^{r-1} p_\alpha \partial_\alpha$ at $\sum p_\alpha = 0$ if applied to the $q_r$-independent function $\hat{I}$, we find after integration over $q_j$

$$J_1 = \int \prod_1^r dp_i \, \tilde{E}(-p_1, \ldots, -p_r) \, \delta^4\left(\sum p_i\right) \sum_{\alpha=1}^{r-1} p_\alpha \partial_\alpha \hat{I}(p_1, \ldots, p_{r-1})\Big|_{p_j=0} \,,$$

which is the $\rho = 1$ subtraction in $p$-space.

Let now $G$ be a sector graph with external variables $x_1, \ldots, x_n$, internal variables $u_1, \ldots, u_m$, and $I_G$ its unrenormalized integrand. We define its UV finite part $I_G^f(x_1, \ldots, u_m)$ as the Fourier transform of its $p$-space form (10.41). It reads exactly the same,

$$I_G^f = \sum_U \prod_{\Gamma_i \in U} (-t_{\Gamma_i}) \, I_G \,, \tag{10.47}$$

but with all expressions now referring to $x$-space. We claim that this expression is free of UV divergences. More explicitly this means the following.

160   10. Renormalization

**Theorem 10.3.** *The expression $I_G^f(x_1, \ldots, x_n, u_1, \ldots, u_m)$ of (10.47) defines a tempered distribution.*

In order to prove this theorem we show that the integral of $I_G^f$ over any test function $f(x_1, \ldots, u_m) \in \mathcal{S}$ exists. We will not show explicitly that this integral depends continuously on $f$, the proof of this fact by an extension of our arguments not presenting any serious difficulties. For the sake of simplicity we consider only test functions $f$ with compact support. From the explicit form of the propagators it is clear that they have no strong increase at infinity, so that no catastrophic behaviour at infinity is to be feared: if $I_G^f$ is a distribution, then it is tempered. Moreover it has been shown after Theorem 10.2 that the forest formula applied to a sector graph factorizes with respect to sectors. We need therefore only prove the theorem for one-sector graphs. More exactly we assume that the vertices of $G$ and the lines connecting them form a single sector, e.g. a $T^+$ sector, while the external lines may be cross lines or amputated lines as introduced in connection with $N$-products. We prove the theorem by induction with respect to the number $V$ of vertices of $G$. The theorem is obviously true for $V = 1$, because in that case there exist no R-parts. We must show that the theorem holds for $V$ vertices if it holds for fewer than $V$ vertices.

We first prove a geometrical lemma.

**Lemma 10.4.** *Let $U = \{u_1, \ldots, u_m\}$ be the set of vertex variables of $G$, $|u_i - u_j|$ the Euclidean distance between $u_i$ and $u_j$. Consider the manifold*

$$S_\lambda = \left\{ U : \sum_{i<j} |u_i - u_j|^2 = \lambda^2, \quad \lambda > 0 \right\}. \tag{10.48}$$

*Choose $U \in S_\lambda$ and let $\delta$ be a real number with*

$$0 < \delta < \frac{1}{2} \binom{m}{2}^{-\frac{1}{2}} \frac{1}{m-1}.$$

*Then there exists a proper subset $U' \subset U$ such that*

$$|u_i - u_j| > \delta\lambda \quad \forall \quad u_i \in U', \, u_j \in U^c = U \backslash U'. \tag{10.49}$$

**Proof.** Condition (10.48) means that

$$\sum_{\mu=0}^{3} \sum_{i<j} |u_i^\mu - u_j^\mu|^2 = \lambda^2.$$

Hence we have

$$\sum_{i<j} |u_i^\mu - u_j^\mu|^2 \geq \frac{\lambda^2}{4} \tag{10.50}$$

## 10.3 Configuration Space

for at least one value of $\mu$. We order the corresponding components $u_i^\mu$ according to their size. Assume e.g. that this ordering happens to be the natural ordering: $u_1^\mu \geq u_2^\mu \geq \cdots \geq u_m^\mu$. Assume that $u_i^\mu - u_{i+1}^\mu \leq \delta\lambda$ for all $i$. Then $u_i^\mu - u_j^\mu \leq (m-1)\delta\lambda$ for all $i > j$, and thus $\sum_{i<j} |u_i^\mu - u_j^\mu|^2 \leq \binom{m}{2} \delta^2\lambda^2 (m-1)^2$ in contradiction to (10.50). Hence there exists a $u_I$ such that $u_I - u_{I+1} > \delta\lambda$. Define $U' = \{u_i : i \geq I\}$. Then $u_i^\mu - u_j^\mu > \delta\lambda$ for $u_i \in U'$, $u_j \in U^c$, and thus $|u_i - u_j| > \delta\lambda$, which proves the lemma.

**Proof of Theorem 10.3.** For the proper subset $U' \subset U$ and $\lambda > 0$ define

$$S_\lambda^{U'} = \{\{u_i\} : |u_i - u_j| > \delta\lambda \ \forall \ u_i \in U', u_j \in U^c\} \tag{10.51}$$

with $U^c = U \setminus U'$. From Lemma 10.4 we find that the union $\bigcup_{U'} S_\lambda^{U'}$ covers the full $U$-space. Take the case $\lambda = 1$. We can find $C^\infty$ functions $\chi_{U'}(u_1 - u_2, \ldots, u_{m-1} - u_m)$ with support in $S_1^{U'}$ and $|\chi_{U'}| \leq 1$, which form a decomposition of the unity on $S_1$:

$$\sum_{U'} \chi_{U'}(u_1 - u_2, \ldots, u_{m-1} - u_m) \equiv 1 . \tag{10.52}$$

Defining $\lambda$ by $\lambda^2 = \sum_{i<j} |u_i - u_j|^2$ and $\psi_{U'}$ by

$$\psi_{U'}(u_1 - u_2, \ldots, u_{V-1} - u_V) = \chi_{U'}\left(\frac{u_1 - u_2}{\lambda}, \ldots, \frac{u_{V-1} - u_V}{\lambda}\right) \tag{10.53}$$

we obtain a decomposition of the unity on the full $U$-space. The $\psi_{U'}$ are bounded by 1, and they are $C^\infty$ except at the origin $u_1 = u_2 = \cdots = u_V$ of the difference space. There they are discontinuous but still bounded. For any fixed value of $\lambda > 0$ the support of $\psi_{U'}$ lies in $S_\lambda^{U'}$. We multiply $I_G^f(X, U)$ with this decomposition of the unity and consider the $U'$ contribution

$$i_{U'} = I_G^f(X, U)\, \psi_{U'}(U) \tag{10.54}$$

outside the points of full coincidence $u_i = \bar{u} \ \forall \ i$. Applying the forest formula (10.47) to $i_{U'}$ we find that $t_{\Gamma_i}$ vanishes if $\Gamma_i$ contains points of both $U'$ and $U^c$ because $\psi_{U'}$ vanishes at the corresponding subtraction point. Thus the forest formula factorizes,

$$I_G^f \psi_{U'} = \sum_{W'} \prod_{\Gamma_i \in W'} (-t_{\Gamma_i}) \sum_{W^c} \prod_{\Gamma_i \in W^c} (-t_{\Gamma_i})\, I_G\, \psi_{U'} \, ,$$

where $W'$ and $W^c$ are forests of the subgraphs consisting of the $U'$ and $U^c$ vertices respectively. But this expression exists as a distribution according to the inductive assumption.

This proves the Theorem if the interior $G^o$ of $G$, i.e. $G$ without its external lines, is not itself a R-part. If $G^o$ is a R-part we are still saddled with a possible divergence at the points of full coincidence. Let $X$ be the set of external variables of $G$, $U$ that of the external vertex variables, i.e. those which are directly connected to an external point, $W$ that of the fully internal variables. Let $I^{f'}$ be the integrand subtracted with respect to all R-parts except $G^o$ itself, $f(X, U, W)$ a test function with compact support, and $\psi(U, W)$ one of the auxiliary functions introduced above. We want to prove the existence of

$$J = \int dX\, dU\, dW\, (1 - t_{G^o}) \prod P(x_a - u_a)\, I^{f'}(U, W)\, f(X, U, W)\, \psi(U, W)\,. \tag{10.55}$$

The $P(x_a - u_a)$ are the propagators of the external lines. Let $\hat{U}$ be a set of independent variables in 1–1 correspondence to $U$. The function

$$g(U, \hat{U}, W) = \int dX \prod P(x_a - u_a)\, f(X, \hat{U}, W) \tag{10.56}$$

is $C^\infty$ and of compact support in $\hat{U}$ and $W$. The $u$-derivations in $t_{G^o}$ act on the propagators $P$ and thus on $g$. We can write

$$J = \int dU\, dW\, \psi(U, W)\, I^{f'}(U, W)\, (1 - t_{G^o}) g(U, \hat{U}, W) I \Big|_{\hat{U} = U}, \tag{10.57}$$

the identification $\hat{U} = U$ being effected after the application of $t_{G^o}$. But $t_{G^o}$ is constructed such that $(1 - t_{G^o}) g(U, \hat{U}, Z)|_{\hat{U} = U}$ vanishes at the coincidence points $\{u_i = \bar{u} \,\forall\, i\}$ sufficiently strongly, so that the singularity of the product $\psi I^{f'} (1 - t_{G^o}) g$ at those points is integrable. The existence of $J$ is thus demonstrated.

It remains to be seen that renormalization does not destroy the properties of the $\mathcal{W}$-functions that have been established in Sect. 9.4 in the unrenormalized case. The demonstrations given there can essentially be taken over without problems. There are only two points which need some elaboration. The first one concerns the equations of motion, in particular the argument that a factor $T^+(\bar{z}, z)$ in a $\mathcal{W}$-function can be replaced by $\bar{\psi}(z)\psi(z)$ without changing anything. This has been shown to be true at the level of singular functions. But the N-vertex standing for $T^+(\bar{z}, z)$ may be included in a R-part, while at first sight this is not the case for $\bar{\psi}(z)\psi(z)$, because the two factors belong to different sectors. This appearance is deceptive: the identification of two external variables may lead to UV divergences even if these variables are taken from different sectors. In $p$-space language: identification of external points in different sectors can create new UV divergent loop integrals. Since the two considered expressions are identical as singular functions, their singularities are also identical and are therefore rendered integrable by the same subtraction procedure. Hence the desired identity of $T^+(\bar{\psi}\psi)$ and $\bar{\psi}\psi$ obtains if we define the finite part of the latter product in the same way as for the former, which is a legitimate prescription. This prescription leads also to equality of the said product with the commutator actually occurring in (8.2.), and the same argument applies to the product $A\!\!\!/\,\psi$. Notice that

## 10.3 Configuration Space

in graphs in which $z$ and $\bar{z}$ are directly connected, this isolated 2-point subgraph is a R-part. Its subtraction by our prescription annihilates these graphs without recourse to the somewhat arbitrary definition (5.72).

The second problem concerns the proof of Lemma 9.2. It has been shown that the sum in (9.29) vanishes as a singular function. Hence the singularities of the various terms add up to zero. And therefore the subtracted finite parts add to zero, provided that each term is subtracted in the same way. That this is the case can be seen as follows. Let $\Gamma'$ be a R-part of the subgraph $\Gamma$ considered in Lemma 9.2. It gives rise to subtractions in the terms of (9.29) in which $\Gamma' \subset \Gamma_L$ or $\Gamma' \subset \Gamma_R$. Such a subtraction is a sum of terms of the form
$$DE(X, \bar{y}, \ldots, \bar{y}) \, I^A(Y, Z)$$
in the notation introduced before (10.44). $A$ stands for $L$ or $R$. $I^A$ is the integrand of $\Gamma'$, including the $\hat{y}$-products in (10.44). $DE$ is a derivative of the integrand of $\Gamma \backslash \Gamma'$ taken at $y_i = \bar{y}$. Since $DE$ depends on the $y_i$ only in the combination $\bar{y}$, we can consider it as the integrand of a new graph with a "reduced vertex" with vertex variable $\bar{y}$ and the $Y$–$Z$-dependent vertex factor $I^A$. Applying the proof of the Lemma to these reduced graphs we find that the $L$–$R$-sum vanishes if $I^L = -I^R$. Assuming that the Lemma is correct in lower orders and using it for the subgraph $\Gamma'$ we obtain $I^L + I^R = -\sum I^{LR}$, where the right-hand side sums the $\Gamma'$-subtractions from graphs in which $\Gamma'$ extends over both sectors. We have seen above that these $I^{LR}$ are UV harmless. But they do not vanish, so that the desired relation $I^L = -I^R$ is not satisfied locally. However, we are eventually only interested in the integral over all the variables of $\Gamma'$, and this integral of $I^{LR}$ will be shown to exist and be zero in the next chapter. This corresponds to the $p$-space result discussed after Theorem 10.2, according to which subtractions of proper subgraphs extending over two sectors vanish. But this has been shown ignoring possible IR problems, hence the postponement of the final proof.

The cluster property involves the long-distance behaviour of the graph integrands in an essential way, and its exact proof must therefore also be postponed to the next chapter.

The proofs of current conservation and the charge identity given in Sect. 9.5 can be taken over, again introducing reduced vertices like in the proof of Lemma 9.2, which have the same algebraic properties as the original vertices.

# Chapter 11

# The IR Problem for Wightman Functions

In this chapter we discuss what may be called the "mild" IR problem: a problem that occurs in the calculation of the $W$-functions as explained in Chaps. 9 and 10. In these calculations the external momenta are not restricted to the mass shell. The "hard" IR problem, which is more severe, will be encountered in Part III in connection with the calculation of scattering cross sections, which necessitates the restriction of external momenta to the mass shell.

## 11.1 The Problem

Let $G$ be a sector graph, $I_G^f(x_1, \ldots, x_n, u_1, \ldots, u_m)$ its renormalized integrand with external variables $\{x_i\}$ and internal variables $\{u_j\}$. According to Theorem 10.3 $I_G^f$ is a tempered distribution. But this does not guarantee the existence of the integral

$$J_G^f = \int \prod_i dx_i \prod_j du_j f(x_1, \ldots, x_n) I_G^f(x_1, \ldots, u_m) \qquad (11.1)$$

for $f \in \mathcal{S}$, because the internal variables are not cut off at large distances by a strongly decreasing test function. This problem of existence at infinity is the IR problem. That the problem is real, i.e. that there exist graphs for which $J_G^f$ diverges even after renormalization, is best seen in $p$-space.

The problem has already appeared in Sect. 10.1 in the form of problem A connected with the 2-point functions in second order. Consider first the 2-photon case: the graphs shown in Fig. 10.1, where now the renormalized finite form of the loop integrals is used. These loop integrals of the graphs

(b) and (c) are of the form

$$L_b^{\mu\nu}(p) = -L_c^{\mu\nu}(p) = p^2 K^{\mu\nu}(p) ,$$

with $K^{\mu\nu}$ smooth functions. In graph (b) we find the product $(p^2 + i\varepsilon)^{-1} p^2 \delta_+(p)$ multiplied with a regular function. This product is ambiguous. The same holds for graph (c). But it has been shown that these ambiguities disappear in the sum of the two graphs, which remark solves the problem. This particular instance is not, properly speaking, an IR problem: it occurs in the same way in purely massive theories like the $\varphi^4$ theory with a positive mass. But its solution is the same as that of the proper IR problem, hence we subsume it under that term.

The true IR problem shows up in the analogous discussion of the fermionic 2-point function, the graphs of Fig. 10.4. There the renormalized loop integrals of graphs (b) and (c) are of the form

$$(p^2 - m^2) \log(p^2 - m^2) K_{b,c}(p) ,$$

with $K$ being regular functions. Hence, in evaluating e.g. graph (b) we are faced with a product $(\Delta + i\varepsilon)^{-1} \Delta \log \Delta \delta_+(\Delta)$ with $\Delta = p^2 - m^2$. This product is not only ambiguous because it is non-associative, it is even divergent if starting the multiplication from either end. This aggravation relative to the photon case is a true IR effect: the logarithmic singularity is due to the vanishing photon mass. Nevertheless it has been shown in Sect. 10.1 that the problem is solved in the same way as in the photon case: it disappears in the sum over the three graphs of Fig. 10.4.

The IR problem in its purest form, unencumbered by problem-A ambiguities, presents itself in another simple example, that of the function $\widetilde{W}_2(p_1, p_2, +|\bar{q}_1, \bar{q}_2, -)$ in second order. The connected graphs contributing to this expression are shown in Fig. 11.1. Consider graph (a). Its $\delta$-reduced integrand contains the singular factors

$$SF = (-2pk + k^2 + i\varepsilon)^{-1} (2qk + k^2 + i\varepsilon)^{-1} (k^2 + i\varepsilon)^{-1} \delta_+(p) \delta_+(q) .$$

Except at $k = 0$ the five singularity manifolds are in general position in the 12-dimensional $p$–$q$–$r$-space: their normals are linearly independent. Hence SF is well defined as a distribution outside an arbitrarily small neighbourhood of $\{k = 0\}$. But at $k = 0$ this is no longer true. There SF has a singularity of order $|k|^{-4}$ which is neither integrable in the usual sense of the word nor defined as a distribution. Hence the corresponding integral diverges unless one of the omitted regular factors happens to vanish there. This divergence is a pure IR divergence. It would not occur if the photon had a positive mass $\lambda$, so that the factor $\frac{1}{k^2}$ were replaced by $\frac{1}{k^2 - \lambda^2}$. The other three graphs are divergent for the same reason.

We claim that these divergences cancel in the sum of the four graphs. We note first that since the divergences are concentrated at $k = 0$ we may

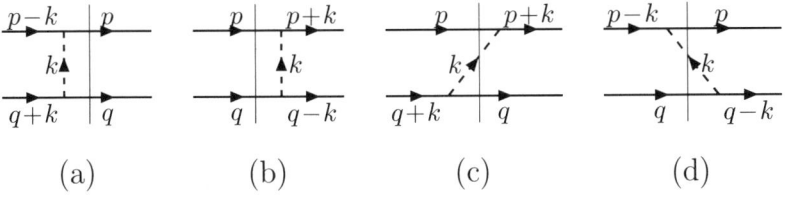

**Fig. 11.1a–d.** Graphs of $\tilde{\mathcal{W}}_2(p_1, p_2, +|\bar{q}_1, \bar{q}_2, -)$.
The thin vertical line in each graph separates the two sectors of $\tilde{\mathcal{W}}$. The relation between the external variables $p_i$, $q_j$, and the $\delta$-reduced graph variables $p$, $q$, $k$, differs from graph to graph

neglect the terms $k^2$ in the first two factors of SF compared to $pk$ or $qk$.[1] Second, the regular factors $f(p,q,k)$ not shown in SF are different for the different graphs, but they coincide at $k=0$, except for some constants which are different because of the different character of the $k$-lines. Cancellation of the IR divergences is then equivalent to their cancellation in the SF-sum, properly including those different constants. We obtain

$$\sum \text{SF} = \delta_+(p)\,\delta_+(q)\,\frac{1}{2\,pk-i\varepsilon}\,\frac{1}{2\,qk+i\varepsilon}$$
$$\times \left[-\frac{i}{2\pi}\frac{1}{k^2+i\varepsilon}+\frac{i}{2\pi}\frac{1}{k^2-i\varepsilon}+\delta_+(k)+\delta_-(k)\right].$$

Using $(k^2-i\varepsilon)^{-1}-(k^2+i\varepsilon)^{-1}=2\pi i\,\delta(k^2)$ we find that the square bracket vanishes.

Hence we have again found that the IR divergences present in individual graphs vanish if summed over all graphs with the same scaffolding. To show that this holds generally is the subject of the following section.

## 11.2 The Solution

For the solution of the mild IR problem we use methods developed in [Ste 93] for the massless $\varphi_4^4$ theory.

The IR divergence of a graph in $x$-space is due to an insufficiently fast decrease of its integrand at infinity. The behaviour at infinity is determined by the asymptotic behaviour of the propagators $D_\pm$, $D_F$, $S_\pm$, $S_F$. All these functions can be expressed in terms of $D_+(\xi)$ or $\Delta_+(\xi)$, respectively, as defined in (5.102) and (5.105). The asymptotic behaviour of these functions for large $\xi$ has been indicated in Sect. 5.4. $D_+(\xi)$ is explicitly given by

---

[1] This argument may not look convincing because the scalar products $pk$ and $qk$ vanish at some non-zero values of $k$. But at the moment we are not concerned with rigour. We only wish to indicate the general mechanism underlying the cancellation of IR divergences.

(5.107). It decreases like $|\xi|^{-2}$ in spacelike and timelike directions, like $|\xi|^{-1}$ in lightlike directions. The latter statement is meaningless on the light cone itself, because there $D_+$ is singular. As a function of the complex 4-vector $\zeta$, $D_+(\zeta)$ decreases like $|\zeta|^{-2}$ if $\zeta$ tends to infinity in a complex direction inside the backward tube $\mathcal{T}_-$. The analytic function $\Delta_+(\zeta)$ with $m > 0$ decreases in $\mathcal{T}_-$ exponentially if $\zeta \to \infty$ in a complex direction. Its boundary value $\Delta_+(\xi)$ at real $\xi$ also decreases exponentially in spacelike directions. In timelike directions it behaves asymptotically as $(\xi^2)^{-3/4} \exp\left(\pm im\sqrt{\xi^2}\right)$. In this case its absolute value decreases slower than that of $D_+(\xi)$. These statements hold also for the fermion propagator $S_+(\xi)$. This decrease in real directions is not sufficient to make the general $G$-graph convergent at infinity. We will improve the situation by shifting integration paths into complex domains in such a way that the more favourable asymptotics in complex directions can be used. This is not possible for most individual graphs. But the method is applicable to the sum of the graphs with the same scaffolding.

Let $G$ be a sector graph and $I_G(x_1, \ldots, x_n, u_1, \ldots, u_\sigma)$ its renormalized integrand, with $X = \{x_i\}$ its external and $U = \{u_j\}$ its internal variables. We consider $I_G$ as a function of $U$ for fixed values of the external variables $X$. More exactly, in order to avoid sitting precisely on a $X$-singularity of the $U$-integral, we consider the integral

$$I'_G(U) = \int dX\, f(X)\, I_G(X, U) \tag{11.2}$$

over a test function $f(X)$ with compact support, again relying on the argument given in connection with Theorem 10.3 that if $\int dU\, I_G(X,U)$ exists as a distribution in $X$, then this distribution cannot help being tempered. For a given $f \in \mathcal{D}$ there exists a positive constant $A$ such that only values of $x_i$ with $|x_i^0| \le A$ contribute to $I'_G$. We need discuss the behaviour of $I'_G$ for large values of the $u_j$. Now, for large spacelike $\xi$ the asymptotic behaviour of the propagators $D_+(\xi)$, $S_+(\xi)$, ..., is optimal and does not cause any divergence problems. The IR divergences are due to their less-than-optimal behaviour in timelike or lightlike directions: the divergence of $\int dU\, I'_G(U)$ is due to the behaviour of $I'_G$ if one or several $u_i^0$ become large.

We therefore integrate $I'_G$ first over the time components $u_i^0$ at fixed values of $\vec{u}_i$. This low-dimensional integral is still IR convergent: the decrease of the propagators in timelike directions is sufficient to make it so. It is convenient to separate the variables $u_i^0$ into harmless bounded ones and dangerous large ones. This we achieve by splitting $I_G(X, U)$ into a finite number of pieces. Let $\beta(s)$ be a $C^\infty$ function of the real variable $s$ with support in the interval $[-2A, 2A]$, and let $\alpha_+(s)$, $\alpha_-(s)$, be $C^\infty$ functions with support in $[\frac{3}{2}A, \infty]$ and $[-\infty, -\frac{3}{2}A]$ respectively, such that

$$\alpha_+(s) + \alpha_-(s) + \beta(s) \equiv 1\,. \tag{11.3}$$

We multiply $I_G$ with the decomposition of unity

$$1 = \prod_{1 \leq i < j \leq \sigma} [\alpha_+(u_i^0 - u_j^0) + \alpha_-(u_i^0 - u_j^0) + \beta(u_i^0 - u_j^0)]$$
$$\times \prod_{\substack{1 \leq i \leq n \\ 1 \leq j \leq \sigma}} [\alpha_+(u_j^0 - x_i^0) + \alpha_-(u_j^0 - x_i^0) + \beta(u_j^0 - x_i^0)] \quad (11.4)$$

and consider each term in this decomposition separately. Select such a term $T$. Its product $I_G^T$ with $I_G$ is still a distribution. We partition the zero-components $N = \{x_i^0, u_j^0\}$, into classes called "clusters" as follows. Two variables $z_i^0, z_j^0 \in N$, either external or internal, are said to be "connected" if there exists a sequence $z_{\alpha_1}^0 = z_i^0, z_{\alpha_2}^0, \ldots, z_{\alpha_r}^0 = z_j^0$, of variables in $N$ such that each difference $z_{\alpha_i}^0 - z_{\alpha_{i+1}}^0$ of adjacent variables occurs as argument of a $\beta$-factor in $T$. The "$X$-cluster" is defined to consist of the external $x_i^0$ and the internal $u_j^0$ connected to a $x_i^0$. If $u_i^0, u_j^0$ do not belong to the $X$-cluster, they are said to be elements of the same "internal" cluster if and only if they are connected. A variable $u_j^0$ not occurring in a $\beta$-factor forms a cluster by itself. These clusters form a partition of $N$ into mutually non-overlapping sets. And there exists a positive number $B < \infty$ such that $|z_i^0 - z_j^0| < B$ if $z_i^0$ and $z_j^0$ belong to the same cluster. (Remember that $|x_i^0| \leq A$.) Furthermore, to any $z_i^0$ in a cluster with at least two elements there exists another $z_j^0$ in the same cluster with $|z_i^0 - z_j^0| \leq 2A$. On the other hand, if $z_i^0$ and $z_j^0$ are in different clusters, then $|z_i^0 - z_j^0| > A$. Because of these properties the clusters are temporally ordered:

**Lemma 11.1.** *Let $C_1, C_2$, be different clusters. Then in the support of $T$ all $z_i^0 \in C_1$ are later than all $z_j^0 \in C_2$, or vice versa.*

**Proof.** Assume that there is a $z_i^0 \in C_1$ such that

$$\min_{C_2} z_j^0 - A \leq z_i^0 \leq \max_{C_2} + A .$$

Then there exists a $z_j^0 \subset C_2$ with $|z_i^0 - z_j^0| \leq A$, which is impossible. Therefore any $z_i^0 \in C_1$ is either larger by at least $\frac{3}{2}A$ than all $C_2$ variables or smaller by the same amount than all $C_2$ variables. But variables $\notin C_2$ with that property lying on different sides of $C_2$ cannot be connected, hence belong to different clusters.

Let $U_0$ be the set of $u_i^0$'s in the $X$-cluster, $U_+$ ($U_-$) those in later (earlier) clusters. We replace the $u_i^0$ as integration variables by difference variables as follows. In the $X$-cluster we choose arbitrarily a difference $t_0 = x_i^0 - u_j^0$ between an external and an internal variable, and a maximal number of linearly independent differences $t_\alpha = u_{i_\alpha}^0 - u_{j_\alpha}^0$ as new variables. Likewise we choose a maximal independent set of such internal differences $t_\beta$ in each of the other clusters. Finally, if the clusters $C_{(-M_-)}, \ldots, C_0 = C^X, \ldots, C_{M_+}$, $M_\pm \geq 0$, are arranged in temporal order, we introduce inter-cluster differences $\omega_\rho =$

170   11. IR Problem

$u_{i_\rho}^0 - u_{i_{\rho-1}}^0$ with $u_{i_\rho}^0$ an arbitrarily selected element of $C_\rho$. $\rho$ takes the values $-M_- + 1, -M_- + 2, \ldots, M_+$. In the support of $T$ we have $\omega_\rho > A$. The $t_\alpha$ and $\omega_\rho$ together form a complete set of variables replacing the $u_i^0$. The $t_\alpha$-integrations extend in $T$ only over compact regions, hence propagators whose arguments are linear combinations of only such variables are IR harmless. On the other hand, the UV problem is fully contained within the clusters, i.e. it is exclusively due to the $t_\alpha$-integrations and is dealt with by the methods explained in Chap. 10.

The IR problem thus is due to the propagators of lines connecting points in different clusters. Such a propagator contains a factor $D_+(a)$ or $S_+(a)$,[2] with an argument of the form

$$a = \pm \sum_{\rho=m_1}^{m_2} \omega_\rho + \sum_{\alpha \in R} \pm t_\alpha \;. \qquad (11.5)$$

$m_1$ and $m_2$ are integers with $-M_- \leq m_1 < m_2 \leq M_+$ and $R$ is a set of $t$-indices. Since $a$ is a difference between variables in different clusters we have $a > A$ or $a < -A$ depending on the sign in front of the $\rho$-sum. If a given $\omega_\rho$ occurs always with the same sign in these arguments, then the product of the propagators containing it is, as a function of $\omega_\rho$, a boundary value of an analytic function in $\Im \omega_\rho \lessgtr 0$, as the case may be. We can then evaluate the integral over $\omega_\rho$ by shifting the integration path into the complex, thus possibly improving its IR behaviour. This is in general not possible for individual graphs. For, if a $U_+$ variable $u_1$ belongs to a non-extremal sector, it may be connected by lines to a $U_0$ or $U_-$ variable $u_2$ in a lower sector and another such variable $u_3$ in a higher sector. The resulting product $P_+(u_2 - u_1) P_+(u_1 - u_3)$ is in $u_1$ not a boundary value of an analytic function, and therefore also not in the $\omega_\rho$ connecting the $u_1$-cluster to the next-lower one. But we will now demonstrate that this situation does not occur in the sum of all integrands belonging to the same scaffolding. We still consider only the contributions of a particular term $T$ in the decomposition (11.4). We prove the following lemma.

**Lemma 11.2.** *Let $\mathcal{G}$ be the set of graphs contributing to $\mathcal{W}_\sigma(X_1, s_1| \cdots |X_N, s_N)$ with the same given scaffolding. Let $U = U_+ \cup U_- \cup U_0$ be a partition of the set $U$ of internal variables of this scaffolding into three non-overlapping subsets, the empty set being admitted. Consider the sum*

$$I_\mathcal{G}(X, U) = \sum_{G \in \mathcal{G}} I_G(X, U)$$

*in a region in which the sets are temporally ordered: $U_+$ is later than $U_0 \cup X$, and $U_0 \cup X$ is later than $U_-$. Then all contributions to $I_\mathcal{G}$ cancel except those in which all $U_\pm$ variables belong to the extremal sectors $S_1$, $S_N$, such that $S_1$*

---

[2] Henceforth we call such a propagator $P_+(a)$ if its type (photon or electron) is unimportant. $P_-$ and $P_F$ are defined analogously.

contains no $U_-$ ($U_+$) variables if $s_1 = +(-)$, and $S_N$ no $U_+$ ($U_-$) variables if $s_N = +(-)$.

According to our rules there exists in $G \in \mathcal{G}$ an internal $T^-$-sector, possibly empty, between two adjacent external $T_+$-sectors, and vice versa. No internal sectors are present between adjacent external sectors of different type. This means that in all cases there exists a regular alternation of $T_+$- and $T_-$-sectors. Consider two neighbouring sectors $S^1$ and $S^2$ of type $T_+$ and $T_-$, respectively. Let $U^\rho = U_+^\rho \cup U_0^\rho \cup U_-^\rho$ be the partitioning of the internal variables $U^\rho$ of $S^\rho$ according to the given partitioning of $U$. Let $S_\pm^\rho$ be the subgraph of $S^\rho$ generated by the $U_\pm^\rho$-vertices, $S_0^\rho$ that generated by $U_0^\rho$ and the external variables $X^\rho$ in $S^\rho$. Because of the assumed temporal order we can write $S^1 S^2$ as a product

$$S^1 S^2 = S_+^1 S_0^1 S_-^1 S_-^2 S_0^2 S_+^2 . \tag{11.6}$$

The $S_I^1$ are $T^+$-sectors, the $S_I^2$ $T^-$-sectors. We use the same notation as in Lemma 9.2: the products in (11.6) include the propagators of the cross lines connecting the factors. Some of the factors in (11.6) may be empty. The proof of the decomposition (11.6) is a corollary to the proof of the time-ordering relations given in Sect. 9.4. If we now sum over all graphs which differ from $G$ at most in the partition of $U^{12} = U_-^1 \cup U_-^2$, we obtain an expression containing the sum $\sum U_-^1 U_-^2$ over all these partitions. By Lemma 9.2 this sum vanishes unless $U_-^{12} = \emptyset$. If $S^1$ is $T^-$ and $S^2$ is $T_+$ we find in the same way that the contribution of $G$ to the $\mathcal{G}$-sum is cancelled by other contributions except if $U_+^1 = U_+^2 = \emptyset$. Lemma 11.2 is an immediate consequence of these results.

Consider now $I_\mathcal{G}^T(X, U)$, the summed integrand of Lemma 11.2 multiplied with the term $T$ in the decomposition (11.4). Its subsets $U_0$, $U_\pm$, satisfy the assumptions of Lemma 11.2. We evaluate the integral

$$J_\mathcal{G}^T = \int dX \, dU \, I_\mathcal{G}^T(X, U) \, f(X)$$

by integrating first over the time components $x_i^0$, $u_j^0$, and then over the spatial components. But we replace the $u_j^0$ as integration variables by the cluster variables $t_\alpha$ and the inter-cluster variables $\omega_\rho$. The temporal integral is still IR convergent, and the integrations can be carried out in an arbitrary order. We integrate first over the bounded variables $x_i^0$ and $t_\alpha$ which are not involved in the IR problem, then over the $\omega_\rho$. The support of $I_\mathcal{G}^T$ lies in $\{\omega_\rho \geq A\}$. We define $\int_A^\infty d\omega_\rho \cdots = \lim_{R_\rho \to \infty} \int_A^{R_\rho} d\omega_\rho \cdots$.

Consider first the case in which the two extremal factors in $\mathcal{W}_\sigma$ are of the same type, e.g. $s_1 = s_N = +$. The $U_+$ variables occur in $I_\mathcal{G}^T$ only in the first sector $S_1$, $U_-$ variables only in the last sector $S_N$. Using

$$P_F(\xi) = \theta(\xi^0) P_+(\xi) + \theta(-\xi^0) P_+(-\xi)$$

we find that any $\omega_\rho$ occurs in the arguments of the $P_+$ functions always with a positive sign if it occurs at all. The product of these $P_+$ is then analytic in $\omega_\rho$ in $\{\Im\omega_\rho < 0\}$, and the factor $\alpha_\pm(\pm\omega_\rho)$ is constant in $\omega_\rho \geq 2A$. We can therefore deform the integration contour of $\omega_\rho$ into a complex contour consisting of three straight lines: the compact real interval $[A, 2A]$, the straight line $D_\rho$ leading from $2A$ to $E_\rho = R_\rho - i(R_\rho - 2A)$, and the straight line $V_\rho$ leading from $E_\rho$ to the real point $R_\rho$. We assume $R_\rho > 2A$. Moreover we assume that $R_\rho$ is already so large that for $\omega_\rho \geq R_\rho$ none of the local singularities of $I_G^T$ is encountered for the given values of $\vec{x}_i$ and $\vec{u}_j$. The first of the three contributions is not involved in the IR problem, hence can be omitted. The remaining $\omega_\rho$-integral we split into two parts: its values over $D_\rho$ and $V_\rho$. This results in $M^2$ contributions to the $2M$-dimensional $\prod d\omega_\rho$ integral, with $M = M_+ + M_-$ the number of $\omega$'s.

We claim that any such contribution with at least one $V_\rho$ vanishes in the limit $R_\rho \to \infty \forall \rho$. The proof is not given in full but indicated far enough so that filling in the missing details should not be difficult. Consider such a contribution. Let $\Omega$ be the set of $\omega_\rho$'s integrated along $V_\rho$, $\Omega'$ the set of $\omega_\sigma$'s integrated along $D_\sigma$. Let $r$ ($r'$) be the number of elements in $\Omega$ ($\Omega'$). We have $r + r' = M$, and we assume $r > 0$. Let $\alpha$ be the number of propagators $P_+$ depending on $\Omega$ but not on $\Omega'$, $\beta$ that of $P_+$ depending on both types of variables, and $\gamma$ that of $P_+$ depending on $\Omega'$ but not on $\Omega$. The graph rules imply that from each internal cluster at least two lines lead to other clusters, and that each internal cluster is directly or indirectly connected by lines to the $X$-cluster. From this we obtain the estimates $\alpha + \beta \geq r+1$, $\beta + \gamma \geq r'+1$ if $r' \neq 0$, $\alpha + \beta + \gamma \geq r + r' + 1 = M + 1$. We wish to show the vanishing for $R_\rho \to \infty$ of

$$J = \prod_\Omega \left(\int_{V_\rho} d\omega_\rho\right) \prod_{i=1}^\alpha P_+(a_i) \prod_{\Omega'} \left(\int_{D_\sigma} d\omega_\sigma\right) \prod_{j=1}^\beta P_+(a_j + a'_j)$$
$$\times \prod_{k=1}^\gamma P_+(a'_k) \,. \qquad (11.7)$$

Here the $a_i$ ($a'_j$) are non-empty partial sums of $\Omega$ ($\Omega'$) variables. The dependence of the propagators on the bounded variables $t_\alpha, x_i, \vec{u}_j$ is irrelevant at the moment and has been suppressed. We propose that $J$ vanishes if all $R_\rho$ tend to $\infty$ simultaneously: $R_\rho = \lambda_\rho R$, $\lambda_\rho > 0$, $R \to \infty$. A more detailed analysis along the same lines shows that $J$ vanishes also if the $R_\rho$ tend to $\infty$ at different rates or one after the other. Consider first the $\Omega'$ integral $J'(\Omega)$ in (11.7). Along $D_\sigma$ both photon and electron propagators decrease at least like $|\omega_\sigma|^{-1}$ for $|\omega_\sigma| \to \infty$.[3] Hence we have

$$|J'(\Omega)| \leq \text{const} \prod_{\Omega'} \int_{2A}^{\lambda_\sigma R} d\omega_\sigma \prod_1^\beta |b_j R + a'_j|^{-1} \prod_1^\gamma |a'_k|^{-1} \,,$$

---
[3] We use this less-than-optimal estimate for later purposes.

with $b_j = \frac{a_j}{R} > 0$. Substituting $\omega_\sigma = R\kappa_\sigma$, $a'_j = R v_j$, this becomes

$$|J'| \leq \text{const}\, R^{r'-2(\beta+\gamma)} \prod_{\Omega'} \left(\int_{2A/R}^{\lambda_\sigma} d\kappa_\sigma\right)^\beta \prod_1^\beta |b_j + v_j|^{-1} \prod_1^\gamma |v_j|^{-1} .$$

Using $\beta + \gamma \geq r' + 1$ we find for $R \to \infty$:

$$\begin{aligned} |J'| &= \mathcal{O}(R^{-\beta} \log R) \text{ if } \gamma \geq r' , \\ |J'| &= \mathcal{O}(R^{r'-(\beta+\gamma)}) \text{ if } \gamma < r' . \end{aligned} \tag{11.8}$$

This result we insert into (11.7) and estimate the remaining $\Omega$-integral. On $V_\rho$ both types of propagator decrease at least of order $\mathcal{O}(R^{-1})$ for $R \to \infty$. On the other hand, the volume of the domain of integration $\otimes_\Omega V_\rho$ increases of order $\mathcal{O}(R^r)$. Hence we obtain the estimate

$$|J| \leq \text{const}\, R^{r-\alpha} |J'| . \tag{11.9}$$

For $\gamma \geq r'$ this becomes

$$|J| \leq \text{const}\, R^{r-\alpha-\beta} \log R \leq R^{-1} \log R \to 0 . \tag{11.10}$$

For $\gamma < r'$ we obtain

$$|J| \leq \text{const}\, R^{r-\alpha+r'-(\beta+\gamma)} \leq \text{const}\, R^{-1} , \tag{11.11}$$

which converges to 0.

As a result we find that the integral of $I_\mathcal{G}^T$ over the variables $t_\alpha$, $\omega_\rho$ is completely given by the $\Omega = \emptyset$ contribution: we may replace the $\omega_\rho$-integral over the real interval $[A, \infty]$ by the sum of the integrals over $[A, 2A]$ and $D_\rho$ for $R_\rho = \infty$. This integral exists because each of the at least $M + 1$ propagators containing $\omega$'s decreases at least of first order at large values of $\omega$.

This result applies at first only to the case of extremal sectors of equal type: $s_1 = s_N$. But the other case, e.g. $s_1 = +$, $s_N = -$, can be reduced to that one, because from Lemma 9.2 we find that

$$\mathcal{W}(\cdots | X_N, -) = \sum_{X_N^R} \pm \mathcal{W}(\cdots | X_N^L, -|X_N^R, +) , \tag{11.12}$$

the sum extending over all partitions of the set $X_N$ into two complementary subsets $X_N^L$, $X_N^R$ with $X_N^R \neq \emptyset$.

We can now carry out the full integration over all variables, including the spatial ones, with the understanding that the variables $\omega_\rho$ are integrated over the complex paths $D_\rho$. The local singularities of $I_\mathcal{G}^T$ are confined to a compact set $K$. We claim that the integral of $I_\mathcal{G}^T$ over the set $S = \mathbf{R}^{4(|X|+|U|)} \setminus K$ is absolutely convergent. Remember that we integrate over a test function $f(X)$

with compact support, so that we need worry only about large values of the $U$-variables. Convergence at infinity obtains if the integrand decreases at least like $R^{-4|U|-1}$ for $R = \sqrt{\sum |u_j|^2} \to \infty$ in generic directions. The decrease may be slower along sufficiently low-dimensional subspaces, e.g. in directions in which some of the variables are kept constant. We study the decrease of the integrand along the line $u_i = \hat{u}_i + \lambda n_i$ for $\lambda \to \infty$, $\hat{u}_i$ and $n_i$ being fixed 4-vectors. We also define such $n_j$ for the external variables $x_j$, setting $n_j = 0$ for them. A set of variables with identical $n_i$ is called a "preswarm". Inside a preswarm the distances between points remain constant for $\lambda \to \infty$. All external variables belong to the same preswarm called the "$X$-preswarm". The other preswarms are called "internal". Let there be $N$ internal preswarms. The following lemma suffices to guarantee the claimed absolute convergence.

**Lemma 11.3.** *If $N$ is the number of internal preswarms, then $I_{\mathcal{G}}^T(X, U)$ decreases for $u_i = \hat{u}_i + \lambda n_i$, $\lambda \to \infty$, at least of order $\lambda^{-2-4N}$.*

As in the case of the clusters considered before, we know that from each preswarm issue at least two lines, and all preswarms are connected to one another directly or indirectly. If two preswarms are directly connected by an electron line, its propagator decreases exponentially for $\lambda \to \infty$. This exponential decay of a single propagator already implies the lemma. Therefore we can combine preswarms connected by electron lines into a "swarm", and we need prove the lemma only for such swarms.

Let $N$ be the number of internal swarms. There are at least $N + 1$ lines connecting different swarms. These lines are photon lines. By Furry's theorem the number of lines joining a swarm must be even. Let us assume for the moment that there are no swarms for which this number is 2. Then there are $P \geq 1 + 2N$ lines connecting different swarms, and their propagators lead to a decrease at least like $\lambda^{-2-4N}$, as claimed in the lemma. If swarms with only two external (to the swarm) lines exist, then this estimate for $P$ does not hold. But such a 2-swarm has the form of a self-energy part. Let $C_{\mathcal{G}}^T(u, v)$ be the value of this part integrated over its internal variables ($u$ and $v$ are the variables of the points joined by the external lines), and let $C_G(u, v)$ and $C_{\mathcal{G}} = \sum_{G \subset \mathcal{G}} C_G(u, v)$ be the corresponding factors in $I_G$ and $I_{\mathcal{G}}$ before the $T$-splitting. According to the results of Chap. 10, the Fourier transforms of $C_G$ and therefore of $C_{\mathcal{G}}$ are of the form

$$\tilde{C}_{\mathcal{G}}(^{\mu}p, ^{\nu}q) = \delta^4(p+q) \left[ p_\rho q^\rho g^{\mu\nu} A(p^2) + p^\mu q^\nu B(p^2) \right] . \tag{11.13}$$

We transfer the $p$- and $q$-factors to the adjacent propagators. Transforming back into $x$-space and only then effecting the $T$-splitting and all the rest of our proceedings, we arrive at the same sort of swarm picture as before, but with the photon propagators connected to 2-swarms being once differentiated. But $\partial_\mu D_+(\zeta)$ decreases like $|\zeta|^{-3}$ in the relevant directions, and this improvement in the decrease offsets the lowering of $P$, so that the lemma still holds.

What we have shown is this: if $f(X) \in \mathcal{S}$, then $f'(X, U) \equiv f(X)$ is an admissible test function for the distribution $I_\mathcal{G}(X, U)$. The value of $I_\mathcal{G}$ on $f'$ is defined by first integrating over the time components $x_i^0, u_j^0$, then over $\vec{x}_i, \vec{u}_j$. This means that

$$J_\mathcal{G}(X) = \int dU \, I_\mathcal{G}(X, U) , \qquad (11.14)$$

if defined as explained, is a tempered distribution. In this way of formulating the result we are again talking of an integral over real variables. The shift to complex paths was helpful for proving the result, but it is not needed for stating it.

The Fourier transform $\tilde{J}_\mathcal{G}(P)$ of $J_G(X)$ is given by the renormalized graph rules in $p$-space explained in Chaps. 9 and 10, summed over a scaffolding. We sketch a proof of this fact without going into all the details. It is known that any distribution can be approximated arbitrarily well by a smooth function with arbitrarily strong decrease at infinity.[4] This means that for any distribution $T(X)$ there exist $C^\infty$-functions $T^\nu(X)$, $\nu = 1, 2, \ldots$, with a good behaviour at infinity, such that

$$\int dX \, T(X) \, f(X) = \lim_{\nu \to \infty} \int dX \, T^\nu(X) \, f(X) \qquad (11.15)$$

for all test functions $f(X)$. Let $P_\alpha(\xi)$ be any of the propagators occurring in our graphs (see the definitions (9.30), and (9.31)). Let $\delta_\nu(\vec{\xi}) \in \mathcal{D}$ be a $\delta$-sequence, i.e. a sequence of test functions with compact support converging for $\nu \to \infty$ to $\delta^3(\vec{\xi})$ in the sense of (11.15), and let $\chi(\vec{\xi}) \in \mathcal{D}$ with $\chi(0) = 1$. Then

$$P_\alpha^\nu(\xi) = \chi\left(\frac{\vec{\xi}}{\nu}\right) \left(\delta_\nu(\vec{\xi}) * P_\alpha(\xi^0, \vec{\xi})\right) \qquad (11.16)$$

is an approximating sequence for $P_\alpha$ in the sense of (11.15), as can easily be ascertained. The symbol $*$ denotes convolution in the 3-dimensional $\vec{\xi}$-space. The $P_\alpha^\nu(\xi)$ are $C^\infty$, have compact support in $\vec{\xi}$, and decrease for $|\xi^0| \to \infty$ at least like $|\xi^0|^{-\frac{3}{2}}$. Replacing $P_\alpha$ by $P_\alpha^\nu$ for all propagators of the graph $G$ we obtain an integrand $I_G^\nu(X, U)$ which is $C^\infty$. It is invariant under translations. Written as a function of one of its variables, e.g. $x_1$, and a complete set of difference variables, it has compact support in the spatial components of the latter and decreases in their time components more than sufficiently fast to make it integrable. $I_G^\nu(X, U)$ clearly defines a tempered distribution, and so does

$$J_G^\nu(X) = \int dU \, I_G^\nu(X, U) . \qquad (11.17)$$

---

[4]In fact, it can be approximated even by test functions with compact support (see [Sch], Chap. III, § 3).

Summing over the graphs with a given scaffolding $\mathcal{G}$ and following the existence proof for $J_\mathcal{G}(X)$ step by step, one finds, defining $J_\mathcal{G}^\nu = \sum_{G \in \mathcal{G}} J_G^\nu$, that

$$J_\mathcal{G}(X) = \lim_{\nu \to \infty} J_\mathcal{G}^\nu(X) \tag{11.18}$$

in the sense of tempered distributions. Notice that the definition (11.16) of $P_\alpha^\nu$ does not touch its $\xi^0$-dependence, so that the steps of the existence proof involving time ordering (the decomposition (11.4), Lemma 11.2, the shift of integration paths into the complex) are not affected. This fact is also helpful, though not vital, for the rigorization of the formal verifications of Sect. 9.4, in particular of the time-ordering relations and of Lemma 9.2.

The Fourier transforms $\tilde{P}_\alpha^\nu(k)$ of the $P_\alpha^\nu(\xi)$ are $C^\infty$ in $\vec{k}$, at least Hölder continuous in $k_0$, and strongly decreasing in all directions. The convolution theorem and the other known properties of the Fourier transform are applicable to $J_G^\nu(X)$ without problem. The resulting Fourier transform $\tilde{J}_G^\nu(P)$ is given by the rules of Chaps. 9 and 10, but with the singular propagators $\tilde{P}_\alpha(k_i)$ replaced by by the well-behaved functions $\tilde{P}_\alpha^\nu(k_i)$. The $\delta^4$ at the vertices are still present. But their arguments are independent, so that their product is well defined. In the $\delta$-reduced rules there survives a $\delta^4$ for each connected component of the graph. Summing over $\mathcal{G}$ we obtain the Fourier transform $\tilde{J}_\mathcal{G}^\nu$ of $J_\mathcal{G}^\nu$. Since $J_\mathcal{G}^\nu$ converges to $J_\mathcal{G}$ for $\nu \to \infty$, the sequence $\tilde{J}_\mathcal{G}^\nu$ converges to $\tilde{J}_\mathcal{G}$. In this limit the summed integrand takes the form specified in Chaps. 9 and 10. The products of singular factors in this integrand may be defined by the above limiting procedure, though in practice they can usually be evaluated directly. A direct proof of the existence of our $p$-space expression for the special case of time-ordered functions can be found in [Zi 71]. It applies to the individual sectors of sector graphs. The IR problems associated with the cross lines can in principle be handled directly in $p$-space with the methods to be developed later in Chaps. 15 and 16, where only the simple case using (16.57) is relevant. For the IR existence the summation over $\mathcal{G}$ is essential.

In this way it is established that our renormalized graph rules in $p$-space are indeed the correct Fourier transforms of the $x$-space rules.

## 11.3 The W-Properties and Other Conditions

### The W-Properties

We have completed the proof that $\mathcal{W}_\sigma(\cdots | X_i, s_i | \cdots)$ exists if defined properly. But it must yet be established that it satisfies all the postulated properties. We start from the formal proofs given in Sect. 9.4 and amended at the end of Sect. 10.3. Some of these proofs rely exclusively on local properties of the integrands. They are therefore unrelated to the IR problems and remain proofs, even after renormalization, which is a local operation (see the last part of Sect. 10.3). This holds for the permutation symmetry inside $T^\pm$-factors, the time-ordering relations (8.27), the sign independence (9.22), hermiticity, and for the equations of motion if interpreted as explained near the end of Sect. 10.3.

Covariance is not obvious, because our way of integrating first over $u_i^0$, then over $\vec{u}_i$, is frame-dependent. Covariance under rotations and the parity operation is clearly not affected, so that we need only concern ourselves with boosts. But the boosts in the $i$-direction form a Lie group, so that it suffices to prove covariance under infinitesimal boosts. What must be shown is this: let $I_A(X,U)$ be the sum over integrands to a given scaffolding, with $A = \{\rho_1, \cdots, \rho_N\}$ the set of external vector and spinor indices, and consider the integral

$$J_A(X) = \int dU \, I_A(X,U) \,, \tag{11.19}$$

whose existence as a distribution has just been established. Covariance under an infinitesimal boost in the $i$-direction means

$$\sum_{x_j \in X} \Lambda_{0i}^j J_A(X) = \sum_{A'} B_{AA'}^i J_{A'}(X) \,, \tag{11.20}$$

where

$$\Lambda_{0i}^j = x_j^0 \frac{\partial}{\partial x_j^i} + x_j^i \frac{\partial}{\partial x_j^0} \,,$$

and $B^i$ is the representative of the infinitesimal $i$-boost in the finite-dimensional product representation $\otimes_j R^{(j)}$, i.e. the sum over the boost generators of all fields in $\mathcal{W}_\sigma$. The formal argument of Sect. 9.4 went like this: the integrand $I_A(X,U)$ satisfies

$$\left( \sum_{x_j \in X} \Lambda_{0i}^j + \sum_{u_k \in U} \Lambda_{0i}^k \right) I_A(X,U) = \sum_{A'} B_{AA'}^i I_{A'}(X,U) \,, \tag{11.21}$$

which differs from (11.20) by the $u_k$ terms on the left. But these are derivatives of $I_A$ with respect to $u_k^0$ or $u_k^i$, hence they give a vanishing contribution to the $U$-integral. This argument assumes that $I_A(X,U)$ decreases rapidly enough for any $|u_j| \to \infty$. This is the case for the $I_{\mathcal{G}}^T$ as defined before, integrated along $D_\rho + V_\rho$. Derivation does not invalidate our estimates, and in all relevant estimates (11.9)–(11.11) and in Lemma 11.3 there is at least one order to spare, which can accomodate the extra factors $u^i$ or $u^0$ in $\Lambda_{0i}$. Once covariance is shown, locality follows then also, as shown in Sect. 9.4.

To demonstrate the spectral condition we argue as follows. The $p$-space integrands of individual graphs are not necessarily distributions, but their sum over a given scaffolding is. But if in certain domains the individual terms *are* defined, e.g. as being zero due to support properties, then their sum also vanishes there. This is the case for the complement of the spectral support.

And since the condition only involves the external variables, while the UV problem is due to internal integrations, renormalization does not destroy this result: for values of the external variables for which the integrand vanishes identically there is clearly no UV problem.

The heuristic argument for the cluster property (9.39) can now be expanded into a proof. We must show that the contributions to the left-hand side of (9.39), which come from graphs in which at least one $x_i$ is connected to at least one $y_j$ by a sequence of lines, vanish for $\lambda \to \infty$. Consider a scaffolding $\mathcal{G}$ with this connectedness and the corresponding integrands $I_{\mathcal{G}}^T$ integrated along the paths $D_\rho$. We know that these integrals are absolutely convergent at large $U$. For simplicity we assume again that the functions $f(X)$, $g(Y)$, and $\hat{\gamma}$ in (9.38) and (9.39) have compact supports. This is no serious restriction of generality but shortens the proof considerably. The differences between the $X$ variables are then bounded uniformly in $\lambda$, as are the differences between the $Y$ variables. Therefore we introduce now two external swarms, the $X$-swarm and the $Y$-swarm. Their distance is still uniformly bounded in the time direction, but in spatial directions it is bounded by a quantity of order $\lambda$. Hence there are now three more potentially large inter-swarm variables to be integrated over, compared to the case of Sect. 11.2. But the estimate $P \geq 2N+1$ for the number of inter-swarm lines given there still holds. ($N$ is the number of internal swarms.) Using the uniform boundedness of $\lambda^3 f_\lambda(X)$ in $\lambda$ and $X$ we find by the methods of Sect. 11.2 that

$$\int dX\, dY\, dU\, \lambda^3 f_\lambda(X)\, g(Y)\, I_{\mathcal{G}}^T(X, Y, U, )$$

diverges for $\lambda \to \infty$ at most of order $\lambda$. This is more than offset by the additional factor $\lambda^{-3}$ in $f_\lambda$. Hence the vanishing of this integral for $\lambda \to \infty$ is demonstrated.

### The Charge Identity

Besides the W-properties we must also rigorize certain aspects of the renormalization conditions that have not yet been dealt with adequately. This concerns the smoothness assumptions for the 2-point functions and the charge identity as treated in Sect. 9.5. We start with the charge identity. The proof of current conservation given in Sect. 9.5 and amended in Sect. 10.2 can easily be transformed into $x$-space. Since the argument is purely local, it applies to the integrands of individual $x$-space graphs, which are well defined as distributions.

We prove that $Q_C \Omega = 0$ by showing that $(\Omega, \mathcal{M} Q_C \Omega)_\sigma = 0$ for every field monomial $\mathcal{M}$. $Q_C$ is defined by (7.14):

$$(\Omega, \mathcal{M} Q_C \Omega)_\sigma = \lim_{R \to \infty} \int dx\, \alpha(x^0)\, \chi(R^{-1}\vec{x})\, (\Omega, \mathcal{M} j^0(x) \Omega)_\sigma\,. \qquad (11.22)$$

## 11.3 W-Properties

We choose the $j^0$-sector and the first sector in $\mathcal{M}$ to be $T^+$. We write $\alpha(x^0) = -\frac{d\beta(x^0)}{dx^0}$ with $\beta(x^0) = \int_{x^0}^\infty du\,\alpha(u)$. If $\text{supp}\,\alpha \subset [-A, A]$ then $\beta$ vanishes for $x^0 > A$ and is $\equiv 1$ for $x^0 < -A$. We insert this form of $\alpha$ into (11.22). A legitimate integration by parts moves the $x^0$-derivation over to $j^0(x)$. Using current conservation we replace $\partial_0 j^0$ by $-\partial_i j^i$ and then again transfer the differentiation to the test function $\chi$, obtaining

$$(\Omega, \mathcal{M} Q_C \Omega)_\sigma = \lim_{R \to \infty} \frac{1}{R} \int dx\, \beta(x^0)\, \chi_i(R^{-1}\vec{x})\, (\Omega, \mathcal{M} j^i(x)\,\Omega)_\sigma \quad (11.23)$$

with $\chi_i(\vec{u}) = \partial_i \chi(\vec{u})$. This expression is calculated for finite $R$ like in the proof of IR convergence, by introducing a $T$-decomposition and summing over the set $\mathcal{G}$ of graphs. $x^0$ now also varies over an unbounded set, so it is treated like an internal $u_j^0$: it is included in the $T$-decomposition. But since it assumes only large negative values, not positive ones, it can only belong to $U_-$, not to $U_+$. This is fortunate because $x$ occurs necessarily in the highest sector, which in the $\mathcal{G}$-sum contains all the $U_-$ variables. Notice also that $\beta(x^0)$ is constant for $x^0 < -A$, so that its presence does not impede the shifting of integration paths to $D_\rho$. The estimate of Lemma 11.3 is now modified by the fact that an internal swarm[5] containing $x$ is connected to other swarms by an odd number of lines, because only two lines are incident at the $x$-vertex. Hence, if $x$ belongs to an internal swarm, the relation between the number $P$ of interswarm lines and the number $N$ of internal swarms is changed to $2P \geq N$, so that the $\lambda$-decrease of Lemma 11.3 is changed to $\lambda^{-4N}$. This would not suffice for convergence, were it not for the fact that in this case the strongly decreasing function $\chi_i(R^{-1}\vec{x})$ contains the dangerous variable. For $R \to \infty$ the integral (11.23) may diverge logarithmically, but this is overcome by the factor $R^{-1}$ in front. This proves the vanishing of the $R$-limit in (11.23).

Theorem 7.1 is then applicable. It implies that $(\Omega, \mathcal{N} Q_C \mathcal{M} \Omega)_\sigma$ exists for all field monomials $\mathcal{M}, \mathcal{N}$. In its expression in terms of $W$-functions including a factor $j^0(x)$ the limit $R \to \infty$ exists. That is, we can integrate over $\vec{x}$ without a cutoff function, or in $p$-space set the corresponding variable $\vec{k}$ equal to zero in the sum over all contributing graphs. And the result is independent of the choice of the smearing function $\alpha(x^0)$. Hence we can again use the trick of Sect. 9.5: that of introducing the scaled family $\alpha_\lambda(x^0) = \lambda \alpha(\lambda x^0)$ and considering the limit $\lambda \to 0$.

To prove the charge identity we start from the formal proof given in Sect. 9.5 which applies to the integrand $I_G$ of an individual graph in $p$-space. If the current vertex lies in an R-part, we apply the argument separately to the original unsubtracted part and to the subtractions. This is meaningful because these contributions make sense separately for the integrands. The subtractions do not depend on the external variables of the R-part, or at the worst this dependence is polynomial. They can therefore be treated as reduced

---

[5] An internal swarm is now a swarm not containing $\mathcal{M}$-variables.

vertices as introduced near the end of Sect. 10.3, and the $\alpha$-integration can be studied in the corresponding reduced graph. This argument has as yet ignored the IR problem: the integrands $I_G$ in $p$-space are not defined as distributions because of the possible existence of bad local singularities. But these singularities cancel in the sum over a scaffolding. Hence consider such a sum. Applying our procedure to each term in this sum gives again an existing scaffolding sum for a function without the $k$-vertex, hence the procedure is legitimate, provided this handling of the IR problem and the former handling of the UV problem are not at cross-purposes. This is not the case. From the proof of IR cancellations it follows that time-ordered functions, whose graphs consist of one sector only, are IR convergent graph by graph. This is then true for the individual sectors of a multi-sector graph: the IR singularities of $I_G$ necessarily involve cross lines, and these cannot occur inside a R-part. Hence our handling of subtractions by means of reduced graphs is in order: IR cancellations occur separately for the nonrenormalized integrands and the reduced integrands.

This reasoning may become more transparent in Part III, where we will have occasion to study the explicit $p$-space mechanism of IR cancellation in a different context.

## Smoothness Conditions

The last conditions to be discussed are the smoothness conditions for the 2-point functions $\tilde{W}_\sigma(\mathsf{p},\mathsf{q})$ and $\tilde{W}_\sigma(p,\bar{q})$ as formulated in Sect.s 8.1 and 8.2. Either of these functions can in $\{p^0 > 0\}$ be represented as

$$\tilde{W}_\sigma(p,q) = \tilde{\tau}_\sigma^+(p,q) + \tilde{\tau}_\sigma^-(p,q) , \qquad (11.24)$$

$p$ and $q$ now denoting arguments of arbitrary type. Hence the smoothness of $\tilde{W}_\sigma$ can be derived from that of $\tilde{\tau}_\sigma^\pm$. We study more generally the singularities of

$$\tilde{\tau}_\sigma^+(p_1,\ldots,p_{n+1}) = \delta^4(\sum p_i) t^+(p_1,\cdots,p_n)$$

with $t^+$ considered as a *function* (not a distribution) of the external variables $p_i$, which may be of any type. The results obtained will also be essential in Part III. The problem is quite involved, and our treatment does not aspire to anything approaching full rigour. We give only indications of how to solve a problem that would deserve a cleaner analysis. The following strategy is used. We start from the known singularities of low-order graphs, examine what new singularities are produced by these initial ones if they occur inside graphs of higher order, and iterate this procedure until no new types of singularities emerge any longer.

Consider first a 1-particle reducible graph, e.g. the one shown in Fig. 11.2. We use the $\delta$-reduced graph rules. The momenta $P_L$ of the cuttable lines ($p_i$, $P_\alpha$, in our example) are linear combinations of the independent external variables $p_1,\ldots,p_n$. These lines contribute singular factors $\left(P_L{}^2 - M_L{}^2\right)^{-1}$

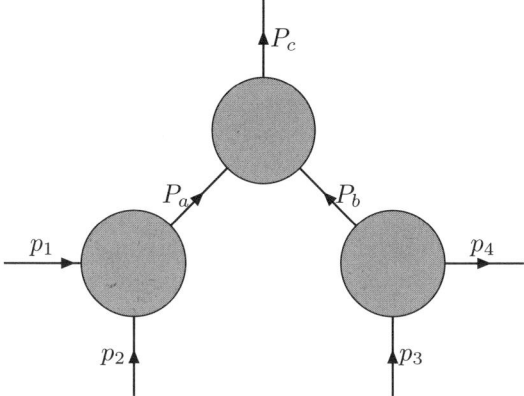

**Fig. 11.2.** A 1-particle reducible graph.
The blobs denote 1PI subgraphs

to the graph, with $M_L = m$ or 0. Any other singularities are due to the 1PI subgraphs whose "external" lines are the cuttable lines. Hence we need only discuss 1PI graphs any further. In particular this means that the external lines of the graphs have been "amputated", i.e. they have been multiplied with $p^2$ for photon lines, with $p^2 - m^2$ for fermion lines. The vertex factors and the numerators of the propagators are regular functions which are immaterial for the general form of the emerging singularities. They are therefore suppressed.

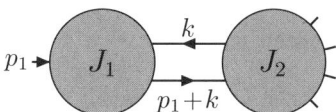

**Fig. 11.3.** A composite graph.
$J_1$ and $J_2$ are connected subgraphs. The lines are of arbitrary type. The amputated external lines are shown as stumps

Let $G$ be a 1PI graph contributing to a $t(p_1, \ldots, p_n)$ which is not a R-part. Split $G$ into the largest R-part $J_1$ containing the $p_1$-vertex and the rest $J_2$. If the $p_1$-vertex does not belong to a R-part, then $J_1$ is that vertex by itself. A typical configuration is shown in Fig. 11.3. If $p_1$ is a photon variable, $J_1$ may have an additional external photon line or an additional line connecting it to $J_2$. We discuss only the case of two connecting lines explicitly. The case of $J_1$ being a 4-photon part can be handled along similar lines. $J_1$ is 1PI by assumption. Assume at first that $J_2$ is also 1PI and that both $J_1$ and $J_2$ are real-analytic functions of their external variables. We must integrate

$$J_G(p_1,\ldots,p_n)$$
$$= \int dk\, J_1(\ldots,k)\, J_2(\ldots,k)\, \frac{1}{k^2 - M_1^2 + i\varepsilon}\, \frac{1}{(p_1+k)^2 - M_2^2 + i\varepsilon}\,. \tag{11.25}$$

At least one of the masses $M_i$ is the electron mass $m$, the other one may be $m$ or 0. The $k$-integral is UV convergent. $J_G$ is then analytic at all $p$-values for which the product of the two singular factors in (11.25) is a distribution in $k$. This is the case if the two manifolds $k^2 = M_1^2$ and $(p_1+k)^2 = M_2^2$ meet everywhere under a finite angle, i.e. they are tangential for no value of $k$. Tangentiality means linear dependence of the gradients $k$ and $p_1 + k$ of the denominators at a point of intersection. This condition is satisfied for all $k$ if $p_1 = 0$. However, in that case the product $\prod_i (k^2 - M_i^2)^{-1}$ is still defined as a distribution even if both $M_i$ are $m$. Therefore nothing bad happens at $p_1 = 0$. But a potentially dangerous tangentiality exists also where $k$ and $p_1$ are parallel, i.e. at the "thresholds" $p_1^2 = 4m^2$ if $M_1 = M_2 = m$ or $p_1^2 = m^2$ if one of the connecting lines is a photon line. In the latter case we are confronted with an integral of a type already studied in Sect. 10.1 (see (10.23) and (10.24)). Like there we find that $J_G(P)$ is of the form

$$J_G(P) = A(P) + B(P)\,(p_1^2 - m^2)\,\log(p_1^2 - m^2 + i\varepsilon)\,, \tag{11.26}$$

with $A$ and $B$ regular. Since the $B$-term vanishes at the mass shell, $J_G$ is there still $C^\infty$ in the directions tangential to the mass shell. If $p_1$ is a photon variable we find by a similar calculation a square root singularity

$$J_G(P) = A(P) + B(P)\sqrt{p_1^2 - 4m^2 + i\varepsilon} \tag{11.27}$$

at the electron-positron threshold $p_1^2 = 4m^2$. In both cases the result is still continuous, even Hölder continuous, and in the case (11.27) $J_G(P)$ is still regular in a neighbourhood of the mass shell $p_1^2 = 0$. If $J_1$ is a 4-photon part with two external variables $p_1$, $p_2$, we obtain the threshold singularity

$$J_G(P) = A(P) + B(P)\log\left[(p_1+p_2)^2 + i\varepsilon\right]\,, \tag{11.28}$$

if three photon lines connect $J_1$ and $J_2$ the singularity

$$J_G(P) = A(P) + B(P)\,p_1^2\log(p_1^2 + i\varepsilon)\,. \tag{11.29}$$

In this latter case the dangerous $B$-term again vanishes at the mass shell, hence $J_G$ is there $C^\infty$ in the tangential directions. In the case (11.28) it can be shown that $B(P)$ vanishes at $p_1 = p_2 = 0$ at least of second order after summing over all $J_1$-graphs of a given perturbative order. Define $p' =$

$p_1+p_2$. A closer inspection of the integral (11.25) reveals that the logarithmic singularity at $p'^2 = 0$ is due to $k$'s which are parallel to $p'$ and with $k^0 \in [0, p'^0]$. This set shrinks to the point $k = 0$ for $p' = 0$, in particular for $p_1 = p_2 = 0$. But we know from Sect. 10.2 that $J_1$ vanishes at $p_1 = p_2 = k = 0$, due to the WT identities. Since there exist no covariant first-order polynomials in these variables, the power expansion of $J_1$ starts with a term of at least second order.[6] A $k$-factor in $J_1$ improves the behaviour of the integrand at $k = 0$ sufficiently to remove the IR divergence at $p' = 0$. Only terms containing exclusively $p_1$ and $p_2$ retain the logarithm. This means that the coefficient $B(P)$ vanishes at $p_1 = p_2 = 0$ together with its first derivatives. If all external variables of $J_2$ are photonic too, then both $A(P)$ and $B(P)$ vanish at the origin (see footnote 4 of Sect. 10.2).

Clearly, the tentative assumption that $J_1$ and $J_2$ are regular is not generally satisfied. What happens if we admit in them threshold singularities of the above forms? A singularity of the type (11.26) in $k$ (or $p_1 + k$) replaces the corresponding pole factor by the less singular factor $\log(k^2 - m^2 + i\varepsilon)$. This generates new but *weaker* singularities at the thresholds, so that nothing worse than is already known will occur. A $e^+$-$e^-$ threshold is separated from the pole in the same variable, so that these singularities do not reinforce each other. By an extension of the former calculations one finds a buildup of successively weaker singularities at the higher thresholds $p_1^2 = n^2 M^2$, for $n$ positive integers. In every case $J_G$ is still Hölder continuous at the threshold.

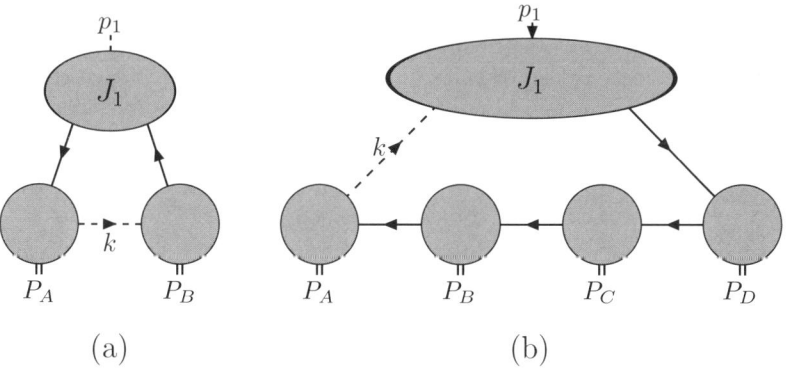

**Fig. 11.4a,b.** Graphs generating singularities.
The double stumps stand for sets of amputated external lines

We consider now the additional problems occurring if $J_2$ is 1-particle reducible. Then new singularities are generated in graphs of the types shown in

---

[6] As a matter of fact, the lowest polynomial satisfying the WT identities is of 4th order.

Fig. 11.4. Internal lines are shown distinguished according to type. The external variables $P_A, P_B, \ldots$ of $J_2$ may be partial sums of individual $p_j$'s. Graph (a) is locally divergent if $P_A{}^2 = P_B{}^2 = m^2$, because in this case we have at $k = 0$ the singularity $(k^2 + i\varepsilon)^{-1}[(P_A, k) + i\varepsilon]^{-1}[(P_B, k) - i\varepsilon]^{-1}$ which is not a distribution. This leads to a logarithmic singularity at $P_A{}^2 = P_B{}^2 = m^2$ which is a boundary value of an analytic function in the same way as the original integral: the integral as a function of $P_A^0(1+i\varepsilon)$, $P_B^0(1+i\varepsilon)$, is analytic in $\varepsilon > 0$, and the boundary value is defined as a distribution by taking the limit $\varepsilon \downarrow 0$ after integrating over a test function. If we return to the original case of $J_2$ being 1PI, but assume that $J_1$ or $J_2$ have singularities of the new kind, we find the following. If these singularities involve only external variables of the full graph, they obviously survive the $k$-integration. If one or both of the lines between the two $J_i$ are involved, the new singularities can cause problems only at the thresholds, where they may lead to additional factors $\log(p_1{}^2 - m^2 + i\varepsilon)$ or $\log(p_1{}^2 - 4m^2 + i\varepsilon)$ in the $B$-terms. The occurrence of these new terms may be contingent on some external partial sums $P_A$ lying on the mass shell. In any case, these new singularities are not significantly stronger than the original threshold singularities. Similar effects occur if the two fermion lines shown explicitly in graph (a) are connected by more than one photon line attached to vertex parts on both sides. The results are qualitatively the same. Self-energy insertions into the explicitly shown lines do not change the result either, because they do not significantly change the strength of the propagator singularities. This holds also for graph (b).

We call a function $f(u)$ "weakly singular" at $u = 0$ if it possesses there a singularity which is weaker than any power singularity, so that $u^\eta f(u)$ is bounded at $u = 0$ for all $\eta > 0$. For such a function we use the generic notation $\ell(u)$. The above iteration then results in a replacement of the logarithms in (11.26)–(11.29) by $\ell(p_1{}^2 - m^2 + i\varepsilon)$, etc., and the multiplication of the square root in (11.27) by $\ell(p_1{}^2 - 4m^2 + i\varepsilon)$.

For the graph (b) of Fig. 11.4 the $k$-integrand is not defined as a distribution at $k = 0$ if $P_A = P_B + P_C = 0$. Assuming at first that $P_B{}^2 = P_C{}^2 \neq m^2$, $P_D{}^2 \neq m^2$, we have at $k = 0$ a singularity of order $|k|^{-6}$. It leads to a singularity of the integral of second order at $P_A = P_B + P_C = 0$. But this manifold has codimension 8 in the full $p$-space, so that the new singularity is integrable, and so are its derivatives of order $\leq 5$. If $P_B{}^2 = P_C{}^2 = m^2$ and/or $P_D{}^2 = m^2$, then the fermion lines between $B$ and $C$ and/or between $D$ and $J_1$ contribute additional singularities at $k = 0$. The singularity of the integral in the external variables is then in the worst case only three times integrably differentiable. If the external variables are restricted to the mass shell $P_B{}^2 = P_C{}^2 = P_D{}^2 = m^2$ the codimension of the singularity manifold is reduced by 3, so that the singularity in the remaining variables is still integrable but its derivatives are not. The result is Hölder continuous in the transversal directions to the mass shell. This means that $\prod(P_L{}^2 - m^2 + i\varepsilon)^{-1} I_G(P)$ is defined as a distribution.

Consider now $k$-loops of the form of graph (b) but of arbitrary length, e.g. $n$ photon lines alternating with $n-1$ fermion lines of the $P_B - P_C$ configuration, and ending in a $P_D$ blob. In that case we have at $k = 0$ a singularity of order $3n$ if all sums $P_L$ of the external variables of the individual $J_2$ blobs lie on the mass shell and $P_{B_i} + P_{C_i} = 0$ for each of the $B$–$C$ pairs. In this worst case the $k$-integration yields a $P$-singularity of order $3n - 4$,[7] of lower order if $P_L{}^2 \neq m^2$ for one or several $L$. The codimension of the singularity manifold is $3(n-1)$ within the mass shell, higher within the full $P$-space. As a result we find again that $I_G(P) \prod_L (P_L{}^2 - m^2 + i\varepsilon)^{-1}$ is a distribution. Weakening the original assumptions on $J_2$ by admitting the presence of such singularities does not significantly alter these results. The threshold singularities remain what they were, still Hölder continuous and therefore multiplicable with the corresponding external propagators. There may be graph (a) singularities which prevent restricting $I_G(P)$ to the mass shell in certain partial $p$-sums, but these singularities are weak. And there may be graph (b) type infinities where a certain number of non-overlapping partial sums of external variables vanish. But these singularities are integrable even over the mass shell (if they do not coincide with (a)-type singularities), and they and a certain number of their derivatives are integrable over the full $P$-space.

Until now we have assumed that the full graph $G$ is not a R-part. If it is, we define $J_1$ to be the largest R-part inside $G$ which contains $p_1$. $J_2$ depends then on at most three external variables. If this number is 3, then all external variables are photonic. Because we are at present concerned with the local singularities of $I_G$ we may consider the unsubtracted $k$-integrand and the subtractions separately. About the former we notice that $I_G$ does not contain enough external variables to produce type (b) singularities. Singularities of type (a) occur in the vertex function but not in the 2-point functions in which we are especially interested. Take first the 2-point function $t^+(\mathbf{p})$. We might expect it to possess a threshold singularity of the form $p^2 \ell(p^2 + i\varepsilon)$ at $p^2 = 0$ originating from graphs in which $J_1$ and $J_2$ are 4-photon parts. But we know from Sect. 10.2 that the sum of such parts over all graphs of a given order vanishes of second order at the origin, possibly up to weakly singular factors. Using this fact and retracing the derivation of the threshold singularity, we find that the dangerous terms vanish at $p = 0$ of $6^{\text{th}}$ order, possibly up to a weak singularity. By covariance it must then be of the form

$$(p^2)^2 \left( a\, p^2 g^{\mu\nu} + b\, p^\mu b^\nu \right) \ell(p^2) , \qquad (11.30)$$

which after deamputation, i.e. division by $(p^2)^2$, is at most weakly singular at $p^2 = 0$. The remainder $r^+(p)$ of $t^+(\mathbf{p})$, i.e. the generalization of the $A$-term in (11.29), is regular up to the $e^+$–$e^-$ threshold at $p^2 = 4m^2$. The same holds for

---

[7] A negative order means that no singularity is present, order 0 indicates a weak (typically logarithmic) singularity.

the corresponding anti-time-ordered expression $r^-(p)$. Moreover, $r^+ + r^- = 0$ for $p^2 < 0$ on account of the spectral condition, hence by analytic continuation this holds for $p^2 < m^2$. Deamputation gives in $\tilde{w}_\sigma(\mathbf{p})$ the contribution

$$\theta(p_0)\, r(p) \left\{ \frac{1}{(p^2+i\varepsilon)^2} - \frac{1}{(p^2-i\varepsilon)^2} \right\},$$

which is a sum of a $\delta(p^2)$ and a $\delta'(p^2)$ term. The good behaviour of $t^+$, $t^-$ at $p = 0$ ensures also that the necessary subtraction to second order at that point is possible. The subtraction terms being polynomials, they do not introduce new singularities. Rather, as has been discussed in Chap. 10, they cancel the $\delta(p^2)$ and $\delta'(p^2)$ found above. And thus the desired form (8.7) of $\tilde{w}_\sigma(p)$, including the smoothness of $\Xi$, is established.

For the fermionic 2-point function $t^+(p)$ the situation is simpler. The subtractions are effected at $p = 0$, which does not lie on the mass shell, hence they contain no non-integrable local singularities. The singularities of $t^+(p)$ are entirely due to the unsubtracted integrand, which yields the known threshold singularity $(p^2-m^2)\,\ell(p^2-m^2+i\varepsilon)$. After deamputation it becomes $(p^2-m^2+i\varepsilon)^{-1}\ell(p^2-m^2+i\varepsilon)$, plus a second-order pole $(p^2-m^2+i\varepsilon)^{-2}$ coming from the regular $A$-part in (11.26), which is removed by mass renormalization. The same holds for $t^-(p)$, and we find that $\tilde{W}_\sigma(p,\bar{q})$ contains at $p^2 = m^2$ a singularity which is slightly stronger than a first-order pole and is Hölder continuous elsewhere. This singularity is one of the causes of the hard IR problem that will be met in Part III.

In the same way one finds that the vertex function $\Lambda(p,-\bar{q})$, i.e. the contribution of 1PI graphs to the amputated $t^+(p,-\bar{q})$, has a weak singularity at $p^2 = q^2 = m^2$, while it is smooth elsewhere. The weak singularity is present in graphs in which $J_1$ is itself a vertex part with $p_1$ its photon variable. In the other cases $\Lambda$ is smooth at the mass shell.

# Chapter 12

# Physical States

## 12.1 The General Strategy

The field theory whose perturbation theory has been developed in Chaps. 8–11 is not yet the QED we want. Firstly, the Lorentz condition (6.15) has been ignored. Therefore the true Maxwell equation (6.12) is not satisfied on the state space $\mathcal{V}$. Secondly, the Maxwell charge $Q_M$ (see (7.9) and (7.11)) vanishes identically on $\mathcal{V}$: $\mathcal{V}$ contains no charged physical states. Thirdly, there is the question of the positivity of the scalar product. But positivity is a somewhat elusive property in perturbation theory, and its discussion will therefore be postponed to the end of the chapter.

The problem cannot be solved with the methods used in Sect. 5.3. for the free electromagnetic field. We need to evolve a more elaborate strategy, as vaguely indicated at the end of Sect. 6.1. It is not the strategy usually employed in textbooks. Compared to this, it has the advantage of not introducing unnecessary new structures, such as a second scalar product, into the formalism. And it yields a clear, compelling prescription for how to renormalize QED in physical gauges, e.g. the Coulomb gauge.

Before explaining the strategy, a few general remarks are in order. Our way of proceeding looks quite arbitrary: it is not arrived at by arguing from first principles. But this is not objectionable, as is seen by looking at the problem in the wider context of the local observables formalism. A general description of this formalism is found in [H], a discussion of our particular problem, that of constructing suitable Hilbert space representations of QED, in a paper by Buchholz [Bu 82] (see also [Fr 79]). For us the salient point of this analysis is the following. As has been pointed out in a seminal paper by Haag and Kastler [HK 64], *any* faithful[1] representation of the algebra of observables is in principle sufficient to describe the full physical content of

---
[1] An operator representation of an algebra is called faithful if different elements of the algebra are represented by different operators.

the theory. This is so because it is never possible to measure more than a finite number of observables, localized in a bounded space-time region, and with a finite accuracy. But under the standard assumptions of the local observables formalism – which are admittedly to some extent motivated by mathematical convenience – it can be shown (Fell's theorem) that such a set of measurements can be described in any faithful representation. More explicitly: if both the theory and the experiments are correct, then there exist in any faithful representation of the observables states which are compatible with the given (or contemplated) measurements. The choice of representation is thus a matter of convenience. In fact, in QED one might do with a representation containing only uncharged states. If in a laboratory on earth some experiments on a charged system are carried out, we can interpret them as experiments concerning a larger system of total charge zero, which consists of the system of actual interest plus compensating charges at such a large distance ("behind the moon") that their influence on the actual experiment is negligible. But this is hardly a convenient way of handling charged systems, and therefore the quest for a formalism describing charged systems in a more direct way is justified.

In the search for such a formulation we first let ourselves be guided by the general ideas of the Gupta–Bleuler (henceforth called GB) formalism, which we have used with success in the free case. From the p-Maxwell equation (6.14) and current conservation it follows that $B(x) = \partial_\mu A^\mu(x)$ satisfies the free wave equation

$$\Box B(x) = 0 , \qquad (12.1)$$

so that $B$ can be split as in the free case into a positive-frequency part $B_+(x)$ and a negative-frequency part $B_-(x)$. We look for a state space $\mathcal{V}_1$ which is in some sense an extension of $\mathcal{V}$, on which $B_\pm$ are still defined, and such that the kernel $\mathcal{V}' \subset \mathcal{V}_1$ of $B_+$ has the properties used in Sect. 5.3 for the construction of a physical state space for the free electromagnetic field. For $\mathcal{V}_1$ we choose the set of states of the form $\mathcal{F}(\psi, \bar\psi, A)\Omega$ with $\mathcal{F}$ a "sufficiently general" functional of the basic fields. In order that the previous perturbative results can be used we assume that $\mathcal{F}$ can be formally expanded into a power series in $e$, such that the term of order $\sigma$ is a field polynomial whose coefficients are, however, not necessarily tempered test functions, but may be of slow decrease at infinity, and locally they may be distributions. For the moment we set all questions of existence aside.

From the perturbative expressions for the $W$-functions we find that on $\mathcal{V}$ the following relations hold:

$$[B(y), A_\mu(x)] = -i\frac{\partial}{\partial x^\mu} D(y-x) , \qquad (12.2)$$

$$[B(y), \psi(x)] = -e\, D(y-x)\, \psi(x) , \qquad [B(y), \bar\psi(x)] = e\, D(y-x)\, \bar\psi(x) . \quad (12.3)$$

Relation (12.2) is easily derived by using that only graphs in which the $y$-line is disconnected from the rest of the graph contribute to a $W$-function containing $B(y)$. Using the WT identities (9.41) for the sectors, one finds similarly that to $(\Omega, \cdots [B(y), \psi(x)] \cdots \Omega)$ contribute only the graphs in which the $y$-line is directly attached to a vertex on the $x$-trajectory in the $x$-sector, and these graphs sum to the result given in (12.3).

Let $f(y)$ be a real test function and define $C_f = \int dy\, f(y)\, B(y)$. The relations (12.2) and (12.3) yield

$$[C_f, A_\mu(x)] = -i\partial_\mu G(x) ,$$
$$[C_f, \psi(x)] = -e\, G(x)\, \psi(x) , \qquad [C_f, \bar\psi(x)] = e\, G(x)\, \bar\psi(x) , \qquad (12.4)$$

with

$$G(x) = \int dy\, f(y)\, D(y-x) .$$

$G(x)$ is $C^\infty$ and decreases at least like $|x|^{-1}$ for $|x| \to \infty$. Equation (12.4) shows that $C_f$ defines an infinitesimal gauge transformation. The operators $U(\alpha) = e^{i\alpha C_f}$ represent a one-parameter family of gauge transformations with gauge functions $\alpha\, G(x)$:

$$U(\alpha)\, A_\mu(x)\, U^*(\alpha) = A_\mu(x) + \alpha \partial_\mu G(x) ,$$
$$U(\alpha)\, \psi(x)\, U^*(\alpha) = \psi(x)\, e^{-i\alpha G(x)} . \qquad (12.5)$$

Such a gauge transformation with a gauge function vanishing at infinity we call "localized".

We conclude that $B(y)$ commutes with the functional $\mathcal{F}(A, \psi, \bar\psi)$ for all $y$ if $\mathcal{F}$ is invariant under localized gauge transformations. The same is then true for $B_+(y)$, so that

$$B_+(y)\, \mathcal{F}(A, \psi, \bar\psi)\, \Omega = 0 \qquad (12.6)$$

for such $\mathcal{F}$, since $B_+ \Omega = 0$ by the spectral property. Hence states of the form $\mathcal{F}\Omega$ with $\mathcal{F}$ invariant under localized gauge transformations are good candidates for the physical states forming $\mathcal{V}_1$. Notice that it would be too restrictive to admit only fully gauge invariant $\mathcal{F}$'s, since this would imply invariance under gauge transformations of the first kind, hence $Q_G \mathcal{F} \Omega = 0$: such a $\mathcal{V}_1$ would still not contain charged states.

We propose that a sufficiently large set of locally gauge invariant $\mathcal{F}$'s is formed by polynomials in suitable new "physical" fields $\mathcal{A}_\mu$, $\Psi$, $\bar\Psi$. Let $r^\mu(x)$ be four real functions or distributions satisfying

$$\partial_\mu r^\mu(x) = \delta^4(x) , \qquad (12.7)$$

or in $p$-space

$$p_\mu \tilde r^\mu(p) = i , \qquad (12.8)$$

with

$$\tilde r(p) = \int dx\, e^{ipx}\, r^\mu(x) .$$

## 12. Physical States

The desired physical fields are defined by an ansatz originally proposed by Dirac,

$$\Psi(x) = \exp\left[i\,e \int dy\, r^\rho(x-y)\, A_\rho(y)\right] \psi(x) ,$$
$$\bar\Psi(x) = \Psi^*(x)\,\gamma^0 ,$$
$$\mathcal{A}_\mu(x) = A_\mu(x) - \frac{\partial}{\partial x^\mu} \int dy\, r^\rho(x-y)\, A_\rho(y) . \qquad (12.9)$$

These expressions are invariant under localized gauge transformations of the original fields $A, \psi, \bar\psi$, provided that the gauge functions decrease sufficiently rapidly at infinity to allow integration by parts in the $y$-integrals. Hence (12.6) is satisfied. Under gauge transformations of the first kind ($G = $ const) $\Psi$ transforms as $\Psi \to \Psi\, e^{-iG}$, as it behoves a charged field. Distribution-theoretical problems are ignored in this formal discussion.

Notice that (12.9) themselves have the form of a gauge transformation with the operator-valued gauge function

$$G(x) = -\int dy\, r^\rho(x-y)\, A_\rho(y)$$
$$= -(2\pi)^{-\frac{3}{2}} \int dp\, e^{-ipx}\, \tilde r^\rho(p)\, \tilde A_\rho(p) . \qquad (12.10)$$

They effect (formally) the transition from the GB space to a physical space. But at first we try to stick to the original idea of defining the fields (12.9) as non-local composite fields on an extension $\mathcal{V}_1$ of $\mathcal{V}$. $\Psi$ is not a polynomial in the GB fields and $r^\mu$ is not a test function. Hence the new fields do not map $\mathcal{V}$ into itself. But they might be definable on an extension $\mathcal{V}_1 \supset \mathcal{V}$. We could then define a subspace $\mathcal{V}' \subset \mathcal{V}_1$ as the space generated from the vacuum by polynomials in the new fields. And on $\mathcal{V}'$ the Maxwell equations hold because the matrix elements of $B_+$ between states of $\mathcal{V}'$ vanish. In a vague sense the exponential factor in $\Psi$ might be construed as supplying the "cloud of soft photons" that necessarily accompanies the "charged particle" described by $\psi$ according to the standard lore. But interacting fields should better not be considered to be closely associated with particles. In any case, the idea fails because $\Psi$ is not defined on $\mathcal{V}$, in particular not on $\Omega$, hence the new fields do not generate an extension of $\mathcal{V}$. This is easily seen in perturbation theory: the scalar product $(\Omega, \psi(x)\,\bar\Psi(y)\,\Omega)$ diverges in order $e^2$ even after renormalization (see Sect. 12.3), hence the vectors $\psi^*(x)\Omega$ and $\bar\Psi(y)\Omega$ have a divergent scalar product and cannot be elements of the same vector space $\mathcal{V}_1$. We must of course insist on the scalar product of $\mathcal{V}$ being extensible to $\mathcal{V}_1$, otherwise the construction would not be of much use.

We will therefore not pursue the extension idea, but rather consider (12.9) as a gauge transformation leading to a new gauge with a new state space called $\mathcal{V}'$. This $\mathcal{V}'$ is constructed from $\mathcal{V}$ via a detour through the reconstruction theorem. It turns out, at least in perturbation theory, that our previous rules allow the calculation of the vacuum expectation values of products of new fields as finite quantities, even though this is not possible for mixed old–new products. More exactly, factors $A_\mu$ are admissible, but not $\psi$ and $\bar\psi$. These vacuum expectation values define then by the reconstruction theorem a field theory with the basic fields $\Psi, \bar\Psi, \mathcal{A}_\mu$, and the state space $\mathcal{V}'$, satisfying the W-axioms with the exception of locality and manifest Lorentz invariance, but including positivity of the scalar product and the validity of the Maxwell equations. The Dirac equations, being gauge invariant, remain valid under the transformation (12.9). But there is no simple direct connection between $\mathcal{V}$ and $\mathcal{V}'$, i.e. no physically meaningful 1–1 mapping between $\mathcal{V}$ or a subspace thereof and $\mathcal{V}'$.

The condition (12.8) implies that at least one $\tilde{r}^\mu(p)$ must have a singularity of at least first order at the origin $p = 0$, and that at least one $\tilde{r}^\mu$ cannot decrease faster at infinity than $|p|^{-1}$. Correspondingly, at least one $r^\mu(x)$ must not decrease faster than $|x|^{-2}$ for $|x| \to \infty$. That $\Psi(x)$ is not defined as an operator on $\mathcal{V}$ is due to this slow decrease. We can therefore approximate $\Psi$ etc. by fields $\Psi_\xi$ etc. defined on $\mathcal{V}$ by means of a cutoff of $r^\mu$ at large distances. Let $\chi(u)$ be a test function on $R^4$ with compact support and with $\chi(u) \equiv 1$ in a neighbourhood of the origin. Define

$$r_\xi^\mu(x) = \chi(\xi x)\, r^\mu(x) \qquad (12.11)$$

for $0 < \xi < \infty$, and define $\Psi_\xi, \mathcal{A}_\xi^\mu$, by (12.9) with $r^\mu$ replaced by $r_\xi^\mu$. Then it can be verified in perturbation theory that under suitable assumptions on the $r^\mu$ these fields exist on $\mathcal{V}$, after a slight elaboration of the renormalization procedure, and that the corresponding $W$-functions $W_\xi(\cdots)$ converge for $\xi \to 0$ to distributions $W(\cdots)$ satisfying the W-properties except locality and Lorentz covariance.

It remains to specify the "suitable assumptions" just mentioned. According to the earlier remarks about the wide freedom we have in choosing a particular representation of the theory, the choice of the $r^\mu$ is largely a matter of convenience. We let ourselves be guided by the following considerations. The exponential factor in the definition of $\Psi$ becomes a polynomial in every finite order of perturbation theory. This allows a perturbative calculation of the new $W$-functions. But the degree of this polynomial increases with increasing $\sigma$, so that graphs with an ever larger number of external sectors must be considered. In order to minimize this inconvenience we choose the $r^\mu(x)$ such that their supports are contained in the 3-plane $x^0 = 0$ in a given frame of reference. This is harmless as far as relativistic invariance is concerned because manifest invariance cannot be achieved in any case. But it means that the $r^\mu$ are distributions, not functions: they contain a factor $\delta(x^0)$. The advantage of this choice is that the ordinary product (12.9) can be replaced by a time-ordered product

$$\Psi(x) = \sum_{n=0}^\infty \frac{(ie)^n}{n!} \int \prod_{j=1}^n \{dy_j\, r^{\mu_j}(x - y_j)\}\, T^\pm\left(\cdots A_{\mu_j}(y_j) \cdots \psi(x)\right), \qquad (12.12)$$

at least at the present formal level. Then for every $\Psi$ or $\bar\Psi$ in a $W$-function there is only one external sector of a type whose selection is again a matter of convenience. Apart from the introduction of this temporal $\delta$-factor we demand that $r^\mu$ and $\tilde r^\mu$ be as well behaved as is compatible with the conditions (12.7) and (12.8). More exactly this means that $\tilde r^\mu(p)$ depends only on $\vec p$, not on $p_0$, as a result of the support condition for $r^\mu$. Condition (12.8) implies then that $\tilde r^0 \equiv 0$. The $\tilde r^i$ shall be $C^\infty$ except at $\vec p = 0$. There at least one $\tilde r^i$ must have a singularity, which we assume to be of first order, and the $\tilde r^i$ shall decrease for $|\vec p| \to \infty$ not slower than $|\vec p|^{-1}$. Slight relaxations of these conditions, i.e. only boundedness for every $\varepsilon > 0$ of $|\vec p|^{1-\varepsilon}|\tilde r^i(\vec p)|$ for large $|\vec p|$, and of $|\vec p|^{1+\varepsilon}|\tilde r^i(\vec p)|$ for small $|\vec p|$, are acceptable. We say then that $\tilde r$ is of "essential order" $|\vec p|^{-1}$ both for $|\vec p| \to 0$ and $|\vec p| \to \infty$. Correspondingly, $r^i(x)$ shall be of the form

$$r^i(x) = \delta(x^0)\,\hat r^i(\vec x)\,, \qquad (12.13)$$

with $\hat r^i \in C^\infty$ for $\vec x \neq 0$, and $|\vec x|^2 \hat r(x)$ should be bounded. Also, the $n^{\text{th}}$ derivatives $\partial^n \hat f^i$ shall decrease like $|\vec x|^{-2-n}$ for $|\vec x| \to \infty$ and have at most a singularity of order $2+n$ at $\vec x = 0$ for $n \leq N$ sufficiently large. Again, slight deviations of logarithmic type from these boundedness conditions are acceptable both for large and small $|\vec x|$. The simplest choice satisfying these conditions is

$$\tilde r_C^j(\vec p) = -i\,\frac{p^j}{|\vec p|^2}\,,\quad \tilde r_C^0(\vec p) = 0\,. \qquad (12.14)$$

The resulting fields $\mathcal{A}_C, \Psi_C, \bar\Psi_C$ are those of the Coulomb gauge: it is easily seen that $p_j \tilde{\mathcal{A}}_C^j(p) = 0$. Hence our construction directly yields the Wightman functions of the Coulomb gauge or of other physical gauges.

## 12.2 Unrenormalized Perturbation Theory

It is our aim to derive a graphical representation for the $W$-functions of the physical fields $\mathcal{A}^\mu, \Psi, \bar\Psi$ starting from the known sector graphs of the GB fields $A^\mu, \psi, \bar\psi$. The notation is taken over from that case. The way of proceeding is as in the GB case. In the present section unrenormalized graphs are introduced in a formal way. In the next section their UV divergences are removed by renormalization, and it will be shown that the sum of graphs of a given order contributing to a given $W$ is IR convergent. It is only in this last step that the $\xi$-limit becomes essential, so that at present we may work directly with the correct $r^\mu$ without cutoff.

We start with the unrenormalized theory. The new $W$-functions are obtained from the old GB ones by expressing the physical fields with the help of the definition (12.9) for $\mathcal{A}^\mu(x)$ and the expansion (12.12) for $\Psi(x)$ and the corresponding expansion for $\bar\Psi(x)$. In order $\sigma$ only the terms with $n \leq \sigma$ of these expansions occur. Let $\Psi_n$ be the $n^{\text{th}}$ term in (12.12). It is treated as a composite field as $j^\mu$ and $A\!\!\!/\,\psi$ were in Sect. 10.2. The $W$-functions containing

$\Psi_n$ as a factor involve the GB functions

$$\mathcal{W}_\sigma^{GB}(\cdots|y_1,\ldots,y_n,x,\pm|\cdots),$$

which are evaluated according to the rules of Chap. 9.

Let $\mathcal{W}_\sigma(X_1,s_1|\cdots|X_N,s_N)$ be defined as before, with the $X_i$ now being sets of arguments of physical fields. We write down the graph rules for $\mathcal{W}_\sigma$ obtained as just explained. The graphs differ from the GB graphs by a more complicated prescription concerning the external points associated with the external variables. In the GB case they were points in which exactly one line of the graph ended, this line carrying a standard propagator. This is now different. With respect to external photon lines the geometrical form of the graphs is not changed. But if $_\mu x$ is an external photon variable, $_\nu z$ the variable at the other end of its line, then the corresponding propagator is changed to

$$i\left[g_{\mu\nu}D_\alpha(x-z) - \frac{\partial}{\partial x^\mu}\int dy\, r_\nu(x-y)\,D_\alpha(y-z)\right], \tag{12.15}$$

where $D_\alpha$ is $D_F$, $-D_F^*$, or $D_+$, as the case may be (see Sect. 9.3), and $r_\nu = g_{\nu\rho}r^\rho$.

With the $\Psi$ variable $x$ is associated a whole set of external vertices, one for each $\Psi_n$. A $\Psi_n$ vertex is counted of order $n$: it contributes the value $n$ to the total order $\sigma$ of the graph. Incident at a $\Psi_n$ vertex are a fermion line directed away from it and $n$ photon lines (see Fig. 12.1). The fermion line carries the usual fermion propagator, the photon lines the propagators

$$-\int du\, r^{\nu_j}(x-u)\,D_{\alpha_j}(u-z_j), \tag{12.16}$$

where $^{\nu_j}z_j$ is the variable at the other end of the line. $D_{\alpha_j}$ is one of the invariant $D$-functions according to the general rules. The vertex itself carries no vertex factor. A $\bar{\Psi}$ factor is represented in the same way by $\bar{\Psi}_n$ vertices, $n = 0, 1, 2, \ldots$, which have the same structure as the $\Psi_n$, except that the fermion line points into the vertex and the factor $-1$ is replaced by $+1$.

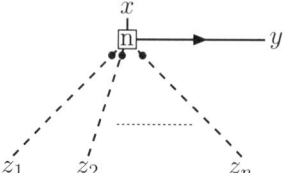

**Fig. 12.1.** A $\Psi_n$-vertex.
The dots near the ends of photon lines denote convolution with $r^\mu$

The internal vertices and lines, i.e. lines not attached to external vertices, retain the same topological structure and the same vertex factors and propagators as in the GB case. Finally, the integrand of the graph $G$ must be divided by its symmetry number $N_G$: the number of permutations of lines and internal vertices which leave $G$ topologically unchanged. Such permutations do not exist in the GB case because all the lines joining a given vertex are distinct by type or direction. But the $\Psi_n$ and $\bar{\Psi}_n$ vertices are joined by $n$ indistinguishable photon lines.

For the verification of the *field equations* (Maxwell and Dirac) external vertices representing the composite fields $j^\mu(x) = \bar{\Psi}(x)\gamma^\mu\Psi(x)$ and $\mathcal{A}(x)\Psi(x)$ must be introduced. We start with the current. From

$$\Psi(x) = \sum_n \frac{(-ie)^n}{n!} G^n(x)\psi(x)$$

we obtain formally, noting that $G$ is real and commutes with the Dirac matrices,

$$\begin{aligned}j^\mu(x) &= \bar{\psi}(x)\gamma^\mu \sum_{N=0}^\infty (ie)^N G^N(x) \sum_{n=0}^N \frac{(-1)^n}{n!(N-n)!}\psi(x) \\ &= \bar{\psi}(x)\gamma^\mu\psi(x) \; .\end{aligned} \quad (12.17)$$

This means that $j^\mu(x)$ can be represented by a vertex of the old GB form. This can be directly verified from the graph rules, where in the graphs of a finite order the current is represented as a finite sum over products of $\bar{\Psi}_n(x)$ and $\Psi_{N-n}(x)$ vertices with identical variables, i.e. vertices coalesced into one vertex. A similar trivial summation as in (12.17) shows that only the term $N=0$ survives in the sum over graphs with identical structure but different assignments of the photon lines joining the $x$-vertex to the two factors. The consideration of Sect. 9.4 showing that the ordinary product (12.17) (more exactly the commutator $\frac{1}{2}[\bar{\psi}(x),\gamma^\mu\psi(x)]$) may be replaced by a time-ordered product then becomes applicable.

This result allows the verification of the *Maxwell equations*. The field tensor $F^{\nu\mu}$ is given by

$$F^{\nu\mu}(x) = \partial^\nu \mathcal{A}^\mu(x) - \partial^\mu \mathcal{A}^\nu(x) \; . \quad (12.18)$$

The homogeneous Maxwell equation is automatically satisfied. The inhomogeneous equation $\partial_\nu F^{\nu\mu} = j^\mu$ remains to be proved. Consider an external point representing $F^{\nu\mu}(x)$ of a graph for a $W$-function and the corresponding photon line. Let its other end be a vertex with variable $^\rho u$. The propagator of the $x$–$u$ line is obtained from the $\mathcal{A}$-propagator (12.15) using the definition (12.18). It is found that the $r$-term in (12.15) drops out, and the propagator takes the form

$$i[g^{\mu\rho}\partial^\nu - g^{\nu\rho}\partial^\mu]D_\alpha(x-u) \; . \quad (12.19)$$

Applying $\partial_\nu$ yields

$$i[g^{\mu\rho}\Box - \partial^\mu\partial^\rho]D_\alpha(x-u) \quad (12.20)$$

as propagator of the $\partial_\nu F^{\nu\mu}$-line. Consider first the case that the $u$-vertex lies in the $x$-sector, which for definiteness we assume to be of type $T^+$. Then the contributions of the $\partial^\mu \partial^\rho$-term to a suitable set of such graphs cancel as in the proof of current conservation in Sect. 9.5. $D_\alpha$ is $D_F$, and $\Box D_F(x-u) = \delta^4(x-u)$: the wave operator of the $g^{\mu\rho}$-term amputates the $x$-line, and what remains is a $j^\mu(x)$ vertex. These contributions yield the desired right-hand side of the Maxwell equation. If the $x$-line is a cross line ending in an internal sector or another external photon sector, then the $\partial^\mu \partial^\rho$-terms cancel between graphs for the same reason as before. But now $\Box D_\alpha = \Box D_\pm$ vanishes too: we obtain no contribution. A vanishing contribution obtains also if the $x$-line ends at another external $\mathcal{A}$-point, as can be easily ascertained. In that case the $F$–$\mathcal{A}$ propagator still contains a factor of the form (12.15) coming from the $\mathcal{A}$-end, and the $\partial^\mu \partial^\rho$ term in (12.20) applied to that factor vanishes because of condition (12.7). There remains the case of the $x$-line ending in a $\Psi$- or $\bar\Psi$-sector. For definiteness we assume this to be a $\Psi$-sector of type $T^+$ standing to the right of the $x$-sector in $W$. The other cases are handled in the same way. Again, the $\Box$-term in (12.20) gives a vanishing contribution, and the $\partial^\mu \partial^\rho$ contributions to almost all graphs cancel each other as before. The only exceptions are graphs in which $x$ is directly connected to a $\Psi_n$-vertex or to the first inner vertex on the fermion trajectory starting at a $\Psi_n$-vertex (including the case $n=0$). Let the $\Psi$-variable be $z$. In the first case the operator $\partial^\mu_x \partial^\rho_x$ acts on a propagator factor of the form (12.16) with $\nu_j = \rho$ and yields the new propagator

$$i\partial^\mu_x \partial^\rho_x \int dy\, r_\rho(z-y)\, D_+(x-y) = -i\, \partial^\mu_x D_+(x-z)\,.$$

In the second case we find from the WT identity (9.41), or rather from its proof, that the relevant term surviving the previous cancellations excises the fermion line between $z$ and $u$, so that the $x$-line is joined directly to the $z$-vertex with propagator $i\partial^\mu_x D_+(x-z)$. In the sum $\sum_n \Psi_n$ the two types of contributions cancel.

Since the transformation (12.9) from GB to physical fields is formally a gauge transformation, we expect the gauge invariant *Dirac equations* to remain valid.[2] But the simple product form of (12.9) will not survive renormalization. It is therefore desirable to check the Dirac equations explicitly already at the present non-renormalized level. The external $\mathcal{A}\Psi$-vertices are obtained by the fusion of a $\mathcal{A}_\mu$-vertex, a factor $\gamma^\mu$, and a $\Psi_n$-vertex into a single vertex, analogously to what has been done for $j^\mu$-vertices. From such a vertex (variable $x$) issues a fermion line (end variable $y$), $n$ photon lines (end variables $u_1,\ldots,u_n$) with propagators (12.16), and a photon line (end variable $v$) with a propagator (12.15) contracted with $\gamma^\mu$. This last propagator consists of a "GB term" and the $r$-dependent "gauge term". For verifying

---

[2] The same transformation turns the non-invariant p-Maxwell equation into the gauge invariant true Maxwell equation. This means simply that the mapping $\psi \to \Psi$ etc. is not invertible.

the Dirac equation we must find out what happens if the Dirac operator $\mathcal{D}$ is applied to a $\Psi_n$-vertex. Assume $\Psi_n$ to lie in a $T^+$-sector. The derivation $\partial\!\!\!/$ acts on the joining propagators according to the product rule. The contribution from the fermion line is combined with the mass term of $\mathcal{D}$. The fermion propagator is then annihilated if this line is a cross line; the line is excised and its ends coalesced into a single vertex if it is an sector line. This new vertex is the part of a $\mathcal{A}\Psi$-vertex containing the GB term of the exceptional $\mathcal{A}$-line, this line being the line formerly joining the end vertex of the excised fermion line. The operator $i\partial\!\!\!/$ applied to a photon propagator at a $\Psi_n$-vertex with the appropriate connections to other vertices supplies the missing gauge term of the exceptional line at the $\mathcal{A}\Psi$-vertex.

In *p-space* the graph rules are changed as follows with respect to the GB rules. The factor $g_{\mu\nu}$ in an external $\mathcal{A}_\mu$-propagator with momentum $\ell$ is replaced by

$$g_{\mu\nu} + i\,\ell_\mu\,\tilde{r}_\nu(\ell) \ . \tag{12.21}$$

A $\tilde{\Psi}_n(p)$-vertex carries the factor $i^n(2\pi)^{-\frac{3}{2}n}\delta^4(p - q - \sum k_i)$, where $q$ is the momentum of its fermion line, $k_i$ those of the photon lines, and all lines are directed out of the vertex. In the $k_i$-propagators the factor $g_{\mu\nu_i}$ of the GB form is replaced by $\tilde{r}_{\nu_i}(k_i)$. The vertex factor of a $\tilde{\bar{\Psi}}_n(p)$-vertex is $(-i)^n(2\pi)^{-\frac{3}{2}n}\delta^4(p+q-\sum k_i)$, where now the fermionic $q$-line is pointing into the vertex but the photonic lines still away from it. The photon propagators are the same as in the $\Psi_n$-case. The same symmetry number as in $x$-space must be included. Otherwise the rules are as in the GB case. Notice that all cross lines carry $\delta_\pm$-factors as before, so that the spectral condition for $\tilde{W}$ is trivially satisfied.

The rules for the general $\mathcal{W}$-functions are changed accordingly, both in $x$- and $p$-space, by replacing the external $A$-points by $\mathcal{A}$-points and adding $\Psi_n$- and $\bar{\Psi}_n$-vertices.

Let us call the graph rules just introduced the "A-rules". It is sometimes more convenient to work with an equivalent set of rules, the "B-rules". They are the rules obtained from the canonical formalism or from path integrals, starting directly in a physical gauge like the Coulomb gauge (see e.g. [BD], [He 79]). In these rules the graphs have exactly the same form as in the GB case. No $\Psi_n$-vertices are introduced. But for all photon lines, external and internal, the propagator $i\,g_{\mu\nu}D_\alpha(z_i - z_j)$ (see Sect. 9.3) is changed to

$$D_\alpha^{\mathrm{ph}}(z_i - z_j) = i\left(\delta_\mu^\kappa - \frac{\partial}{\partial z_i^\mu}\int du_1 r^\kappa(z_i - u_1)\right)g_{\kappa\lambda}$$
$$\times \left(\delta_\nu^\lambda - \frac{\partial}{\partial z_j^\nu}\int du_2 r^\lambda(z_j - u_2)\right)D_\alpha(u_1 - u_2) \ .$$
$$\tag{12.22}$$

In $p$-space this means that the $g_{\mu\nu}$ in the photon propagators are changed to

$$\left(\delta_\mu^\kappa + i\ell_\mu \tilde{r}^\kappa(\ell)\right) g_{\kappa\lambda} \left(\delta_\nu^\lambda - i\ell_\nu \tilde{r}^\lambda(-\ell)\right) , \tag{12.23}$$

the momentum $\ell$ flowing from the $\mu$-end to the $\nu$-end. The signs of the $r$-terms are independent of the type of sector in which the respective end of the line lies.

The proof of the equivalence of the two sets of rules follows the by-now familiar route. Starting from the B-rules, working in $p$-space, we look at the $r$-term in the left-hand factor of (12.23). If its line end is an external $\mathcal{A}$-point, then this is just the correct factor of the A-rules. If the end is attached to a trajectory, we contract $\ell_\mu$ with the $\gamma^\mu$ of the corresponding vertex and sum over all possible positions of that vertex on the trajectory. As usual we obtain almost complete cancellation of these terms, except for the extremal vertices on the trajectory, where contributions of $\Psi_1$-type survive. Carrying out this procedure for all ends of all photon lines, we recover the A-rules. The same procedure can be carried out in the reverse direction, from A to B.

The necessary properties of the W- and $\mathcal{W}$-functions that have not yet been checked can be easily established with both the A- and B-rules, following the proofs given in Sect. 9.4 for the GB case. This is obvious for the permutation symmetry within sectors and for hermiticity. Lemma 9.2 remains correct for the physical rules because the vital equation (9.32) is still correct even for the modified photon propagators. Here the support condition $\xi^0 = 0$ for $r^\mu(\xi)$ becomes essential! As a result, the sign independence of $W_\sigma$ remains satisfied. Current conservation remains intact too, because the $j^\mu$-vertices have the same form as in the GB case, and the exact form of the photon propagator is immaterial for the proof in Sect. 9.5. The only not entirely trivial point is that of proving the time-ordering relations using the A-rules. In the GB case the proof started from (9.28), comparing the two expressions $I_1$ and $I_2$ defined there. The generalization of this to the case that the specially selected variable $x$ is a $\Psi_n$ variable is not straightforward. And we are forced to consider that case e.g. in purely fermionic $\mathcal{W}$'s. But the generalization is simple: single out the fermion line leaving the $\Psi_n$-vertex, call its other endpoint $u$, change the explicitly shown propagator function $D_\alpha$ to $S_\alpha$, replace $\mathcal{W}_{\sigma-1}$ by $\mathcal{W}_{\sigma-n-1}$, and include in the set $X$ $n$ photon variables x belonging to the photon lines joining $\Psi_n$. Then everything goes through as before. If an external photon variable is present in $X'$, the GB proof can of course be taken over as it stands, only replacing the $D_\alpha(x-u)$ by their modified form (12.15).

## 12.3 The UV Problem

The UV problem is most easily solved for the A-rules, which therefore will be used exclusively from now on. We work in $p$-space, directly using the full

$\tilde{r}^\mu$-factors, not their cutoff version $\tilde{r}_\xi$. The $\xi$-limit will only become important in the discussion of the IR problem.

We follow the procedure of Sect. 10.2. Nothing is changed relative to the GB case for R-parts containing only internal vertices. The change of the external photon propagators is UV irrelevant because such propagators do not occur in 1PI subgraphs. But new UV divergences are introduced by the $\Psi_n$- and $\bar{\Psi}_n$-vertices. These new divergences reflect the fact that the composite field $\Psi_n(x)$ is as usual defined as a product of distribution-valued factors, but with the roughness of the photonic factors being softened by their convolution with $r^\mu$ (see (12.16)). By our assumptions, $\tilde{r}^\mu(\vec{p})$ decreases at least like $|\vec{p}|^{-1}$ for $|\vec{p}| \to \infty$. That means that the photon propagators decrease at infinity of order 3: the factor $\tilde{r}^\mu$ improves the UV behaviour by one order. That such an additional decrease is not present for $p_0 \to \infty$, $\vec{p} = $ const, does not invalidate this assessment, because these exceptional directions are sufficiently low-dimensional not to matter greatly. Hence the dimension $d(\Gamma)$ of a subgraph $\Gamma$ containing such a photon line is diminished by 1 relative to the standard value (10.37). The new R-parts involving $\Psi_n$-lines are of one of the two types shown in Fig. 12.2. Type (a) consists of a photon line possibly interrupted by SEP's, connecting the $n$-vertex to itself. Such a subgraph is logarithmically UV divergent (it contains two $\tilde{r}$-factors). But the integrand does not depend on any external variables. Hence the standard subtraction at vanishing external momenta completely removes that graph: graphs containing subgraphs of type (a) must be omitted. Note, however, that this loop integral is also IR divergent because at least one $\tilde{r}^\mu$ has a singularity of order 1 at the origin. This IR divergence is then at first also removed by renormalization. But it is needed later on for cancelling other IR divergences, and we will therefore be forced to partially retract this subtraction. But at the moment this need not concern us.

The subgraphs of type (b) are also logarithmically divergent if situated inside a sector. They are rendered finite by the standard prescription: subtract the integrand at $q = 0$, then integrate. In this case the subtraction is not IR divergent, so that the procedure is well defined. However, the subgraph (b) is IR divergent for $q$ lying on the mass shell $q^2 = m^2$ because there the integrand has a singularity of $4^{\text{th}}$ order at $k = 0$. But the resulting mass-shell singularity is only logarithmic. Together with the pole of the adjacent $q$-propagator it is still of allowed strength, so that no additional mass renormalization is required.

The 2-photon R-parts remain exactly the GB ones, so that the condition (8.7) remains satisfied except for two additional external factors of the form (12.21). For the same reason the Maxwell equations remain satisfied. The Dirac equation is a more delicate problem. As has been remarked before, the gauge form (12.9) of $\Psi$ is modified by the UV subtractions introduced above. Hence the validity of the Dirac equation is no longer assured. The problem shows up as follows. The procedure used to prove the unrenormalized Dirac

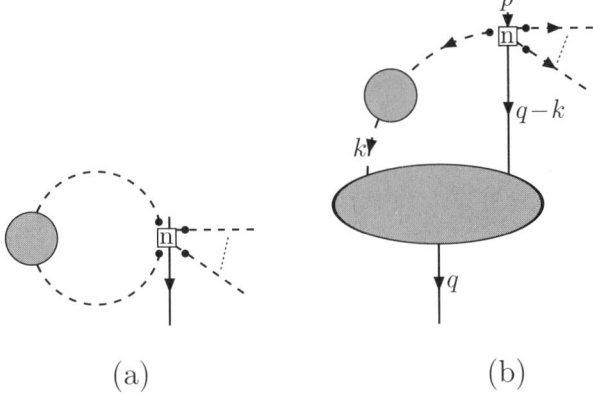

**Fig. 12.2a,b.** R-parts involving $\Psi_n$-vertices.
The grey blobs are renormalized subgraphs. The 2-line blobs may be absent, the 3-line blob may be a single vertex

equation amounts in $p$-space to this: the factor $(\not{p} - m)$ applied to a $\tilde{\Psi}_n(p)$ vertex is written as $(\not{q} - m) + \sum \not{k}_i$, where $q$ is the momentum of the electron line attached to the vertex, $k_i$ those of the photon lines. The various terms are then multiplied with the respective propagators. This turns the external factor $(\not{p} - m)$ into internal factors of subgraphs, increasing their dimension by 1, thereby creating new R-parts and raising the necessary number of subtractions in already existing ones. These "unnecessary" subtractions (they are not needed if the factor is left outside) do not cancel between the various resultant graphs, though their sum is UV convergent. Nevertheless the Dirac equation can be saved by an appropriate renormalization prescription for its right-hand side $\mathcal{A}\Psi$. Consider a $\mathcal{A}\Psi_n$-vertex as introduced in the formal proof of the Dirac equation, but now in $p$-space. The photon lines of the factor $\Psi_n$ are UV improved by one order by the factor $\tilde{r}^\mu$. This is not true for the photon line coming from $\mathcal{A}$, because in that case the decreasing $\tilde{r}(k)$ is multiplied with an increasing $k_\mu$. We split this extra line again into its GB part and its gauge part. R-parts containing the latter are subtracted as usual, by subtracting the integrand to sufficiently high order at vanishing external momenta. But in the graphs containing the GB part we insert the decomposition of unity,

$$\mathbf{1} = (\not{p} - \sum \not{k}_i)(\not{p} - \sum \not{k}_i - m)^{-1} - m(\not{p} - \sum \not{k}_i - m)^{-1},$$

and treat the terms separately, again subtracting the ensuing R-parts by the standard prescription. The $k_i$ are the momenta of the $\Psi_n$ photon lines, not including the $\mathcal{A}$-line. Note that the second term in this decomposition has a better UV behaviour than the first one, so that it gives rise to fewer subtractions. It can be shown that with this definition of $\mathcal{A}\Psi$ the Dirac equation is still satisfied. We refrain from giving the proof here. It is rather cumbersome and not particularly instructive. It might seem that this definition of the interaction term is rather arbitrary. Indeed it amounts at least in part (but not fully: see the unrenormalized case) to turning the Dirac equation into a definition of its right-hand side. But the Dirac equation is not an equation between observables, its physical meaning is not clear, and we can therefore afford to be generous in its exact formulation.

These renormalization prescriptions are hard to translate into the B-rules, and we will not endeavour to do so. One might, however, be tempted to start directly from the B-rules. Noting that the photonic propagators with numerator (12.23) have the same UV behaviour as in the GB case, it is found that $d(\Gamma)$ is equal to its GB value and the R-parts are the same in both cases. One might then try to render them finite by taking over the GB prescription of subtracting at vanishing external momenta. However, in order to preserve the physically essential moderate strength of the mass-shell singularity of the 2-point electron function it would then be necessary to introduce an additional finite mass renormalization (i.e. additional to the GB one) which is *not* a Lorentz scalar. This would really destroy the validity of a sensible Dirac equation, and it would also change the photonic $W$-functions in unfortunate ways. For example, a vacuum expectation value of a product of observable fields $F^{\mu\nu}$, $j^\mu$ would no longer be covariant, which is not acceptable (see remark at the end of this chapter).

In $x$-space the photon propagator (12.16) of a $\Psi_n$-line takes the form

$$P_\alpha^\nu(\xi) = -\int d^3u \, \hat{r}^\nu(\vec{u}) \, D_\alpha(\xi^0, \vec{\xi}-\vec{u}), \qquad (12.24)$$

with $\xi = x - z$. $\hat{r}^\nu$ has an integrable singularity at $\vec{u} = 0$ and is $C^\infty$ elsewhere. The decrease of the integrand at large $\vec{u}$ is sufficiently fast to guarantee existence of the integral at infinity. It follows immediately that $P^\nu$ exists and is $C^\infty$ off the light cone. For $\xi^2 = 0$ but $\xi \neq 0$ the $\hat{r}$-singularity sits on top of the $D_\alpha$-singularity. But the $\hat{r}$-singularity is of essential order 2, the $D_\alpha$-singularity of order 1, which results in a divergence of $P_\alpha(\xi)$ of at most order 1: the light-cone singularity is weaker than that of $D_\alpha$. This is true too for the singularity at the origin $\xi = 0$. For the special case (12.14) of the Coulomb gauge this is easily seen by using a scaling argument. For the general case we find a singularity of essential order 4 at $\vec{u} = \vec{\xi} = 0$, hence $P(\xi)$ diverges there of order $|\xi|^{-1}$ in contrast to the $|\xi|^{-2}$ divergence of $D_\alpha$. The $r$-term in the $\mathcal{A}$-propagator (12.15) is of the form $\partial_\mu P_\nu(\xi)$. The derivative worsens the singularities at the light cone and the origin by one order, so that they become of the same strength or marginally stronger than those of the $g_{\mu\nu}$-term. The convolution with $\hat{r}$ destroys the analyticity of $P_\alpha(\xi)$ in the spatial components $\vec{\xi}$, but it preserves its $\xi^0$-analyticity: $P_+(\xi)$ is as a function of $\xi^0$ analytic in $\Im\xi^0 < 0$, $P_F(\xi)$ in $\Im\xi^0 \{{}^<_>\} 0$ for $\Re\xi^0 \{{}^>_<\} 0$.

These properties of the physical propagators are sufficient to make the contents of Sect. 10.3 – the definition of BPHZ renormalization in $x$-space and the proof of its success (Theorem 10.3) – work. The restricted analyticity of $P_\alpha(\xi)$ in $\xi^0$ only is quite sufficient for that purpose.

## 12.4 The IR Problem and the Final Verifications

The IR existence of the physical $\mathcal{W}_\sigma$ can be shown with the methods developed in Chapt. 11. By way of example, consider first the 4-point function $\tilde{\mathcal{W}}_2(p_1, p_2, +|\bar{q}_1, \bar{q}_2, -)$ of Sect. 11.1, but now for the physical fields. Its connected part is given by the graphs of Fig. 11.1. We use the B-rules which are unproblematic in this case without UV problems. The photon propagators carry now the numerators (12.23) (with $\ell$ replaced by $k$). This expression is $C^\infty$ everywhere except at $\vec{k} = 0$. At $\vec{k} = 0$ but $k_0 \neq 0$ it has an integrable singularity of essential order $|\vec{k}|^{-1}$, at $k = 0$ a singularity of essential order 0 (i.e. a bounded discontinuity or a singularity of logarithmic type), which gives together with the factor $\delta_\pm(k)$ or $(k^2 \pm i\varepsilon)^{-1}$ a singularity of order 2, which is responsible for the IR divergences of the individual graphs. But these singularities cancel in the sum of the four integrals exactly like in the GB case, the numerator (12.23) being the same in all four cases.

However, this cancellation does *not* happen in the mixed GB-physical function

$$\left(\Omega,\, T^+\left(\tilde{\psi}(p_1)\,\tilde{\psi}(p_2)\right) T^-\left(\tilde{\bar{\Psi}}(q_1)\,\tilde{\bar{\Psi}}(q_2)\right)\Omega\right)$$

because in that case only the propagator ends lying in the right-hand sector give rise to an $r$-term, so that the $r$-terms are different in the various graphs. This proves the earlier statement that $\Psi$ and $\bar{\Psi}$ are not defined on the state space $\mathcal{V}$ of the GB formalism.

The IR existence of $\mathcal{W}_\sigma$ is shown as in Sect. 11.2 for the GB functions. The A-rules are used. And for simplicity we work directly with the full $\hat{r}^\mu$, not with the cutoff $\hat{r}^\mu_\xi$. But more extensive cancellations of IR divergences between graphs are needed than in the GB case: not only between graphs with the same scaffolding. This makes summing the integrands before integration impractical. It is therefore understood that one should calculate at first with the $r_\xi$, and the limit $\xi \to 0$ must be taken in the end in sums over suitable classes of graphs.

We need to study the long-distance behaviour of the new photon propagators $P_\alpha^\nu(\xi)$ in the relevant directions. The relevant cases are $P_+(\lambda\xi)$ and $P_F(\lambda\xi)$, $\lambda \to \infty$, for $\xi$ spacelike or timelike, and $P_+(\lambda\zeta)$, $\lambda \to \infty$, for $\zeta = (\zeta^0, \vec{\xi})$ with $\Im\zeta^0 < 0$ but $\vec{\xi}$ real: we are not interested in lightlike directions. For the Coulomb gauge it is immediately established with the help of a scaling argument that $P_\alpha(\lambda\xi)$ decreases in these directions like $\lambda^{-1}$:

worse by one degree than the GB propagator. This can be shown to be true also for the more general $r^\mu$ satisfying our conditions. Notice that for $\xi^2 \neq 0$ the singularities of $\hat{r}^\mu(\vec{u})$ and $D_\alpha(\xi^0, \vec{\xi} - \vec{u})$ in (12.24) are separated. We are interested in

$$t(\lambda) = \lambda \int d^3u\, \hat{r}^\nu(\lambda \vec{u})\, D_\alpha(\xi^0, \vec{\xi} - \vec{u}) \,. \tag{12.25}$$

For $\xi = \zeta$ complex, the $D_\alpha$-factor is regular in $\vec{u}$. The $\vec{u} = 0$ singularity of $\hat{r}^\nu$ is integrable, hence the integral (12.25) is absolutely convergent and we find

$$|t(\lambda)| \leq \lambda \int d^3u\, |\hat{r}^\nu(\lambda \vec{u})|\, |D_\alpha(\zeta^0, \vec{\xi} - \vec{u})|$$

$$\leq \lambda^{-1+\varepsilon} \int d^3u \left\{ \frac{c_1}{|\vec{u}|^{2+\varepsilon}} + \frac{c_2}{|\vec{u}|^{2-\varepsilon}} \right\} |D_\alpha(\zeta^0, \vec{\xi} - \vec{u})|$$

for any arbitrarily small positive $\varepsilon$ and $\lambda$ sufficiently large. $c_1$ and $c_2$ are positive constants depending on $r^\nu$ and $\varepsilon$. The $u$-integral exists, hence this estimate proves our contention. If $\xi$ is real but not lightlike, the $D_\alpha$ in (12.25) is singular for $(\vec{u} - \vec{\xi})^2 = (\xi^0)^2$, i.e. on a sphere with center $\vec{\xi}$ and radius $|\xi^0|$ which does not contain the origin $\vec{u} = 0$. We can therefore find a test function $\chi(\vec{u})$ with compact support, which is $\equiv 1$ in a neighbourhood of that sphere and $\equiv 0$ in a neighbourhood of $\vec{u} = 0$. We introduce the decomposition of unity $1 = \chi + (1 - \chi)$ into the integrand of (12.25). The second term excludes the $D_\alpha$-singularity from the region of integration and the resulting integral can be estimated like in the complex case. In the $\chi$-part of the integral, which does not contain the $\hat{r}$-singularity, we use the fact that any distribution can be written as a derivative of a continuous function. For instance, we can write

$$(2\pi)^4 D_F(\xi) = (1 - \triangle_\xi)^2 \int dk\, e^{-ik\xi} \frac{1}{k^2 + i\varepsilon} \frac{1}{(1 + |\vec{k}|^2)^2}$$

$$=: (2\pi)^4 (1 - \triangle_\xi)^2 F(\xi)$$

where $F$ is continuous. The $\chi$-part of $t(\lambda)$ becomes

$$\lambda \int d^3u\, \chi(\vec{u})\, \hat{r}^\nu(\lambda \vec{u})\, (1 - \triangle_{\vec{u}})^2\, F(\xi^0, \vec{\xi} - \vec{u}) \,.$$

We transfer the $\triangle$-operator onto $\chi \hat{r}^\nu$ through integration by parts. Using the boundedness of $F$ in the compact support of $\chi$ and the assumed decay properties of the derivatives of $\hat{r}^\nu$ we can again estimate the absolute value of this integral, finding a $\lambda^{-1+\varepsilon}$-decrease.

The derivative in the $r$-term of the $\mathcal{A}$-propagator (12.15) restores the $\lambda^{-2}$-decay of the GB case, as can be shown in the same way. This propagator has the same long-distance behaviour as the GB propagator up to logarithmic-type terms.

With these results the proof of IR convergence given in Sect. 11.2 can be adapted to the new situation. The situation is changed (a) by the weaker decrease of certain photon propagators at large distances, as just discussed, and (b) by the presence of $\Psi_n$- and $\bar\Psi_n$-vertices. We follow exactly the procedure of Sect. 11.2. We integrate first over the time components, decompose this integral according to (11.4) into cluster terms, etc. Note that all external points including the $\Psi_n$- and $\bar\Psi_n$-vertices belong to the $X$-cluster. Lemmas 11.1 and 11.2 are not affected by the changes. The estimates (11.8)–(11.11) remain valid because in their derivation only a decrease of first order of the photon propagators has been assumed. A problem occurs for the first time with Lemma 11.3 whose proof assumes the decrease of the GB propagators. If $r$-lines occur as inter-swarm lines, the estimate of Lemma 11.3 is no longer necessarily true. Consider such an $r$-line starting at a $\Psi_n$-vertex. This starting point lies in the $X$-swarm. So does the fermion trajectory starting at that vertex, because fermion lines occur only inside swarms. Hence the end point of that trajectory, a $\bar\Psi_m$-vertex, lies in the $X$-swarm too. Hence there exists another graph which is identical to the original one, except that the considered $r$-line starts at this $\bar\Psi_m$-vertex instead of the $\Psi_n$-vertex. This change from $\Psi(x)$ to $\bar\Psi(y)$ also involves a change of sign, so that in the sum of the two integrands occurs the difference $P_+(x-u) - P_+(y-u)$ with $u$ the variable of the internal end point of the line. But this difference decreases like $|u|^{-2+\varepsilon}$ for $u \to \infty$ in a complex direction with $x, y$, remaining bounded. Thus Lemma 11.3 remains true if we sum over pairs of graphs for each inter-swarm $r$-line. 2-line internal swarms are of the GB form and can be handled like in the GB case. The possibility of the $X$-swarm being a 2-line subgraph has already been admitted in the estimate $P \geq 1 + 2N$ of the proof of Lemma 11.3. Hence the proof of that lemma goes through.

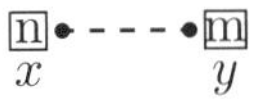

$\boxed{n}\bullet\text{-}\text{-}\text{-}\text{-}\bullet\boxed{m}$
$\quad x \qquad\qquad y$
**Fig. 12.3.** An IR divergent photon line

But there remains an additional difficulty concerning photon lines connecting two $\Psi_n$- or $\bar\Psi_n$-vertices, i.e. lines of the form shown in Fig. 12.3. The corresponding propagator is of the form

$$Q(x-y) = \int d^3u\, d^3v\, \hat r^\mu(\vec u)\, \hat r_\mu(\vec v)\, D_\alpha(x^0 - y^0, \vec x - \vec y - \vec u + \vec v)\,.$$

For $\vec u, \vec v \to \infty$ the $\hat r$ decrease of second order and so does $D_\alpha$. This is not sufficient to make the 6-dimensional integral converge at infinity. The line leads to an IR divergence even though it stays completely inside the $X$-swarm. But $x$ and $y$ are bounded, hence the divergent part of the integral is independent of them. Similarly to the earlier treatment of inter-swarm $r$-lines we find that there is also a graph in which the line in question connects the $x$-vertex not to the $y$-vertex but to the other end of the $y$-trajectory, all else being equal, and we obtain cancellation of the divergences between these two graphs. This does not work, however, if the $x$- and $y$-vertices are the end-vertices of one and the same trajectory. In that case we have in the unrenormalized rules also the graphs in which this particular line connects the $\Psi_n(x)$-vertex or the $\bar\Psi_m(y)$-vertex to itself (see case (a) of Fig. 12.2). These contributions have a different sign from the original one because the original line connects vertices of different type. And the new contributions are divided by the symmetry number 2. Hence we get again cancellation. But these cancelling graphs were unfortunately removed by renormalization. We must re-introduce their IR parts, which can be done by a UV finite but IR divergent (for $\xi = 0$) renormalization of the $\Psi$- and $\bar\Psi$-fields:

$$\begin{aligned}\Psi(x) &\longrightarrow C_{\mathrm{IR}}\,\Psi(x)\,,\\ \bar\Psi(x) &\longrightarrow C_{\mathrm{IR}}\,\bar\Psi(x)\,.\end{aligned} \qquad (12.26)$$

Let $\kappa(\vec w)$ be a real test function with compact support, which is invariant under the inversion $\vec w \to -\vec w$ and which is $\equiv 1$ in an open neighbourhood of the origin. We define

$$C_{\mathrm{IR}} = \frac{1}{2}\int d^3 u\, d^3 v\, d^3 w\, \kappa(\vec w)\,\hat r^\mu(\vec u)\,\hat r_\mu(\vec v)\, i\, D_\alpha(0, \vec w - \vec u + \vec v)\,. \qquad (12.27)$$

Notice that all $D_\alpha(\xi)$ are equal for spacelike $\xi$. This $C_{\mathrm{IR}}$ is real and independent of the type of sector in which it occurs. The $\kappa$-cutoff ensures the UV convergence of $C_{\mathrm{IR}}$. Its IR divergence cancels the IR divergence of the $r$–$r$-line we started from. It is especially at this point that the necessity of a preliminary $\xi$-regularization imposes itself in order to make the argument rigorous. The dependence of $C_{\mathrm{IR}}$ on the choice of $\kappa$ is unproblematic since the normalization of $\Psi$ has not been fixed by any physical requirement.

The verification of the W-properties given in Sect. 11.3 can be taken over to the new situation with some obvious changes patterned after the changes in the foregoing proof of IR existence, of course with the exception of locality and covariance[3]. On the other hand, positivity of the scalar product is now included among the demonstrable W-properties. But since this is a novel feature, we postpone its discussion and continue with the other properties treated in Sect. 11.3 for the GB case.

No new problems occur in the proofs of current conservation and the charge identity. Concerning the smoothness of the $\tilde\tau$-functions we must study the influence of $\Psi_n$-vertices on the arguments of Sect. 11.3 (see (11.24) and further). We use the notations of that section. Let the $p_1$-vertex be a $\bar\Psi_n$-vertex. Then the subgraphs $J_1$ and $J_2$ are connected by one fermion line and an arbitrary number of photon lines, possibly including $r$-lines starting from the $p_1$-vertex. But the more such lines are present, the more integrations are carried out and the smoother is the result. We need therefore only consider the case (11.25) with one fermion and one photon line, which we assume to be an $r$-line. Let $k$ be its momentum. Then the additional factor $\tilde r^\mu(\vec k)$ is

---

[3] See the end of this section for a discussion of observational Lorentz invariance.

present in the integrand. It has a $|\vec{k}|^{-1}$ singularity which is integrable if taken by itself. Its influence in conjunction with the propagator singularities can be studied by the methods used in the GB case. One finds that problems occur only for $p_1^2 = m^2$, at $k = 0$. They result in an expression of the form (11.26) but without the factor $(p_1^2 - m^2)$ in the log-term. However, (11.26) was derived for an amputated graph, in which the external $p_1$-propagator was removed through multiplication with $(p_1^2 - m^2)$. The same multiplication restores the missing factor in the $\Psi_n$-case, because in this case no external $p_1$-line is present in the first place. From there on the argument goes through like in the GB case. We find that the singularity of the 2-fermion function is of the same strength as in the GB case. It has already been remarked that the 2-photon function is not changed from its GB form apart from the trivial replacement of its external propagators by the new, physical, ones.

We come to the *positivity* of the scalar product on $\mathcal{V}'$. In a rigorous QED positivity means the following. Let $\mathcal{P}(\Psi, \bar{\Psi}, \mathcal{A})$ be a field polynomial and (see (9.36))

$$\mathcal{P}^*(\Psi, \bar{\Psi}, \mathcal{A}) = \prod_\Psi \gamma^0 \mathcal{P}^\leftarrow(\bar{\Psi}, \Psi, \mathcal{A}) \prod_{\bar\Psi} \gamma^0 \qquad (12.28)$$

its adjoint. For simplicity we assume the test functions in $\mathcal{P}$ to be real. The left arrow on the right-hand side means that the ordering of the field operators in each term of $\mathcal{P}^\leftarrow$ is the reverse of the original one. Then positivity means that

$$N := (\Omega, \mathcal{P}^* \mathcal{P} \Omega) \geq 0 \ . \qquad (12.29)$$

But in this form the condition cannot be checked in perturbation theory: it has no clear consequences for the individual terms of a perturbative expansion of $N$, except that the lowest non-vanishing term must be positive. E.g. the function $e^{-g^2}$ is positive for all real $g$. But how does this property manifest itself in the power series $1 - g^2 + g^4/2 - \cdots$ ? We use therefore a different formulation which we call "perturbative positivity". We say that $N$ is positive if it can be expressed in the sense of formal power series as

$$N = \sum_{i,j} A_i^* K_{ij} A_j \ , \qquad (12.30)$$

where $K$ is a positive matrix and the $A_i$ are defined as formal power series. More explicitly this means

$$N_\sigma = \sum_{i,j} \sum_{\tau=0}^{\sigma} A_{i,\tau}^* K_{ij} A_{j,\sigma-\tau} \qquad (12.31)$$

if $N_\sigma$ and $A_{i,\sigma}$ are the terms of order $\sigma$ in the expansions of $N$ and $A_i$. The $i$–$j$-sum is to be understood in a general sense: $i$ and $j$ may stand for possibly infinite sets of indices, some of which may be continuous, in which case the sum becomes an integral. The exact formulation is given in (12.36).

For studying the perturbative form of $N$ it suffices to study the expression

$$(\Omega, \mathcal{M}_\tau^* \mathcal{N}_{\sigma-\tau} \Omega) \,, \tag{12.32}$$

$\mathcal{M}$ and $\mathcal{N}$ being field monomials contained in $\mathcal{P}$ and $\mathcal{M}^*$ being obtained from $\mathcal{M}$ like in (12.28). We represent the fields in $\mathcal{N}$ by $T^+$-sectors, those in $\mathcal{M}^*$ as $T^-$-sectors, so that no internal sector exists between the two factors. A $(\Omega, \mathcal{M}^* \mathcal{N} \Omega)$ graph is a $(\Omega, \mathcal{M}^{\leftarrow} \mathcal{N} \Omega)$ graph (in obvious notation) with factors $\gamma^0$ tacked onto the external $\mathcal{M}$-points. We multiply the fermionic cross lines connecting $\mathcal{M}$- with $\mathcal{N}$-sectors by $\gamma^{0^2} = 1$ at their $\mathcal{M}$-ends. One of the $\gamma^0$ is then multiplied into the corresponding vertex. Together with the external $\gamma^0$'s this turns the $\mathcal{M}^{\leftarrow}$-part of the graph into the adjoint of an $\mathcal{M}$-part, as is shown in the same way as hermiticity has been proved in Sect. 9.4. The second $\gamma^0$ is multiplied into the cross propagator, yielding the product

$$\pm \gamma^0 (\ell + m) \delta_\pm(\ell) = \pm (\ell_0 + \gamma^0 \gamma^i \ell_i + m) \delta_\pm(\ell) \,, \tag{12.33}$$

which is a positive matrix in spin space. These results lead to the desired form (12.31).

Concerning the photonic $\mathcal{M}$–$\mathcal{N}$-lines we can go over to the B-rules. We replace their propagators by the B-form (12.23):

$$M_{\mu\nu}(\ell) \, \delta_+(\ell) := - \left( \delta_\mu^\alpha + i \, \ell_\mu \tilde{r}^\alpha(\ell) \right) g_{\alpha\beta} \left( \delta_\nu^\beta - i \, \ell_\nu \overline{\tilde{r}^\beta(\ell)} \right) \delta_+(\ell) \tag{12.34}$$

and drop the graphs in which such lines are directly attached to $\Psi_n$- or $\bar{\Psi}_n$-vertices. This is shown to be equivalent to the original rules like in the proof of the formal equivalence of the A- and B-rules. But now the equivalence holds rigorously: the change of rules for these special lines only is unproblematic because cross lines do not occur in R-parts. The propagator (12.34) is positive. We show this first for the special case of the Coulomb gauge, forming $M_{\mu\nu}^C$ with the $\tilde{r}_\mu^C$ defined in (12.14). A simple calculation shows that on the mass shell $\ell_0 = |\vec{\ell}|$ we have

$$M_{0\nu}^C = M_{\mu 0}^C = 0 \quad , \quad M_{ij}^C(\ell) = \delta_{ij} - \frac{\ell_i \ell_j}{|\vec{\ell}|^2} \,. \tag{12.35}$$

$M_{ij}^C$ is a positive matrix. Using that $\left( \delta_\mu^\alpha + i \, \ell_\mu \tilde{r}^\alpha(\ell) \right) \ell_\alpha = 0$ as a consequence of the condition (12.8) we can replace the factor $-g_{\mu\nu}$ in (12.34) by $M_{\alpha\beta}^C(\ell)$, obtaining a positive hermitian matrix. We can now concretize the expression (12.30). It reads

$$N = \sum_{\alpha,\beta,\gamma=0}^{\infty} \frac{1}{\alpha!\beta!\gamma!} \int \prod_1^\alpha dp_i \prod_1^\beta dq_j \prod_1^\gamma d\ell_h \, A_{\alpha\beta\gamma}^*(P,\bar{Q},L)$$

$$\times \prod_1^\alpha K(p_i) \prod_1^\beta \bar{K}(q_j) \prod_1^\gamma \mathcal{K}(\ell_h) \, A_{\alpha\beta\gamma}(P,\bar{Q},L) \,. \tag{12.36}$$

$A_{\alpha\beta\gamma}$ is the formal sum over all possible subgraphs containing the external points of any of the monomials in $\mathcal{P}$ and the end points $p_i$, $q_j$, $\ell_h$, of cross lines, $p_i$ for electron lines pointing into the subgraph, $q_j$ for electron lines pointing out of it, $\ell_h$ for photon lines. $K(p)$ and $\bar{K}(q)$ are of the form (12.33) with the upper sign for $K$, the lower one for $\bar{K}$. $\mathcal{K}(\ell)$ is the propagator (12.34). All these kernels are positive. The factorials take care of the fact that the numbering of the cross lines has not been fixed. It must be remarked that in finite orders of perturbation theory the individual terms in the sum in (12.36) do in general not exist because they are IR divergent. The known cancellations of these divergences involve graphs with different numbers of cross lines. On the other hand, in the unlikely case that the formal sum defining $A_{\alpha\beta\gamma}$ converges, it is expected to vanish on the mass shell of the $p_i$, $q_j$, so that the individual terms for $(\alpha, \beta) \neq (0,0)$ vanish. In this case an initial IR regularization is needed to make sense of (12.36) as a limit obtained by removing that regularization after summing.

Last but not least, the *relativistic invariance* of the theory must be discussed. The fact that the physical $W$-functions do not transform covariantly under the Lorentz group must not be interpreted as a breakdown of observational invariance. We discuss this problem in the local observables framework, though not at the level of mathematical rigour that one is wont to from the practitioners of that approach. The formalism is based on the observation that in the last resort what is measured in actual experiments are always expectation values of observables localized in a finite space-time region: measurements take place in the limited confines of a laboratory and are of finite duration. The operators representing observables in QED must be gauge invariant, in particular not depending on the choice of the auxiliary functions $r^\mu$ if expressed as functions of the physical fields. The problem will not be discussed in depth. Only a rather formal overview will be given, disregarding subtleties like the need for renormalization and such-like.

In order to see what becomes of expectation values of observables, let us first consider a manifestly covariant theory: the pseudo-QED of the GB formulation. It is best to explain this for an exemplary observable with a simple Lorentz behaviour, the current operator $j^\mu(x)$. Let $\mathcal{M}(Y)$, $\mathcal{N}(Z)$, be two field monomials, and $m(Y) = \mathcal{M}(Y)\Omega$, $n(Z) = \mathcal{N}(Z)\Omega$, monomial states as introduced at the beginning of Chapt. 7. Any state in $\mathcal{V}$ is a finite superposition of monomial states integrated over tempered test functions. It suffices to consider the matrix elements $(m(Y), j^\mu(x) n(Z))$ under the Lorentz transformation $\Lambda \in L$. In the new frame of reference the state $m$ is seen as

$$m_\Lambda(Y) = \prod_{y_i \in Y} R^{(i)}(\Lambda)\, m(\Lambda Y)\,. \tag{12.37}$$

$n_\Lambda(Z)$ is defined in the same way. The matrix element of $j^\mu$ becomes (see (4.40))

$$(m_\Lambda(Y),\, j^\mu(x)\, n_\Lambda(Z)) = \Lambda^\mu{}_\nu \left(m(Y),\, j^\nu(\Lambda^{-1}x)\, n(Z)\right)\,, \tag{12.38}$$

the expected behaviour of a vector observable.

This consideration can be generalized to physical gauges. In preparation we note that our conditions for the choice of the functions $r^\rho(u)$ were unnecessarily restrictive. Let $r^\rho$ satisfy these conditions and let $\Lambda \in L$. Then

$$r_\Lambda^\rho(u) = \left(\Lambda^{-1}\right)^\rho{}_\sigma r^\sigma(\Lambda u) \tag{12.39}$$

satisfies condition (12.7). Its support is no longer contained in $\{u^0 = 0\}$ but in a general spacelike 3-plane through the origin. The variables of the GB fields in the definitions (12.9) of $\Psi(x)$ and $\mathcal{A}_\mu(x)$ are then located in a spacelike plane through $x$ which is parallel to this $r_\Lambda$-support. The planes of localization of two such fields at different points $x_1$ and $x_2$ are time ordered in the sense that every point of one of them is either spacelike or positive timelike relative to every point of the other. This is sufficient for our purposes. All arguments relying on the special support of $r^\mu$ can be generalized to the transformed situation without inconvenience. The gauge function $G(x)$ defined in (12.10) becomes

$$\begin{aligned}G_\Lambda(x) &= -\int du\, r_\Lambda^\rho(u)\, A_\rho(x - u) \\ &= -U^*(\Lambda)\, G(\Lambda x)\, U(\Lambda)\,.\end{aligned} \tag{12.40}$$

The unitary $L$-representation $U$ drops out from $\mathcal{W}$-functions in which all fields are transformed under $\Lambda$. Let $\Psi_\Lambda(x)$, $\mathcal{A}_\Lambda^\mu(x)$, be defined by (12.9) but using $r_\Lambda^\rho$ instead of $r^\rho$. Then the relation (12.38) remains true if (12.37) is replaced by the more complicated formula

$$m_\Lambda(Y) = \prod R^{(i)}(\Lambda)\, m(\Lambda\, Y_\Lambda)\,, \tag{12.41}$$

the subscript $\Lambda$ at the variable $y_j$ indicating that it is the argument of a field $\Psi_\Lambda$, $\bar{\Psi}_\Lambda$, or $\mathcal{A}_\Lambda^\mu$. Hence the correct transformation laws are preserved if a more elaborate transformation law for states is introduced.

This is not yet the full story. The state $m_\Lambda(Y)$ is a state in the state space $\mathcal{V}'_\Lambda$ constructed with the auxiliary functions $r_\Lambda^\rho$. It does *not* lie in the original space $\mathcal{V}'$. Conventionally speaking, the two observers work with different gauges. This should not be necessary. Our claim is that the space $\mathcal{V}'$ is already sufficiently large to describe all possible experimental situations irrespective of the state of motion of the observer. To substantiate this claim it must be shown that any state $\Phi_\Lambda$ in $\mathcal{V}'_\Lambda$ can be approximated arbitrarily well in the sense of Fell's theorem by states in $\mathcal{V}'$. That is, for any normalized $\Phi_\Lambda \in \mathcal{V}'$ and any finite set $\mathcal{O}_1, \ldots, \mathcal{O}_n$ of local observables there must exist a normalized state $\Phi \in \mathcal{V}'$ such that the differences

$$(\Phi_\Lambda,\, \mathcal{O}_i \Phi_\Lambda) - (\Phi,\, \mathcal{O}_i \Phi)$$

are smaller than the given experimental uncertainties. In a rigorous formulation of QED, if such a thing exists, this is guaranteed by Fell's theorem.

But in our perturbative context it would be desirable to find an explicit construction of such a state. To the best of my knowledge, such a construction has not yet been worked out. I can only roughly indicate a way of proceeding that has a good chance of attaining the goal. It will be explained by means of a simple example: the expectation value

$$E = (\Omega, \Psi(f)^* j^\mu(x) \Psi(f) \Omega) \tag{12.42}$$

of a current component in a one-$\Psi$ state $\int dy\, f(y)\, \Psi(y)\, \Omega$ with a test function $f$ with compact support. $\Psi(y)$ is written in the form (12.9): $\Psi(y) = e^{-ieG(y)}\psi(y)$ with $G(y) = -\int du\, r^\rho(y-u)\, A_\rho(u)$. Choose a $C^\infty$ function $\sigma(\vec{u})$ such that the points $(y^0, \vec{u})$ with $y^0$ in the $y^0$-support of $f$ and $\vec{u} \in \operatorname{supp}\sigma$ are spacelike relative to $x$ and to $\operatorname{supp} f$, and such that $\sigma(\vec{u}) \equiv 1$ for $|\vec{u}|$ sufficiently large. Then the expression

$$\exp\left[i e \int du\, \sigma(\vec{u})\, \delta(y^0 - u^0)\, \hat{r}^\rho(\vec{u})\, A_\rho(u)\right]$$

commutes with $j^\mu(x)$ and $\Psi(f)$: in $E$, this factor multiplied with $\Psi(f)$ and its adjoint with $\Psi(f)^*$ cancel. Hence we can replace $\Psi(y)$ by

$$\Psi'(y) = \psi(y) \exp[-ie \int du\, s^\rho(y, u)\, A_\rho(u)] , \tag{12.43}$$

with

$$s^\rho(y, u) = r^\rho(y - u) - \sigma(\vec{u})\, \delta(y^0 - u^0)\, \hat{r}^\rho(\vec{u}) . \tag{12.44}$$

Under our assumptions on $\hat{r}^\rho$ and its derivatives we find that $s^\rho$ decreases like $|\vec{u}|^{-3}$ for $|\vec{u}| \to \infty$ uniformly in $y \in \operatorname{supp} f$. This decrease suffices to make $\Psi'(y)$ definable on $\mathcal{V}$, at least in perturbation theory. But Lorentz invariance is no problem in $\mathcal{V}$. We define

$$\Psi'_\Lambda(y) = \psi(y) \exp[-ie \int du\, s^\rho_\Lambda(y, u)\, A_\rho(u)] , \tag{12.45}$$

where $s^\rho_\Lambda$ is formed from $r^\rho_\Lambda$ in analogy to (12.44), and we obtain

$$\Psi'(y) = \Psi'_\Lambda(y) \exp\left\{-ie \int du[s^\rho(y, u) - s^\rho_\Lambda(y, u)]\, A_\rho(u)\right\} . \tag{12.46}$$

$\Psi'_\Lambda(y)$ can again be replaced in $E$ by $\Psi_\Lambda(y) = e^{-ieG_\Lambda(y)}\psi(y)$ without changing $E$. We obtain

$$E = (\Omega, \hat{\Psi}_\Lambda(f)^* j^\mu(x) \hat{\Psi}_\Lambda(f) \Omega) , \tag{12.47}$$

with

$$\hat{\Psi}_\Lambda(f) = \int dy\, f(y)\, \psi(y)\, e^{-ieG_\Lambda(y)}\, U(y) \tag{12.48}$$

where $U(y)$ is the unitary exponential factor in (12.46). $\hat{\Psi}_\Lambda(f)\Omega$ can be expected to be defined as a vector in $\mathcal{V}'_\Lambda$. The unitary factor $U(y)$ should not create any IR problems on account of the sufficiently strong decrease of $s^\rho$ and $s^\rho_\Lambda$. The $A_\rho$ are defined as fields on $\mathcal{V}'$ and $\mathcal{V}'_\Lambda$, only $\psi$ is not, so there is also no trouble on that count. Of course, the perturbative formulation of these considerations, defining the exponentials as power series, needs to be worked out properly. In particular, the UV problems which have been ignored in the above outline must be properly dealt with.

This procedure can only work if applied to observables localized in a given bounded set. If this set is enlarged, the function $\sigma$ must be changed accordingly. But this was to be expected. It is not possible to find for any given $\Phi_\Lambda \in \mathcal{V}'_\Lambda$ a state $\Phi \in \mathcal{V}'$ such that $(\Omega, \mathcal{O}\Omega) = (\Phi_\Lambda \mathcal{O} \Phi_\Lambda)$ for all local observables $\mathcal{O}$ irrespective of their localization region. This is due to the existence of "observables at infinity", gauge invariant quantities which may vanish identically on states constructed with certain $r^\rho$'s but not for different $r^\rho$. For a discussion of this problem, which lies outside the scope of the present work, see [Bu 82].

# Part III

# Particles and Their Reactions

In this last part it is shown how the phenomenologically important notion of particles emerges from QED, a field theory. In Chap. 13, the standard view of this problem is briefly discussed. A more directly phenomenological approach is developed in Chaps. 14 and 15 and applied to the description of scattering events in Chaps. 16 and 17.

# Chapter 13

# The Standard View

The word "particle" and particle names: electron, positron, photon, have occasionally been mentioned in Parts I and II, although without much emphasis. The theory has been developed strictly as a field theory, a theory of a continuum. But the fields $\psi$ and $\bar{\psi}$ are never observed, electrons and positrons are. Therefore, in order to make contact with reality, particles must be incorporated into the theory. The prevalent view seems to be that particles and fields are complementary aspects of one and the same situation. This view is based on the Fock space description of free particles, which can be obtained from a free field theory as described in Sect. 5.1 for the free scalar field. As has been mentioned in the Introduction, this view is more problematic in the interacting case. We will therefore talk about particles only as approximate notions which are useful for the phenomenological description of situations which are close to the free case. More exactly, this means that the state under scrutiny is examined in a regime in which it behaves sensibly like a state of a free theory. Empirically one finds that the states that are successfully described by perturbative QED can be characterized as *"scattering states"*. These are states that at a sufficiently early time are experimentally indistinguish- able from states of a finite number of well-separated particles (electrons, positrons, photons) behaving classically, i.e. following well-defined straight orbits with constant velocity. These particles then approach one another closely enough to start interacting, forming a conglomerate the examination of whose detailed behaviour is beyond the capabilities of present-day experimental techniques. Finally, this mess separates again into entities recognizable as particles, which after a sufficiently long time are no longer distinguishable from free particles following classical paths. This experience gives rise to the field-theoretical picture of such events that has already been introduced in Sect. 6.3: suitably averaged field operators converge for large negative or positive times to free fields generating two asymptotic Fock spaces. And the particles of the particle interpretation of these Fock spaces are what we observe. This idea is described more exactly, though without proofs, in Sect. 13.1 for the case in which it can be implemented without problems: the case of a purely massive field theory. Unfortunately, these methods are not immediately applicable to QED because of the vanishing photon mass and the ensuing IR problems. These difficulties and the traditional ways of overcoming them are discussed in Sect. 13.2.

## 13.1 Massive Theories

Let $P_\mu$ be the momentum operators defined in (4.25). Under a massive theory we understand a theory in which the spectrum of the mass operator $M^2 = P_\mu P^\mu$ consists of a non-degenerate eigenvalue 0 with the vacuum $\Omega$ as sole eigenstate, a continuum starting at $m_c^2 > 0$, and one or several discrete eigenvalues $m^2$ in the open interval $(0, m_c^2)$. The simplest but typical example is a theory of a single hermitian scalar field $\varphi(x)$ with $M^2$ having the single nonvanishing eigenvalue $m^2 > 0$, the continuum starting at $m_c^2 = 4m^2$. The restriction of the $L$-representation $U(\Lambda)$ to the $m^2$-eigenspace is assumed to be irreducible. The Wightman axioms are assumed to hold in the strict version with $\mathcal{V}$ a Hilbert space. This model is sufficient for explaining what needs be explained. The complications introduced by the presence of several fundamental fields and particles and of spin degrees of freedom are merely notational and immaterial for our purpose. Under the mentioned assumptions the condition (6.33) is satisfied and the LSZ asymptotic conditions can be proved. We will not prove them here, but only formulate them in a more rigorous way than in Sect. 6.3. Let $f(x)$ be a smooth positive-frequency solution of the Klein–Gordon equation as given by (6.28). The averaged field $\varphi(f,t)$ is in general undefined because $\varphi(x)$ and $\dot\varphi(x)$ cannot be restricted to a fixed value of $x^0$. But this problem can be solved like the similar problem encountered in the definition of the charge as an integral over the charge density. Let $\alpha(x^0)$ be a test function with a compact support centered at the origin, which satisfies $\int dx^0\, \alpha(x^0) = 1$. We define

$$\varphi(f,t) = -i \int dx\, \alpha(x^0 - t)\, f(x)\, \overleftrightarrow{\partial}_0 \varphi(x) . \tag{13.1}$$

Then it can be proved (see [Hp 65] or Chap. 13 of [BLOT]) that there exist free fields $\varphi_{\text{in}}(x)$, $\varphi_{\text{out}}(x)$ defined on $\mathcal{V}$, such that

$$\lim_{t \to \pm\infty} \left( \Psi, \varphi^{(*)}(f,t)\, \Phi \right) = \left( \Psi, \varphi^{(*)}_{\substack{\text{out}\\ \text{in}}}(f)\, \Phi \right) \tag{13.2}$$

for all "non-overlapping asymptotic states" $\Psi$, $\Phi$, a notion that we refrain from defining here. The superscript $(*)$ means that the equation holds both for $\varphi$ and for $\varphi^*$. $\varphi_{\text{ex}}$ with $ex = out, in$, is defined by inserting $\varphi_{\text{ex}}(x)$ instead of $\varphi(x)$ into (13.1). For free fields this expression does not depend on $t$ and $\alpha$. In $p$-space it takes the simple form

$$\varphi_{\text{ex}}(f) = \int \frac{d^3p}{2\omega(\vec{p})}\, \hat f(\vec p)\, a^*_{\text{ex}}(\vec p) , \tag{13.3}$$

where $a^*_{\text{ex}}$ is the creation operator introduced in (5.34). We assume that the Fock spaces $\mathcal{F}_{\text{ex}}$ of these free fields are identical to the full Hilbert space $\mathcal{V}$ of the theory:
$$\mathcal{F}_{\text{in}} = \mathcal{F}_{\text{out}} = \mathcal{V} \ . \tag{13.4}$$
This property is called *"asymptotic completeness"*. It signifies that any state in $\mathcal{V}$ can be approximated arbitrarily well by scattering states as defined at the beginning of this chapter.

Under the stated assumptions the in- and out-fields are related through the *"S-matrix"*: a unitary operator on $\mathcal{V}$ defined by
$$S\Omega = \Omega \ , \quad \varphi_{\text{out}}(x) = S^* \varphi_{\text{in}}(x) S \ . \tag{13.5}$$
Its "matrix elements"
$$\begin{aligned} S(\vec{p}_1, \ldots, \vec{p}_F; \vec{q}_1, \ldots, \vec{q}_I) &= \left( \Omega, \prod_{i=1}^F a_{\text{in}}(\vec{p}_i) \, S \prod_{j=1}^I a^*_{\text{in}}(\vec{q}_j) \, \Omega \right) \\ &= \left( \Omega, \prod_{i=1}^F a_{\text{out}}(\vec{p}_i) \prod_{j=1}^I a^*_{\text{in}}(\vec{q}_j) \, \Omega \right) \end{aligned} \tag{13.6}$$
are obtained from the Green's functions of the interacting field $\varphi$ via the *LSZ reduction formula*
$$S(\vec{p}_1, \ldots ; \ldots, \vec{q}_I) = \tilde{\tau}^{\text{amp}}(p_1, \ldots, p_F; -q_1, \ldots, -q_I)|_{\text{MS}} \ . \tag{13.7}$$
$\tilde{\tau}^{\text{amp}}$ is defined as
$$\tilde{\tau}^{\text{amp}}(p_1, \ldots, p_n) = \prod_{j=1}^n \left( 2\pi i (p_j^2 - m^2) \right) \tilde{\tau}^+(p_1, \ldots, p_n) \ . \tag{13.8}$$
In perturbation theory this means simply that the external propagators of the $\tau^+$ graphs are removed, hence the term "amputation". The subscript MS signifies that the arguments $p_i$, $q_j$, are restricted to the positive mass shell: $p_i^0 = \omega(\vec{p}_i)$, $q_j^0 = \omega(\vec{q}_j)$. More exactly, the formula (13.7) holds in this simple form only if no $\vec{p}_i$ is equal to a $\vec{q}_j$ ("non-forward directions"). This condition is met in the cases of experimental interest.

Knowledge of the $S$-matrix allows to calculate the probabilities of specific scattering events taking place; it embodies the theoretical information on such events. Consider the initial state
$$\Phi_{\text{in}} = \prod_{j=1}^I \left( \int d^3 q_j \, f_j(\vec{q}_j) \, a^*_{\text{in}}(\vec{q}_j) \right) \Omega \tag{13.9}$$
prepared at an early time as a state of $I$ free particles with wave functions $f_j$. What is the probability that eventually, at large times, the state
$$\Phi_{\text{out}} = \prod_{i=1}^F \left( \int d^3 p_i \, g_i(\vec{p}_i) \, a^*_{\text{out}}(\vec{p}_i) \right) \Omega \tag{13.10}$$

is present? If both states are normalized to 1, then this probability is the absolute square of the transition amplitude

$$T(\Phi_{\text{in}} \to \Phi_{\text{out}}) = \int \prod \left( \frac{d^3 p_i}{2\omega(\vec{p}_i)} g_i^*(\vec{p}_i) \right)$$
$$\times \prod \left( \frac{d^3 q_j}{2\omega(\vec{q}_j)} f_j(\vec{q}_j) \right) S(\ldots, \vec{p}_i, \ldots; \ldots, \vec{q}_j, \ldots). \qquad (13.11)$$

A realistic experiment differs from this idealized situation in various respects. First, the initial state is a 2-particle state, since it is for all practical purposes impossible to get more than two particles close enough together to have them all interacting with one another. Second, one does not shoot two particles at each other, but two beams of particles or, more frequently, a beam is directed at a stationary target. Third, the wave functions of the outgoing particles are not determined. Measured are only their 4-momenta within some experimental accuracy. The quantity of actual interest for the comparison of theory and experiment is the *reaction cross section* which is defined for the case of a fixed target by

$$d\sigma = \frac{n_f(\vec{q}_1, \ldots, \vec{q}_F) \prod_1^F dq_j}{N_T D}, \qquad (13.12)$$

where the numerator gives the number of events/sec with $F$ particles emerging with momenta in the infinitesimal region $\prod d^3 \vec{q}_j$, $N_T$ is the number of particles in the target, and $D$ is the current density at the target of the impinging beam. This cross section is expressible in terms of the $S$-matrix (see [IZ] Sect. 5-1, or [GW], Sect. 3.3) as

$$d\sigma = \frac{\pi^2}{|\vec{p}|\,\omega(\vec{p})} \left| \hat{S}(\vec{p}; \vec{q}_1, \ldots, \vec{q}_F) \right|^2 \prod \frac{d^3 q_j}{2\omega(\vec{q}_j)} \delta^4 \left( p + p_T - \sum q_j \right). \qquad (13.13)$$

Here $p$ is the momentum of a particle in the in-beam, which is supposed to be known very accurately, $p_T = (m, \vec{0})$ is the 4-momentum of a target particle, and $\hat{S}$ is defined by

$$S(\vec{p}, \vec{p_T}; ..,\vec{q}_j, ..) = \delta\left(\omega(\vec{p}) + m - \sum \omega(\vec{q}_j)\right) \delta^3\left(\vec{p} + \vec{p}_T - \sum \vec{q}_j\right) \hat{S}(\vec{p}; .., \vec{q}_j, ..). \qquad (13.14)$$

Finally an important point should be stressed which is not always given sufficient attention to in the literature. The expressions for transition amplitudes and cross sections given above are derived from asymptotic conditions referring to the limits of infinitely large positive and negative times. But actual measurements are carried out at finite times. Therefore the theoretical expressions are not strictly applicable to the experiment. The solution of this dilemma seems obvious: measurements are always carried out with a finite experimental uncertainty. And within this uncertainty the asymptotic theoretical expressions are accurate if only we measure sufficiently long before and after the scattering takes place. However, a complete theory must be able to tell us just how long is "sufficiently long". This information is *not* contained in the S-matrix. The often heard statement that $S$ embodies all the relevant information on scattering experiments must therefore be taken with a grain of salt.

## 13.2 The IR Problem

In the description of the LSZ formalism in the previous section it has been assumed that the 2-point function of the field $\varphi$ contains a $\delta_+(p)$ at the mass shell, which is isolated both from the vacuum value $p = 0$ and the continuum $p^2 \geq m_c^2$. This condition of isolation is not necessary. LSZ-like asymptotic conditions can be proved if the 1-particle mass shell is not isolated from the rest of the spectrum, provided that the mass shell singularity is still a $\delta_+$-function. This has been shown in [Hb 71] for the massive case $m > 0$, in [Bu 75, Bu 77] for $m = 0$. But in QED such a $\delta_+$-singularity is not present for the electron field, so that the LSZ formalism in its pure form is not applicable.

The current solutions of this problem grow from two roots: the treatment of Coulomb scattering in non-relativistic quantum mechanics, and the Bloch–Nordsieck model. Let us first look at the non-relativistic situation. Let $V(\vec{x})$ be a smooth static potential with strong decrease at infinity, meaning that it decreases for $|\vec{x}| \to \infty$ faster than $|\vec{x}|^{-1}$ in all directions. Then there exists for each momentum $\vec{k}$ a "scattering solution" of the time-independent Schrödinger equation

$$-\frac{1}{2m}\triangle \psi_{\vec{k}}(\vec{x}) + V(\vec{x})\,\psi_{\vec{k}}(\vec{x}) = E\,\psi_{\vec{k}}(\vec{x}) \qquad (13.15)$$

with energy $E = \frac{|\vec{k}|^2}{2m}$, which behaves for $r = |\vec{x}| \to \infty$ asymptotically like

$$\psi_{\vec{k}}(\vec{x}) \sim e^{i\vec{k}\vec{x}} + h_{\vec{k}}(\theta,\phi)\,\frac{e^{ikr}}{r} \ . \qquad (13.16)$$

$r, \theta, \phi$, are spherical coordinates with central axis in direction $\vec{k}$, and $k = |\vec{k}|$. This solution describes the scattering from the potential $V$ of a beam of particles with momentum $\vec{k}$. The plane-wave part represents the incoming beam. The second term on the right-hand side is an outgoing spherical wave describing the scattered particles. The current density of the *in*-contribution is $\vec{j}_{\text{in}} = \vec{k}/m$. Let $d\Omega = \sin\theta\,d\theta\,d\phi$ be a small element of solid angle in a non-forward direction $\theta > 0$. Then the outflux $j_{\text{out}}\,d\Omega$ in this range of directions converges for $r \to \infty$ to

$$j_{\text{out}}\,d\Omega = \left|h_{\vec{k}}(\theta,\phi)\right|^2 d\Omega \ ,$$

yielding the "differential cross section"

$$\frac{d\sigma}{d\Omega} = \frac{j_{\text{out}}}{|\vec{j}|} = \left|h_{\vec{k}}(\theta,\phi)\right|^2 \ . \qquad (13.17)$$

This means that $h_{\vec{k}}$ is an $S$-matrix element in the sense of the preceding section. At macroscopic distances $r$ from the scattering center a scattered particle behaves like a classical free particle with a velocity proportional to its momentum, so that the angles $\theta$, $\phi$ fix the direction of its momentum, its magnitude being determined by energy conservation.

The similarity to the situation of Sect. 13.1 is seen better if we go over to the time-dependent scattering formalism. Let $a(\vec{k})$ be a smooth function with a small compact support whose convex hull does not contain the origin. Then

$$\psi(\vec{x},t) = \int d^3k\, a(\vec{k})\, \psi_{\vec{k}}(\vec{x})\, e^{-iEt} \tag{13.18}$$

is a $L_2$-solution of the time-dependent Schrödinger equation. The following asymptotic conditions can be proved.

**In-condition.** The function

$$\psi_{\text{in}}(\vec{x},t) = \int d^3k\, a(\vec{k})\, e^{i\vec{k}\vec{x}} e^{-iEt} \tag{13.19}$$

solves the free Schrödinger equation, and

$$\lim_{t \to -\infty} \|\psi(\vec{x},t) - \psi_{\text{in}}(\vec{x},t)\| = 0 , \tag{13.20}$$

where $\|\phi(\vec{x},t)\| = \left(\int d^3x\, |\phi(\vec{x},t)|^2\right)^{1/2}$ is the $L_2$-norm of $\phi$.

**Out-condition.** Let $K^f \subset R^3$ be the cone with apex 0 which is spanned by the support of $a(\vec{k})$. It contains the forward directions, i.e. the directions in which a classical non-interacting particle with a momentum in the support of $a$ would fly. Let $K$ be another cone with apex 0 and $K \cap K^f = \{0\}$. Then there exists a solution $\psi_{\text{out}}(\vec{x},t)$ of the free Schrödinger equation such that

$$\lim_{t \to \infty} \left\| \psi_{\text{out}}(\vec{x},t) - \int d^3k\, a(\vec{k})\, h_{\vec{k}}(\theta,\phi)\, \frac{e^{ikr}}{r}\, e^{-iEt} \right\| = 0 \tag{13.21}$$

and

$$\lim_{t \to \infty} \int_K d^3x\, |\psi(\vec{x},t) - \psi_{\text{out}}(\vec{x},t)|^2 = 0 . \tag{13.22}$$

These two conditions are non-relativistic versions of the LSZ conditions of Sect. 13.1. That (13.22) holds only in non-forward directions is a feature that is also present in the LSZ case. Remember that condition (13.2) is only claimed to hold on a dense set of states, not on all of $\mathcal{V}$.

Let now the potential $V(\vec{x})$ be slowly decreasing. More exactly, we demand that $V$ should at large distances approach a Coulomb potential, i.e.

$$V(\vec{x}) - \frac{\gamma}{mr} \to 0 \quad \text{for} \quad r \to \infty \tag{13.23}$$

sufficiently fast, with $\gamma$ a real constant. Then the results just quoted are no longer true. But only slightly more complicated ones are. Equation (13.16) is replaced by

$$\psi_{\vec{k}}(\vec{x}) \sim \exp\left\{i\vec{k}\vec{x} + i\gamma \log[kr(1-\cos\theta)]\right\}$$
$$+ h_{\vec{k}}(\theta,\phi)\frac{1}{r}\exp\left\{i\,k\,r - i\gamma\log[kr(1-\cos\theta)]\right\}\,, \quad (13.24)$$

which holds for $r \to \infty$ in all directions with $\theta > 0$. The first term in this expression is no longer a free solution, but the additional logarithmic phase drops out from $\vec{j}_{\text{in}}$ at large $r$: the term still describes asymptotically free incoming particles. The same holds for $j_{\text{out}}$ as calculated from the last term. Equation (13.17) remains true, hence $h_{\vec{k}}$ can still be considered an $S$-matrix element, albeit defined with the help of more complicated asymptotic conditions. The condition (13.20) still holds, but with $e^{i\vec{k}\vec{x}}$ being replaced in the definition (13.19) by the in-term in (13.24). And $\psi_{\text{out}}$ in (13.22) is now defined by

$$\psi_{\text{out}}(\vec{x},t) = \int d^3k\, a(\vec{k})\, h_{\vec{k}}(\theta,\Phi)\,\frac{e^{ikr}}{r}\, e^{-i\gamma\log[kr(1-\cos\theta)]}\,. \quad (13.25)$$

Neither $\psi_{\text{in}}$ nor $\psi_{\text{out}}$ approach solutions of the free equation for $|t| \to \infty$, but they deviate from such solutions in a fixed way independent of the local behaviour of $V$. This deviation can be considered a consequence of an explicitly constructible non-local and $t$-dependent interaction Hamiltonian describing the essence of the long-distance effects of the potential (see [Do 64] for details and generalizations).

These facts suggest a generalization to QED. Perhaps asymptotic conditions and the $S$-matrix can be saved if we do not insist on the asymptotic fields being free fields acting on a Fock space of free particle states, but allow them to interact via a truncated interaction incorporating the essential long-range aspects of the true interaction. It must, of course, be possible to write down the modified asymptotic fields and states explicitly. Formulations of this type have been proposed by various authors working at various levels of rigour (see e.g. [Ki 68], [KF 70], [Zw 75]). The results cannot be said to be very appealing. They are too complicated to be of much practical use either for the calculation of cross sections or for providing insights into the underlying structures of the theory.

The second approach to the IR problem in scattering theory goes back to the classical paper [BN 37] of Bloch and Nordsieck. In this paper the authors introduced an approximation to QED which is much better suited than perturbation theory for the investigation of the IR problem, especially by treating the emission of low-energy photons by accelerated charges in a simple way. Essentially the approximation neglects at first the recoil of the charged particle in this emission, which makes the problem solvable by methods known from classical electrodynamics. The probability of emission of a single photon or of a finite number of them is found to be zero. This contrasts sharply with what is obtained by an uncritical application of the LSZ reduction formula in perturbation theory, which produces divergent probabilities. But at first sight the zero-result seems to be equally inacceptable, since the emission of photons by accelerated charges is an observational fact. The solution proposed by Bloch and Nordsieck is this: all photon detectors have threshold energies below which photons are not registered. The final state of an observed scattering event may therefore contain an arbitrary, possibly infinite, number of unobserved low-energy photons with a finite total energy. For calculating the probability of observing what is in fact observed, the probabilities of emission of $n$ soft photons must be summed over $n$. This can be done and results in a finite but non-vanishing value. It has been realized later on that the corresponding effect occurs also in perturbation theory, using the LSZ reduction formula. Let $\sigma_n$ be the scattering cross section for an $n$-photon final state which is compatible with the observations. It is formally obtained from (13.13) by integrating over the final momenta according to the experimental situation: observed momenta are fixed at the observed value, the momenta of unobserved soft photons are integrated over all values with energies below the deSect. threshold. The factor $\hat{S}$ in (13.13) is defined by (13.14) and the reduction formula (13.7). $\sigma_n$ vanishes in low orders (depending on $n$) but is in general IR divergent in higher orders. However, in any finite order of perturbation theory these divergences are found to cancel in the sum $\sum_n \sigma_n$ over all final configurations which are compatible with the experimental setup. In this naked form the statement clearly makes no sense. In order to give it a meaning, one must first introduce an IR regularization removing the divergences of the individual terms. This may consist in giving the photons a small non-vanishing mass $\lambda$, or in cutting off photon energies below a small positive value $\epsilon$. The $\sigma_n$ become then finite, their sum can be formed, and the limit $\lambda \to 0$ or $\epsilon \to 0$ of this sum can be shown to exist.[1] Satisfying though this procedure may be from a practical point of view, it leaves something to be desired as far as the theoretical understanding of the situation is concerned. A result of direct physical relevance that is only arrived at via a detour through an unphysical, regularized theory can hardly be claimed to have been understood. There is also the question whether the two limits of going to the mass shell and of removing the regularization may really be interchanged. In the next three chapters we will therefore develop a more physical approach aiming directly at the observable inclusive cross sections, without ever introducing an $S$-matrix or other transition amplitudes, and without using IR regularizations. A result of this approach is that the exchange of limits just mentioned is indeed a potential source of errors. For a closer discussion see Chaps. 15 and 16, especially Sect. 16.5.

---

[1] The standard reference is still [YFS 61]. See also [W], Chap. 13.

# Chapter 14

# Particle Probes

The fermionic fields $\psi$ and $\bar{\psi}$ of QED do not satisfy the conditions that are assumed in the derivation of the LSZ asymptotic conditions. Therefore the experimentally observed electrons and positrons are not particles in the sense of the LSZ formalism, they are expected to be "infraparticles" in the sense of Schroer [Sc 63]. Characteristic for them is a mass-shell singularity in the 2-point functions of their fields which is weaker than the $\delta$-singularity of an ordinary particle. Typically, an infraparticle singularity is of the branch point type $\theta(p^2 - m^2)(p^2 - m^2)^{-1+\alpha}$ with $m$ the observed mass of the particle and $\alpha$ a small positive number. As a consequence of this weak singularity there exists no 1-particle subspace of the state space, which is an eigenspace of the mass operator $M^2 = P_\mu P^\mu$. But the spectrum of $M^2$ still shows a strong concentration near the mass shell: loosely speaking, the density of states is large there. This fact might be used as a starting point for the search of a scattering formalism for infraparticles, by looking for a generalization of the LSZ formalism which is capable of handling this situation. As it turns out, it is more promising to apply a different strategy which is closer to experiment. In this approach a particle is not characterized by possessing a sharp mass but by its spatial localization. This notion of particles describes an observed lumpiness of quantum reality which becomes apparent at low matter densities.

Such an approach to particle scattering has first been described in the framework of local observables by Araki and Haag [AH 67]. They consider local observables representing detectors registering the presence of a particle in a given region of space during a given time interval. Positioning a number of such detectors at macroscopic distances but susceptible to triggering during the same time interval, as is done in a coincidence arrangement of a real scattering experiment, a theoretical expression can then be derived directly for a scattering cross section without introducing an $S$-matrix. The proofs of Araki and Haag were still based on the assumptions underlying the LSZ formalism. Hence their results are not directly applicable to QED.[1] But it will be shown in this and the following chapters that their results can be generalized quite naturally to QED, in marked contrast to the LSZ approach. This generalization is, however, not just based on a few simple assumptions like the axioms of axiomatic field theory or of the local observables formalism. It relies on a detailed knowledge of the dynamics and is therefore given in perturbation theory, using the results of Part II.

Due to this necessity of arguing dynamically, we can only work with quite simple local observables with little similarity to real detectors. They will therefore be given the less specific denomination *"particle probes"*. The present chapter is devoted to the description of these probes and their connection with particles in the free field case. Chapter 15 contains the extension of the notion to interacting fields, and in Chaps. 16 and 17 it will be used to describe scattering events and to calculate reaction cross sections from the $\mathcal{W}$-functions of QED.

The contents of this and the remaining chapters are not intended to be the final word on the subject. They just suggest a promising (in my opinion) new point of view, which may lead to a better understanding of what a particle is in a relativistic quantum field theory, giving the IR aspect of the problem its full due.

## 14.1 Mathematical Preliminaries

In this section we prove some mathematical results which lie at the heart of our approach. They are generalizations of a classical theorem by Ruelle on the large-time behaviour of smooth solutions of the Klein–Gordon equation, which is an important ingredient in the proof of the conventional asymptotic conditions.

We first introduce a definition patterned after the definition of Hölder continuous functions. A "truncated cone" $K = (\mathcal{C} \cap B_r) \subset R^N$ is the intersection of a non-empty open cone $\mathcal{C}$ with apex at the origin and an open ball $B_r = \{u \in R^N : |u| < r < \infty\}$. We define:

**Definition 14.1.** *Let $f(u)$, $u \in R^N$, be an integrable function on $R^N$. We say that $f$ is "H-integrable of index $\alpha$" with $0 < \alpha \leq 1$ if there exists a truncated cone $K \subset R^N$ and a non-negative integrable function $h(u)$ such that*

$$\left| \frac{f(u+a) - f(u)}{|a|^\beta} \right| \leq h(u) \qquad (14.1)$$

---

[1] A possible extension of these results to infraparticles is discussed in Chap. VI of [H] and in [Bu 94]. Our way of proceeding may be considered an implementation of these ideas in a concrete field theoretical setting, though the conditions on the "detectors" are relaxed, in particular using a very weak definition of localization.

## 14.1 Mathematical Preliminaries

for all $\beta \in [0, \alpha)$ and $a \in K$. $f$ is said to be "of H-type $(n, \alpha)$" if all its derivatives of order $n$ exist and are H-integrable of index $\alpha$. If the index is not specified (H-type $n$), then the function is of H-type $(n, \alpha)$ for every $\alpha \leq 1$.

An H-integrable function of index $\alpha$ is of H-type $(0, \alpha)$. A Hölder continuous function of index $\alpha$ with compact support is H-integrable of index $\alpha$. A one-dimensional example of direct relevance is $f(u) = \theta(u)\,\theta(1-u)\log u$, which is H-integrable. This is seen as follows. As the truncated cone $K$ we choose the open interval $(0,1)$, and we define $v = a/u$. Then

$$\frac{1}{a^\beta}\left(\log(u+a) - \log u\right) = \frac{1}{u^\beta}\left[\frac{1}{v^\beta}\log(1+v)\right].$$

The square bracket is a bounded function of $v$ on $[0, \infty]$ and $u^{-\beta}\theta(u)\,\theta(1-u)$ is integrable for $\beta < 1$, hence our contention is proved.

A number of statements about the asymptotic behaviour of the Fourier transforms of H-integrable and related functions will now be proved.

**Lemma 14.1.** *Let $f(u)$, $u \in R^N$, be of H-type $(n, \alpha)$ and have a compact support. Let $\nu$ be an arbitrary unit vector in $R^N$. Then there exists for every $\beta \in [0, \alpha)$ a positive constant $C_{\beta n}$ such that*

$$\lambda^{n+\beta}\left|\int du\, e^{-i\lambda(\nu,u)} f(u)\right| \leq C_{\beta n} \tag{14.2}$$

*for all $\nu$ and all $\lambda \geq 0$.*

The assumption of a compact support of $f$ is unnecessarily restrictive but adequate for our purposes.

**Proof.** We first prove the Lemma in the case $n = 0$. Let $K = \mathcal{C} \cap B_R$ be the truncated cone in the characterization of $f$ as being H-integrable. To each unit vector $\nu \in R^N$ we can find a unit vector $\eta_\nu \in \mathcal{C}$ such that $|(\eta_\nu, \nu)| \geq C_1 > 0$ for a positive constant $C_1$ which does not depend on $\nu$. Then

$$g(u) = \frac{f(u)}{e^{i(\eta_\nu,\nu)} - 1}$$

is H-integrable of index $\alpha$ too: there exists an integrable function $h(u)$ independent of $\nu$ with

$$\left|\frac{g(u+a) - g(u)}{|a|^\beta}\right| \leq h(u)$$

for all $\beta \in [0, \alpha)$ and all $a \in K$. We find

$$\left|\int du\, \frac{g(u+a) - g(u)}{|a|^\beta} e^{-i\lambda(\nu,u)}\right| \leq \int du\, h(u) =: C_\beta.$$

The integral on the left-hand side can be written

$$\lambda^\beta \int du\, g(u)\, \frac{e^{i\lambda(\nu,a)} - 1}{|\lambda a|^\beta} e^{-i\lambda(\nu,u)}.$$

Choosing $a = A\,\eta_\nu$ we obtain

$$\lambda^\beta \left| \int du\, g(u)\, \frac{e^{i\lambda A(\eta_\nu,\nu)} - 1}{(\lambda A)^\beta}\, e^{-i\lambda(\nu,u)} \right| \leq C_\beta\ .$$

The expression (14.2) is clearly bounded for $\lambda \to 0$, hence we need only study its behaviour for large $\lambda$, e.g. for $\lambda > R$. With $A = 1/\lambda$ our inequality becomes

$$\lambda^\beta \left| \int du\, f(u)\, e^{-i\lambda(\nu,u)} \right| \leq C_\beta\ ,$$

the claimed result.

The case of positive $n$ can be reduced to the case $n = 0$ by an $n$-fold integration by parts. Define $\partial_i = \frac{\partial}{\partial u_i}$. Then $\partial_i^n f(u)$ is H-integrable of index $\alpha$ by assumption. Hence

$$\lambda^{n+\beta} \left|\nu_i^n \int du\, f(u)\, e^{-i\lambda(\nu,u)}\right| = \lambda^\beta \left|\int du\, \partial_i^n f(u)\, e^{-i\lambda(\nu,u)}\right| \leq C_\beta$$

and thus

$$\lambda^{n+\beta} \left|\int du\, f(u)\, e^{-i\lambda(\nu,u)}\right| \leq \min_i \frac{C_\beta}{|\nu_i|^n} \leq C_\beta N^{-\frac{n}{2}} =: C_{\beta n}\ ,$$

which proves the Lemma.

The following lemmas are applications of Lemma 14.1 to situations of physical interest with $N = 4$. The first two are Ruelle's original result and a generalization of it to the zero-mass case.

**Lemma 14.2.** *Let $f(\vec{p})$, $\vec{p} \in R^3$, be a $C^n$ function with compact support, $n \geq 2$, and $\nu \neq 0$ a 4-vector with $\nu^0 \geq 0$. Define*

$$I(\lambda\nu) = \int dp\, \delta_+(p,m)\, f(\vec{p})\, e^{-i\lambda(\nu,p)}\ , \qquad (14.3)$$

*with $\delta_+(p,m) = \theta(p^0)\,\delta(p^2 - m^2)$, $m > 0$. Then, if the ray from the origin in direction $\nu$ intersects the mass shell at a point $P_\nu$ with $\vec{P}_\nu \in \mathrm{supp} f$, $I(\lambda\nu)$ behaves for $\lambda \to \infty$ as*

$$I(\lambda\nu) \sim -\left(\lambda\sqrt{(\nu,\nu)}\right)^{-\frac{3}{2}} \pi^{\frac{3}{2}} \sqrt{m}\, (1+i)\, e^{im\lambda\sqrt{(\nu,\nu)}}\, f(\vec{P}_\nu) + \mathcal{O}(\lambda^{-2})\ . \qquad (14.4)$$

*In all other directions $\nu$, $I(\lambda\nu)$ decreases at least of order $\mathcal{O}(\lambda^{-n+\varepsilon})$ for every $\varepsilon > 0$.*

**Proof.** Consider first a timelike $\nu$. We can choose a system of reference such that $\nu = (\nu^0, \vec{0})$. Then

$$I(\lambda\nu) = \int \frac{d^3p}{2\omega(\vec{p})}\, e^{-i\lambda\nu^0\omega(\vec{p})}\, f(\vec{p})$$

$$= 2\pi \int_m^\infty d\omega\, \sqrt{\omega - m}\, \sqrt{\omega + m}\, \bar{f}(\sqrt{\omega^2 - m^2})\, e^{-i\lambda\nu^0\omega}\ ,$$

where $\bar{f}(r) \in C^n$ is the mean value of $f$ over the sphere $|\vec{p}| = r$. We introduce $\tau = \omega - m$ as new variable of integration and split

$$\sqrt{\tau + 2m}\, \bar{f}(\sqrt{\tau(\tau + 2m)}) = \sqrt{2m}\, \bar{f}(0)\, e^{-\tau}$$
$$+ \left[\sqrt{\tau + 2m}\, \bar{f}(\sqrt{\tau(\tau + 2m)}) - \sqrt{2m}\, \bar{f}(0)\, e^{-\tau}\right]. \quad (14.5)$$

The contribution to $I$ of the first term in this decomposition is

$$I_1(\lambda \nu) = 2\pi \sqrt{2m}\, e^{-i\lambda \nu^0 m} f(0) \int_0^\infty d\tau\, \sqrt{\tau}\, e^{-\tau(1+i\lambda \nu^0)}$$
$$= 2\pi \sqrt{2m}\, e^{-i\lambda \nu^0 m} f(0)\, (\lambda \nu^0)^{-\frac{3}{2}} \int_0^\infty du\, \sqrt{u}\, e^{-u((\lambda \nu^0)^{-1}+i)}.$$

The $u$-integral has the value (see [GR], formula 3.381)

$$\Gamma(\tfrac{3}{2}) \left(1 + \frac{1}{(\lambda \nu^0)^2}\right)^{-\frac{3}{4}} \exp(-\tfrac{3}{2} i \arctan \lambda \nu^0),$$

which after insertion into $I_1$ gives the leading term in (14.4) plus a term decreasing at least like $\lambda^{-\frac{5}{2}}$. Calling the square bracket in (14.5) $B(\tau)$ we obtain as its contribution

$$I_2(\lambda \nu) = 2\pi\, e^{-i\lambda \nu^0 m} \int_0^\infty d\tau\, \sqrt{\tau}\, B(\tau)\, e^{-i\lambda \nu^0 \tau}.$$

$B(\tau)$ vanishes at $\tau = 0$ of order 1, hence $\sqrt{\tau}\, B(\tau)$ is of H-type (1,1) if $n = 2$, of H-type $(2, \tfrac{1}{2})$ if $n > 2$, and by Lemma 14.1 $I_2$ decreases at least like $\lambda^{-2+\varepsilon}$.

If the $\nu$-ray does not meet the support of the integrand in (14.3) we can introduce $u_1 = (\nu, p)$, $u_2 = (p, p)$, and $u_3$, $u_4$, two suitably chosen components $p_i$ as new integration variables. The Jacobian $J = \frac{\partial(p)}{\partial(u)}$ is then regular in the support of the integrand, and $I$ can be written as

$$I(\lambda \nu) = \int du_1\, e^{-i\lambda u_1}\, F(u_1),$$

with

$$F(u_1) = \int du_2\, \delta(u_2 - m^2) \int du_3\, du_4\, J(u)\, f(\vec{p}(u))$$

a $C^n$ function. It is then a standard result that $I(\lambda \nu) = \mathcal{O}(\lambda^{-n})$ for $\lambda \to \infty$.

**Lemma 14.3.** *Let $f(\vec{p})$ be a $C^n$ function with compact support, $n \geq 2$, and $\nu \neq 0$ a 4-vector with $\nu^0 \geq 0$. Define*

$$I(\lambda \nu) = \int dp\, \delta_+(p, 0)\, f(\vec{p})\, e^{-i\lambda(\nu, p)}. \quad (14.6)$$

*Then, if $(\nu, \nu) = 0$,*

$$I(\lambda \nu) = \frac{-i\pi}{\lambda |\vec{\nu}|} \int_0^\infty ds\, f(s\hat{\nu}) + \mathcal{O}\left(\lambda^{-\frac{3}{2}+\varepsilon}\right) \quad (14.7)$$

for $\lambda \to \infty$, with $\hat{\nu} = \frac{\vec{\nu}}{|\vec{\nu}|}$ and $\varepsilon > 0$ arbitrarily small. For timelike or spacelike $\nu$ the stronger estimate

$$I(\lambda \nu) = \mathcal{O}\left(\lambda^{-2+\varepsilon}\right) \tag{14.8}$$

holds.

**Proof.** Let $\nu$ be spacelike. We can find a coordinate system in which $\nu^0 = 0$. Then

$$I(\lambda \nu) = \frac{1}{2} \int d^3p \, \frac{f(\vec{p})}{|\vec{p}|} \, e^{i\lambda(\vec{\nu},\vec{p})} \, .$$

Since the integrand is integrable, the region of integration can be split into the half-spaces $(\vec{\nu}, \vec{p}) \{\substack{>\\<}\} 0$. Choosing for $K$ a truncated cone in the respective half-space with apex at 0 we find in both cases that $\frac{f(\vec{p})}{|\vec{p}|}$ is of H-type $(1,1)$, hence (14.8) follows from Lemma 14.1. For timelike $\nu$ we can choose $\nu = (\nu^0, \vec{0})$ without restricting generality. Let $\bar{f}(r) \in C^n$ be the mean value of $f$ over the sphere $|\vec{p}| = r$. Then

$$I(\lambda \nu) = 2\pi \int_0^\infty r \, dr \, \bar{f}(r) \, e^{-i\lambda \nu^0 r} \, .$$

$r \bar{f}(r)$ is in $[0, \infty)$ of H-type $(1,1)$, from which fact the estimate (14.8) follows. Both for spacelike and timelike $\nu$ the decrease of $I$ is actually of order $\lambda^{-n}$ if $\vec{p} = 0$ does not lie in the support of $f$.

For $\nu$ lightlike we may choose the typical value $\nu = (1, 1, 0, 0)$ without restricting generality. Then

$$I(\lambda \nu) = \frac{1}{2} \int d^3p \, \frac{f(\vec{p})}{|\vec{p}|} \, e^{-i\lambda(p^0 - p^1)} = \frac{1}{2} \int du \, g(u) \, e^{-i\lambda u} \tag{14.9}$$

with

$$g(u) = \int d^3p \, \frac{f(\vec{p})}{|\vec{p}|} \, \delta(u - |\vec{p}| + p^1) \, . \tag{14.10}$$

$g$ vanishes for negative $u$ because $|\vec{p}| - p^1 \geq 0$. For $u > 0$ the $p$-support of the $\delta$-function is a smooth manifold not containing the dangerous point $\vec{p} = 0$. Hence $g \in C^\infty$ for $u > 0$. The behaviour of $I$ for $\lambda \to \infty$ is thus determined by the singularity of $g$ at $u = 0$. Define $\rho = \sqrt{(p^2)^2 + (p^3)^2}$, $z = p^1$, and let $\bar{f}(z, \rho) \in C^n$ be the mean value of $f(\vec{p})$ over the circle $\rho = \text{const}$, $z = \text{const}$. We obtain

$$g(u) = 2\pi \int_{-\infty}^\infty dz \int_0^\infty \rho \, d\rho \, \frac{\bar{f}(z,\rho)}{\sqrt{z^2 + \rho^2}} \, \delta\left(u - \sqrt{z^2 + \rho^2} + z\right)$$

$$= 2\pi \int_{-\frac{u}{2}}^\infty dz \, \bar{f}\left(z, \sqrt{u^2 + 2uz}\right) \, .$$

This function is continuous on $[0, \infty)$ and

$$g(0) = 2\pi \int_0^\infty dp^1 \, f(p^1, 0, 0) \, . \tag{14.11}$$

Its derivative is, with $f_2(z,\rho) = \frac{\partial \bar{f}(z,\rho)}{\partial \rho}$,

$$g'(u) = \pi f(p^1, 0, 0) + \frac{2\pi}{\sqrt{u}} \int_{-\frac{u}{2}}^{\infty} dz\, f_2\left(z, \sqrt{u^2 + 2uz}\right) \frac{u+z}{\sqrt{u+2z}}\ .$$

The $z$-integral remains bounded for $u \downarrow 0$, hence $g'(u)$ has at $u = 0$ a singularity $u^{-1/2}$. This means that $g'$ is H-integrable of index $1/2$. Inserting $g$ into (14.9) and integrating by parts we find

$$\begin{aligned}I(\lambda \nu) &= \frac{i}{2\lambda} \int_0^\infty du\, g(u) \left(e^{-i\lambda u}\right)' \\ &= \frac{-i}{2\lambda} g(0) - \frac{i}{2\lambda} \int_0^\infty du\, g'(u)\, e^{-i\lambda u} \\ &= -\frac{i\pi}{\lambda} \int_0^\infty dp^1\, f(p^1, 0, 0) + \mathcal{O}\left(\lambda^{-\frac{3}{2}+\varepsilon}\right)\end{aligned}$$

by Lemma 14.1, which is the desired result (14.7).

**Definition 14.2.** *Let*

$$H_\pm = \{p \in R^4 : p^0 = \pm\omega(\vec{p})\,,\ m \geq 0\} \tag{14.12}$$

*be the* $\left\{\begin{matrix}positive\\negative\end{matrix}\right\}$ *mass shell. Let* $\mathcal{R} \subset R^4$ *be a compact set with a smooth boundary, such that* $\mathcal{R}$ *intersects* $H_+$ *but not* $H_-$. *Let the function* $S(p)$ *be defined on* $\mathcal{R}$. *Write it as a function of the variables* $p_- = p^0 - \omega(\vec{p}), p^1, p^2, p^3$. *Assume that for all* $\vec{p}$-*derivatives* $D_{\vec{p}}$ *of arbitrary order* $D_{\vec{p}} S(p_-, \vec{p})$ *is* $C^\infty$ *in* $\vec{p}$, *and* $C^\infty$ *in* $p_-$ *except at* $p_- = 0$ *where* $\left|p_-^{s+\varepsilon}\left(\frac{\partial}{\partial p_-}\right)^s D_{\vec{p}} S(p_-, \vec{p})\right|$ *is bounded for all non-negative integers* $s$ *and all* $\varepsilon > 0$. *Then* $S(p)$ *is called "weakly singular at* $H_+$*".*

**Lemma 14.4.** *Let* $H_\pm$, $\mathcal{R}$, $S(p)$, *satisfy the conditions of Definition 14.2. In addition, assume that* $S(p_-, \vec{p})$ *is in* $p_-$ *a boundary value of an analytic function in* $\{0 \geq \Im p_- \geq -A\}$ *for a positive* $A$. *Let* $f(p)$ *be* $C^\infty$ *with support in* $\mathcal{R}$, *and* $\nu \neq 0$ *a 4-vector with* $\nu^0 \geq 0$. *Then*

$$I(\lambda \nu) = \int dp_-\, d^3p\, \frac{S(p)}{(p_- - i\varepsilon)^r}\, f(p)\, e^{-i\lambda(\nu,p)}\ , \tag{14.13}$$

$r \in Z_+$, *is a continuous function of* $\lambda$ *which decreases for* $\lambda \to \infty$ *stronger than any negative power* $\lambda^{-N}$, $N \in Z_+$.

**Proof.** Existence and continuity of $I$ are obvious, because $\frac{S(p)}{(p_- - i\varepsilon)^r}$ is a distribution under our assumptions. $S$ and $f$ can be written as functions of $p_-, \vec{p}$. $f(p_-, \vec{p})$ is $C^\infty$, $S(p_-, \vec{p})$ has the properties stated in Definition 14.2. The function $F(p_-, \vec{p}) = f(p_-, \vec{p})\, e^{-\vec{p}^{\,2}}$ is $C^\infty$ and of compact support.

Defining $F^\alpha(p_-,\vec{p}) = \left(\frac{\partial}{\partial p_-}\right)^\alpha F(p_-,\vec{p})$ we expand

$$F(p_-,\vec{p}) = \sum_{\alpha=0}^{n} \frac{1}{\alpha!} F^\alpha(0,\vec{p})\, p_-^\alpha + R'(p)$$

for $n$ any positive integer. $R'$ vanishes at $p_- = 0$ of order $n$. For $f$ we obtain

$$f(p) = \sum_{\alpha=0}^{n} \frac{1}{\alpha!} F^\alpha(0,\vec{p})\, p_-^\alpha e^{-p_-^2} + R(p) , \qquad (14.14)$$

with $R(p) = R'(p)\, e^{-p_-^2}$. All the terms in this expansion are strongly decreasing for $|p| \to \infty$. We insert the expansion into the $p$-integral in (14.13). The $\alpha$-term is

$$\int d^3p\, F^\alpha(0,\vec{p})\, e^{i\lambda(\vec{\nu},\vec{p})} e^{-i\lambda\nu^0 \omega(\vec{p})} \int dp_- \frac{S(p_-,\vec{p})}{(p_- - i\varepsilon)^{r-\alpha}} e^{-p_-^2} e^{-i\lambda\nu^0 p_-} .$$

Under our assumptions we can shift the $p_-$ integration path in a neighbourhood of $p_- = 0$ into the lower complex half-plane, where the integrand is regular. Setting $p_- = p' - i p''$, $p'' \geq 0$, we find $e^{-i\lambda\nu^0 p_-} = e^{-i\lambda\nu^0 p'} e^{-\lambda\nu^0 p''}$ which decreases strongly for $\lambda \to \infty$ if $\nu^0$ and $p''$ are positive. Hence the $p_-$-integral decreases strongly. For $\nu^0 = 0$ it is a smooth function of $\vec{p}$, so that the remaining $\vec{p}$-integral, with $\vec{\nu} \neq 0$, decreases strongly for $\lambda \to \infty$. The contribution of $R$ to (14.13) is

$$\int dp\, S(p) \frac{R(p)}{p_-^r} e^{-i\lambda(\nu,p)} .$$

The quotient vanishes at $p_- = 0$ of order $n-r$ if $n > r$, hence the integrand is an integrable function. Like in the proof of Lemma 14.3 for spacelike $\nu$ we split the region of integration into the sets $\{p_- \geq 0\}$ and $\{p_- \leq 0\}$. Choosing suitable truncated cones we find in both cases that $S\,R\,(p_-)^{-r}$ is of H-type $(n-r, 1)$, hence $I(\lambda\nu)$ decreases at least like $\lambda^{-n+r-1+\varepsilon}$ for every $n > r$.

Clearly this method gives similar estimates if $f$ is only $n$ times differentiable with $n$ sufficiently large.

**Lemma 14.5.** *Let $\mathcal{R}$, $S(p)$, $f(p)$, $\nu$, be defined like in Lemma 14.4, except that $S$ is not assumed to be analytic in $p_-$. Then*

$$I(\lambda\nu) = \int d^3p\, \mathrm{P.f.} \int dp_- \frac{S(p_-,\vec{p})}{p_-} f(p_-,\vec{p})\, e^{-i\lambda(\nu,p)} \qquad (14.15)$$

*is continuous in $\lambda$ and decreases for $\lambda \to \infty$ at least like $\lambda^{-\frac{3}{2}+\varepsilon}$ if $m > 0$, like $\lambda^{-1+\varepsilon}$ if $m = 0$, for every $\varepsilon > 0$.*

The finite-part prescription P.f. is a generalization of the prescription in (10.26). There exists on $\{p_- \geq 0\}$ a weakly singular function $T(p_-,\vec{p})$ such that $S(p_-,\vec{p}) = \frac{\partial}{\partial p_-}T(p_-,\vec{p})$. Then, for $f$ a test function with support in $\mathcal{R}$ we define

$$\text{P.f.} \int_0^\infty dp_- \, \frac{S(p_-,\vec{p})}{p_-} f(p_-,\vec{p}) = -\int_0^\infty dp_- \, T(p_-,\vec{p}) \frac{\partial}{\partial p_-} f(p_-,\vec{p}) \,. \quad (14.16)$$

This means that we integrate by parts and drop the divergent boundary term at $p_- = 0$. This prescription is known to define a distribution.

**Proof of Lemma 14.5.** $I(\lambda \nu)$ is given by

$$I(\lambda \nu) = -\int d^3p \, e^{i\lambda(\vec{\nu},\vec{p})} \, e^{-i\lambda\nu^0 \omega(\vec{p})}$$
$$\times \int_0^\infty dp_- \, T(p_-,\vec{p}) \frac{\partial}{\partial p_-} \left[ f(p_-,\vec{p}) e^{-i\lambda\nu^0 p_-} \right] \,. \quad (14.17)$$

The integrand is an integrable function, so that the orders of integration can be freely interchanged. Also, differentiations with respect to $\vec{p}$ of the $p_-$-integral can be carried out under the integral sign. Thus the $p_-$-integral $J(\vec{p}, \lambda \nu^0)$ is $C^\infty$ and has a compact support in $\vec{p}$. For $\nu^0 = 0$, $I(\lambda \nu)$ is a Fourier integral of a test function from $\mathcal{D}$, hence strongly decreasing. For $\nu^0 > 0$ the derivative of the square bracket in (14.17) is $\frac{\partial}{\partial p_-} f \exp(\cdots) - i\lambda\nu^0 \exp(\cdots)$. The second term is clearly the more critical one for the $\lambda \to \infty$ behaviour. As a function of $p_-$, $Tf$ is H-integrable of index 1, hence

$$\lambda^{1-\varepsilon} \int_0^\infty dp_- \, T f \, e^{-i\lambda\nu^0 p_-} =: \lambda^{1-\varepsilon} J_c(\vec{p}, \lambda \nu^0)$$

remains bounded for $\lambda \to \infty$ uniformly in $\vec{p}$. The same holds for its $\vec{p}$-derivatives. But

$$I_c(\lambda \nu^0) = -\int d^3p \, e^{i\lambda(\vec{\nu},\vec{p})} \, e^{-i\lambda\nu^0 \omega(\vec{p})} \, J_c(\vec{p}, \lambda \nu^0)$$

is of a form discussed in the proofs of Lemmas 14.2 and 14.3, apart from the explicit $\lambda$-dependence of $J_c$. Notice that in the case $m = 0$, $J_c$ is still $C^\infty$ at $\vec{p} = 0$ because the origin $p = 0$ lies outside the support of $f$. Due to the favourable $\lambda$-behaviour of $J_c$ the estimates used in those proofs still apply and yield the result claimed in Lemma 14.5.

## 14.2 Free Particles

Under a *"particle probe"* we understand an observable registering the presence of a particle with a given 4-momentum in a given region of space at a given time. The underlying particle picture is the classical one of an object which is well localized at all times and follows a well defined world line, which is a straight line if no external forces act on the particle. This corresponds to the experimentalist's notion of a particle, because this is what he observes with his detectors. Of course, this picture can only be approximately true. The notion of a sharp world line makes no sense in quantum mechanics, because position and velocity of a particle are subject to uncertainty relations which preclude fixing them simultaneously with arbitrary precision. But in a scattering experiment the uncertainties of the position measurements are macroscopic, i.e. large compared to atomic dimensions. The same applies to the thickness of the observed trajectories and therefore to the momenta deduced from them: the limits set by the uncertainty relations are safely surpassed. Also, the relevant observations are made at macroscopic 4-distances from the scattering event, and this fact washes out a possible influence of the precise form of the involved wave functions: directly measured are only the "classical" variables position, momentum, and polarization.

We propose a theoretical implementation of this phenomenological picture, at first in a theory of free fields. Spin is an inessential complication in this context, hence we explain the formalism for the scalar free field of Sect. 5.1.

Let $\varphi(x)$ be a hermitian scalar field solving the Klein–Gordon equation (5.1). Let $f(x)$ be a test function, $\tilde{f}(p)$ its Fourier transform, with the following properties. $\tilde{f}(p)$ has a small compact support contained in the half-space $\{p^0 > 0\}$ and centred around a momentum $P$ on the positive mass shell $\{(P,P) = m^2,\ P^0 > 0\}$. Moreover, $\tilde{f}(p)$ is smooth in the sense that it has no strong oscillatory behaviour anywhere, i.e. its derivatives are roughly as small as is compatible with the size of the support and the maximal value of $|\tilde{f}|$. This implies that $f(x)$ is concentrated around the origin of $x$-space in the sense that it is negligibly small outside a small neighbourhood of the origin. What "negligibly small" means depends on the concrete experimental situation. It would clearly be preferable to work with test functions with compact supports in both spaces. But such functions do not exist. That we prefer strict localization in $p$-space rather than in $x$-space is reasonable in view of an embarrassing feature of relativistic quantum mechanics: well-defined self-adjoint momentum operators exist but position operators with all the properties generally associated with particle localization do not. The localization of a particle is an unavoidably vague and ill-defined concept, and there is no point in wasting one's energy on trying to render it strict.

We represent a probe located in this vague sense at the origin and susceptible to particles with momentum $\simeq P$ by the operator

$$D = \varphi(f)^* \varphi(f) , \qquad (14.18)$$

$$\varphi(f) = \int dp\, \tilde{\varphi}(p)\, \tilde{f}(p) = \int dy\, \varphi(y)\, f(y) . \qquad (14.19)$$

$\varphi(f)$ is an annihilation operator, its adjoint a creation operator. A probe located at an arbitrary point $a$ of Minkowski space is represented by

$$D(a) = T(a)\, D\, T^*(a) = \varphi(f_a)^* \varphi(f_a) , \qquad (14.20)$$

where $T(a)$ is the familiar unitary representation of the translation group and $f_a(x) = f(x-a)$.

It must now be discussed in what way this operator acts as a particle probe. A first trivial remark is that $D(a)\Omega = 0$. Hence we have $(\Omega, D(a)\Omega) = 0$: $D$ sees no particle in the vacuum, which would be fatal to our claim. Let $g(x)$ be another test function with the same properties as $f$ except that the support of $\tilde{g}$ may reach into the lower half-space $\{p^0 < 0\}$ (this relaxation of requirements is important only for massless particles). Then

$$\Phi_g = \varphi(g)^* \Omega \tag{14.21}$$

is a 1-particle state in the conventional sense: it is the 1-particle Fock state

$$\int \frac{d^3p}{2\omega(\vec{p})}\, \overline{\hat{g}(\vec{p})}\, a^*(\vec{p})\, \Omega$$

with wave function $\hat{g}(\vec{p}) = \tilde{g}(\omega(\vec{p}), \vec{p})$, which we assume to be normalized: $\int \frac{d^3p}{2\omega(\vec{p})} |\hat{g}(\vec{p})|^2 = 1$. We claim that $\Phi_g$ can be interpreted as a 1-particle state in our meaning of the word, which has been prepared at time $t \simeq 0$ in a neighbourhood of the origin with a momentum distribution described by $\hat{g}$. More exactly, this means that if monitored by a probe $D(a)$ it behaves as such a state in the approximate sense indicated above and hopefully clarified by what follows. The preparation region of this state we call the *source* of the particle.

The expectation value

$$E(a) = (\Phi_g, D(a)\Phi_g) \tag{14.22}$$

can be calculated using the cluster expansion (5.13):

$$E(a) = (\Omega\,\varphi(g)\,\varphi(f)^*\,\Omega)\,(\Omega\,\varphi(f)\,\varphi(g)^*\,\Omega)$$
$$= \left| \int dp\, e^{-ipa}\, \hat{f}(\vec{p})\, \overline{\hat{g}(\vec{p})}\, \delta_+(p) \right|^2 . \tag{14.23}$$

Call the $p$-integral in this expression $I(a)$. It vanishes if the supports of $\hat{f}$ and $\hat{g}$ do not overlap, in accordance with the contention that the probe registers only particles with momenta in the support of $\hat{f}$. For overlapping supports we study $I(a)$ for macroscopic separations $a$ of the probe from the source $O$. We write $a = \lambda\nu$ with $\nu^0 \geq 0$, and $\lambda$ large. The behaviour of $I$ for $\lambda \to \infty$ is given by Lemmas 14.2 and 14.3. Consider first the case $m > 0$. If the $\nu$-ray intersects the positive mass shell at a point $P$ with $\vec{P}$ in the support of $\hat{f}\hat{g}$, then the asymptotically dominant term is

$$E(\lambda\nu) \sim \frac{1}{\lambda^3}\frac{1}{(\nu,\nu)^{3/2}} \pi^3 (2m)^{3/2} |\hat{f}(\vec{P})|^2 |\hat{g}(\vec{P})|^2 . \tag{14.24}$$

If $m = 0$, $(\nu, \nu) = 0$, and the $\nu$-ray intersects the support of $\hat{f}\,\hat{g}$, then

$$E(\lambda\nu) \sim \frac{1}{\lambda^2} \frac{\pi^2}{|\vec{\nu}|^2} \left| \int_0^\infty ds\, \hat{f}(s\vec{\nu})\, \overline{\hat{g}(s\vec{\nu})} \right|^2. \tag{14.25}$$

$E(\lambda\nu)$ decreases faster in all other directions, the speed of decrease depending on the smoothness of $\hat{f}\hat{g}$. These results, which are valid at macroscopic distances, can be interpreted in terms of classical particles. For $m > 0$ a particle issuing from a small neighbourhood of the origin, possessing the velocity $\vec{v} = \frac{\vec{p}}{m}$ with probability $\frac{1}{2\omega}|\hat{g}(\vec{p})|^2$, will hit a probe located in a small neighbourhood of the distant point $\lambda\nu$ with a probability decreasing like $\lambda^{-3}$ for increasing $\lambda$. This is a purely geometrical result. Remember that "small" refers to space *and* time: the probe does not only have a spatial diameter which is small relative to its distance from the origin, it also is only receptive to a particle during a time interval which is small relative to the particle's time of flight. This latter aspect is clearly not very realistic, but this is of little consequence for our discussion of principal ideas. The factor $|\hat{g}(\vec{P})|^2$ in (14.24) describes the probability distribution of the momentum of the observed particle, the factor $|\hat{f}|^2$ characterizes the efficiency of the probe.

A similar reasoning applies to the case $m = 0$. The slower decrease of $E$ in this case is due to the fact that all particles travel at the same speed, namely the velocity of light, irrespective of their momenta, so that there is no spreading of the wave packet in longitudinal directions.

In the directions not corresponding to this classical picture the counting probability decreases so fast that it is negligibly small for times at which the leading terms (14.24) or (14.25) are still of measurable size where they are present. Notice that $\hat{f}(\vec{p})$ vanishes in a neighbourhood of $\vec{p} = 0$ by our assumptions, so that this non-classical decrease is strong if $\hat{f}$ and $\hat{g}$ are $C^\infty$. In the spirit of the LSZ approach one might consider $\lim_{\lambda \to \infty} \lambda^{(3 \text{ or } 2)} E(\lambda\nu)$ the relevant asymptotic quantity. But this does not quite correspond to reality, real measurements being effected at finite times.

For later purposes it is convenient to replace the factor $(\Omega, \varphi(x)\,\varphi(y)\,\Omega)$ in $I(a)$ by $(\Omega, T^+(x,y)\,\Omega)$. This is legitimate because the difference between the two expressions is

$$\Delta(x,y) = \left(\Omega, \left(\varphi(x)\,\varphi(y) - T^+(x,y)\right)\Omega\right) = \theta(y^0 - x^0)\left(\Omega, [\varphi(x), \varphi(y)]\,\Omega\right), \tag{14.26}$$

with support $S$ in $(x - y) \in \bar{V}_+$. On this support the test function $f(x - \lambda\nu)\,g(y)$ vanishes for $\lambda \to \infty$ in the topology of $\mathcal{S}$ for the $\nu$ of interest to us. As a result, the relevant Schwartz seminorms vanish on $S$ sufficiently fast to make $\int dx\, dy\, f(x - \lambda\nu)\, g(y)\, \Delta(x,y)$ vanish strongly for $\lambda \to \infty$. This can be seen more explicitly. The replacement of $\varphi(x)\,\varphi(y)$ by $T^+(x,y)$ in $I(\lambda\nu)$ results in a replacement of $\delta_+(p)$ in (14.23) by

$$\frac{i}{2\pi}\frac{1}{p^2 - m^2 + i\varepsilon} = \delta_+(p) + \frac{i}{(p_0 - i\varepsilon)^2 - \omega(\vec{p})^2}. \tag{14.27}$$

## 14.2 Free Particles

The first term on the right restores the original expression. The second term leads to an expression satisfying the assumptions of Lemma 14.4, hence gives a strongly decreasing contribution to the new $I$.

We also need to study the reaction of probes to several-particle states. It suffices to consider 2-particle states, the generalization to more particles not presenting any problems.

The particles observed in the initial and final states of a scattering experiment are located at macroscopic distances from one another and have different momenta. The particles present before the event (in-case) originate in spatially separated sources, while those present after the event (out-case) all originate in the same small region where the reaction took place. We study only states conforming to one of these patterns. Let $g_1(x)$, $g_2(x)$, be two test functions with the properties described before, with $\hat{g}_1(\vec{p})$ and $\hat{g}_2(\vec{p})$ having non-overlapping supports and being normalized to 1. We define $g_{i,b}(x) = g_i(x - b)$ with $b = (0, \vec{b})$, and we consider the 2-particle state

$$\Phi_b = \varphi(g_1)^* \, \varphi(g_{2,b})^* \, \Omega \;, \tag{14.28}$$

where $\vec{b}$ is macroscopic in the in-case, $\vec{b} = 0$ in the out-case. $\Phi_b$ is a state prepared at time 0 with two particles at $\simeq 0$ and $\simeq b$ with momentum wave functions $\hat{g}_i$.

We analyse this state at first with a single probe, considering

$$E(a, b) = (\Phi_b, D(a) \, \Phi_b) \;, \tag{14.29}$$

with $D(a)$ defined as before. This expression must be evaluated for macroscopic $a$ and either macroscopic or vanishing $b$. We set $a = \lambda \nu$ as before and $b = \lambda \beta$, $\beta = (0, \vec{\beta})$, $\vec{\beta} \neq 0$ in the in-case, $\vec{\beta} = 0$ in the out-case, and study the asymptotic behaviour of $E'(\lambda) = E(\lambda \nu, \lambda \beta)$ for large values of the scaling parameter $\lambda$. Using the cluster expansion (5.13) we find

$$E'(\lambda) = E_1(\lambda \nu, g_1) + E_1(\lambda(\nu - \beta), g_2) \;, \tag{14.30}$$

with $E_1(\lambda \nu, g)$ the 1 particle expectation value (14.23). The asymptotically dominant $\lambda^{-3}$ term agrees again with a classical particle interpretation: there are two particles present which start at time 0 from the locations O and $\lambda \beta$ with 3-momenta in the supports of $\hat{g}_1$ and $\hat{g}_2$. The term is present if at least one of the particles moving as a classical free particle reaches the probe during its active period. It is then registered as the corresponding 1-particle state would be. If the 4-dimensional geometry of the arrangement is such that both particles arrive at the probe in its active period, then their effects add. This cannot happen in the out-case. In the in-case it happens if the straight lines through O and $b$ in directions $P_1$ and $P_2$, respectively, meet approximately at $\nu$. This is impossible if $\nu^0$ is sufficiently small, depending on $\vec{b}$, $\vec{P}_1$, $\vec{P}_2$, certainly if $\nu^0 < \frac{1}{2} |\vec{b}|$. In this case the two particles do not have enough time to approach each other by the time $\lambda \nu^0$.

For a coincidence arrangement of two probes $D(a_1)$, $D(a_2)$, with $a_i = \lambda \nu_i$, $\nu_1^0 = \nu_2^0 > 0$, $\vec{\nu}_1 \neq \vec{\nu}_2$, we must evaluate

$$E'(\lambda) = \left(\Phi_{\lambda\beta}, D(\lambda\nu_1) D(\lambda\nu_2) \Phi_{\lambda\beta}\right). \tag{14.31}$$

We define $\beta_1 = 0$, $\beta_2 = \beta$, and

$$I_{ij}(\lambda) = \int dp \, e^{-i\lambda(p,\nu_i - \beta_j)} \, \hat{f}_i(\vec{p}) \, \overline{\hat{g}_j(\vec{p})} \, \delta_+(p). \tag{14.32}$$

$E'$ is again computed with the help of the cluster expansion. There emerge non-vanishing mixed terms in which either $f_i$ occurs with both $g_1$ and $g_2$, for example

$$\overline{I_{11}(\lambda)} \, \overline{I_{22}(\lambda)} \, I_{12}(\lambda) \, I_{21}(\lambda). \tag{14.33}$$

If both $\hat{f}_i$-supports intersect both $\hat{g}_j$-supports, then every factor in this product contains a dominant $\lambda^{-3/2}$-term in appropriate directions. But there is no "appropriate direction" common to all factors. In the out-case the term is present in $I_{11}$ if the $\nu_1$-ray intersects the $\hat{g}_1$-support, in $I_{12}$ if the $\nu_1$-ray intersects the $\hat{g}_2$-support. These conditions are incompatible, hence the product (14.33) decreases strongly. In the in-case the factors $I_{11}$ and $I_{12}$ also cannot be slowly decreasing together because the rays in direction $\nu_1$ through O and through $b$ never meet.

The remaining asymptotically relevant terms in $E'$ are

$$E'(\lambda) \sim E_1(\lambda\nu_1, f_1, g_1) \, E_1(\lambda\nu_2, f_2, g_2) + E_1(\lambda\nu_1, f_1, g_2) \, E_1(\lambda\nu_2, f_2, g_1), \tag{14.34}$$

with $E_1$ the 1-particle expectation value from before. The first term exhibits the dominant $\lambda^{-6}$ behaviour if the $a_1$-probe is hit by a classical particle passing through O with a 3-momentum in supp $\hat{g}_1$, and the $a_2$-probe by a particle through $\lambda b$ with momentum in supp $\hat{g}_2$. In the second term the two particles are interchanged. The $\lambda^{-6}$ decay is fully explained by geometry. If neither of the two conditions is satisfied, then $E'(\lambda)$ decreases strongly and is negligible at macroscopic distances.

In the same way as in the 1-particle case it can be shown that $E'(\lambda)$ may be replaced by

$$E''(\lambda) = \left(\Omega, \, T^-\!\left(\varphi(g_1)\,\varphi(g_{2,b})\,\varphi(f_{1,a_1})^*\,\varphi(f_{2,a_2})^*\right) \right.$$
$$\left. \times T^+\!\left(\varphi(f_{1,a_1})\,\varphi(f_{2,a_2})\,\varphi(g_1)^*\,\varphi(g_{2,b})^*\right) \Omega\right) \tag{14.35}$$

without changing the dominant term, and similarly for the 1-probe case (14.29).

The above methods and results can be transferred without problems to the fields of free QED. But a few details need discussing. Consider first the spinors $\psi$, $\bar\psi$. A 1-electron state $\Phi_g^-$ is defined in analogy to (14.21) by

$$\Phi_g^- = \psi(\bar{g})\,\Omega \quad , \quad \psi(\bar{g}) = \int dx \, \bar{g}(x)\,\psi(x). \tag{14.36}$$

Here $\bar{g}(x) = g(x)^* \gamma^0$ and $g(x)$ is a spinor-valued function whose four components have the properties of the function $g(x)$ of the scalar case. An electron probe is given accordingly by

$$D^-(a) = \bar\psi(f_a)^* \bar\psi(f_a) \quad, \quad \bar\psi(f_a) = \int dy\, \bar\psi(y)\, f(y-a)\,, \tag{14.37}$$

where $f$ is a spinor function with the same properties as $g$. In analogy to (14.23) we obtain

$$E^-(a) = \left(\Phi_g^-,\, D^-(a)\, \Phi_g^-\right) = |I(a)|^2\,,$$
$$I(a) = \int dp\, e^{-ipa}\, \delta_+(p)\, \tilde{\bar g}(-p)\,(\not{p}-m)\,\tilde f(p)\,. \tag{14.38}$$

This expression can be simplified by a convenient choice of $f$ and $g$. We split

$$\tilde{\bar g}(-p) = \frac{1}{2m}\left(\tilde{\bar g}(-p)(\not{p}+m) - \tilde{\bar g}(-p)(\not{p}-m)\right)\,.$$

The factor $(\not{p}+m)$ in the first term annihilates the product $(\not{p}-m)\delta_+(p)$ in (14.38), so that this term does not contribute: we can assume without restriction of generality that this term is not present, i.e. that $\tilde{\bar g}(-p)\not{p} = -m\tilde{\bar g}(-p)$ on the mass shell. Analogously we assume that $\not{p}\,\tilde f(p) = -m\,\tilde f(p)$ on the mass shell. Then $I(a)$ becomes

$$I(a) = -2m \int dp\, e^{-ipa}\, \delta_+(p)\, \hat g(\vec p)^*\, \gamma^0\, \hat f(\vec p)\,, \tag{14.39}$$

and the asymptotic behaviour of $E^-$ is given as in (14.24) by

$$E^-(\lambda\nu) \sim \frac{1}{\lambda^3}\frac{1}{(\nu,\nu)^{3/2}}\pi^3(2m)^{\frac{7}{2}}|\hat g(\vec p)^*\gamma^0\hat f(\vec p)|^2 \tag{14.40}$$

if the $\nu$-ray intersects the mass shell at $(\omega(\vec p),\vec p)$ in the support of $\tilde g(p)\tilde f(p)$.

From the theory of the Dirac equation it is known that a 1-electron state with a given momentum $p$ can have two polarizations. A convenient basis of the polarization space is formed by the helicity eigenstates. The "$helicity$" is the value of the spin component in direction $\vec p$. From the spin operators $\Sigma$ defined in (2.47) we construct the helicity operator

$$h(\vec p) = \frac{1}{2|\vec p|}\begin{pmatrix}\vec p\vec\sigma & 0 \\ 0 & \vec p\vec\sigma\end{pmatrix}\,. \tag{14.41}$$

$h(\vec p)$ commutes with $\not{p}$ and with $\gamma^0$. The projection operators

$$H_\pm = \tfrac{1}{2} \pm h(\vec p) \tag{14.42}$$

project onto the states with helicity $\pm\tfrac{1}{2}$. One finds that the $4\times 4$ matrices $(\not{p}+m)H_\pm$ are of rank 1, so that their four columns are parallel 4-vectors.

They form eigenstates of $h$ with eigenvalues $\pm\frac{1}{2}$. This is easily verified for the special case $p_1 = p_2 = 0$, $p_3 \neq 0$, hence it holds generally by rotation invariance if $\vec{p} \neq 0$. Let $\eta_\pm(p)$ be such columns, where care must be taken not to choose columns that vanish identically. We may normalize them such that

$$\eta_+^* \eta_+ = \eta_-^* \eta_- = 1 , \qquad \eta_+^* \eta_- = 0 . \tag{14.43}$$

It suffices to consider test spinors $f$ and $g$ of the form

$$\tilde{f}_\pm(p) = \eta_\pm(p)\, \alpha(p) , \qquad \tilde{g}_\pm(p) = \eta_\pm(p)\, \beta(p) , \tag{14.44}$$

with $\alpha$ and $\beta$ being scalar test functions with the desired smoothness and support properties. $\gamma^0 \tilde{f}_\pm(p)$ are then eigenstates of $h(\vec{p})$ too, because $\gamma^0$ commutes with $h$. Hence the factor $|\hat{g}^* \gamma^0 \hat{f}|$ in (14.40) vanishes if $f$ and $g$ have different helicity: the $f_\pm$-probe sees only particles with helicity $\pm\frac{1}{2}$.

The helicity is not defined if $\vec{p} = 0$. This is the case for the target particles in a fixed-target experiment. In this case one may replace $h$ by the spin component $\Sigma_n = (\vec{\Sigma}, \vec{n})$ in a fixed direction $\vec{n}$, $|\vec{n}| = 1$. For $\vec{p} \neq 0$ such a choice is more problematic because $[\Sigma_n, \slashed{p}] \neq 0$ unless $\vec{p} = 0$.

The sum of the two helicity probes with the same $\alpha$ describes a probe which does not discriminate between polarizations.

A 1-positron state is written as

$$\Phi_g^+ = \bar{\psi}(g^p)\,\Omega , \qquad g^p(x) = g(-x) , \tag{14.45}$$

a positron probe is

$$D^+(a) = \psi(f_a^p)^* \psi(f_a^p) , \qquad \psi(f_a^p) = \int dy\, \bar{f}(-y+a)\, \psi(y) . \tag{14.46}$$

The sign changes in the test spinors $f^p$ and $g^p$ have the effect that their Fourier transforms are centred at a point on the negative mass shell. Exactly as in the electron case we find

$$E^+(a) = |I(a)|^2 ,$$
$$I(a) = \int dp\, e^{-ipa} \delta_+(p)\, \tilde{f}(p)^* \gamma^0 (\slashed{p} + m)\, \tilde{g}(p)$$
$$= 2m \int dp\, e^{-ipa}\, \tilde{f}(p)^* \gamma^0\, \tilde{g}(p) . \tag{14.47}$$

The latter equality holds if $\tilde{g}$ satisfies the Dirac equation $(\slashed{p} - m)\, \tilde{g}(p) = 0$ on the mass shell. The asymptotic expression (14.40) holds for $E^+(\lambda\nu)$ with $\hat{f}$ and $\hat{g}$ interchanged.

Concerning photons we are only interested in physical states. According to Lemma 5.3, the physical state space of the free electromagnetic field is generated from the vacuum by polynomials in the electric field strengths $E_i = F^{i0}$. Moreover, one of the Maxwell equations states that

$$\vec{p} \cdot \vec{E}(\vec{p}) = 0 \,. \tag{14.48}$$

Hence a sufficiently general 1-photon state is

$$\Phi_g^\gamma = F(g)^* \Omega \,, \qquad F(g) = \int dy\, \vec{E}(y) \cdot \vec{g}(y) \,, \tag{14.49}$$

where the $g^i$ are test functions with the by now familiar properties, and $\nabla \vec{g} = 0$. This last condition is satisfied if

$$\tilde{g}^i(p) = e^i(\vec{p})\, \alpha(p) \tag{14.50}$$

with $\vec{e}(\vec{p})$ a possibly complex "polarization vector" which is orthogonal to $\vec{p}$ (="transversal"). $\alpha(p)$ is a scalar test function with the usual properties. A real $\vec{e}$ means linear polarization. A photon probe is formed with the same operators:

$$D^\gamma(a) = F(f_a)^* F(f_a) \,. \tag{14.51}$$

Notice that under our conditions the origin $p = 0$ is not contained in the support of $\tilde{f}$, in accordance with the fact that photons with zero energy are unobservable: all photon detectors have positive threshold energies below which they are insensitive.

From (5.98) we find

$$(\Omega,\, \tilde{E}_i(p)\, \tilde{E}_j(q)\, \Omega) = \delta^4(p+q)\, \delta_+(p)\, (|\vec{p}|^2 \delta_{ij} - p_i p_j) \,.$$

From this we obtain like in the scalar case, using the transversality of $\tilde{f}$ and $\tilde{g}$:

$$E^\gamma(a) = \left(\Phi_g^\gamma,\, D^\gamma(a)\, \Phi_g^\gamma\right) = |I(a)|^2 \,,$$
$$I(a) = \int dp\, e^{-ipa}\, \delta_+(p)\, |\vec{p}|^2\, \hat{f}^i(p)\, \delta_{ij}\, \overline{\hat{g}^j(p)} \,. \tag{14.52}$$

Like in (14.25), the asymptotic behaviour of $I(\lambda \nu)$ is given by

$$E^\gamma(\lambda \nu) \sim \frac{\pi^2}{\lambda^2} \left| \int_0^\infty ds\, s^2\, \hat{f}^i(s\vec{\nu})\, \delta_{ij}\, \overline{\hat{g}^j(s\vec{\nu})} \right|^2 \,. \tag{14.53}$$

The generalization of these results to the case of several particles and probes presents no difficulties.

The large-distance behaviour of expectation values of probes that has been derived above for free particles we consider to indicate the presence of particles also in the interacting case. This is the definition of "particles" that we will use.

# Chapter 15

# Interacting Particles

## 15.1 Probes for Interacting Particles

In the interacting case the IR problem constitutes an essential complication. We therefore directly consider QED. We define particle probes like in the free case by (14.37), (14.46), and (14.51), though the photon case will be shown to need some elaboration. The fields in these definitions are now the interacting ones. In the fermion probes $D^\pm$ we use the physical fields $\Psi$, $\bar{\Psi}$, because we are only interested in physical states. There is a problem here: even though formed with the physical fields, the probe

$$D^-(a) = \bar{\Psi}(f_a)^* \bar{\Psi}(f_a)$$

is not a gauge invariant operator, hence not a true observable. The same holds for $D^+$.[1] But we do not claim the probes to represent realistic detectors. We use them only as a convenient means of motivating and describing the concept of a particle as a localized object. Therefore the use of these probes is justified despite their lack of gauge invariance. In any case, in the asymptotic region in which we are interested, the interacting fields behave sensibly as free fields. And in the free theory there are no gauge transformations of the second kind, hence the problem does not occur. A more satisfying way of handling the problem would be to work with genuine observables, for example, with $D(a) = \int dx\, f(x-a)\, j^\mu(x)$, where $f$ is a real test function which is essentially concentrated around the origin both in $x$- and $p$-space. Similar methods as for our choice could then be employed to analyze the expectation value of $D(a)$ in suitable states, with similar results. But the detailed calculations are even more complicated than with our "simple" definition without really giving more insight. We therefore feel entitled to work with the slightly objectionable $D^\pm$.

---

[1] This objection does not apply to $D^\gamma$ which is formed with the gauge invariant $F_{\mu\nu}$.

240   15. Interacting Particles

In this chapter we study the expectation values of the $D$'s in "one-particle states" also defined, provisionally, like in the free case by (14.36), (14.45), and (14.49), with the interacting fields replacing the free ones. It will be shown that the expectation values

$$E^-(\lambda\nu) = \left(\Phi_g^-,\, D^-(\lambda\nu)\,\Phi_g^-\right) \tag{15.1}$$

etc. still exhibit the asymptotic behaviour of the free case which is interpretable in terms of free classical particles. This justifies speaking of $\Phi_g^\pm$ as 1-particle states, while for $\Phi_g^\gamma$ the situation will turn out to be slightly more complicated. The derivation of this appealing result is unfortunately lengthy and tedious. This cannot be helped.

We start with $E^-$. It reads explicitly

$$E^-(a) = \int dx\,dy\,dx'\,dy'\,g(x')\,\bar{f}(y'-a)\,f(y-a)\,\bar{g}(x)$$
$$\times \left(\Omega,\,\bar{\Psi}(x')\,\Psi(y')\,\bar{\Psi}(y)\,\Psi(x)\,\Omega\right), \tag{15.2}$$

where $\bar{g}(x) = g(x)^*\gamma^0$, $\bar{f}(y) = f(y)^*\gamma^0$ and $f$ and $g$ are the spinor functions already used in the free case. The spinor indices in $\bar{g}(x)$ are contracted with those in $\Psi(x)$ and similarly for the other pairs. The $W$-function is the interacting one given by the graph rules of Part II. Since 2-sector graphs are simpler than 4-sector ones we replace it by

$$\left(\Omega,\,T^-\left(\bar{\Psi}(x')\,\Psi(y')\right)T^+\left(\bar{\Psi}(y)\,\Psi(x)\right)\Omega\right),$$

which does not change the asymptotic behaviour of $E^-(\lambda\nu)$ for $\lambda \to \infty$ if $\nu^0 > 0$. The argument is the same as in (14.26). Because of the non-locality of $\Psi$, $\bar{\Psi}$, it cannot be used without further elaboration for $\nu^0 = 0$, hence we exclude that case from consideration. This is physically justified: the state $\Phi_g$ is prepared at time 0 and is then analyzed with the probe $D^-$ at time $\lambda\nu^0$. Clearly, this makes sense only if $\lambda\nu^0 > 0$. Notice that $\nu$ may still be spacelike.

The actual discussion of $E^-(\lambda\nu)$ is carried out in momentum space where it reads

$$E^-(\lambda\nu) = \int dp\,dq\,dp'\,dq'\,e^{-i\lambda(\nu,q'+q)}\,\tilde{g}(p')\tilde{\bar{f}}(q')\,\tilde{f}(q)\,\tilde{\bar{g}}(p)$$
$$\times \tilde{\mathcal{W}}(p',q',-|\bar{q},p,+)\,. \tag{15.3}$$

The test functions $\tilde{g}(p)$, $\tilde{f}(q)$, have small supports centred at points on the positive mass shell. "Small" means that the maximal diameter of the support is small compared to the mass $m$. $\tilde{\bar{g}}(p) = \tilde{g}^*(-p)\,\gamma^0$ and $\tilde{\bar{f}}(q)$ have small supports centred on the negative mass shell. These properties have an important consequence for the sector graphs of $\tilde{\mathcal{W}}$ that contribute to $E^-$. Such graphs are of the general form shown in Fig. 15.1. The external momenta of the $T^-$-sector add to a momentum $p' + q'$ flowing into the sector, with $(p' + q')^2 \ll m^2$. The cross momenta flow from $T^-$ to $T^+$. Because momentum is conserved, only photon lines can occur as cross lines, and they carry momenta which are small compared to $m$. Since no fermion trajectory can cross the sector boundary, the $q$- and $p$-lines are connected by a trajectory inside the $T^+$-sector and so are the $p'$- and $q'$-lines inside the $T^-$-sector. We call these two trajectories the "basic trajectories" of the graph.

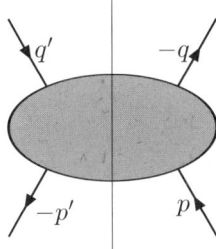

**Fig. 15.1.** Graph contributing to $\tilde{\mathcal{W}}(\bar{p}', q', -|\bar{q}, p, +)$. The thin vertical line is a sector boundary

Let $G$ be a graph of this form, not containing composite $\Psi_n$- or $\bar{\Psi}_n$-vertices whose influence will be considered later. The asymptotic behaviour of its contribution $E_G$ to $E^-$ is determined by the singularities of $\tilde{\mathcal{W}}_G$ in the variable $q + q'$ after integration over the internal variables and a complete set of other external variables. As a first result we establish that these singularities are associated with cuts through $G$ which separate the final variables $q$, $q'$, from the initial $p$, $p'$ as shown in Fig. 15.2.

Let $s_1, \ldots, s_N$, be the momenta of the cut lines. Then $\sum \pm s_i = -q - q'$ by momentum conservation. The sign is positive if the $s_i$-line points upwards. Each propagator $P(s_i)$ contains a singular factor $(s_i^2 - m_i^2 \pm i\varepsilon)^{-1}$ or $\delta_\pm(s_i)$. Integrating the product of these singular factors over the $s_i$ yields the usual threshold singularity in $q + q'$. We need not determine this singularity explicitly since its contribution to $E_G$ can be found in a more direct way. Namely, it can be written as

$$E_G(\lambda \nu) = \int \prod_{j+1}^{N} \left[ ds_j \, e^{\pm i\lambda(\nu, s_j)} P(s_j) \right] U(s_1, \ldots, s_N) \, L(s_1, \ldots, s_N) \quad (15.4)$$

where $U$ and $L$ are the contributions of the upper and lower subgraph integrated over the test functions $\tilde{f}$ etc. Assuming at first that $U$ and $L$ are smooth functions, the behaviour of $E_G$ for $\lambda \to \infty$ can be determined from the lemmas of Sect. 14.1, since the essential, singular, part of the integrand is a product of 1-variable factors. That $UL$ does not factorize correspondingly is immaterial. If the $s_i$-singularity is $\delta_\pm(s_i)$, then Lemma 14.2 or Lemma

242    15. Interacting Particles

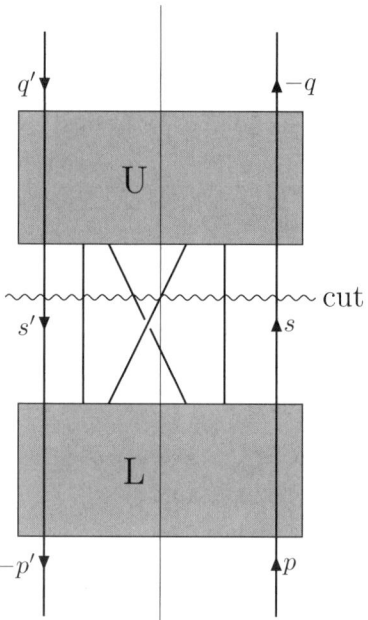

**Fig. 15.2.** A cut graph.
The cut lines other than those on the basic trajectories may be of any type

14.3 applies. It gives a decrease like $\lambda^{-3/2}$ or $\lambda^{-1}$ if $\nu$ is as stated there, stronger for other $\nu$. If the singularity is $(s_i^2 - m_i^2 \pm i\varepsilon)^{-1}$, the $i\varepsilon$-sign according with the sign of $s_i$ in the exponent, we split

$$\frac{1}{s_i^2 - m_i^2 \pm i\varepsilon} = \frac{1}{(s_i^0 \pm i\varepsilon)^2 - \omega(\vec{s}_i)^2} \mp 2\pi i\, \delta_\mp(s_i) \,. \tag{15.5}$$

The first term on the right produces a strong $\lambda$-decrease by Lemma 14.4, while the $\delta_\mp$-term gives again a $\lambda^{-3/2}$ or $\lambda^{-1}$ for favorable $\nu$, a stronger decrease otherwise. An analogous method applies if the two signs disagree. In all cases the asymptotically relevant $s_i$-values are such that in the $T^-$-sector energy is flowing out of the $L$-part. The only inflowing energy is that of the $p'$-line. Momentum conservation then tells us that the asymptotically relevant cut lines can only be photonic with the exception of one line belonging to the basic trajectory. The same holds for the $T^+$-sector. But $\nu$ cannot be both timelike and lightlike. Therefore at least one $s_i$-factor produces a strong decrease in $E_G$ unless the only cut lines are two lines of the two basic trajectories. This fact forms the basis of the subsequent analysis.

Let $s, s'$, be the momenta of the cut trajectory lines flowing in the direction of the trajectory. $U$ and $L$ are functions of $s$ and $s'$ which are smooth under our provisional assumption. An asymptotically relevant contribution to $E^-(\lambda\nu)$ is then given by Lemma 14.2 as

$$E_G(\lambda\nu) \sim \frac{1}{\lambda^3}\frac{1}{(\nu,\nu)} \cdot 2\pi^3 m\, U(s,s')\, L(s,s')|_{s=s'=-\hat{\nu}}\,, \qquad (15.6)$$

with $\hat{\nu} = \nu/\sqrt{(\nu,\nu)}$. This holds if $\nu$ is time-like and the product $UL$ does not vanish at $s = s' = -\hat{\nu}$. Otherwise the decrease is faster. $UL$ is in general not real for a given graph, but the terms (15.6) summed over the graphs with the same scaffolding *are* real. As in the free case, this result agrees with the geometrical behaviour of a classical free particle. But it differs from its free equivalent by the dynamics-dependent factors $U$ and $L$ which have, however, also a physical explanation: $L$ and $U$ describe the effect of unobserved low-energy photons that have been set free in the preparation procedure or by bremsstrahlung of the observed electron in its encounter with the probe. They also describe the efficiency of the source and the probe.

But these considerations ignore the IR problem, as a consequence of which $U$ and $L$ are not smooth. In zeroth order there are no subgraphs $U$ and $L$, or to put it differently: $U = L = 1$. This is the free case in which the claimed result is clearly correct. If the subgraph $U$ consists of a single cross line with momentum $k$, then $U$ is given by

$$U(s,s') = \int dk\,\delta_+(k)\, \frac{1}{(s'+k)^2 - m^2 - i\varepsilon}\, \frac{1}{(s+k)^2 - m^2 + i\varepsilon}\, \tilde{f}(s'+k)\,\tilde{f}(s+k)\,,$$

up to smooth factors. This integral diverges logarithmically at $k = 0$ if $s^2 = s'^2 = m^2$: $U(s,s')$ is weakly singular at the mass shell. Thus $E_G$ decays slightly slower than $\lambda^{-3}$ and this destroys the classical interpretation. The same effect occurs if a cross line is not directly connected to the basic trajectories but to vertex parts (VPs), i.e. 3-line 1PI subgraphs, on them, or if there are several cross lines connected to VPs on the two trajectories. This is the first cause of the hard IR problem.

Some possible generalizations of this trouble are fortunately harmless. A cross-line may be interrupted by self-energy parts (SEPs) before meeting the trajectory or a 1PI subgraph intersecting it. This adds a term of the type $\rho\, k^\mu k^\nu \delta_+(k)$ of (8.7) to the cross propagator. This must not be forgotten in the final result. But because of the $k$-factors this term creates no IR problems; hence it can be disregarded as far as these problems are concerned. IR divergences also do not occur if two or more cross lines are attached to a 1PI subgraph intersecting e.g. the $p - q$ trajectory as shown in Fig. 15.3. According to the results of Sect. 11.3, the 1PI subgraph of Fig. 15.3 may be weakly singular if both $s$ and $q$ are restricted to the mass shell. But only $s$ is so restricted while $q$ is integrated over the 4-dimensional test function $\tilde{f}(q)$.

Some vital questions are still left open by this discussion, and they must be answered.

First question: for a given graph there exist in general different possibilities of splitting it into two parts by cutting two trajectory lines. E.g. the graph

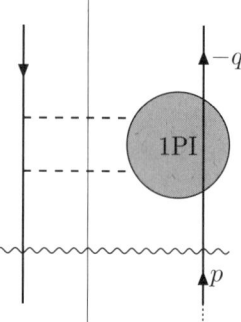

**Fig. 15.3.** An IR harmless graph

of Fig. 15.2 could also be decomposed by cutting the external $q$- and $q'$-lines or the $p$- and $p'$-lines. Clearly there are graphs with even more possibilities of cutting. The question is: do these various ways of cutting produce the same asymptotic term or do their contributions add? The answer is that the latter is true, with a proviso to be noted presently. The argument is this: the expression (15.6) derives from the singularity of the integrand at $s^2 = s'^2 = m^2$. It is not changed if the integral (15.4) is taken only over an arbitrarily small neighbourhood of this manifold. In the same way the leading term from a $q$–$q'$-cut is due to the contribution of $\{q^2 = q'^2 = m^2\}$ to the integral. But the sum $K$ of the cross momenta in $U$ is timelike so that $s^2 = (q+K)^2 \neq m^2$ if $q^2 = m^2$ except for $K = 0$. The singularity manifolds responsible for the two terms are disjoint except at $K = 0$ where they intersect but remain different manifolds nevertheless. This is essential because for instance in the case of a single cross line with momentum $k$ the behaviour at $k = 0$ is responsible for the logarithmic deviation from the desired $\lambda^{-3}$ behaviour. The cancellation of the various IR terms will be shown to occur separately for each one. This makes sense only if several terms are indeed present.

An exception to this additivity occurs if a SEP is inserted in one of the trajectories. Then the trajectory momentum above this insertion is the same as that below it, and it makes no difference whether the one or the other of these lines are cut. Also, such insertions introduce weak singularities which are the second cause of the hard IR problem.

Second question: does the irrelevance of cuts through more than two lines remain true if R-parts are cut? The answer is yes. It has been shown (see (15.5)) that only $\delta_+$-parts of the cut propagators are asymptotically relevant. This relevant contribution to the R-part vanishes if all external variables of the part are set equal to zero, like in an R-part intersecting several sectors. Hence it does not contribute to the renormalizing subtractions and thus creates no problems.

Third question: might combinations of propagator singularities other than those defined by cuts give rise to asymptotically relevant terms? That this is not the case is seen as follows. Consider a small neighbourhood in the space of external and internal variables in which a certain number of propagators, called S-propagators, are singular, the others called N-propagators, not. Assume that there is a generalized trajectory possibly including photon lines but not including any S-lines, connecting one of the final $q$–$q'$-lines with one of the initial $p$–$p'$-lines. We choose one of its momenta, called $\ell$, as one of the basic momenta in the $\delta$-reduced graph rules and let it flow through that trajectory. That is, we choose the other basic momenta such that only the trajectory propagators depend on $\ell$. Then the integrand is on the considered set smooth in $\ell$. The graph may be cut as before, including the $\ell$-line among the cut ones. The previous argument gives a strongly decreasing factor from the $\ell$-integral while the other factors increase polynomially at worst, all singularities being of finite order. Hence an arbitrary combination of singularities can compete successfully only if no such trajectory exists: each generalized top-to-bottom trajectory must contain an S-line. Cutting all S-lines then decomposes the graph into several components. And a subset of S-lines can be found which contains exactly one line of each generalized trajectory. Cutting the lines of this subset only is then a cut as considered before, resulting in the asymptotic behaviour found there.

**Fig. 15.4.** A $\Psi_n$ selfenergy part

Fourth question: how does the presence of $\bar{\Psi}_n$- or $\Psi_n$-vertices affect these considerations? A trivial remark is this: if for example the $p$-point is a $\Psi_n$-vertex, then there is no external $p$-line to be cut. But there is also a less trivial point to be considered. The propagators of the photon lines incident at a $\Psi_n$-vertex are more singular at vanishing momentum than the ordinary photon propagators. Besides the problems dealt with in Sect. 12.4 this produces new IR problems of the type encountered above. Consider an SEP containing a $\Psi_n$-vertex e.g. as shown in Fig. 15.4. This part is weakly singular at $s^2 = m^2$, hence if this $s$-line is cut (notice that $s = p$) it produces a factor with a weaker decay than the canonical $\lambda^{-3/2}$. If a photon line connects two composite vertices, its propagator leads already to an IR divergent integral by itself without the cooperation of other propagators. This problem has been solved in Sect. 12.4 by showing these divergences to cancel between various graphs. The relevant graphs are shown in Fig. 15.5 for an example with a single dangerous line. For the handling of lines connecting a $\Psi_n$-vertex to itself see (12.26). It has been shown in Sect. 12.4 that the IR divergences cancel between graphs (a) and (b) as well as between (c) and (d) and the mirrored graph (d), and similarly for (e) and (f). This is not what is needed now because a cut of our kind will cut the photon line in (b) but not that in (a) and similarly for the other combinations, so that the cancellation clashes with the $\lambda$-asymptotics. But a close inspection of the situation reveals that the divergent parts of the various integrals depend only on the location of the vertices involved, but not on the variables of the other lines joining them. It can then be shown that in our case cancellation occurs in different combinations too: the divergence of (a) cancels those of (d) and (f), the divergences of (b) and its mirrored form cancel those of (c) and (e). And these cancellations are not beset with the problem just mentioned.

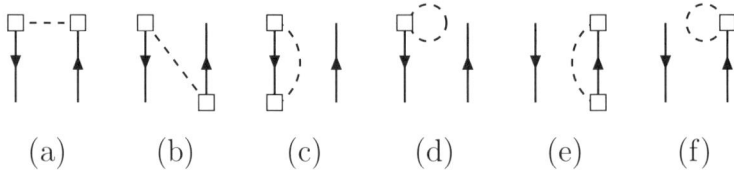

(a)   (b)   (c)   (d)   (e)   (f)

**Fig. 15.5a–f.** IR divergent physical graphs.
Only the dangerous photon line and the basic trajectories are shown. Other relevant graphs are obtained by reflection at a horizontal line. The omitted internal parts are the same in all cases

## 15.2 IR Cancellations

In this section it will be shown that the undesirable non-geometrical terms in $E^-(\lambda\nu)$ cancel in the sum over all graphs of a given order. The proof of this cancellation is very intricate. We will therefore first exhibit its basic mechanism in a simple but typical situation: $2^{\text{nd}}$-order perturbation theory. Since the IR problems are due to the singularities of the propagators, not to their regular numerators, we consider a purely scalar theory with a scalar massive "electron"-field $\psi$ and a scalar massless "photon"-field $A$ with interaction $e\psi^*\psi A$. And we will partly argue heuristically, using approximations without full justification. We are only concerned with timelike $\nu$ which we normalize to $(\nu,\nu) = m^2$.

The second-order graphs of this model contributing to the factor $\tilde{\mathcal{W}}$ in the expression (15.3) are shown in Fig. 15.6. Consider first the asymptotic contributions of graph (c). There are two of them, depending on whether the $q$-lines or the $p$-lines are cut. The first contribution comes from the singularities at $q^2 = q'^2 = m^2$ of the integrand of

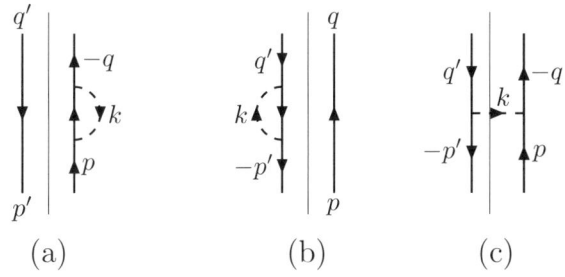

**Fig. 15.6a–c.** Second-order contributions to $E^-$

$$E_1(\lambda\nu)$$
$$= \int dq\, dq'\, \overline{\tilde{f}(-q')}\, \tilde{f}(q)\, e^{-i\lambda(\nu, q+q')} \frac{1}{q^2 - m^2 + i\varepsilon} \frac{1}{q'^2 - m^2 - i\varepsilon}$$
$$\times \int dk\, \delta_+(k) \frac{1}{(q'-k)^2 - m^2 - i\varepsilon} \frac{1}{(q+k)^2 - m^2 + i\varepsilon}\, \tilde{g}(-q'+k)\, \overline{\tilde{g}(q+k)}\, .$$
(15.7)

The $k$-integral in this expression is logarithmically divergent at $k = 0$ if $q^2 = q'^2 = m^2$: considered as a function of $q$ and $q'$ it is logarithmically singular on the mass shell. This singularity is responsible for the undesirable $(\log \lambda)$-term in the asymptotic expansion of $E_1$. For the proof of cancellation we must isolate the IR relevant part of the integrand in a simple form. We note that only an arbitrarily small neighbourhood of $k = 0$ is relevant. Therefore we need evaluate the $k$-integral only over a suitable neighbourhood $\mathcal{K}$ of the origin, which we choose to be the cylinder

$$\mathcal{K} = \{k \in R^4 : |\vec{k}| \leq R\}\, .$$
(15.8)

Note that $\mathcal{K}$ is invariant under the reflection $k \to -k$. Furthermore, we may set $k = 0$ in the factors which depend smoothly on $k$, that is, in the test functions $\tilde{g}$. Finally, we neglect the $k^2$-term in the denominator $(q+k)^2 - m^2 = q^2 - m^2 + 2(q, k) + k^2$ compared to the first order term $(q, k)$, and analogously for the $q'$-denominator. Similar approximations apply to the dependence on $q$ and $q'$ because only the vicinity of $q^0 = \omega(\vec{q}),\ q'^0 = -\omega(\vec{q'})$ is of interest. That is, we replace $q^2 - m^2$ by $\Delta = 2\omega(\vec{q})\left(q^0 - \omega(\vec{q})\right)$ and the factor $q$ in $(q, k)$ by its mass shell projection $\hat{q} = (\omega(\vec{q}), \vec{q})$, and analogously $q'^2 - m^2$ by $\Delta' = -2\omega(\vec{q'})(q'^0 + \omega')$ and $q'$ by $\hat{q}' = (-\omega', \vec{q'})$ with $\omega' = \omega(\vec{q'})$. Replacing $q^0$, $q'^0$, by $\Delta$, $\Delta'$, as integration variables and writing

248    15. Interacting Particles

$$(\nu, q) = \alpha \Delta + (\nu, \hat{q}), \qquad \alpha = \frac{\nu^0}{2\omega},$$

$$(\nu, q') = -\alpha' \Delta' + (\nu, \hat{q}'), \qquad \alpha' = \frac{\nu^0}{2\omega'}, \qquad (15.9)$$

we obtain

$$\begin{aligned}&E_1(\lambda \nu)\\&= \int d^3q\, d^3q'\, e^{-i\lambda(\nu,\hat{q}+\hat{q}')} \int \frac{d\Delta}{\Delta + i\varepsilon} \frac{d\Delta'}{\Delta' - i\varepsilon} e^{-i\lambda(\alpha\Delta - \alpha'\Delta')}\\&\quad \times F(\Delta, \Delta', \vec{q}, \vec{q}') \int_K dk\, \delta_+(k) \frac{1}{\Delta + 2\hat{q}k + i\varepsilon} \frac{1}{\Delta' - 2\hat{q}'k - i\varepsilon},\quad (15.10)\end{aligned}$$

with $F$ a smooth function. After the rescaling $\lambda \Delta = \delta$, $\lambda \Delta' = \delta'$, $\lambda k = u$, the $\Delta$–$\Delta'$–$k$-integral becomes

$$\begin{aligned}&\int \frac{d\delta}{\delta + i\varepsilon} \frac{d\delta'}{\delta' - i\varepsilon} e^{-i\delta\alpha} e^{i\delta'\alpha'} F\left(\frac{\delta}{\lambda}, \frac{\delta'}{\lambda}, \vec{q}, \vec{q}'\right)\\&\quad \times \int_{\lambda K} du\, \delta_+(u) \frac{1}{\delta + 2\hat{q}u + i\varepsilon} \frac{1}{\delta' - 2\hat{q}'u - i\varepsilon}. \quad (15.11)\end{aligned}$$

In the argument of $F$ we can replace $\frac{\delta}{\lambda}, \frac{\delta'}{\lambda}$, by their limits 0 without destroying the existence of the $\delta$–$\delta'$-integral at infinity. At $\delta = 0$ or $\delta' = 0$ the $\delta$-dependent factors are singular, but their product is defined as a distribution by the $i\varepsilon$-prescription. Therefore the only critical $\lambda$-dependence is that of the region of integration $\lambda K$. The $u$-integrand decreases for $|u| \to \infty$ only of fourth order, which does not suffice to make the $u$-integral exist for $\lambda \to \infty$. The divergent part of the integral can be written as a surface integral in which we may neglect $\delta$ and $\delta'$ relative to $2\hat{q}u$ and $2\hat{q}'u$. In this way we obtain

$$\int_{\lambda K} du \cdots = \log \lambda\, A(\hat{q}, \hat{q}') + B(\delta, \delta', \hat{q}, \hat{q}', \lambda), \quad (15.12)$$

with

$$A(\hat{q}, \hat{q}') = -\int_{\partial K} d^3\Omega\, \delta_+(u) \frac{1}{2\hat{q}u + i\varepsilon} \frac{1}{2\hat{q}'u + i\varepsilon} \quad (15.13)$$

a smooth function. $d^3\Omega$ is the surface element on the 3-sphere $\partial K$. $(\delta + i\varepsilon)^{-1}(\delta' - i\varepsilon)^{-1} B$ is smooth in $\hat{q}, \hat{q}'$, a well defined distribution in $\delta, \delta'$, and its $\lim_{\lambda \to \infty}$ exists and has the same properties. Inserting $A$ into (15.11) yields an existent integral. Evaluating the remaining $\vec{q}$–$\vec{q}'$-integral in (15.10) by the stationary phase method gives the familiar $\lambda^{-3}$-decrease, but now multiplied with the $\log \lambda$ factor of the $A$-term. The same method applied to $B$ gives the desired $\lambda^{-3}$-decay: the IR problem is localized in the $A$-term. More exactly, $\lambda^{-3} \log \lambda$ is multiplied by

$$A_1(\nu) := A(\nu, -\nu) = -\int_{\partial K} d^3\Omega\, \delta_+(u) \frac{1}{4(\nu, u)^2} : \quad (15.14)$$

## 15.2 IR Cancellations

only these diagonal values are IR relevant. Note that $(\nu, u) \neq 0$, so that the $i\varepsilon$-terms are irrelevant.

The contribution $E_2(\lambda \nu)$ of graph (c) cut through the $p$–$p'$-lines is evaluated in exactly the same way. Its $\lambda^{-3} \log \lambda$ term differs from the $E_1$ one only by the replacement of $A_1$ by

$$A_2(\nu) = -\int_{\partial K} d^3\Omega\, \delta_-(u) \frac{1}{4(\nu, u)^2}\,, \tag{15.15}$$

which is however identical to $A_1$, as is seen by the change of variable $u \to -u$.

The dangerous contribution of graph (a) is contained in

$$\begin{aligned} E_3(\lambda\nu) \\ = -\frac{1}{2\pi i} \int dq\, dq'\, \overline{\tilde{f}(-q')}\, \tilde{f}(q)\, \tilde{g}(-q)\, \overline{\tilde{g}(q)}\, e^{-i\lambda(\nu, q+q')} \frac{1}{q^2 - m^2 + i\varepsilon} \\ \times \frac{1}{q'^2 - m^2 - i\varepsilon} \int dk\, \frac{1}{k^2 + i\varepsilon}\, \frac{1}{(q+k)^2 - m^2 + i\varepsilon}\, \frac{1}{2qk + k^2 + i\varepsilon}\,. \end{aligned} \tag{15.16}$$

The numerical factors that have been omitted in $E_1$ and $E_2$ are omitted here too. The remaining factor $(2\pi i)^{-1}$ was not present in $E_{1,2}$. The soft-photon contribution to mass renormalization is included. The same operations as before yield again a $\lambda^{-3} \log \lambda$ term that is identical to the one in $E_1$ except for the replacement of $A_1(\nu)$ by

$$A_3(\nu) = -\frac{1}{2\pi i} \int_{\partial K} d^3\Omega\, \frac{1}{u^2 + i\varepsilon}\, \frac{1}{(2\nu u + i\varepsilon)^2}\,. \tag{15.17}$$

And graph (b) contributes a similar term with

$$A_4(\nu) = \frac{1}{2\pi i} \int_{\partial K} d^3\Omega\, \frac{1}{u^2 - i\varepsilon}\, \frac{1}{(2\nu u - i\varepsilon)^2}\,. \tag{15.18}$$

By the substitution $u \to -u$ the last denominator becomes the same as in (15.17). In (15.14) and (15.15) we can also introduce an (irrelevant) $+i\varepsilon$ in the denominator factors. But

$$-\frac{1}{2\pi i} \frac{1}{u^2 + i\varepsilon} + \frac{1}{2\pi i} \frac{1}{u^2 - i\varepsilon} + \delta_+(u) + \delta_-(u) = 0\,,$$

hence

$$A_1(\nu) + \cdots + A_4(\nu) = 0\,, \tag{15.19}$$

which proves the cancellation of the non-geometrical terms in the asymptotics of $E^-(\lambda\nu)$.

In this simple example the form of the relevant singularities at $\Delta = \Delta' = 0$ can be found explicitly. It is instructive to do so for the contribution of graph

250   15. Interacting Particles

(c) cut through the $q$-lines. More specifically, let us calculate the $k$-integral in (15.10) at the relevant points $\hat{q} = -\hat{q}' = \nu$ for the typical case $\nu = (m, \vec{0})$. It is

$$\int_K dk\, \delta_+(k) \frac{1}{\Delta + 2m|\vec{k}| + i\varepsilon} \frac{1}{\Delta' + 2m|\vec{k}| - i\varepsilon}$$

$$= 2\pi \int_0^R dr\, r \frac{1}{\Delta + 2mr + i\varepsilon} \frac{1}{\Delta' + 2mr - i\varepsilon},$$

which can be calculated explicitly. Its IR relevant terms are

$$\frac{\pi}{m(\Delta - \Delta') + i\varepsilon} \left[\Delta' \log(\Delta' - i\varepsilon) - \Delta \log(\Delta + i\varepsilon)\right]$$

$$= \frac{\pi}{2m} \frac{\Delta + \Delta'}{\Delta - \Delta'} \log \frac{\Delta'}{\Delta} - \frac{\pi}{2m} (\log \Delta + \log \Delta') . \qquad (15.20)$$

The last term in the second expression multiplied with the factor $\frac{1}{\Delta\Delta'}$ of (15.10) is the term of excessive singularity which is cancelled in $E^-$ by corresponding terms from its other contributions. The first term on the right-hand side of (15.20), which also occurs with the same sign in the contribution from the $p$–$p'$ cut, is not cancelled by anything. This is all right at first sight, since the term has the proper scaling behaviour to produce a $\lambda^{-3}$-contribution to $E^-$, not anything of weaker decay. But the contribution is nonzero, and this is an awkward fact: this term does not occur if the $\lambda^{-3}$ coefficient is found by the standard method of first introducing an IR regularization (like a small-momentum cutoff or a small nonvanishing photon mass), computing the desired coefficient by restricting the appropriate expression to the mass shell $\Delta = \Delta' = 0$, and then removing the regularization.

Let us look at this point in more detail. Consider our scalar model, but with a positive photon mass $\mu > 0$. This means that the photon propagators are $\theta(\pm k_0)\delta(k^2 - \mu^2)$ and $\mp \frac{i}{2\pi}(k^2 - \mu^2 \pm i\varepsilon)^{-1}$. There are no IR problems in this model. Take again the graphs of Fig. 15.6 and their four contributions to the asymptotics of $E(\lambda)$. Each of them gives a leading term $C_i \lambda^{-3}$ with $i = 1, \ldots, 4$, the numbers introduced above in $A_i$. The $C_i$ are found like in (15.6). Define

$$D_{1,2}(\Delta, \Delta') = \int dk\, \delta_\pm^\mu(k) \frac{1}{\Delta + 2(\hat{\nu}, k) + k^2 + i\varepsilon} \frac{1}{\Delta' + 2(\hat{\nu}, k) + k^2 - i\varepsilon}$$

$$D_3(\Delta, \Delta') = \frac{-i}{2\pi} \int \frac{dk}{k^2 - \mu^2 + i\varepsilon} \frac{1}{\Delta + 2(\hat{\nu}, k) + k^2 + i\varepsilon} \frac{1}{2(\hat{\nu}, k) + k^2 + i\varepsilon}$$

$$D_4(\Delta, \Delta') = \frac{i}{2\pi} \int \frac{dk}{k^2 - \mu^2 - i\varepsilon} \frac{1}{\Delta' + 2(\hat{\nu}, k) + k^2 - i\varepsilon} \frac{1}{2(\hat{\nu}, k) + k^2 - i\varepsilon} .$$

$$(15.21)$$

Then the $C_i$ are, up to a common factor, the restrictions $C_i = D_i(0, 0)$ of the $D_i$ to the mass shell $\Delta = \Delta' = 0$. The $D_i$ are continuous at this point, hence the $C_i$ are well defined. The second-order coefficient of the $\lambda^{-3}$-term in $E(\lambda)$ is given by $C = \sum C_i$ multiplied by the omitted common factor. In the limit $\mu \to 0$ the individual $C_i$ diverge logarithmically on account of IR divergences developing at $k = 0$ in the integrals defining them. But their divergent parts $C_i^d$ can again be isolated in a simple form. Define $D_i^d(\Delta, \Delta')$ by (15.21), except that the terms $k^2$ in the denominators of the electron propagators are dropped and the $k$-integral is taken over $\mathcal{K}$ only. Then it is easy to see that $B_i = \lim_{\mu \to 0}(C_i - C_i^d)$ exists. Moreover, one finds that $\sum_i C_i^d = 0$, hence $C = \lim_{\mu \to 0} \sum C_i = \sum B_i$ exists. $C$ gives the asymptotically leading term in $E(\lambda)$ of massless QED as obtained by an uncritical application of the conventional method of first introducing the IR regulator $\mu$, then finding the leading term in $E$, and then removing the regularization. This result does not agree with what we have arrived at earlier by working directly in the true theory with a massless photon. The correct result is related to the case $\mu > 0$ as follows. The $D_i$ and the $D_i^d$ converge for $\mu \to 0$ in the sense of distributions to the corresponding expressions of the massless theory that have been used in our derivation. Moreover, $D_i - D_i^d$ is continuous in $\Delta, \Delta'$ and remains so for $\mu \to 0$. Its restriction to the mass shell $\Delta = \Delta' = 0$ remains meaningful even in this limit, where it becomes $B_i$. Writing $\sum D_i = \sum(D_i - D_i^d) + \sum D_i^d$, taking the limit $\mu \to 0$, and only then determining the asymptotically leading term in $E(\lambda)$, we find a contribution from the first term in the right-hand side which agrees with the result obtained previously with the regulator method. But the last term $\sum D_i^d$ does *not* vanish for $\mu \to 0$, but produces the unexpected term in (15.20) and its $p$–$p'$ analogue, which give an additional contribution to $E(\lambda)$ that is missed in the regulator method. The physical relevance of this fact and its generalizations to be encountered later on will be discussed in Chap. 17.

The procedure of this "simple" low-order example must now be generalized to QED in all orders of perturbation theory.

The IR deviation from the geometrical $\lambda^{-3}$ decrease of the contribution $E_G(\lambda \nu)$ of the given graph $G$ to $E^-$ is caused by photon cross lines attached directly to the basic trajectories or to VPs situated on them, and by SEPs in the basic trajectories. Both cases are relevant only if occurring in dangerous proximity to the cut line, i.e. not separated from it by larger 1PI subgraphs of the section in question. Moreover, if SEPs are present next to the cut line, not even separated from it by VPs, then the position of the cut is ambiguous. Such a situation is shown in Fig. 15.7. The three cuts indicated there lead to equivalent asymptotic expressions. Adding their contributions would lead to multiple counting.

We address this ambiguity problem first. It could be solved by arbitrarily requiring the cut to be effected at the first possible position on the trajectory. But this asymmetric procedure is unsuitable for the intended proof of IR cancellations. We need a more complicated prescription. Note that SEPs containing $\Psi_n$-vertices are not involved in the ambiguity since a cut is only

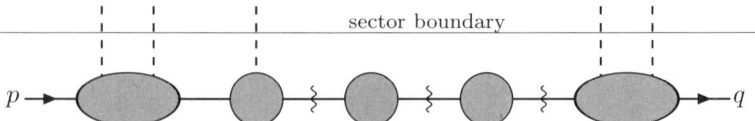

**Fig. 15.7.** A dangerous interval on a basic trajectory. The wavelets indicate equivalent cut positions

possible on one side of them. We need only consider SEPs of GB type. A chain of SEPs is a contribution to the "clothed" electron propagator $t^+(s)$. We replace it by

$$t^+(s) = t^+(s)\,\frac{1}{t^+(s)}\,t^+(s)\;, \tag{15.22}$$

which expression must of course be perturbatively expanded. The two factors $t^+$ are included in the considered graph on either side of the cut. The quotient in the middle remains standing by itself. It may be considered to be associated with the cut. Notice that the singularities $(s^2 - m^2 + i\varepsilon)^{-1}$ in the three factors combine into the original simple pole.

It must be shown that the undesirable slowly decreasing terms in the asymptotic expansions cancel in a sum over sufficiently many graphs of a given order. In contrast to the mild IR problem of Chap. 11 it does not suffice to sum over graphs with the same scaffolding. A larger class of graphs is necessarily involved, as indicated by the proposed solution of the ambiguity problem and by our paradigmatic example. Since this analysis is lengthy and not overly transparent, we will first state the pertinent results and only then plunge into the frightful details of the proofs.

Consider the 2-cut graphs of the form shown in Fig. 15.2 with $s$, $s'$, the momenta of the cut lines, and their contribution $E(\lambda)$ to $E^-(\lambda\nu)$ as given in (15.4) amended by the trick (15.22): [2]

$$E(\lambda) = \int ds\,ds'\,e^{i\lambda(\nu,s-s')}U'(s,s')\,\frac{1}{t^-(s')\,t^+(s)}\,L'(s,s')\;. \tag{15.23}$$

$U'$ and $L'$ are 4-point functions like in Fig. 15.2 but including the cut propagators and integrated over the external variables $q$, $q'$, and their test spinors, and formally summed over all orders of perturbation theory. Such formal sums will be generally used in the remainder of this section: unless noted otherwise, the results stated and proved hold in the sense of formal power series, which implies that they hold in every finite order of perturbation theory. This involves inserting the necessary powers of the coupling constant $e$ in finite-order expressions. The factors $U'$, $L'$, $1/t^\pm$ carry spinor indices which

---

[2] We consider timelike $\nu$ normalized to $(\nu,\nu) = m^2$. For spacelike or lightlike $\nu$ the decrease is known to be fast.

are partially contracted according to the ordering of the factors on the basic trajectories. This ordering does not correspond to the ordering of the factors in (15.23). The formal sums mentioned do probably not converge. We are, in fact, only interested in the contributions of given finite order to the integrand. But these are clearly not easy to write down explicitly in terms of the expansions of $U'$, $L'$, $t^\pm$, wherefore we use the expedient of introducing the formal sum as a generating function.

The behaviour of $E$ for $\lambda \to \infty$ is determined by the singularity of the integrand at $s = s' = -\nu$. We define

$$
\begin{aligned}
s^+ &= s_0 + \omega, & s'^+ &= s'_0 + \omega', \\
\hat{s} &= (-\omega, \vec{s}), & \hat{s}' &= (-\omega', \vec{s}'), \\
\Delta &= 2\omega s^+, & \Delta' &= 2\omega' s'^+,
\end{aligned}
\qquad (15.24)
$$

with $\omega = \omega(\vec{s})$, $\omega' = \omega(\vec{s}')$. $\hat{s}$ and $\hat{s}'$ are projections of $s$ and $s'$ onto the negative mass shell. Replacing $s_0$, $s'_0$, by $\Delta$, $\Delta'$, as variables of integration, we obtain

$$
E(\lambda) = \int \frac{d^3s}{2\omega} \frac{d^3s'}{2\omega'} e^{i\lambda(\nu, \hat{s}-\hat{s}')}
$$

$$
\times \int d\Delta\, d\Delta'\, e^{i\lambda\nu^0(s^+ - s'^+)} I(\Delta, \Delta', \vec{s}, \vec{s}'), \qquad (15.25)
$$

with $I$ the product occurring in (15.23) written as a function of the new variables. Since only the points $\hat{s} \sim \hat{s}' \sim -\nu$ are asymptotically relevant, we can introduce in the integrand a cutoff function $\chi(\vec{s}, \vec{s}')$ with $\chi(-\vec{\nu}, -\vec{\nu}) = 1$ and an arbitrarily small support around this point. Off the mass shell, i.e. for $\Delta$ and $\Delta'$ non-zero, $I$ is a smooth function of $\vec{s}$ and $\vec{s}'$. Even on the mass shell its behaviour at $\vec{s} \neq \vec{s}'$ is not worse than at $\vec{s} = \vec{s}'$. We are therefore justified in setting $\vec{s} = \vec{s}' = -\vec{\nu}$ in the argument of $I$ without changing the asymptotically leading term in $E_G$. This is understood in the sequel and the arguments $\vec{s}$, $\vec{s}'$ are dropped.

Let $I_\sigma$ be the sum over all these contributions of order $e^\sigma$. Define

$$
D = \sqrt{\Delta^2 + \Delta'^2}. \qquad (15.26)
$$

The following theorem will be proved.

**Theorem 15.1.** *The function $I_\sigma(\Delta, \Delta')$ can be written as*

$$
I_\sigma = \frac{1}{D^2} I^0_\sigma(\Delta, \Delta') + \mathcal{O}\left(\frac{1}{D^{1+\varepsilon}}\right) \qquad (15.27)
$$

*for $1 > \varepsilon > 0$, where $I^0_\sigma$ is homogeneous of order 0:*

$$
I^0_\sigma(\rho\Delta, \rho\Delta') = I^0_\sigma(\Delta, \Delta') \qquad \text{for} \qquad \rho > 0. \qquad (15.28)
$$

$I_\sigma^0$ is obtained from $I_\sigma$ as

$$I_\sigma^0(\Delta, \Delta') = \lim_{D \to 0} D^2 I_\sigma(\Delta, \Delta') . \tag{15.29}$$

Note that this amounts to finding the value of $D^2 I_\sigma$ at the mass shell $\Delta = \Delta' = 0$: it replaces the "restriction to the mass shell" of amputated functions found in the conventional formalism. However, in our case this limit turns out to be direction-dependent even if the factor $D^2$ is replaced by the conventional $\Delta\Delta'$.

By the scaling transformation $\Delta = \lambda\delta$, $\Delta' = \lambda\delta'$, we find that the $\Delta$–$\Delta'$-integral in (15.25) converges for $\lambda \to \infty$ to

$$J = \int \frac{d\delta\, d\delta'}{\delta^2 + \delta'^2} e^{-in(\delta-\delta')} I^0(\delta, \delta') , \tag{15.30}$$

with $n = \frac{\nu^0}{2\omega(\vec{\nu})} > 0$. This integral is defined as $\lim_{\varepsilon \downarrow 0}$ of the expression obtained by the replacement $\delta \to \delta - i\varepsilon$, $\delta' \to \delta' + i\varepsilon$ in the integrand. The remaining $\vec{s}$–$\vec{s'}$-integral in (15.26) can then be evaluated at large $\lambda$ with the help of Lemma 14.2, giving

$$E(\lambda) \sim \lambda^{-3} \frac{2\pi^3 m}{(\nu,\nu)} J . \tag{15.31}$$

This is the $\lambda^{-3}$ behaviour typical for a 1-particle state.

In theories without IR problems, for example massive QED (i.e. with a positive photon mass), the coefficient of the $\lambda^{-3}$-term in $E$ is given by (15.6). Apart from kinematical factors it is a product of two terms which can be considered to be associated with the probe and the source respectively. Such a factorization would be desirable for the scattering formalism to be established in the remaining chapters. We must therefore investigate in how far it also obtains in true, massless QED. The answer is: not completely but to an acceptable extent.

We start by showing that the $\Delta$–$\Delta'$-singularity responsible for the leading term in $E$ can be written in the desired product form. In the product $U'L' \frac{1}{t^+ t^-}$ in the integrand of (15.23), $U'$ and $L'$ get obviously assigned to the upper and lower side of the cut, respectively. The numerator we try to split evenly among the sides, assigning to either a factor $(t^+ t^-)^{-1/2}$. But $t^\pm$ are spin matrices, hence their square roots are undetermined. Consider the case $t^+$. We are only interested in its singularity at $s^2 = m^2$, $s^0 < 0$. According to Chap. 10, $t^+$ can be written as

$$t^+(s) = \frac{N(s)}{\slashed{s} - m} , \qquad N(s) = A(s^2) + (\slashed{s} - m) B(s^2) ,$$

where $A$ and $B$ are scalars with at most weak singularities at the mass shell. The relevant singularity is clearly due to the $A$-term. A simple calculation shows that $A$ can be expressed in terms of $t^+$ as

$$A(s) = (\not{s} - m) t^+(s) - \frac{\not{s} - m}{4s^2} \operatorname{Tr}\left(\not{s}(\not{s} - m) t^+(s)\right). \tag{15.32}$$

The $\frac{1}{s^2}$ singularity is cancelled by the vanishing of the trace at $s^2 = 0$. It lies in any case outside the momentum region of interest to us. For our purposes, the denominator factor $t^+(s)$ can then be replaced by $-\Delta^{-1} A(s) (\not{s} + m)$, whose inverse is

$$\frac{-\Delta}{\not{s} + m} \frac{1}{A(s)}. \tag{15.33}$$

But $U$ and $L$ contain factors $(\not{s} + m)$ from their $s$-lines. One of them cancels the one in the denominator of (15.33), and is then re-introduced by writing the remaining one as $\not{s} + m = \frac{1}{2m}(\not{s} + m)^2 + \mathcal{O}(\Delta)$. This means that $\frac{1}{t^+}$ can be replaced by

$$\frac{1}{t'^+(s)} := \frac{-\Delta + i\varepsilon}{2mA(s)} \tag{15.34}$$

without changing the $\lambda$-asymptotics, and the root of this scalar expression can be drawn without problem. There is of course a sign ambiguity, but this does not affect the final result if the same branch of the square root is consistently chosen everywhere.

The same method applies to $t^-$. The integrand of (15.23) without the exponential can then be factorized as

$$\hat{U}(s, s') \hat{L}(s, s'), \tag{15.35}$$

with

$$\hat{U}(s, s') = \frac{1}{\sqrt{t'^+(s)}} \frac{1}{\sqrt{t'^-(s')}} U'(s, s'),$$

$$\hat{L}(s, s') = \frac{1}{\sqrt{t'^+(s)}} \frac{1}{\sqrt{t'^-(s')}} L'(s, s'). \tag{15.36}$$

$\hat{U}$ and $\hat{L}$ still carry spinor indices which must be summed over appropriately.

We define

$$L(\Delta, \Delta') = \hat{L}(s, s')\big|_{\hat{s} = \hat{s}' = -\nu} \tag{15.37}$$

written as a function of $\Delta$, $\Delta'$, and $U(\Delta, \Delta')$ in the same way. Then the following more detailed version of Theorem 15.1 holds.

**Theorem 15.2.** *The function $U(\Delta, \Delta')$ can be written as*

$$U = \frac{1}{D} U^0(\Delta, \Delta') + \mathcal{O}(D^{-\varepsilon}) \tag{15.38}$$

*for every $\varepsilon \in (0, 1)$, where $U^0$ is homogeneous of order zero. The same holds for $L(\Delta, \Delta')$*

Theorem 15.1 is clearly a consequence of Theorem 15.2.

256    15. Interacting Particles

The decomposition $I^0 = L^0 U^0$ occurs in (15.30) under an integral and does therefore not yet lead to a factorization of the coefficient $J$. The situation is analyzed further with the help of the following result stating that the singularities of $L^0$ and $U^0$ at $D = 0$ are of an universal form not depending on the test spinors $f$ and $g$.

**Theorem 15.3.** *There exist homogeneous functions $N^{U,L}(\Delta, \Delta')$ of degree $-1$ which are spin scalars and do not depend on the test spinors $f$ and $g$, such that*

$$L_{MS} = \lim_{D \to 0} \frac{L(\Delta, \Delta')}{N^L(\Delta, \Delta')}, \qquad U_{MS} = \lim_{D \to 0} \frac{U(\Delta, \Delta')}{N^U(\Delta, \Delta')} \qquad (15.39)$$

*exist and are independent of the direction in which $(\Delta, \Delta')$ tends to zero.*

$N^U$ and $N^L$ are explicitly given by

$$N^U(\Delta, \Delta') = \frac{1}{\Delta - \Delta' - i\varepsilon} \left[ \left(\frac{\Delta' + i\varepsilon}{\Delta - i\varepsilon}\right)^{\frac{1+\xi}{2}} - \left(\frac{\Delta - i\varepsilon}{\Delta' + i\varepsilon}\right)^{\frac{1+\xi}{2}} \right]$$

$$N^L(\Delta, \Delta') = \frac{1}{\Delta - \Delta' - i\varepsilon} \left[ e^{-i\pi\xi} \left(\frac{\Delta' + i\varepsilon}{\Delta - i\varepsilon}\right)^{\frac{1+\xi}{2}} - e^{i\pi\xi} \left(\frac{\Delta - i\varepsilon}{\Delta' + i\varepsilon}\right)^{\frac{1+\xi}{2}} \right],$$

(15.40)

with $\xi = \frac{e^2}{2\pi^2}$. $(\Delta - i\varepsilon)^{\frac{1+\xi}{2}}$ and $(\Delta' + i\varepsilon)^{\frac{1+\xi}{2}}$ are defined to be real for positive real $\Delta, \Delta'$. The asymmetric appearance of the two expressions depends on our definition of $\Delta$ and $\Delta'$. If they were defined with the opposite sign, then $N^U$ would have the more complicated form.

$L_{MS}$ and $U_{MS}$ replace the restriction to the mass shell of the amputated Green's functions occurring in the LSZ reduction formula. Inserting $D L_{MS} N(\Delta, \Delta')$ in place of $L^0(\Delta, \Delta')$ into the formula (15.30) for $J$, and similarly for $U^0$, we find

$$J = L_{MS} U_{MS} \int d\delta \, d\delta' \, e^{-in(\delta - \delta')} N^U(\delta, \delta') N^L(\delta, \delta'). \qquad (15.41)$$

The $f$–$g$-dependence of $J$ is contained in the product in front of the integral. The $\delta$–$\delta'$-integral $I$ depends on $e^2$ but not on $n$ for $n > 0$, for scaling reasons. The factor $L_{MS}$ is assigned to the preparation part of the arrangement, $U_{MS}$ to the probe. $I$ is real and at $e = 0$ it is positive. It cannot be split convincingly into factors connected with the source and the probe, respectively. For the moment we leave this problem open, but will return to it in Chap. 17 as part of a more general problem.

The factor $U'(s, s')|_{\hat{s} = \hat{s}' = -\nu}$ entering our result through (15.36) is

$$\int dq \, dq' \, \tilde{\bar{f}}(q') \, \tilde{f}(q) \left( \Omega, T^- \left( \tilde{\bar{\psi}}(-\nu) \, \tilde{\Psi}(q') \right) T^+ \left( \tilde{\bar{\Psi}}(q) \, \tilde{\psi}(-\nu) \right) \Omega \right). \qquad (15.42)$$

$L(\Delta, \Delta')$ is of a similar form. It might be preferable to replace the fields $\psi$ and $\bar\psi$ in this expression by $\Psi$ and $\bar\Psi$. This is almost possible. For example, a $\Psi_n(s')$ vertex is amputated in $s'$: there is no external $s'$-propagator and therefore as a rule no asymptotically relevant singularity at $s'^2 = m^2$. Unfortunately, there are exceptions to this rule, occurring if the $\Psi_n$ vertex appears in a SEP, i.e. a 1PI subgraph which is connected to the main body of the graph by a single line belonging to the $s'$-trajectory. Summing formally over all these SEPs, keeping the rest of the graph fixed, we find that this results in multiplying the original graph by a matrix-valued factor $Z(s')$. This amounts to a generalized field renormalization. The zero-order term in $Z$ is 1; hence $Z$ is invertible in the sense of formal power series. The $\mathcal{W}$-function in (15.42) can then be replaced by the corresponding $\mathcal{W}$ formed with four physical fields, multiplied with $Z^{-1}(-\nu)$ and the corresponding $\bar\Psi_n$-factor $\bar Z^{-1}(-\nu)$. In the full graph containing $U'L'$, and $\frac{1}{t^+t^-}$, the new factors in $U'$ and $L'$ cancel against the corresponding factors in $t^\pm$ if these functions are also formed with physical fields. Unfortunately this agreeable fact does not carry over to the $U$- and $L$-parts separately, because the trick of forming the square root of the denominator does no longer work. The replacement $\psi \to \Psi$ at the cut positions is therefore of limited usefulness.

It is clear how the foregoing results are transferred to the positron case. Photons will be considered in the next section.

The proof of Theorems 15.1–3 is based on a more explicit analysis of the IR singularities than was given in Chap. 11. This is achieved by splitting a graph integrand into a part called *"web"* of a simple structure but fully containing the IR problem, and the IR harmless but very complicated remainder called *"core"*.[3] We first state this decomposition and then prove that it achieves what it is supposed to do.

Consider a 2-cut graph of the form shown in Fig. 15.2 with $s$, $s'$, the momenta of the cut lines. The decomposition $U' \frac{1}{t^- t^+} L'$ of the graph integrand introduced in (15.23) is used. Each of the four terms in this quotient is separately subjected to a core–web splitting. For definiteness we consider the upper half, the $U$-part of Fig. 15.2 including the $s$- and $s'$-propagators.

A $U'$-web is represented by a graph containing two connected subtrajectories (also called "semitrajectories") of the basic trajectories starting at the cut lines, but no other fermion lines, in particular no closed fermion loops. Photon lines either start and end at one of these trajectories or they connect one of them to a $\Psi_n$- or $\bar\Psi_n$- vertex at the end of one of the trajectories of the original graph. These composite vertices need not belong to one of the semitrajectories included in the web. The momenta of cross lines are defined to be directed from the $T^-$- to the $T^+$-sector. Photon lines connecting a trajectory to itself are directed against the direction of this trajectory. Photon lines with a $\Psi_n$- or $\bar\Psi_n$-end point into the $s$-trajectory, out of the $s'$-trajectory. Photon

---

[3] Our procedure is partly based on the second paper of the series [Ki 68].

lines connecting two composite end vertices do not belong to the web because the IR divergences caused by them have been shown to cancel.

The photon momenta are classified as "soft" or "hard" depending on whether they belong to the cylinder $\mathcal{K}$ defined in (15.8) or to its complement. The constant $R$ characterizing the size of $\mathcal{K}$ need not be small: no terms that vanish together with $R$ will be neglected.

Let $\chi(k)$ be the characteristic function of $\mathcal{K}$. With a web graph as described above an integrand $w_U(s^+, \vec{s}, s'^+, \vec{s}', k_i)$, with $\{k_i\}$ denoting the photon momenta, is associated as follows.

- Each vertex on one of the trajectories carries a factor $-ie\sqrt{\frac{2}{\pi}}\hat{s}^\mu$ or $ie\sqrt{\frac{2}{\pi}}\hat{s}'^\nu$ respectively.

- Each photon line carries its usual propagator as defined in Chaps. 9 and 12, multiplied with $\chi(k)$ if $k$ is its momentum. For each photon line leaving a $\Psi_n$ or $\bar{\Psi}_n$ a factor $i(2\pi)^{-\frac{3}{2}}e$ or $-i(2\pi)^{-\frac{3}{2}}e$ coming from the composite vertex is included in the web.

- An electron line on the $s$-trajectory carries the propagator $\frac{i}{2\pi} \times \frac{1}{-\Delta + 2(\hat{s},K) + i\varepsilon}$, where $K$ is the sum of the photon momenta having flowed into the trajectory before this line, i.e. $s + K$ is the momentum of the line according to the ordinary graph rules. The propagators of the $s'$-trajectory are the complex–conjugates of those with $s'$ replacing $s$.

- On the $s$-trajectory exist 2-line vertices with vertex factors

$$m_W = \frac{e^2 m^2}{2\pi^3} \int \frac{dk}{k^2 + i\varepsilon} \frac{\chi(k)}{2(\hat{s}, k) + i\varepsilon},$$

and similarly for the $s'$-trajectory.

The 2-line vertices of the last item are parts of the mass renormalization vertices which are included in the web in order to prevent the build-up of singularities of high order at $\Delta = 0$ or $\Delta' = 0$. In a summed version of the webs (see below) they result in the replacement of $\Delta$ by $\Delta - m_W$ and of $\Delta'$ by $\Delta' - m_W$ in the formulae.

In the $L$-web $W_L$ we let the cross momenta flow from $s$ to $s'$. The graph rules are then the same as in $W_U$, except that the factors $\delta_+(k)$ of cross propagators are replaced by $\delta_-(k)$. The web integrand $w_+$ of the denominator $t^+(s)$ is defined in the same way, the $s$-line being the first line of the trajectory traversing the graph. No cross lines and no $\Psi_n$ lines are present in this case. The $s'$-denominator integrand $w_-$ is defined accordingly.

Webs of second order have been discussed earlier in this section.

Finally, we define

$$W_U(s, s') = \int \prod_1^\alpha dk_i \prod_1^\beta d\ell_j \prod_1^\gamma d\ell'_h \, w_U(s, s', k_i, \ell_j, \ell'_h) \tag{15.43}$$

and similarly $W_L(s,s')$, $W_+(s)$, and $W_-(s')$. The following result will be established.

**Lemma 15.4.** *In every finite order of perturbation theory, the asymptotically relevant contribution to $E^-(\lambda\nu)$ is a sum of terms of the form*

$$\int ds^+ \, ds'^+ \, d^3s \, d^3s' \, e^{i\lambda\nu^0(s^+ - s'^+)} e^{i\lambda(\nu,\hat{s}-\hat{s}')} C_U(\vec{s},\vec{s}\,')$$

$$\times \frac{1}{C_+(\vec{s})\,C_-(\vec{s}\,')} \, C_L(\vec{s},\vec{s}\,') \, W_U(s,s') \, W_L(s,s') \, \frac{1}{W_+(s)\,W_-(s')}, \quad (15.44)$$

*with smooth functions $C_U, \ldots, C_-$. The core parts $C$ and web parts $W$ of the various terms vary independently of each other: every possible core part occurs combined with every possible web part.*

The $C$-factors carry spin indices which must be arranged as indicated by the trajectory directions and the decomposition (15.22). The Lemma will be proved by explicit construction of the core functions. This is a lengthy procedure with an exceedingly inelegant result. Luckily, the intricate core–web splitting is only needed as a technical tool in the proof of the IR cancellations. It is not needed for the actual calculation of the remaining truly dominant term in $E^-$. More exactly, the form of the web will be needed to be known ecplicitly but not the form of the core.

The rest of the section is devoted to proving Lemma 15.4 and Theorems 15.1–3.

## Proof of Lemma 15.4

We construct the decomposition $U' = C_U W_U$ for the factor $U'$ in (15.23). The other factors are treated analogously. $C_U$ and $W_U$ are defined as integrals over functions $c_U$ and $w_U$ depending on $s$, $s'$, and the internal variables on the $U$-side of the cut of the considered graph

First we remember that 1PI subgraphs of the kind shown in Fig. 15.3 which intersect a trajectory[4] but are not R-parts are IR harmless. They get therefore included in the core. Furthermore, it has been noted before that cross lines connected to VPs[5], which are interrupted by 2-photon parts are IR harmless. They and their end points are included in the core. If a cross line ends on one side in a core vertex, on the other side in a VP on the trajectory, then this line and this VP are included in the core. And all points and internal vertices separated from the cut by core vertices belong themselves to the core, as do all lines joining them. This rule remains valid throughout the following

---

[4] Unless noted otherwise, "trajectory" means in this proof the part of a basic trajectory contained in $U$.

[5] For the time being the term "vertex part" includes simple vertices.

enlargements of the web. If a trajectory does not contain any core vertices, it belongs fully to the web. But the test spinors $\tilde{f}$ and $\bar{\tilde{f}}$ and the numerators $(\slashed{s}+m),\ldots$, of the adjacent propagators belong to the core even in this case.

The remaining putative web consists of connected parts of the two trajectories beginning at the cut, of SEPs lying on the trajectories, and of VPs on the trajectories whose external photon line leads to another such VP or to a $\Psi_n$ or $\bar{\Psi}_n$. In a second step of the construction these contributions are analyzed more closely. We first note that if a photon line with momentum $k$ is connected at one end to a closed fermion loop in a VP, then this VP vanishes at $k=0$. More exactly, this is true for a sum over a suitable set of such VPs. The reason is that the loop part $\ell(k,\ldots)$ of such a graph is, after integration over the loop variable, a smooth function of its external variables, and that by the WT identities we have $k_\rho \ell(^\rho k,\ldots) = 0$. Differentiating with respect to $k$ and setting $k=0$ we find $\ell(0,\ldots)=0$. Such a VP is IR irrelevant and is included in the core. Secondly, consider a SEP none of whose photon lines connects the trajectory to itself. Examples are shown in Fig. 15.8. Such SEPs are called of the "first kind". They vanish at the mass shell $s^2=m^2$ after mass renormalization. Moreover, their first derivatives are continuous at the mass shell: they produce no weak singularity. For the first graph in Fig. 15.8 this is seen as follows. Let the blob be connected to the trajectory by $n$ photon lines. $n$ is even by Furry's theorem. For $n=2$ the blob is an insertion in a photon line (momentum $k$) contributing a factor $\rho k^\mu k^\nu$ whose vanishing at $k=0$ removes any possible IR problem. For $n \geq 4$, introduce $n-1$ photon variables $\ell_1,\ldots$, as independent variables. The blob is a smooth function $b(\ell_1,\ldots)$ by the results of Sect. 11.3. In the $T^+$-sector the $\ell$-integral is at $s^2=m^2$ of the form

$$\Sigma'(s) = \int \prod_1^{n-1} \left\{ \frac{d\ell_i}{\ell_i^2 + i\varepsilon} \frac{1}{\sum_{j=1}^{i}(2(s,\ell_j) + \ell_j^2)} \right\} b(\ell_1,\ldots) \ .$$

This integral converges at $\ell_1 = \cdots = \ell_{n-1} = 0$ because the integrand has there a singularity of order $3(n-1)+2$, which is lower than the number $4(n-1)$ of variables. A derivation with respect to $s$ increases the order of the singularity by 1, at first sight destroying the existence of the integral if $n=4$. But the 4-point function $b(\ell_1,\ell_2,\ell_3)$ vanishes at $\ell_1 = \ell_2 = \ell_3 = 0$ as has been shown in Sect. 10.2. This result involves a summation over graphs. Renormalization does not invalidate the argument because the subtractions are computed at $s=0$, where there is no IR problem, and the mass renormalization involves only $\Sigma'$ itself, not its derivatives. The same sort of estimate works also for the second graph of Fig. 15.8 with a 1-particle reducible blob part. The additional singular propagator is compensated for by the larger number of independent $\ell_j$-variables. The same is true for graphs of higher order with a more extensive tree structure of the blob part. The contribution of graphs of first kind to a SEP is thus of the form

$$\Sigma(s) = -2\pi i\, S\, (\slashed{s} - m) + R(s)\ , \tag{15.45}$$

where $S$ is a finite (in finite order) real constant and $R(s)(\not{s}+m)$ vanishes at the mass shell at least like $(s^2 - m^2)^{2-\varepsilon}$ for every $\varepsilon > 0$. The $R$-contribution is included in the core.

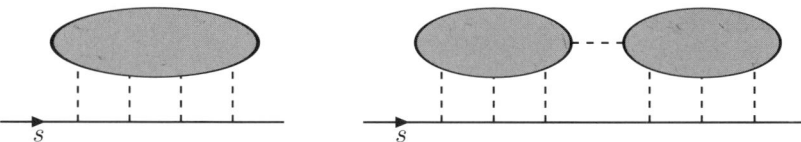

**Fig. 15.8.** SEPs of first kind.
The grey blobs represent 1PI subgraphs

Similar estimates apply to VPs of the first kind, which are generated from SEPs of the first kind by attaching a cross line to the trajectory inside the SEP. In this case it is found that the vertex function $\Lambda(u,v)$, $u$ and $v$ being the adjacent trajectory variables, is finite at $u = v$, $u^2 = m^2$. The WT identities hold for VPs and SEPs of first kind. We obtain

$$(v-u)_\rho \Lambda^\rho(u,v) = \frac{e}{(2\pi)^{3/2}} \left( \Sigma(u) - \Sigma(v) \right) .$$

Inserting the form (15.45) of $\Sigma$ and differentiating with respect to $v_\rho$ gives

$$\Lambda^\rho(u,v) = \frac{ie}{\sqrt{2\pi}} S \gamma^\rho + R(u,v) , \qquad (15.46)$$

with $R$ vanishing like a positive power for $u^2 \to m^2$, $|u-v| \to 0$. Again the $R$-contribution gets added to the core.

SEPs and VPs of the "second kind" are generated from those of the first kind by adding photon lines with both ends on the trajectory but with at least one end lying inside the original first kind part. Examples are shown in Fig. 15.9. Each of the additional lines adds 4 variables of integration and increases the degree of the IR singularity by 4, so that the former estimates remain valid. The new lines can create subintegrals with logarithmic singularities, but these involve internal variables and do not lead to a worse $u$- or $u$-$v$-behaviour. Equations (15.45) and (15.46) hold for these graphs.

The remaing SEPs and VPs are said to be of the "third kind". They may contain SEPs and VPs of the first or second kind as subgraphs. These benign subgraphs are split according to (15.45) and (15.46). The $R$-contributions remove the weak singularities of the SEPs and VPs in which they are imbedded, turning them into benign graphs to be handled like those of first and second kind. The term $-2\pi i\, S\, (\not{u}-m)$, where $u$ is an internal variable, multiplied with the subsequent propagator $\frac{i}{2\pi}(\not{u}-m)^{-1}$ replaces this propagator by the

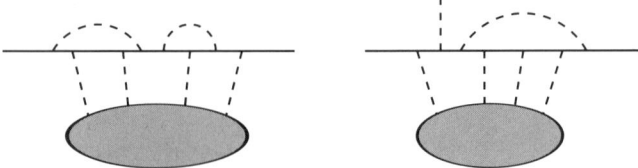

**Fig. 15.9.** A SEP and a VP of second kind

constant factor $S$ which is made part of the numerator of the previous propagator. The presence of a chain of $n$ adjacent SEPs leads in the same way to the replacement of the whole chain by a single propagator with a factor $S^n$ in the numerator. Summation over $n$, including $n = 0$, yields a numerator factor $\frac{1}{1-S}$ in the propagator. But this propagator starts at a single vertex or a VP with web part $\frac{ie}{\sqrt{2\pi}} S \gamma^\rho$ by (15.46). Summing these two expressions gives the factor $-\frac{ie}{\sqrt{2\pi}} \gamma^\rho (1-S)$, whose factor $(1-S)$ cancels the $(1-S)^{-1}$ of the next propagator. A factor remains from the first, cut, line of the trajectory. It is a constant that does not depend on the structure of the web. It can therefore be commuted through to the core and be included in the core.

The result of this procedure is a provisional web consisting of two partial trajectories and photon lines all of whose ends lie on the trajectory.

The SEPs and VPs remaining in the web are UV divergent and must be renormalized. In order to separate the IR from the UV problems we introduce as a third step of the construction a further modification of the web–core splitting. We distinguish between *"soft"* and *"hard"* photons, a photon being soft if its momentum $k$ is contained in the cylinder $\mathcal{K}$ defined in (15.8), hard otherwise. Let $\chi(k)$ be the characteristic function of $\mathcal{K}$ and $\chi'(k) = 1 - \chi(k)$. The photon propagators are decomposed into a soft and a hard part by

$$(k^2 \pm i\varepsilon)^{-1} = \chi(k)(k^2 \pm i\varepsilon)^{-1} + \chi'(k)(k^2 \pm i\varepsilon)^{-1} \qquad (15.47)$$

and analogously for $\delta_\pm(k)$. This produces two kinds of photon lines: soft and hard ones. The soft propagators are UV irrelevant due to their restricted support. Only 1PI subgraphs containing exclusively hard lines can be R-parts. On the other hand, the IR problems are due to the soft lines. This allows a new enlargement of the core. SEPs and VPs of the putative web containing no soft photon lines are considered to be of the first kind and are treated as such. New parts of the second and third kind are then defined as before, the soft lines playing the role of the trajectory→trajectory lines of the previous stage of construction. The subtraction terms are evaluated with the full photon propagators, not only the hard ones. They are included in the R-parts of first and second kind, rendering them UV finite, which those of the third kind are anyhow. Mass renormalization is performed separately

## 15.2 IR Cancellations

in each SEP. The contribution to $\delta m$ of a particular web-SEP depends on $\vec{s}$ because $\mathcal{K}$ is not Lorentz invariant. If a web trajectory ends in a composite external vertex, then the photon lines incident at this vertex are web lines if their other end belongs to the web, core lines otherwise. Its vertex factor is accordingly split into a factor $(\pm i)^\alpha (2\pi)^{-\frac{3}{2}\alpha}$ belonging to the web and a factor $(\pm i)^\beta (2\pi)^{-\frac{3}{2}\beta}$ belonging to the core. And the numerator $g_{\mu\nu}$ of the photon propagator is replaced by $g_{\mu\nu} \pm i k_\mu \tilde{r}_\nu(\pm k)$, as the case may be.

After this splitting the web is still specified by a graph of the same form as before, but containing only soft photon lines. But we are not yet at the end of our travail; we need further simplifications. Consider the $T^+$-trajectory. We write the singular factor in the cut propagator as

$$\frac{1}{s^2 - m^2 + i\varepsilon} = \frac{1}{-\Delta + i\varepsilon} \frac{-2\omega}{s^0 - \omega} , \qquad (15.48)$$

with the notations introduced in (15.24). The last factor is regular at the mass shell $s^+ = 0$ and therefore IR harmless. It is included in the core function. The numerator $\slashed{s} + m$ of the remaining $1/\Delta$ factor is then multiplied with the $\gamma^\mu$ of the first vertex on the trajectory and this product is decomposed as

$$(\slashed{s} + m)\gamma^\mu = 2\hat{s}^\mu - \gamma^\mu(\slashed{s} + \slashed{k}_1 - m) + \gamma_\mu \slashed{k}_1 + 2(s^\mu - \hat{s}^\mu) , \qquad (15.49)$$

with $k_1$ the photon momentum incident at the vertex. The last two terms on the right-hand side are IR harmless because they vanish either at the critical value $k_1 = 0$ or at the mass shell $s^+ = 0$. They are assigned to the core. The second factor of the second term on the right-hand side cancels the singularity $(\slashed{s} + \slashed{k}_1 - m)^{-1}$ of the next trajectory propagator. Therefore it is IR harmless too and is assigned to the core.

The only contribution remaining in the web is $2\hat{s}^\mu$ which replaces the matrix $\gamma^\mu$ of the vertex factor of the original rules. In this contribution we split the next electron propagator as

$$\frac{\slashed{s} + \slashed{k}_1 + m}{(s+k_1)^2 - m^2 + i\varepsilon} = A(s, k_1) + \frac{\slashed{s} + \slashed{k}_1 + m}{-\Delta + 2(\hat{s}, k_1) + i\varepsilon} . \qquad (15.50)$$

The singularity of $A$ at $\{s^+ = 0, k_1 = 0\}$ is lower by one order than that of the full propagator, and this makes $A$ IR harmless. It gets included in the core. This makes the core at first depend explicitly on $k_1$ and on $\hat{s}$, not only on the momentum $s + k_1$ of the line adjacent to it. The numerator of the last term in (15.50) is then multiplied with the vertex factor $\gamma^\nu$ of the next vertex with photon momentum $k_2$, and split

$$(\slashed{s} + \slashed{k}_1 + m)\gamma^\nu = 2\hat{s}^\nu - \gamma^\nu(\slashed{s} + \slashed{k}_1 + \slashed{k}_2 - m) + \gamma^\nu \slashed{k}_2 + 2(s^\nu - \hat{s}^\nu) . \qquad (15.51)$$

The last three terms are IR harmless for the reasons explained before and are included in the core. In the $2\hat{s}^\nu$ term we split the subsequent propagator like in (15.50) with $k_1$ replaced by $k_1 + k_2$, and the $A$-part is included in the core. It is clear how the procedure continues. It stops as soon as a core vertex is reached. Whenever a term is included in the core, then also the later vertices on the trajectory as well as the vertices directly connected to them by photon lines become core vertices. The same procedure is applied to the $s'$-trajectory.

As result we obtain a web which still has the same graph structure as before. But on the $s$-trajectory the vertex factors are $-\sqrt{\frac{\pi}{2}}\, i\, e\, \hat{s}^\mu$ instead of $-\frac{ie}{\sqrt{2\pi}}\gamma^\mu$ and the electron propagators have the simple form

$$\frac{i}{2\pi\bigl(-\Delta + 2(\hat{s}, \sum_1^i k_j) + i\varepsilon\bigr)}$$

if $k_j$ is the photon momentum incident at the $j^{\text{th}}$ vertex counted from the cut. The corresponding rules hold on the $s'$-trajectory.

This result as yet ignores mass renormalization. Accordingly, the presence of SEPs leads to catastrophically strong IR singularities of the form $\bigl(-\Delta + (s, K) + i\varepsilon\bigr)^{-n}$ with $K$ a partial sum (possibly empty) of $k_i$'s. Mass renormalization can therefore not be pushed off completely into the core; its IR relevant part must be performed in the web. Consider a SEP situated on the $s$-trajectory before the first cross vertex. Then its external momentum is the cut momentum $s$ and the subsequent propagator contains the singularity $(-\Delta + i\varepsilon)^{-1}$ which must be removed by mass renormalization. This is achieved by subtracting from the integrand of the SEP its value at $\Delta = 0$. In the special case of a 1-loop SEP with photon momentum $\ell$ this results in the replacement of the internal electron propagator by

$$\frac{i}{2\pi}\left(\frac{1}{-\Delta + 2(\hat{s}, \ell) + i\varepsilon} - \frac{1}{2(\hat{s}, \ell) + i\varepsilon}\right).$$

For a later SEP, let $K$ be the sum of the cross momenta joining the trajectory previous to it. Then the dangerous singularity to be cancelled is the factor $(-\Delta + 2(\hat{s}, K) + i\varepsilon)^{-1}$ of the next electron propagator. In the 1-loop case with photon momentum $\ell$ the integrand of the SEP contains the factor $(-\Delta + 2(\hat{s}, K + \ell) + i\varepsilon)^{-1}$ which is replaced by $(-\Delta + 2(\hat{s}, K + \ell) + i\varepsilon)^{-1} - (2(\hat{s}, \ell) + i\varepsilon)^{-1}$. The SEPs with more than one loop can be shown to add to zero at the "mass shell" $-\Delta + 2(\hat{s}, \ell) = 0$, so that no mass renormalization is required for them. The same procedure is applied for the $s'$-trajectory.

This result is still not final because the core $C_U$ as yet arrived at, integrated over its internal variables, depends on $s, s', k_1, \ldots, k_\alpha$, instead of only on $\vec{s}$ and $\vec{s}'$ as desired. The $k_i$ are the momenta of the cross lines in the web. The final reduction relies on the fact that this $C_U$ is Hölder continuous of order $1 - \varepsilon$ even in the vicinity of the dangerous points $\Delta = \Delta' = k_i = 0$. We first get rid of the $k_i$ as follows. Split

$$C_U(s,\ldots,k_{\alpha-1},k_\alpha) = C_U(s,\ldots,k_{\alpha-1},0)$$
$$+ \left[ C_U(s,\ldots,k_{\alpha-1},k_\alpha) - C_U(s,\ldots,k_{\alpha-1},0) \right] . \tag{15.52}$$

The square bracket vanishes at $k_\alpha = 0$ with a positive power. This renders the $k_\alpha$-propagator harmless: it can be included in the core. The new core is connected to a web of lower order. It, as well as the first term on the right-hand side of (15.52), is then again split in the same way with respect to the variable $k_{\alpha-1}$. Iteration of this procedure yields a sum over web–core expressions with core functions $C'_U(s,s')$ depending only on $s$ and $s'$. These we decompose as

$$C'_U(s,s') = C'_U(\hat{s},\hat{s}') + \left[ C'_U(s,s') - C'_U(\hat{s},\hat{s}') \right] . \tag{15.53}$$

The first term on the right is of the desired form; the square bracket vanishes at the critical values $s^+ = s'^+ = 0$ with a positive power. This more than compensates for the weak singularity present there: this term leads in the final expression to a $\lambda$-decrease which is faster than the dominant $\lambda^{-3}$.

## Proof of Theorems 15.1–3

In a first step we isolate the singularities of the web factors $W_U, \ldots, W_-$, at $\Delta = \Delta' = 0$ which are at least as strong as those of the free theory. As has been argued after (15.25) we may set $\vec{s} = \vec{s}' = -\vec{v}$. Then it will be shown that these separate singularities combine in $I$, $U$, and $L$, to singularities of the form claimed in Theorems 15.1–3. We begin with $W_U$. Consider a web graph contributing to $w_U$, in particular its $s$-trajectory. Let $N$ be the number of its vertices, not counting the end-vertex if it happens to be a $\bar{\Psi}_n$-vertex, and let $k_1, \ldots, k_N$, be the photon momenta flowing into the trajectory at its vertices, numbered consecutively starting from the $s$-line. We abbreviate

$$K_i = 2(\hat{s}, k_i) . \tag{15.54}$$

The lines and vertices of the trajectory contribute to $w_U$ the factor

$$A_N = F_N \prod_{j=0}^{N} \left( \Delta + \sum_{i=1}^{j} K_i \right)^{-1} , \tag{15.55}$$

with

$$F_N = i \left( e \sqrt{\frac{2}{\pi}} \right)^N \left( \frac{1}{2\pi} \right)^{N+1} \prod_{i=1}^{N} \hat{s}^{\mu_i} . \tag{15.56}$$

Using $(x+i\varepsilon)^{-1} = -i \int_0^\infty d\tau\, e^{i\tau(x+i\varepsilon)}$, we obtain after some elementary manipulations

$$A_N = F_N(-i)^{N+1} \int_0^\infty d\sigma_0\, e^{i\sigma_0(-\Delta+i\varepsilon)} \int_0^{\sigma_0} d\sigma_1\, e^{i\sigma_1(K_1+i\varepsilon)}$$
$$\times \cdots \int_0^{\sigma_{N-1}} d\sigma_N\, e^{i\sigma_N(K_N+i\varepsilon)} . \tag{15.57}$$

266  15. Interacting Particles

A permutation of the vertices on the trajectory, leaving the rest of the graph unchanged, leads again to a valid web graph. Summing over these permutations removes the restrictions $\sigma_{i+1} \leq \sigma_i$, replacing them by $\sigma_i \leq \sigma_0$ for $i > 0$. Carrying out the integrations over $\sigma_i$, $i \geq 1$, yields

$$\sum A_N = -i(-1)^N F_N \int_0^\infty d\sigma\, e^{i\sigma(-\Delta+i\varepsilon)} \prod_{j=1}^N \frac{e^{i\sigma K_j}-1}{K_j}. \tag{15.58}$$

But not all permutations lead to new graphs. The exchange of the two end points of a photon line connecting the trajectory to itself amounts merely to replacing its momentum $k$ by $-k$ without producing a new graph. In order to remove these double countings, we must introduce a factor $2^{-L}$ if $L$ is the number of the photon lines in question. A simple renumbering of the variables of these lines does not produce a new graph either. This necessitates an additional factor $(L!)^{-1}$. In the same way, factors $(L_b!)^{-1}$ and $(M_b!)^{-1}$ must be introduced if there are $L_b$ photon lines connecting the trajectory with its end-vertex if this is a composite vertex, and $M_b$ lines connecting it to the end-vertex of the $s'$-trajectory if it is composite.

The $s'$-trajectory is handled in the same way. The quantities occurring in this case are marked by primes: $L'$ is the number of photon lines with both ends on the $s'$-trajectory, and so on. Because of $\hat{s}' = \hat{s}$ we have $K'_i = 2(\hat{s}', k_i) = K_i$. We multiply the contributions from the two trajectories and include yet another combinatorial factor $(M!)^{-1}$, where $M$ is the number of cross lines connecting the two trajectories. This takes care of the fact that a renumbering of their momenta does not produce a new graph. There results the following expression for the sum over all $W_U$-graphs with given numbers $L, L', \cdots$, of photon lines of the various possible types:

$$\Sigma(L,\cdots) = \left(\frac{1}{2\pi}\right)^2 \frac{(-1)^{L+M+L_b+L'_b+M_b}}{L!\,L'!\,M!\,L_b!\,L'_b!\,M_b!\,M'_b!}$$

$$\times \left(\frac{im^2 e^2}{8\pi^4}\right)^{L+L'} \left(\frac{m^2 e^2}{2\pi^3}\right)^M \left(\frac{e^2}{8\pi^4}\right)^{L_b+L'_b} \left(\frac{ie^2}{4\pi^3}\right)^{M_b+M'_b}$$

$$\times \int_0^\infty d\sigma\, e^{i\sigma(-\Delta+i\varepsilon)} \int_0^\infty d\sigma'\, e^{i\sigma'(\Delta'+i\varepsilon)}$$

$$\times \int \prod_L \left\{-\frac{dk_i}{k_i^2+i\varepsilon} \frac{e^{i\sigma K_i}-1}{K_i}\frac{e^{-i\sigma K_i}-1}{K_i}\right\}$$

$$\times \int \prod_{L'} \left\{-\frac{dk_i}{k_i^2-i\varepsilon} \frac{e^{-i\sigma' K_i}-1}{K_i}\frac{e^{i\sigma' K_i}-1}{K_i}\right\}$$

$$\times \int \prod_M \left\{dk_i\, \delta_+(k_i) \frac{e^{i\sigma K_i}-1}{K_i}\frac{e^{-i\sigma' K_i}-1}{K_i}\right\}$$

$$\times \int \prod_{L_b} \left\{ dk_i \, \frac{(\tilde{r}(k_i), \hat{s})}{k_i^2 + i\varepsilon} \, \frac{e^{i\sigma K_i} - 1}{K_i} \right\} \int \prod_{L'_b} \left\{ dk_i \, \frac{(\tilde{r}(-k_i), \hat{s})}{k_i^2 - i\varepsilon} \, \frac{e^{-i\sigma' K_i} - 1}{K_i} \right\}$$

$$\times \int \prod_{M_b} \left\{ dk_i \, \delta_+(k_i) \, (\tilde{r}(k_i), \hat{s}) \, \frac{e^{i\sigma K_i} - 1}{K_i} \right\}$$

$$\times \int \prod_{M'_b} \left\{ dk_i \, \delta_+(k_i) \, (\tilde{r}(-k_i), \hat{s}) \, \frac{e^{-i\sigma' K_i} - 1}{K_i} \right\} . \tag{15.59}$$

The product $\prod_L$ extends over all photon lines of $L$-type, i.e. those connecting the $s$-trajectory to itself, and similarly for the other products. Notice that for a $L$-line the photon momentum flowing into the trajectory is $k_i$ at one end, $-k_i$ at the other end. This explains the extra sign in the corresponding factor. All $k_i$-integrals are taken over the set $\mathcal{K}$ of soft photons.

Summation over the numbers $L, \ldots$, yields the following expression for $W_U$ at $\vec{s} = \vec{s}'$:

$$W_U(\Delta, \Delta') = \frac{1}{(2\pi)^2} \int_0^\infty d\sigma \, e^{-i\sigma(\Delta - i\varepsilon)} \int_0^\infty d\sigma' \, e^{i\sigma'(\Delta' + i\varepsilon)}$$

$$\times \exp \left\{ \frac{e^2 m^2}{2\pi^3} \left[ \frac{i}{4\pi} \int_{\mathcal{K}} \frac{dk}{k^2 + i\varepsilon} \, \frac{e^{i\sigma K} - 1}{K} \, \frac{e^{-i\sigma K} - 1}{K} \right. \right.$$

$$- \frac{i}{4\pi} \int_{\mathcal{K}} \frac{dk}{k^2 - i\varepsilon} \, \frac{e^{-i\sigma' K} - 1}{K} \, \frac{e^{i\sigma' K} - 1}{K}$$

$$\left. \left. - \int_{\mathcal{K}} dk \, \delta_+(k) \, \frac{e^{i\sigma K} - 1}{K} \, \frac{e^{-i\sigma' K} - 1}{K} \right] \right\}$$

$$\times \exp \left\{ -\frac{ie^2}{4\pi^3} \left[ -\frac{i}{2\pi} \int_{\mathcal{K}} dk \, \frac{(\tilde{r}(k), \hat{s})}{k^2 + i\varepsilon} \, \frac{e^{i\sigma K} - 1}{K} \right. \right.$$

$$+ \int_{\mathcal{K}} dk \, \delta_+(k) \, (\tilde{r}(k), \hat{s}) \, \frac{e^{i\sigma K} - 1}{K}$$

$$- \frac{i}{2\pi} \int_{\mathcal{K}} dk \, \frac{(\tilde{r}(-k), \hat{s})}{k^2 - i\varepsilon} \, \frac{e^{-i\sigma' K} - 1}{K}$$

$$\left. \left. - \int_{\mathcal{K}} dk \, (\tilde{r}(-k), \hat{s}) \, \frac{e^{-i\sigma' K} - 1}{K} \right] \right\} . \tag{15.60}$$

The second exponential in this expression is 1 because its exponent vanishes. This is seen as follows. Consider the first term in the square bracket and integrate first over $k_0$. The integration runs over the real axis, going around the poles at $k_0 = \pm |\vec{k}|$ as indicated by the $i\varepsilon$-prescription. The integrand is analytic in $k_0$ everywhere except at these two poles. Because of $\sigma \hat{s}_0 < 0$ we can close the contour with an infinite semicircle in the lower half-plane. The resulting integral is computed with the method of residues. This amounts to replacing the factor $(k^2 + i\varepsilon)^{-1}$ by $-2\pi i \, \delta_+(k)$. The result just cancels the second term in $[\cdots]$. The third and fourth terms cancel in the same way. This means that the IR singularities being studied are gauge independent, at least in the very restricted set of gauges that we are considering.

We look for singularities of $W_U(\Delta, \Delta')$ at

$$D := \sqrt{\Delta^2 + \Delta'^2} = 0 \,. \tag{15.61}$$

We study them by rewriting $W_U$ in terms of the scaled variables $\alpha, \beta, \tau, \tau', y$ defined by

$$\Delta = \alpha D \,, \quad \Delta' = \beta D \,, \quad \tau = \sigma D \,, \quad \tau' = \sigma' D \,, \quad k = yD \,. \tag{15.62}$$

Note that $\alpha^2 + \beta^2 = 1$. We obtain

$$W_U(D) = \frac{1}{(2\pi)^2 D^2} \int_0^\infty d\tau \, e^{-i\tau(\alpha - i\varepsilon)} \int_0^\infty d\tau' \, e^{i\tau'(\beta + i\varepsilon)}$$
$$\times \exp\left\{\frac{e^2 m^2}{2\pi^3}\left[\frac{i}{4\pi}\int_{\mathcal{K}_D} \frac{dy}{y^2 + i\varepsilon} \frac{2 - e^{i\tau Y} - e^{-i\tau Y}}{Y^2}\right.\right.$$
$$-\frac{i}{4\pi}\int_{\mathcal{K}_D} \frac{dy}{y^2 - i\varepsilon} \frac{2 - e^{-i\tau' Y} - e^{i\tau' Y}}{Y^2}$$
$$\left.\left. - \int_{\mathcal{K}_D} dy \, \delta_+(y) \frac{e^{i\tau Y} - 1}{Y} \frac{e^{-i\tau' Y} - 1}{Y}\right]\right\}, \tag{15.63}$$

with $\mathcal{K}_D = \{y \in R^4 : |\vec{y}| \leq \frac{R}{D}\}$ and $Y = 2(\hat{s}, y)$. The $D$-dependence of the exponential is contained exclusively in the domain of integration $\mathcal{K}_D$ which approaches $R^4$ for $D \to 0$.

We write the first term in the square bracket as

$$G_F = \frac{i}{4\pi}\int \frac{dy}{y^2 + i\varepsilon} \frac{(1 - e^{i\tau Y}) + (1 - e^{-i\tau Y})}{Y^2} \,. \tag{15.64}$$

The $Y$-dependent quotient is regular in $y$ even where $Y = 0$. Hence we may replace its denominator by $(Y + 2i\omega\varepsilon)(Y - 2i\omega\varepsilon)$, taking the limit $\varepsilon \downarrow 0$ at the end. We integrate first over $y_0$. The integrand is regular in $y_0$ except for simple poles at $y_0 = \pm(|\vec{y}| - i\varepsilon)$ and—for the individual terms in the numerator—at $y_0 = -\frac{s\vec{y}}{\omega} \pm i\varepsilon$. In the contribution $G_F^1$ from the first bracket in the numerator we may close the integration contour by an infinite semicircle in the lower half-plane. The method of residues gives contributions from the two poles with negative $\varepsilon$-terms. The one from $|\vec{y}| - i\varepsilon$ is obtained by replacing $(y^2 + i\varepsilon)^{-1}$ in the integrand by $-2\pi i \, \delta_+(y)$, yielding

$$\frac{1}{2}X(\vec{s}, \tau) := \frac{1}{2}\int_{\mathcal{K}_D} dy \, \delta_+(y) \frac{1 - e^{i\tau Y}}{Y^2} = \frac{1}{4}\int_{|\vec{y}|\leq \frac{R}{D}} \frac{d^3 y}{|\vec{y}|} \frac{1 - e^{i\tau \hat{Y}}}{\hat{Y}^2} \,, \tag{15.65}$$

with

$$\hat{Y} = -2(\omega|\vec{y}| + (\vec{s}, \vec{y})) \,. \tag{15.66}$$

## 15.2 IR Cancellations

The contribution from the other pole is

$$\frac{\pi}{2\omega} \tau \frac{1}{\left(\frac{\vec{s}\vec{y}}{\omega}\right)^2 - |\vec{y}|^2},$$

which is homogeneous of order $-2$ in $\vec{y}$, so that the remaining $d^3y$ integral over $\{|\vec{y}| \le R/D\}$ gives a term of the form

$$-\frac{i}{2}\frac{\tau}{D}\mu'(\vec{s}) \tag{15.67}$$

with $\mu'$ real. The contribution $G_F^2$ of the second bracket in (15.64) becomes identical to $G_F^1$ by the substitution $y \to -y$. Together we obtain

$$G_F = X(\vec{s},\tau) - i\frac{\tau}{D}\mu'(\vec{s}) . \tag{15.68}$$

The $\mu'$-term produces in the expression (15.63) the factor $\exp[-i\frac{\tau}{D}\mu(\vec{s})]$ with $\mu = \frac{e^2 m^2}{2\pi^3}\mu'$. This factor cancels a factor $\exp[i\frac{\tau}{D}\mu(\vec{s})]$ coming from the mass renormalization vertices in the web which have hitherto been ignored. They lead to the replacement of $\Delta$ in our formulae by $(\Delta - \delta m^2)$ with $\delta m^2 = -\frac{i}{2\pi} m_W$, hence of the exponent $i(-\alpha+i\varepsilon)\tau$ in (15.63) by $i(-\alpha+\frac{\delta m^2}{D}+i\varepsilon)\tau$, and a closer inspection shows that $\delta m^2 = \mu$. The $\mu'$-term in (15.68) can therefore be dropped.

The $(y^2 - i\varepsilon)$-term $G_{\bar{F}}$ in the square bracket of (15.63) is treated in the same way. It is the complex conjugate of $G_F$ with $\tau$ replaced by $\tau'$:

$$G_{\bar{F}} = X(\vec{s}, -\tau') = \overline{X(\vec{s},\tau')} . \tag{15.69}$$

The remaining term in the square bracket is the cross term

$$G_C = -X(\vec{s},\tau) - X(\vec{s},-\tau') + X(\vec{s},\tau-\tau') . \tag{15.70}$$

Together we obtain for the square bracket without the mass renormalization terms:

$$[\cdots] = X(\vec{s},\tau-\tau') \tag{15.71}$$

and thus the factor

$$\exp\left\{\frac{e^2 m^2}{2\pi^3} X(\vec{s},\tau-\tau')\right\} \tag{15.72}$$

in the $\tau$–$\tau'$-integral of (15.63).

The web factor $W_L$ differs from $W_U$ only by the replacement of $\delta_+(y)$ by $\delta_-(y)$ in the cross propagators. This leads to the replacement of $X(\vec{s},\tau-\tau')$ by $X(\vec{s},\tau'-\tau)$ in (15.72) as is seen with the help of the substitution $y \to -y$. The denominator factors $W_\pm$ are of the same general form as $W_U$ but contain only the $G_F$- or $G_{\bar{F}}$-part, respectively. This amounts to a replacement of $X(\vec{s},\tau-\tau')$ by $X(\vec{s},\tau)$ or $X(\vec{s},-\tau')$, respectively, in (15.72).

In order to evaluate the remaining $\tau$–$\tau'$-integrals and their behaviour for $D \to 0$ we need information on the function $X(\vec{s}, T)$ defined in (15.65), with $T$ one of $\tau$, $\tau'$, $\tau - \tau'$. The $T$-independent term of the integrand leads to a logarithmic divergence of the integral for $D \to 0$. We can write

$$X(\vec{s}, T) = c_1(\vec{s}) \log D + c_2(\vec{s}, T) + \mathcal{O}(D) \tag{15.73}$$

with

$$c_2(\vec{s}, T) = \lim_{D \to 0} \left[ \frac{1}{2} \int_{|\vec{y}| \leq \frac{R}{D}} \frac{d^3 y}{|\vec{y}|} \frac{1 - e^{iT\hat{Y}}}{\hat{Y}^2} - c_1(\vec{s}) \log D \right]. \tag{15.74}$$

$c_1$ is real. $c_2$ is continuous in $T$ except at $T = 0$. For $|T| \to \infty$ it shows a logarithmic increase coming from the contribution of a neighbourhood of the origin $\vec{y} = 0$. At $T = 0$ emerges a logarithmic singularity coming from the $e^{iT\hat{Y}}$-term at large values of $\vec{y}$. The resulting exponential in the $\tau$–$\tau'$-integrand is

$$\exp\left[C_1(\vec{s}) \log D\right] \exp\left[C_2(\vec{s}, T) + \mathcal{O}(D)\right] \tag{15.75}$$

with $C_i = \frac{e^2 m^2}{2\pi^3} c_i$. The first factor does not depend on $T$ and can be drawn in front of the integral. It is responsible for the undesirably strong singularity of $W_\alpha$ at $D = 0$. From the properties of $c_2$ we find that after dropping the irrelevant $\mathcal{O}(D)$-term all terms of finite perturbative order of the $T$-dependent factor are locally integrable in $\tau$, $\tau'$, and at most weakly increasing for large $\tau$, $\tau'$. The $\tau$–$\tau'$-integral, for example, in $W_U$:

$$\int_0^\infty d\tau \, e^{-i\tau(\alpha - i\varepsilon)} \int_0^\infty d\tau' \, e^{i\tau'(\beta + i\varepsilon)} \, e^{C_2(\vec{s}, \tau - \tau')},$$

exists then in every finite order as a continuous function of $\alpha$ and $\beta$ if $\varepsilon > 0$. Its limit for $\varepsilon \downarrow 0$ exists as a distribution in $\alpha = \frac{\Delta}{D}$ and $\beta = \frac{\Delta'}{D}$ which is singular at $\alpha = 0$ or $\beta = 0$ but is elsewhere a continuous function. This latter property is due to the smooth local behaviour (no strong oscillations) of $X(\vec{s}, \tau - \tau')$.

The proof of Theorems 15.1 and 15.2 is now immediate. The $D$-singularities not corresponding to these theorems are collected in the factors $\exp[C_1 \log D]$ of the various webs, and these factors cancel in the combinations $(W_U W_L)/(W_+ W_-)$ and $W_U/\sqrt{W_+ W_-}$ occurring in the two theorems.

For the proof of Theorem 15.3 we must determine the exact form of the singularities of the $W_\alpha$ at $D = 0$ after discarding the $\log D$-factor. This singularity is clearly Lorentz invariant. Moreover, since $\Delta = 2\omega(s_0 + \omega) = -(s^2 - m^2) + \mathcal{O}(\Delta^2)$ for $\Delta \to 0$, $\Delta$ may be considered to be Lorentz invariant for the present purpose. That means that the singularity functions $N^a(\Delta, \Delta')$ of the theorem do not depend on $\hat{s}$, hence can be evaluated at the special value $\hat{s} = \hat{s}' = (-m, \vec{0})$. Let us do this for $N^U$. A constant multiplying $N^U$ is irrelevant; $N^U$ need only be determined up to such a factor. We use

## 15.2 IR Cancellations

the symbol $\doteq$ for denoting equality up to a constant factor. We are concerned with the factor $\exp[C_2(T)]$ in (15.75) which is responsible for the looked-for singularity of $W_\alpha$. For our special choice of $\vec{s}$ we have

$$C_2(T) = \lim_{D \to 0} \left\{ \frac{e^2}{2\pi^2} \int_0^{R/D} \frac{dr}{r} \left(1 - e^{-2imTr}\right) - C_1 \log D \right\}, \quad (15.76)$$

with $r = |\vec{y}|$. A constant factor in the exponential is irrelevant, hence so is a constant addition to the exponent. We can therefore replace $C_2(T)$ by any indefinite integral (in the sense of distributions) of its $T$-derivative $\frac{e^2}{2\pi^2} \frac{1}{T-i\varepsilon}$. We find, defining $\xi = \frac{e^2}{2\pi^2}$, that

$$\exp[C_2(T)] \doteq e^{\xi \log(T-i\varepsilon)}. \quad (15.77)$$

In $W_+(\tau)$ this gives the singular factor[6]

$$N_+ = \frac{1}{D} \int_0^\infty d\tau \, e^{-i\tau\alpha} \, \tau^\xi$$
$$\doteq -\frac{i}{D} e^{-i\frac{\pi\xi}{2}} (\alpha - i\varepsilon)^{-1-\xi} \doteq \frac{1}{D} (\alpha - i\varepsilon)^{-1-\xi}. \quad (15.78)$$

The analogous factor in $W_-(-\tau')$ is

$$N_-(-\tau') \doteq \frac{1}{D} (\beta + i\varepsilon)^{-1-\xi}. \quad (15.79)$$

For the corresponding factor $N_U$ of $W_U$ we obtain

$$N_U \doteq \frac{1}{D^2} \int_0^\infty d\tau \, e^{-i\tau\alpha} \int_0^\infty d\tau' \, e^{i\tau'\beta} (\tau - \tau' - i\varepsilon)^\xi.$$

We write

$$(\tau - \tau' - i\varepsilon)^\xi \doteq |\tau - \tau'|^\xi \left( \theta(\tau - \tau') e^{i\frac{\pi\xi}{2}} + \theta(\tau' - \tau) e^{-i\frac{\pi\xi}{2}} \right)$$

and introduce $x = \tau + \tau'$, $y = \tau - \tau'$ as new variables of integration. Using $\int_0^\infty d\tau \int_0^\infty d\tau' = 2 \int_{-\infty}^\infty dy \int_{|y|}^\infty dx$, $N_U$ is found to be

$$N_U \doteq \frac{1}{D^2} \frac{1}{\alpha - \beta - i\varepsilon} \left[ \frac{1}{(\alpha - i\varepsilon)^{1+\xi}} - \frac{1}{(\beta + i\varepsilon)^{1+\xi}} \right]. \quad (15.80)$$

The function $N^U(\Delta, \Delta')$ of Theorem 15.3 is then

$$N^U \doteq \frac{N_U}{\sqrt{N_+ N_-}}$$
$$\doteq \frac{1}{\Delta - \Delta' - i\varepsilon} \left[ \left( \frac{\Delta' + i\varepsilon}{\Delta - i\varepsilon} \right)^{\frac{1+\xi}{2}} - \left( \frac{\Delta - i\varepsilon}{\Delta' + i\varepsilon} \right)^{\frac{1+\xi}{2}} \right], \quad (15.81)$$

---
[6] See Sect. 3.3 of Gel'fand–Shilov [GS].

which is the form claimed in Theorem 15.3. The $N^L$ given there is obtained in the same way. Note the distinction between $N^U$ and $N_U$, and between $N^L$ and $N_L$.

## 15.3 Photons

1-photon states and photon probes are introduced in analogy to the electron case. Since there are no serious IR problems, the description can be shorter than for the electrons. A 1-photon state $\Phi_g^\gamma$ and a photon probe $D^\gamma(a)$ are provisionally defined like in the free case by

$$\Phi_g^\gamma = F(g)^*\Omega \ , \quad D^\gamma(a) = F(f_a)^* F(f_a) \ , \tag{15.82}$$

$$F(g) = \int dy \, F_{i0}(y) \, g^i(y) \ , \tag{15.83}$$

with $f^i$, $g^i$, being test functions with the familiar support properties in $x$- and $p$-space and with $\partial_i f^i(x) = \partial_i g^i(x) = 0$. But the $F_{i0}$ are now interacting.

We wish to determine the behaviour of

$$E^\gamma(\lambda\nu) = \left(\Phi_g^\gamma, D^\gamma(\lambda\nu)\, \Phi_g^\gamma\right) \tag{15.84}$$

$$= \int dp\, dq\, dp'\, dq' \, e^{-i\lambda(\nu,q'+q)} \tilde{g}(p') \, \overline{\tilde{f}(-q')}$$

$$\times \tilde{f}(q) \, \overline{\tilde{g}(-p)} \, \tilde{W}(p', q', -|q, p, +) \tag{15.85}$$

for $\lambda \to \infty$ and $\nu_0 > 0$. $\tilde{g}$ and $\tilde{f}$ are 3-vectors, the arguments of $\tilde{W}$ are $\vec{E}$-variables. $\tilde{f}(p)$ and $\tilde{g}(p)$ have small (i.e. with diameter $\ll m$) compact supports intersecting the forward light cone but not the backward one.

The graphs contributing to the factor $\tilde{W}$ are of the general form shown in Fig. 15.1. The external lines are photon lines which are defined to be directed like in the electron case. For the same reason as in that case, only photon lines can occur as cross lines. Hence electron lines are present only in closed fermion loops inside the sectors. As a consequence of this, any 2-line cut separating the probe variables $q$, $q'$ from the preparation variables $p$, $p'$ is a cut through two photon lines, one in each sector. And, as has been shown in the preceding section, these 2-line cuts determine the asymptotic behaviour of $E^\gamma$.

Let $G$ be such a cut graph, $s$ and $s'$ the momenta of its cut lines, directed like in Fig. 15.2, and let $U(s, s')$ and $L(s, s')$ be the contributions of the halves of $G$ on the probe side and the preparation side, respectively, of the cut. Both halves are integrated over all variables except $s$ and $s'$. Then, if $U$ and $L$ are sufficiently smooth and $(\nu, \nu) = 0$, the asymptotically leading term in $E^\gamma$ is given by

$$E^\gamma(\lambda\nu) \sim \frac{\pi^2}{\lambda^2|\vec{\nu}|^2} \int_0^\infty d\rho' \int_0^\infty d\rho \, U^{\alpha'\alpha}(-\rho'\hat{\nu}, -\rho\hat{\nu}) \, g_{\alpha'\beta'} g_{\alpha\beta} L^{\beta'\beta}(-\rho'\hat{\nu}, -\rho\hat{\nu}) \ , \tag{15.86}$$

with $\hat{\nu} = \frac{\nu}{|\vec{\nu}|}$. This is shown analogously to the corresponding result (15.6) in the electron case, using Lemma 14.3. $\alpha, \cdots, \beta'$, are the Minkowski indices of the cut lines. The decrease is faster if $(\nu, \nu) \neq 0$. The interpretation of this result is similar to the one given for electrons. The $\lambda^{-2}$ decrease is explained by geometry. It agrees with that of a classical free particle of mass zero. The cross lines in $U$ describe unobserved photons that have been knocked out of the probe by the impinging photon, those in $L$ unobserved photons that have been produced together with the observed one in the preparation procedure. And $U$ and $L$ describe the efficiency and other characteristics of the apparatus.

Two points need closer examination. First, the expression (15.86) is not entirely without IR problems. Second, $U$ and $L$ enter not only at the fixed momenta $s = s' = -\hat{\nu}$ like in the electron case. There occurs an integral over multiples of this value, and this destroys factorization of the expression.

We address the IR problem first. From (11.28) we may expect $U$ to show a weak singularity at $(s + s')^2 = 0$, which condition is satisfied for $s = -\rho\hat{\nu}$, $s' = -\rho'\hat{\nu}$. Equation (11.28) was derived for 1-sector graphs, but the conclusion remains valid for the 2-sector graphs contributing to $U$. These singularities are due to 2-line cuts running parallel to the $s$–$s'$-cut. Such a cut is always possible through the $q$- and $q'$-lines if they are not identical to the $s$- and $s'$-lines. But there may be other possibilities, e.g. a cut through the $t$-$t'$-lines in the graph shown in Fig. 15.10. From Sect. 11.3 we know that weak singularities of $U$ at $(s+s')^2 = 0$ arise from points where $t$ and $t'$ are lightlike and parallel, or $q$ and $-q'$ are lightlike and parallel, or both. In the former cases the singularity is a logarithm, in the latter case a square of a logarithm. Graphs with more possible cut positions lead to higher but still weak singularities.

For example, take the singularity caused by the $t$–$t'$-propagators. We are interested in $s$ and $s'$ parallel to $-\nu$. $t$ and $t'$ must also be parallel and lightlike. And we have $t^0 < 0$ because $q^0$ is positive and the momenta $k_1, \ldots, k_n$, of the cross lines between the $s$–$s'$ and $t$–$t'$ cuts are positive lightlike. From momentum conservation we find that these requirements are only met if all the momenta $-s, -t, k_1, \ldots, k_n$, are parallel, positive lightlike, vectors. This helps us in showing that the suspected weak singularity in $U$ is in fact not present. Consider the lower blob in the right-hand sector of Fig. 15.10. Its external variables $k_1, \ldots, k_n, -t, s$, are renamed $p_1, \ldots, p_N$, with $p_N = s$. All of them flow into the blob. $N$ is even by Furry's theorem. Let $\mu_1, \ldots, \mu_N$ be the $A^\mu$-indices of the external lines. Then it can be shown (see at the end of the section for a sketch of the proof) that the blob function $B(p_1, \ldots, p_N)$ can be expanded into a finite series

$$B(p_1, \ldots, p_N) = \sum_A \mathcal{P}_A^{\mu_1 \cdots \mu_N}(p_1, \ldots, p_N) I_A(p_1, \ldots, p_N), \qquad (15.87)$$

where the $\mathcal{P}_A$ are covariant, homogeneous, polynomials formed with the com-

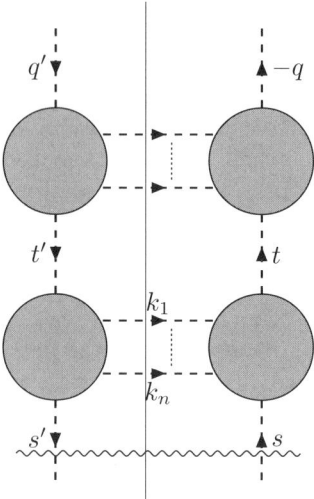

**Fig. 15.10.** IR singularities in $U(s,s')$

ponents $p_i^{\mu_j}$ of the external variables and the metric tensor $g^{\mu_i\mu_j}$, and the $I_A$ are invariant functions with no stronger singularities than $B$ itself. We assume $B$ to have been summed over all graphs of a given order, so that the WT identities

$$p_{i,\mu_i} \mathcal{P}_A^{\cdots\mu_i\cdots} = 0 \qquad (15.88)$$

hold. The same is assumed for the other blobs in the graph, including the smallest subgraph of the $L$-part with only photonic external lines, joined by the $s$-line. Consider a term of (15.87) with $\mathcal{P}_A$ containing at least one factor $p_i^{\mu_j}$. If $j = i$, we contract $p_i^{\mu_i}$ with the next blob along the $p_i$-line, which is annihilated because of (15.88). If the $t$-line is actually the $q$-line and $p_i = -q$, there is no such blob. But in this case $q^\mu$ gets contracted with

$$(g_{\mu h} q_0 - g_{\mu 0} q_h) \tilde{f}^h(q) = q_0 g_{\mu h} \tilde{f}^h(q)$$

because of (15.82) and (15.83), yielding $q_0 q_h \tilde{f}^h(q) = 0$. The same happens if $p_i = s$ and the $s$-line is the $p$-line of the full graph $G$. If $j \neq i$ we write

$$p_i^{\mu_j} = \left(p_i^{\mu_j} - \frac{|p_i|}{|p_j|} p_j^{\mu_j}\right) + \frac{|p_i|}{|p_j|} p_j^{\mu_j} . \qquad (15.89)$$

The last term does not contribute for the reason just described. The bracket vanishes if $p_i$ and $p_j$ are parallel, which they are at the critical points, and this kills the unwanted weak singularity.

There remain the terms in the $A$-sum whose $\mathcal{P}_A$'s consist exclusively of $g^{\alpha\beta}$'s. But in Sect. 10.2 it has been found (see footnote 4 of Chap. 10) that $B$, after integrating away its factor $\delta^4(\sum p_i)$, vanishes at the origin. This must then be the case for the sum over the all-$g$-terms, because for the other $A$ it is achieved by the vanishing of $\mathcal{P}_A$ at the origin. We are interested in the behaviour of $B$ at the points where the independent variables $p_1, \ldots, p_{N-1}$, are all positive lightlike and parallel. For any such point $P$ and an arbitrarily small neighbourhood $\mathcal{N}_0$ of the origin there exist a small neighbourhood $\mathcal{N}$ of $P$ and a Lorentz boost $\Lambda \in L$ such that $\Lambda \mathcal{N} \subset \mathcal{N}_0$. Hence the considered terms are arbitrarily small at the critical points, i.e. they vanish there and thus are not singular. This establishes that $U(s, s')$ is free of dangerous IR singularities. The same consideration applies to $L(s, s')$.

The physical interpretation of the $\rho$–$\rho'$-integral in (15.86) will become important in the following chapters. Consider first the probe factor $U(s, s')$ at $s = -\rho \hat{\nu}$, $s' = -\rho' \hat{\nu}$. The variables $q$ and $q'$, defined as before, are restricted to the supports of $\tilde{f}(q)$ and $\tilde{f}(-q')$. Let $K$ be the total momentum flowing through the cross lines of $U$ from the $T^-$- to the $T^+$-sector. In contrast to the electron case, $K$ need not be small. But it is approximately parallel to $\nu$. We have $-q = s + K$, $q' = s' + K$, hence $s - s' = -q - q'$, which implies that $|s - s'| = |q + q'|$ is small because of the small support of $\tilde{f}$. Therefore $(\rho - \rho')\hat{\nu}$ must be small: only values $\rho \sim \rho'$ contribute to the integral. The admissible size of $|\rho - \rho'|$ can be made arbitrarily small by the choice of $supp\,\tilde{f}$. This means that in factors of the integrand which depend only weakly on $\rho$, $\rho'$, we may set $\rho = \rho'$. This will be used later on. But even so, values of $s \sim s'$ which do not lie in $supp\,\tilde{f}$ but are parallel to such vectors, are asymptotically relevant. Thus, if $s$ is still considered to be the 4-momentum of the photon triggering the probe, then the set of momenta to which the probe is sensitive is no longer determined by the support of $\tilde{f}$. But we have interpreted the cross lines as describing photons that are produced in the probe by its interaction with the impinging particle. And we note that in realistic photon detectors these photons are reabsorbed before leaving the detector, and the total amount $-s^0 = q^0 + K^0$ of deposited energy is measured. Hence we may consider the contribution to (15.86) coming from a given small $\rho$–$\rho'$-interval as being due to photons of energy $-s^0 \sim \rho \hat{\nu}^0$.

For the preparation factor $L(s, s')$ we propose a different interpretation. Like in the $U$-case we find, assuming a small $\tilde{g}$-support, that in order to be relevant $s$ and $s'$ must be roughly equal and parallel to a lightlike vector in $supp\,\tilde{g}$, but are not necessarily themselves contained in this support. This means that the probe can only be triggered if $supp\,\tilde{f}$ and $supp\,\tilde{g}$ contain lightlike vectors which are multiples of each other. But the intersection of the two supports may be empty. Again, the total momentum $K$ of the cross lines in $L$ need not be small. It may be large if it is roughly parallel to $s$. We interpret these facts as follows. The initial state $\Phi_g^\gamma$ is *not* a 1-photon state. It may contain several hard photons with approximately but not exactly parallel 4-momenta, so that they become spatially separated after a sufficiently long time. That is, the state may trigger a coincidence arrangement of several spatially separated probes responsive at the same time. But for our purposes it may nevertheless serve as a preparation procedure for a 1-photon state. If our single probe is triggered and the measured energy is $-s^0$, we say that a photon of momentum $-s$ was present, possibly besides other photons that have not been observed and are of no further interest to us.

A final note: the factor $U^{\alpha\alpha'}$ in (15.86) is calculated from a 4-photon function with physical fields $\tilde{F}_{0\gamma}(q$ or $q')$ and unphysical fields $\tilde{A}^\alpha(s)$, $\tilde{A}^{\alpha'}(-s')$. Like in the electron case, the latter can be replaced by the physical fields $\tilde{F}^{0\alpha}(s), \tilde{F}^{0\alpha'}(-s')$. This is achieved by replacing the factor $g_{\alpha\beta}$ in (15.86) by

$$-\frac{i}{|\vec{s}|}\left(-is^0\delta^\nu_\alpha + is^\nu\delta^0_\alpha\right)g_{\nu\beta} = -\frac{1}{|\vec{s}|}\left(s^0 g_{\alpha\beta} - \delta^0_\alpha s_\beta\right), \quad (15.90)$$

and equivalently for $g_{\alpha'\beta'}$. The $s_\beta$-term in (15.90) annihilates $L^{\beta\beta'}$ for the reason explained before in connection with the IR problem. The quotient $-\frac{s^0}{|\vec{s}|}$ is $=1$ on the negative mass shell: the $g_{\alpha\beta}$-term in (15.90) reproduces the original one. The new $U$ is then constructed from a 4-$F$-function multiplied by $(|\vec{s}||\vec{s'}|)^{-1}$. The same alteration can be applied to $L^{\beta\beta'}$. Equation (15.86) remains valid after these changes.

## Derivation of the Decomposition (15.87)

Though intuitively plausible, the decomposition (15.87) is by no means easy to derive, since one has to beware of producing "kinematical singularities", i.e. a decomposition whose coefficients $I_A$ are more singular than $B$, the unwanted singularities cancelling one another in the sum. A derivation will therefore be indicated, without giving all the details, which avoids this pitfall.[7] What counts for us is only the existence of the decomposition and the strength of the local singularities of the coefficients $I_A$, not their explicit form. To keep the notation simple, constant factors will therefore be discarded. The symbol $\doteq$ is again used to denote equality up to an irrelevant constant factor.

We start with some useful mathematical formulae. The first is

$$\int_{-\infty}^\infty dx\, x^n e^{\pm iax^2} = \sqrt{2}^n e^{\pm i\frac{n\pi}{4}} \int_{-\infty}^\infty d\rho\, \rho^n e^{-2a\rho^2}$$

$$\doteq \begin{cases} 0 & \text{for } n \text{ odd} \\ a^{-\frac{n+1}{2}} & \text{for } n \text{ even}, \end{cases} \quad (15.91)$$

for $a$ a positive real number and $n$ a non-negative integer. This is seen by shifting the integration path from the real axis to the complex line $\rho(1\pm i)$, $-\infty < \rho < \infty$. Let now $k$ be a 4-vector. Then the integral $\int dk\, e^{iak^2}$ factorizes into four integrals of the form (15.91), one with a positive sign in the exponent, three with a negative sign, and we obtain

$$\int dk\, e^{iak^2} \doteq -\frac{i}{a^2}. \quad (15.92)$$

---

[7] We follow the procedure outlined in Sect. 6-2-3 of [IZ] for a scalar theory, generalizing it to our covariant case. Some of the missing details can be found in this reference.

Obviously, we have

$$\int dk\, k_\mu e^{iak^2} = 0 \ . \tag{15.93}$$

The integral

$$J_{\mu\nu} = \int dk\, k_\mu k_\nu e^{iak^2} \tag{15.94}$$

vanishes if $\mu \neq \nu$. For $\mu = \nu$ we have again a product of four 1-dimensional integrals of the form (15.91). Compared to (15.92) we get an additional factor $i/a$ if $\mu = 0$, $-i/a$ if $\mu$ is spatial. The result is

$$J_{\mu\nu} \doteq a^{-3} g_{\mu\nu} \ . \tag{15.95}$$

This result generalizes easily to moments of higher order:

$$J_{\mu_1\cdots\mu_n} = \int dk\, \prod_{j=1}^n k_{\mu_j} e^{iak^2} \tag{15.96}$$

vanishes if $n$ is odd, and

$$J_{\mu_1\cdots\mu_n} \doteq a^{-(2+\frac{n}{2})} \sum_\alpha \prod g_{\mu_\alpha \nu_\alpha} \tag{15.97}$$

for even $n$, the sum extending over all partitions of the index set $\{\mu_1,\ldots,\mu_n\}$, into pairs $(\mu_\alpha, \nu_\alpha)$.

Let $G$ be a 1PI graph with $N$ external photon variables $p_1,\ldots,p_N$, and $A^\mu$-indices $\mu_1,\cdots,\mu_N$, but with no external fermion variables. Let $L$ be the number of lines, $V$ of vertices, of $G$. We use the non-reduced graph rules in which each line $\ell$ carries an independent variable $k_\ell$ and each vertex $v$ carries a $\delta^4$-factor ensuring momentum conservation. Fermion lines occur only in closed loops. Taking the trace of the numerator product of such a loop yields a covariant polynomial $\mathcal{P}^{\nu_1\cdots\nu_r}$ formed with components $k_\ell^{\nu_i}$ of the loop variables and with $g^{\nu_i\nu_j}$'s. The $\nu_i$ are the indices of the loop vertices. Doing this for all fermion loops and contracting with the factors $g_{\alpha\beta}$ of the photon propagators, we obtain a covariant polynomial $\mathcal{P}^{\mu_1\cdots\mu_N}(k_\ell)$ formed with $g^{\mu_i\mu_j}$'s and $k_\ell^{\mu_i}$'s. Let $M = \prod_\ell M_\ell(k_\ell)$ be a monomial term in this polynomial. $M_\ell$ is of degree $|M_\ell| = 0$ or $1$. Factors $g^{\mu_i\mu_j}$ in $M$ can be drawn outside the $k$-integral and will not be shown explicitly in the sequel. For the $k_\ell$-propagator we use the parametric representation

$$\frac{i}{k_\ell^2 - m_\ell^2 + i\varepsilon} = \int_0^\infty d\alpha_\ell\, e^{i\alpha_\ell(k_\ell^2 - m_\ell^2 + i\varepsilon)} \ . \tag{15.98}$$

The $\delta$-factor of the vertex $v$ is written as

$$\delta^4\left(\sum \pm k_\ell\right) \doteq \int dz_v e^{i\left(\sum \pm k_\ell, z_v\right)} \tag{15.99}$$

where $z_v$ is the $x$-space variable associated with $v$ and the $k_\ell$ are the momenta of the lines incident at $v$, counted positive if $\ell$ points into $v$. Define

$$\varepsilon_{\ell v} = \begin{cases} 1 & \text{if } \ell \text{ points into } v \\ -1 & \text{if } \ell \text{ points away from } v \\ 0 & \text{if } \ell \text{ is not incident at } v \end{cases} \quad (15.100)$$

The contribution of the line $\ell$ to the graph integral is

$$C_\ell \doteq \int dk_\ell \, M_\ell(k_\ell) \int_0^\infty d\alpha_\ell \, e^{i\alpha_\ell \left(k_\ell^2 - m_\ell^2 + i\varepsilon\right)} e^{i \sum_v \varepsilon_{\ell v}(k_\ell, z_v)} . \quad (15.101)$$

Exchanging the two integrations, which is allowed in a context of distributions, and substituting

$$u = k_\ell + \frac{1}{2\alpha_\ell} \sum_v \varepsilon_{\ell v} z_v$$

for $k_\ell$, we obtain

$$C_\ell \doteq \int_0^\infty d\alpha_\ell \, e^{-i\alpha_\ell(m_\ell^2 - i\varepsilon)} \exp\left[-\frac{i}{4\alpha_\ell}\left(\sum_v \varepsilon_{\ell v} z_v\right)^2\right]$$

$$\times M_\ell\left(\frac{1}{2\alpha_\ell}\sum_v \varepsilon_{\ell v} z_v\right) \int du \, e^{i\alpha_\ell u^2} . \quad (15.102)$$

The $u$-integral is found by (15.92) to be $\doteq (\alpha_\ell)^{-2}$. Consider the $z_v$ as components of a $V$-dimensional supervector[8] $z'$ and define a $V \times V$ matrix $D^\ell$ by

$$D^\ell_{v_1 v_2} = \frac{1}{\alpha_\ell} \varepsilon_{\ell v_1} \varepsilon_{\ell v_2} . \quad (15.103)$$

Notice that $D^\ell$ is symmetric. $(z', D^\ell z')$ stands for $\sum_{v_1, v_2} D^\ell_{v_1 v_2}(z_{v_1}, z_{v_2})$, and similarly in the future. $C_\ell$ can then be written

$$C_\ell \doteq M_\ell\left(\sum_v \varepsilon_{\ell v} z_v\right) \int_0^\infty d\alpha_\ell \, \frac{e^{-i\alpha_\ell(m_\ell^2 - i\varepsilon)}}{\alpha_\ell^{2+|M_\ell|}} e^{-\frac{1}{4}(z', D^\ell z')} . \quad (15.104)$$

The argument of $M_\ell$ is the difference of the $z_v$ of the end points of the line $\ell$.

We define the polynomial $M_1(z_1, \ldots, z_V)$ of order $|M_1| = \sum_\ell |M_\ell|$ by $M_1 = \prod_\ell M_\ell(\sum \varepsilon_{\ell v} z_v)$, the $V \times V$ matrix $D'$ by $D' = \sum_\ell D^\ell$, and $p_v$ as the sum of external momenta flowing into the vertex $v$, so that $p_v = 0$ if $v$ is an internal point of $G$. The full $G$-integral is a sum of terms of the form

$$J \doteq \int \prod_v dz_v \, e^{i \sum_v z_v p_v} M_1(z')$$

$$\times \prod_\ell \left\{\int d\alpha_\ell \, \frac{e^{-i\alpha_\ell(m_\ell^2 - i\varepsilon)}}{\alpha_\ell^{2+|M_\ell|}}\right\} e^{-\frac{1}{4}(z', D'z')} , \quad (15.105)$$

---

[8] Under a supervector we understand a vector whose components are themselves 4-vectors.

omitting factors $g^{\mu_i \mu_j}$. Both $M_1$ and $(z', D'z')$ are translation invariant because of $\sum_v \varepsilon_{\ell v} = 0$. We introduce $Z = z_V$ and $y_v = z_v - Z$ for $v = 1, \ldots, V-1$, as new variables of integration. $Z$ occurs only in the first exponential, and the $Z$-integration yields the familiar factor $\delta^4(\sum p_i)$. Let $y$ be the $(V-1)$-dimensional supervector with components $y_v$, $p$ the supervector with components $p_1, \ldots, p_{V-1}$, and $D$ the $(V-1) \times (V-1)$ matrix obtained from $D'$ by deleting its $V^{\text{th}}$ line and column. We find

$$J \doteq \delta^4\left(\sum p_i\right) \int \prod_v dy_v\, e^{i(y,p)} M_2(y)$$
$$\times \prod_\ell \left\{ \int_0^\infty d\alpha_\ell \frac{e^{-i\alpha_\ell(m_\ell^2 - i\varepsilon)}}{\alpha_\ell^{2+|M_\ell|}} e^{-\frac{1}{4}(y,Dy)} \right\} \qquad (15.106)$$

with $M_2$ a polynomial of degree $|M_2| = |M_1|$. The determinant $|D|$ of $D$ is a homogeneous polynomial of degree $V-1$ in the variables $\frac{1}{\alpha_\ell}$. It can be calculated to be

$$|D| = \sum_T \prod_{\ell \in T} \frac{1}{\alpha_\ell}\,, \qquad (15.107)$$

the sum extending over all full trees $T$ of $G$, i.e. the connected tree subgraphs of $G$ containing every vertex. Hence $D$ is regular for positive $\alpha_\ell$. The quadratic form $(y, Dy)$ is diagonalizable with only non-vanishing diagonal elements, and the $\alpha$-integral in (15.106) would define a distribution in $z$ if it were not for the singular behaviour of the integrand at $\alpha_\ell = 0$ for one or several $\ell$. This problem is actually the UV problem, which is one of the "details" that will be ignored. Renormalization in the $\alpha$-parametric representation has been treated in Hepp's contribution [Hp 66] to the BPHZ method. At first integrating $\alpha_\ell$ only over $0 < r \leq \alpha_\ell < \infty$, Hepp showed that the limit $r \to 0$ exists after introducing the familiar subtraction terms, which are of the same general structure as the terms discussed here, being described by reduced graphs constructed from $G$ by fusing R-parts into points. Renormalization is a problem concerned with large momenta and does not affect the local singularities in a significant way. We therefore proceed by considering only strictly positive $\alpha_\ell$'s, not worrying about what happens when one or several of them approach zero. We can then again exchange the $y$- and $\alpha$-integrations in (15.106) like we did in (15.101). In the $y$-integral we replace $y$ by $x = y - 2D^{-1}p$ as integration variables, which turns $J$ into a sum of terms of the form

$$J_1 \doteq \int \prod_\ell \left\{ d\alpha_\ell \frac{e^{-i\alpha_\ell(m_\ell^2 - i\varepsilon)}}{\alpha_\ell^{2+|M_\ell|}} \right\} M_3(D^{-1}p)$$
$$\times e^{i(p, D^{-1}p)} \int \prod dx_\nu\, M_4(x)\, e^{-\frac{i}{4}(x, Dx)}\,, \qquad (15.108)$$

where $M_3$ and $M_4$ are monomials with $|M_3| + |M_4| = |M_1|$. Diagonalization of $(x, Dx)$ turns the $x$-integral into a product of integrals of the form (15.96).

$J_1$ becomes a sum of terms of the following form, up to some new factors $g^{\mu_i \mu_j}$ with the indices of $M_4$:

$$J_2 \doteq \int \prod_\ell \left\{ d\alpha_\ell \frac{e^{-i\alpha_\ell(m_\ell^2 - i\varepsilon)}}{\alpha_\ell^{2+|M_\ell|}} \right\} \frac{1}{|D|^2} h_4(\alpha) \, M_3(D^{-1}p) \, e^{i(p, D^{-1}p)} \, . \tag{15.109}$$

$h_4$ is a regular function of $\alpha = \{\alpha_1, \ldots, \alpha_\ell\}$ which is homogeneous of order $\frac{1}{2}|M_4|$. $M_3(D^{-1}p)$ is in $p$ a homogeneous polynomial of order $|M_3|$ with coefficients which are homogeneous functions of $\alpha$ of order 1. Taking one of the terms of $M_3$, drawing its $p$-product in front of the integral while leaving the coefficient inside, and combining the $p$-monomial with the $g$-product already present, we obtain a covariant polynomial with the Minkowski indices $\mu_1, \ldots, \mu_N$, multiplied with an expression of the form

$$J_3 \doteq \int \prod_\ell \left\{ d\alpha_\ell \cdots \right\} \frac{1}{|D|^2} h_4(\alpha) \, h_3(\alpha) \, e^{i(p, D^{-1}p)} \tag{15.110}$$

with $h_3$ homogeneous of order $|M_3|$. $J_3$ depends on $p$ only through the invariant expression $(p, D^{-1}p)$: it is Lorentz invariant as desired. Thus we have arrived at a decomposition of the form (15.87).

All $\alpha$-dependences in the integrand of $J_3$ are regular for $\alpha_\ell > 0$. The matrix elements $D^{-1}_{v_1 v_2}(\alpha)$ are homogeneous of order 1. Diagonalization of $D^{-1}$ gives $J_3$ the form of a Fourier integral. The strengths of its singularities in the external variables $p$ are determined by the behaviour of the integrand

$$\left[ \prod_\ell \frac{e^{-i\alpha_\ell m_\ell^2}}{|\alpha^\ell|^2} \frac{1}{|D|^2} \right] \cdot \left[ \frac{h_4(\alpha) \, h_3(\alpha)}{|M_1|} \right] \tag{15.111}$$

for large $\alpha$. The second square bracket $B_2$ is $= 1$ if all $|M_i| = 0$. In this case, $J_3$ is the contribution of the graph $G$ to a $N$-point function in a QED-like theory with scalar electrons and photons. To this graph the results of Sect. 11.3 are applicable and tell us that at most weak singularities are present if $G$ is 1PI. In general, $B_2$ is homogeneous in $\alpha$ of order $|M_3| + \frac{1}{2}|M_4| - |M_1| = -\frac{1}{2}|M_4|$. Thus this factor decreases like $|\alpha|^{-\frac{1}{2}|M_4|}$ if $\alpha$ tend to infinity in a generic direction. If this were true in all directions, or if at least $B_2$ remained bounded in all directions, then its presence could only weaken the singularities of $J_3$, not strengthen them. At first sight, the behaviour of $B_2$ might be less favorable if some $\alpha_\ell = c_\ell > 0$ are kept constant and only the remaining ones tend to $\infty$. But even in these directions $B_2$ is homogeneous and bounded. Let $\mathcal{L}$ be the set of varying $\alpha_\ell$'s. We start from the graph $G$ but replace the singular propagator factors of the lines $\ell \notin \mathcal{L}$ by the regular factors $e^{ic_\ell(k_\ell^2 - m_\ell^2)}$. The $p$-singularities of this modified graph are entirely due to the $\mathcal{L}$-lines, and they are clearly not worse than those of the original graph. But for this critical $\mathcal{L}$-subgraph we can derive a covariant decomposition of the form (15.87) generalized to the case that the subgraph may contain fermionic external variables, and apply to it the former considerations about the large $\alpha$-behaviour. The generalization causes no problems, nor does the fact that the subgraph may be 1P-reducible or even disconnected. The possibly dangerous subintegrals of (15.110) are therefore actually harmless.

# Chapter 16

# Reactions

In this chapter, interactions between particles are studied and described with the help of the probes introduced in the preceding chapters. It is shown how this leads directly to expressions for the observable reaction rates, without detour through an $S$-matrix. In the first section the basic ideas are explained for a simple model without IR problems, to wit, a theory of a scalar hermitian field whose stable particles are massive and have spin 0. In the following three sections these ideas are generalized to QED. The result is surprising: it does not agree with the customary picture of a particle as a robust object with an unquestionable individuality. It seems that the notion of particles is more subtle than generally believed, in particular that charged particles cannot be defined meaningfully in detachment from an observational situation. This implies that also the notion of reaction cross section takes on a new, intricate meaning. These problems will be discussed in the final Chap. 17.

Our results are not proved to be wrong by differing from those of the conventional approach. But, clearly, they must be viewed with circumspection. Many details of the rather involved calculations would bear closer scrutiny; the derivation does not claim to be fully rigorous.

## 16.1 Massive Scalar Fields

We consider a theory of a hermitian, scalar, interacting field $\varphi(x)$, whose particles as defined in Chap. 15 are of mass $m > 0$ and spin 0. To fix the ideas, we may consider the $\varphi_4^4$ model. The exact form of the interaction is, however, not essential. Nor is the use of perturbation theory essential. Our results hold for an exact $\varphi$-theory, provided it satisfies asymptotic completeness, and the singularities of its $\mathcal{W}$-functions are of the form suggested by perturbation theory. But being obliged to use perturbation theory in the case of QED, we may as well restrict ourselves to a perturbative approach already in the present section.

In actual experiments the incoming state is practically always a 2-particle state, since it is not feasible to bring more than two particles so close together at one time that they are simultaneously within the range of interaction. We start therefore from the "2-particle state"

$$\Phi = \varphi(g_1, b_1)^* \, \varphi(g_2, b_2)^* \, \Omega \qquad (16.1)$$

defined as in (14.28), but with $\varphi$ an interacting field. Remember that this state describes two particles prepared in neighbourhoods of the space-time points $b_1$, $b_2$, with momentum distributions $|\tilde{g}_i(p)|^2$. We assume $b_1^0 = b_2^0$, $\vec{b}_1 \neq \vec{b}_2$. It is frequently convenient to choose coordinates such that $b_i^0 = 0$, or even $b_2 = 0$. The probe operator $D(a)$ is defined by (14.20) with $\varphi$ interacting. We consider the typical case that $\Phi$ is explored with the help of two probes placed at $a_1$, $a_2$, with $a_1^0 = a_2^0 > b_i^0$, $\vec{a}_1 \neq \vec{a}_2$. The generalization to other numbers of probes is straightforward. The assumptions $b_1^0 = b_2^0$ and $a_1^0 = a_2^0$ are only made to simplify notations. They are not essential. It would suffice to assume $a_j^0 > b_i^0$ and $(a_1 - a_2)^2 < 0$, $(b_1 - b_2)^2 < 0$.

We introduce a scale factor $\lambda$ by defining

$$a_j = \lambda \nu_j \, , \quad b_i = \lambda \beta_i \, , \qquad (16.2)$$

and we look for the asymptotic behaviour of

$$E(\lambda) = \bigl(\Phi, \, D(a_1) \, D(a_2) \, \Phi\bigr)$$
$$\sim \Bigl(\Omega, \prod_i \varphi(g_{i,\lambda\beta_i}) \prod_j \varphi(f_{j,\lambda\nu_j})^* \prod_j \varphi(f_{j,\lambda\nu_j}) \prod_i \varphi(g_{i,\lambda\beta_i})^* \Omega\Bigr) \quad (16.3)$$

for $\lambda \to \infty$ with $\nu_i$, $\beta_j$, being kept fixed. Like in (14.35) we may replace the ordinary product of the first four fields in this expression by their $T^-$-product, that of the last four fields by their $T^+$-product, without changing the asymptotically leading term of $E(\lambda)$: the argument applied to (14.26) is easily generalized to more than two local fields, even if they are interacting.

In analogy to (15.3) we have

$$E(\lambda) = \int \prod_{i=1}^{2} \left\{ dp_i \, dp'_i \, e^{-i\lambda(\beta_i, p_i + p'_i)} \tilde{g}_i(-p_i) \, \overline{\tilde{g}_i(p'_i)} \right\}$$
$$\times \prod_{j=1}^{2} \left\{ dq_j \, dq'_j \, e^{-i\lambda(\nu_j, q_j + q'_j)} \tilde{f}_j(q_j) \, \overline{\tilde{f}_j(-q'_j)} \right\}$$
$$\times \tilde{\mathcal{W}}(p'_1, p'_2, q'_1, q'_2, -|q_1, q_2, p_1, p_2, +) \, . \qquad (16.4)$$

The graphs contributing to $\tilde{\mathcal{W}}$ are of the form shown in Fig. 16.1. Like in the 1-particle case, the asymptotically leading terms in $E$ are determined by the strongest singularities of $\tilde{\mathcal{W}}$ in its external variables.

We start with the important special case of disconnected graphs consisting of four 2-point components. Their leading singularities are poles of first

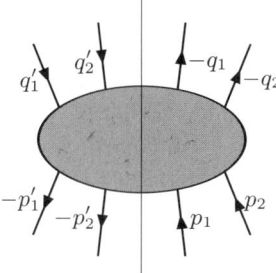

**Fig. 16.1.** Graph contributing to $\tilde{\mathcal{W}}(p'_1,\ldots,-|\ldots,p_2,+)$.
The sign conventions correspond to those of QED

order at the mass shell. In massive theories it is customary to define the field normalization by demanding these poles to have the same residue as in the free case. Then the asymptotically leading term agrees with that of the free case discussed in Sect. 14.2. In analogy to (14.34) we find the possibly dominant terms with a $\lambda^{-6}$-decrease,

$$E(\lambda) \sim E_1\big(\lambda(\nu_1 - \beta_1), f_1, g_1\big) E_1\big(\lambda(\nu_2 - \beta_2), f_2, g_2\big)$$
$$+ E_1\big(\lambda(\nu_1 - \beta_2), f_1, g_2\big) E_1\big(\lambda(\nu_2 - \beta_1), f_2, g_1\big) , \quad (16.5)$$

with $E_1$ the same function as in (14.34). The first term is present in the "forward direction": the supports of $\tilde{f}_1$ and $\tilde{g}_1$ overlap and $\nu_1 - \beta_1$ is parallel to a vector in their intersection, and the same holds for $\tilde{f}_2$, $\tilde{g}_2$, and $\nu_2 - \beta_2$. The second term is dominant if the same condition holds with the two probes interchanged. The decrease is fast if neither of these two conditions is satisfied. This contribution to $E(\lambda)$ corresponds to the case that the two particles propagate freely without taking any notice of each other, e.g. because they never get close enough for one to feel the presence of the other one. In the later analysis of more complicated graphs it will be found that none of them leads to a slower or equally slow decrease in $E(\lambda)$. The free-particle term just described is truly dominant in the forward directions. This justifies the interpretation of $\Phi$ as a 2-particle state: if the geometrical arrangement is such that in the asymptotic classical interpretation no interaction is possible for geometrical reasons, then the state behaves as a state of two free particles. This is the case, in particular, if $\nu_i^0 - \beta_j^0$ is so small that the incoming particles have not sufficient time to meet before hitting the probes.

For describing reactions between the two particles we must consider non-forward directions. Henceforth it will be assumed that neither of the two forward conditions just mentioned is satisfied, in particular that the supports of the functions $\tilde{g}_i$, $\tilde{f}_j$, are mutually non-overlapping. Inclusive reactions (not all emerging particles are detected) are described by connected graphs, exclusive reactions (i.e. elastic scattering in our two-probe example) by disconnected graphs both of whose sectors are connected. It will be found that there is no essential difference between the two cases. We consider graphs of the form of Fig. 16.1, possibly without cross lines. But both sectors are connected. The singularities of such a $\tilde{\mathcal{W}}$-graph can be found with the methods applied in Sect. 11.3 to QED. Of course, there is no IR problem. Singularities may occur where a partial sum $P$ of external variables of $\tilde{\mathcal{W}}$ lies on a threshold $P^2 = n^2 m^2$, $n$ a positive integer. The threshold singularities with $n > 2$ are too weak to be of relevance to us. They correspond to the cuts through more than two lines of Chap. 15. 2-particle threshold singularities with $n = 2$ may exist in graphs which can be decomposed by cutting two lines. A singularity is present if the support of the integrand in (16.4) contains points where the two cut momenta coincide and lie on the mass shell. It can be shown by elementary but somewhat lengthy arguments that only two types of such 2-particle thresholds can occur under our assumptions. The first type occurs in graphs with a sector of the form shown in Fig. 16.2 for the example of the $T^-$-sector. Here there exists a threshold singularity at $(p'_1 + p'_2 - q'_2)^2 = (q'_1 + k)^2 = 4m^2$, provided this condition is compatible with the supports of $\tilde{f}_j$ and $\tilde{g}_i$. Anticipating the results of the subsequent analysis it may be said that this situation corresponds physically to a reaction in which three particles are produced, two of them with identical 4-momenta. As a result, these two "particles" tend to stay together for a long time, thus retarding the time when the asymptotic particle interpretation becomes appropriate. It is then not surprising that such a singularity would complicate our discussion. We will therefore assume that the supports of $\tilde{g}_i$, $\tilde{f}_j$, are such that this type of singularity is not present.

The second type of possibly dangerous thresholds is present at $(p_1 + p_2 + q_1 + q_2)^2 = (p'_1 + p'_2 + q'_1 + q'_2)^2 = 4m^2$ in graphs with two cross

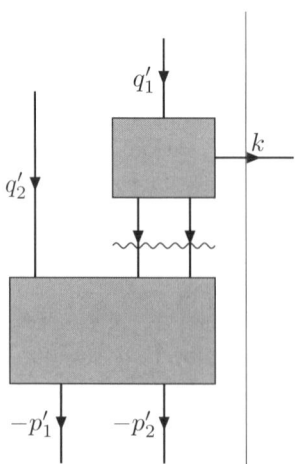

**Fig. 16.2.** Sector leading to a 2-particle threshold. The wavy line cuts the lines responsible for the singularity. The momenta of all the lines shown lie on or near the positive mass shell, flowing in the direction of the arrow

lines. It is due to coinciding cross momenta. But this time both corresponding "particles" are unobserved, so that their dubious claim to particle status is of little consequence. As will be seen, there is no need to exclude this case from consideration.

Cuts involving internal lines of a given graph, like the ones encountered in the 1-particle case of Sect. 15.2, could only contribute to singularities in the relevant variables $\sum(\beta_i, p_i + p'_i)$ or $\sum(\nu_j, q_j + q'_j)$, if they are of the 1-particle form shown in Fig. 15.2. That is, the cut lines would have to occur as pairs $s_\ell, s'_\ell$, each pair being connected on one side to a subgraph whose only other external lines are a pair of $q_j$–$q'_j$-lines or of $p_i$–$p'_i$-lines belonging to the same probe or source. No other combinations are acceptable. But, assuming sufficiently small supports of $\tilde{g}_i, \tilde{f}_j$, such 1-particle $U$- or $L$-parts are impossible at threshold in a massive theory due to momentum conservation. Thus the only asymptotically relevant singularities of $\tilde{\mathcal{W}}$ are the poles of its external lines.

Our task is to find the asymptotic behaviour of $E(\lambda)$. Let $G$ be a $\tilde{\mathcal{W}}$-graph which is by the preceding considerations expected to contribute to the leading term. We introduce some convenient notations. The external variables are renamed

$$r_1 = q_1, \quad r_2 = q_2, \quad r_3 = p_1, \quad r_4 = p_2,$$
$$r'_1 = -q'_1, \quad r'_2 = -q'_2, \quad r'_3 = -p'_1, \quad r'_4 = -p'_2. \quad (16.6)$$

Note that in the expression (16.4) $r_1, r_2, r'_1, r'_2$ are close to the positive mass shell, the remaining variables close to the negative mass shell. Also we define

$$\rho_1 = \nu_1, \quad \rho_2 = \nu_2, \quad \rho_3 = \beta_1, \quad \rho_4 = \beta_2, \quad (16.7)$$

and for the time being we set $\rho_4 = 0$. The contribution of $G$ to $\tilde{\mathcal{W}}$ we write

$$\tilde{\mathcal{W}}_G(-r'_\alpha, -|r_\beta, +) = \delta^4(C + \sum_\beta r_\beta) \prod_{\beta=1}^4 \frac{i}{2\pi(r_\beta^2 - m^2 + i\varepsilon)}$$
$$\times \prod_{\alpha=1}^4 \frac{-i}{2\pi(r'_\alpha{}^2 - m^2 - i\varepsilon)} w(r'_1, r'_2, r'_3, r_1, r_2, r_3; C). \quad (16.8)$$

$C$ is the sum of the cross momenta of $G$. Finally, we define the function $F(r'_1, r'_2, r'_3, r_1, r_2, r_3, C)$ as the product of the factors $\tilde{g}_i$ and $\tilde{f}_j$ in (16.4) expressed in terms of $r'_1, \ldots, r_4$ and setting $r_4 = -C - r_1 - r_2 - r_3$, $r'_4 = -C - r'_1 - r'_2 - r'_3$. The contribution of $G$ to $E$ is then

$$E_G(\lambda) = \int dC \prod_1^3 \left\{ \int dr_\beta \, \frac{i}{2\pi(r_\beta{}^2 - m^2 + i\varepsilon)} \right\} \prod_1^3 \left\{ dr'_\beta \, \frac{-i}{2\pi(r'_\beta{}^2 - m^2 - i\varepsilon)} \right\}$$

$$\times \frac{i}{2\pi[(C + r_1 + r_2 + r_3)^2 - m^2 + i\varepsilon]} \, \frac{-i}{2\pi[(C + r'_1 + r'_2 + r'_3)^2 - m^2 - i\varepsilon]}$$

$$\times w(r'_1, \ldots, r_3; C) \, F(r'_1, \ldots, r_3, C) \prod_1^4 e^{-i\lambda(\rho_\alpha, r_\alpha - r'_\alpha)} \, . \tag{16.9}$$

In the support of $F$, $w$ may contain threshold singularities of higher order in sums of $r_\beta$'s and $r'_\beta$'s. But these are locally of H-type 1 (see Definition 14.1), i.e. $w$ may be said to be "almost twice integrably differentiable" as a function of the $r_\beta$ and $r'_\beta$. As a function of $C$, $w$ is Hölder continuous if $G$ contains at least two cross lines. If there is only one cross line, then $w$ is of the form

$$w(\cdots; C) = \delta_+(C) \, w_L(r'_1, r'_2, r'_3; C) \, w_R(r_1, r_2, r_3; C) \, , \tag{16.10}$$

if there is no cross line, of the form

$$w(\cdots; C) = \delta^4(C) \, w_L(r'_1, r'_2, r'_3) \, w_R(r_1, r_2, r_3) \, . \tag{16.11}$$

In both cases $w_L$ and $w_R$ are of H-type 1. This is all the smoothness we need.

The $\lambda$-dependent exponentials in (16.9) do not depend on $C$: $C$ drops out in the difference $r_4 - r'_4$. The $C$-integration is therefore asymptotically irrelevant, and we may first consider the situation for $C$ fixed. Moreover, the singularities and the exponentials factorize into two expressions containing only primed or unprimed variables, respectively. The essential aspects of the problem factorize. We consider therefore at first the expression

$$E_R(\lambda)$$
$$= \int \prod_1^3 dr_\beta \prod_1^4 \frac{i}{(r_\alpha{}^2 - m^2 + i\varepsilon)} \, w_R(r_1, r_2, r_3) \, F_R(r_1, \ldots, r_4) \prod_1^4 e^{-i\lambda(\rho_\alpha, r_\alpha)}, \tag{16.12}$$

where $r_4 = -C - r_1 - r_2 - r_3$, and $C$ is kept fixed. $F_R$ contains the $r_\alpha$-dependent factors of $F$, and $w_R$ is a function with the smoothness of $w$. The $r_\alpha$-poles are responsible for the asymptotically leading term in $E_R$ for $\lambda \to \infty$. We are therefore only interested in an arbitrarily small open neighbourhood of

$$\mathcal{M} = \bigcap_{\alpha=1}^4 \{r_\alpha{}^2 = m^2\} \, . \tag{16.13}$$

The gradients $r_\alpha$ of the intersecting manifolds, considered as vectors in the 12-dimensional integration space, are linearly independent in the support of $F$. Hence these manifolds are in general position and $\mathcal{M}$ is itself a smooth manifold.

We define
$$t_\alpha := r_\alpha{}^2 \tag{16.14}$$
for $\alpha = 1,\ldots,4$. Let $P \in \mathcal{M}$ be a point at which the sum $T = \sum(\rho_\alpha, r_\alpha)$ in the $\lambda$-exponent is independent of the $t_\alpha$, i.e. the 12-gradient of $T$ is linearly independent of the 12-gradients of the $t_\alpha$. Then we may introduce in a neighbourhood of $P$ the variables $t_1,\ldots,t_4, T$, and seven suitably chosen functions $u_j$ of $r_1, r_2, r_3$, as new variables of integration with a regular Jacobian
$$J = \frac{\partial(r_1, r_2, r_3)}{\partial(t_\alpha, T, u_j)} .$$

Thus
$$E_R(\lambda) = \int dT\, e^{-i\lambda T} \int \prod \frac{dt_\alpha}{t_\alpha + i\varepsilon} \prod du_j\, A(t_\alpha, T, u_j) , \tag{16.15}$$

with $A = J w_R F_R$ a smooth function, if we assume for the moment that no thresholds are present in the support of $F$, i.e. that no partial sum $R$ of $r_\alpha$'s satisfies $R^2 = n^2 m^2$ with $n \geq 2$.[1] We will return to the influence of thresholds later on. The $t$–$u$-integral in (16.15) taken over the considered neighbourhood of $P$ is then a test function in $T$, and its Fourier transform $E_R^P(\lambda)$ decreases strongly.

Asymptotically decisive are therefore the points $(R_1, R_2, R_3)$ on $\mathcal{M} \cap \text{supp}\, F$ where $\text{grad}\, T$ is perpendicular to $\mathcal{M}$, i.e. it is a linear combination of the 12-gradients of the $t_\alpha$. Writing the generic vector in $r$-space as $(r_1, r_2, r_3)$ we have $\text{grad}\, t_1 = 2(r_1, 0, 0), \ldots, \text{grad}\, t_4 = 2(-r_4, -r_4, -r_4)$. The condition of dependence is satisfied iff there exist real numbers $c_1, \ldots, c_4$, such that (remember $\rho_4 = 0$)
$$\rho_1 = c_1 R_1 + c_4 R_4, \quad \rho_2 = c_2 R_2 + c_4 R_4, \quad \rho_3 = c_3 R_3 + c_4 R_4 . \tag{16.16}$$

It is easily seen that this condition has the geometrical interpretation shown in Fig. 16.3. There exists a point $N$ such that the rays starting from $\rho_\alpha$ in direction $-R_\alpha$ meet at $N$. Note that $\xi_\alpha = \rho_\alpha - N$ is timelike. Because of this and the smallness of the sets in which the $R_\alpha$ must lie, $N$ is restricted to a bounded set in Minkowski space.

We show that only a position of $N$ as shown in the figure: $0 < N^0 < \rho_1^0$, corresponding to $c_\alpha > 0\ \forall \alpha$, is asymptotically relevant. The argument is related to that used in Sect. 14.2 for justifying the replacement of the ordinary product $\varphi(x)\varphi(y)$ by the time-ordered product $T^+(x, y)$ (see (14.26) and (14.27)). The variables $t_1, \ldots, t_4$ are independent in the critical region. The product $\prod_\alpha (r_\alpha{}^2 - m^2 + i\varepsilon)^{-1} = \prod_\alpha (t_\alpha - m^2 + i\varepsilon)^{-1}$ is therefore a direct

---
[1] The case $(\sum_1^4 r_\alpha)^2 = n^2 m^2$ leads to a threshold singularity in $C$ which is of no consequence. This case we do not exclude.

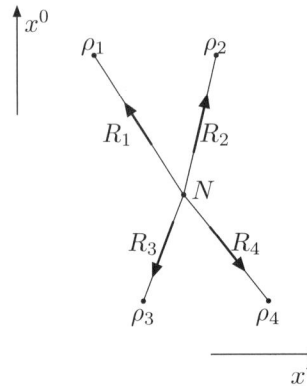

**Fig. 16.3.** Geometry of a scattering event. A 2-dimensional projection is shown

product of four distributions in independent variables. The $\lambda$-exponential can be written

$$e^{-i\lambda \sum \rho_\alpha r_\alpha} = e^{-i\lambda \sum (\rho_\alpha - N, r_\alpha)} e^{i\lambda N C} \ .$$

The $C$-factor is at present of no interest. Consider the asymptotic contribution of a very small neighbourhood of a critical point $\{R_\alpha\}$. We can replace $r_\alpha$ by $R_\alpha$ in the smooth factors in an excellent approximation. In the exponential we write $r_\alpha = \tau_\alpha R_\alpha + \delta_\alpha$ with $\tau_\alpha \approx 1$ a scalar and $(R_\alpha, \delta_\alpha) = 0$. In this neighbourhood we introduce the $\tau_\alpha$ and a sufficient number of other regular variables as new variables of integration. The corresponding Jacobian is smooth and only its value at $\tau_\alpha = 1$ is asymptotically relevant. The only essentially $\tau_\alpha$-dependent factors are then $\left(m^2(\tau_\alpha + i\varepsilon)^2 + (\delta_\alpha, \delta_\alpha)\right)^{-1} \exp(-i\lambda m^2 c_\alpha \tau_\alpha)$. The path of the $\tau_\alpha$-integration can be shifted into the upper half-plane: $\tau_\alpha \to \tau_\alpha + i\tau'_\alpha$, $\tau'_\alpha > 0$, introducing a factor $e^{\lambda m^2 c_\alpha \tau'_\alpha}$ which decreases rapidly for $\lambda \to \infty$ if $c_\alpha < 0$. Hence only the values of $N$ for which all $c_\alpha$ are positive are asymptotically relevant.

In the same way we find a strong decrease for $c_\alpha > 0$ if we replace $+i\varepsilon$ by $-i\varepsilon$ in the $t_\alpha$-pole. For evaluating the asymptotically leading term we may therefore replace the factor $(r_\alpha^2 - m^2 + i\varepsilon)^{-1}$ by

$$\frac{1}{r_\alpha^2 - m^2 + i\varepsilon} - \frac{1}{r_\alpha^2 - m^2 - i\varepsilon} = -2\pi i\, \delta(r_\alpha^2 - m^2) \ ,$$

and then $\delta(r_\alpha^2 - m^2)$ by $\delta_+(r_\alpha)$ for $\alpha = 1, 2$, by $\delta_-(r_\alpha)$ for $\alpha = 3, 4$.

Actually, the set of critical points $\{R_\alpha\}$ is very small. We prove:

**Lemma 16.1.** *The set $S_C$ of points $N$ for which $\rho_\alpha - N = c_\alpha R_\alpha$ with $c_\alpha > 0$ solves the condition (16.16) is discrete.*

This means that the asymptotic behaviour of $E_R$ can be determined by considering contributions to the $r$-integral from domains containing exactly one such critical point. In fact, if the supports of $\tilde{f}_j, \tilde{g}_i$, are chosen sufficiently small, there is at most one such point. We assume that this is the case, more precisely that there is exactly one such point.

**Proof of the Lemma.** Define $\xi_\alpha := \rho_\alpha - N = c_\alpha R_\alpha$. For $c_\alpha > 0$ we have

$$R_\alpha = m \frac{\xi_\alpha}{\sqrt{\xi_\alpha{}^2}}, \qquad (16.17)$$

which is uniquely determined by $N$. Consider the function

$$Z(N) = m \sum_{\alpha=1}^{4} \sqrt{(\rho_\alpha - N)^2} + NC . \qquad (16.18)$$

$Z$ is stationary at $N$ if

$$\operatorname{grad} Z = m \sum_\alpha \frac{\rho_\alpha - N}{\sqrt{(\rho_\alpha - N)^2}} + C = 0 .$$

This equation is implied by $\sum R_\alpha + C = 0$ and (16.17), which is equivalent to the condition (16.16): $Z$ is stationary for the $N$ associated with a solution of (16.16). But these stationary points of $Z$ are isolated maxima. Let $N'$ be a stationary point. We expand $Z$ around $N'$ into a power series

$$Z(N) = Z(N' + \Delta) = Z(N') + Z_2(\Delta) + \cdots \qquad (16.19)$$

with $Z_2$ a form of second order in $\Delta$. The term of first order vanishes because of the stationarity at $N'$. $Z_2(\Delta)$ is the sum of the terms of second order in the individual $\alpha$-terms. Expanding

$$\sqrt{(\xi_\alpha + \Delta)^2} = \sqrt{\xi_\alpha{}^2} \left(1 + \frac{2(\xi_\alpha, \Delta) + \Delta^2}{\xi_\alpha{}^2}\right)^{\frac{1}{2}}$$

into a power series in $\Delta$ we find, using $\xi_\alpha = \sqrt{\xi_\alpha{}^2} \frac{R_\alpha}{m}$:

$$Z_2(\Delta) = \frac{1}{2} \sum_{\alpha=1}^{4} \frac{1}{\sqrt{\xi_\alpha{}^2}} \left(\Delta^2 - m^{-2}(R_\alpha, \Delta)^2\right) .$$

Since $R_\alpha$ is timelike, we have $\Delta^2 - m^{-2}(R_\alpha, \Delta)^2 \leq 0$. For $\Delta \neq 0$ the equality holds only if $\Delta$ is parallel to $R_\alpha$. This can be the case for at most one $\alpha$, because the four $R_\alpha$ are not parallel. Hence $Z_2(\Delta)$ is negative definite, which proves our assertion.

The uniqueness of $N$ and therefore of $R_\alpha$ has two beneficial effects. First, the potentially troublesome factor $e^{i\lambda NC}$ in $E_R$ is cancelled in the generalization of the procedure to $E$ by a corresponding factor coming from the primed variables. Second, if higher thresholds ($n > 2$) are present in the relevant momentum region, then $w_R$ is still continuously differentiable in $r_\alpha$.

Only its value at $r_\alpha = R_\alpha$ is asymptotically relevant, because the difference $w_R(r_\alpha) - w_R(R_\alpha)$ vanishes of order 1 at the critical point. This lowers the degree of the singularity and therefore leads to a stronger, non-dominant, $\lambda$-decrease. We may therefore replace the regular factor $w_R$ in (16.12) by its value at $r_\alpha = R_\alpha$, thus turning $W_R F_R$ into a $C^\infty$ function.

We must, then, determine the asymptotically leading term of

$$I_R(\lambda) = \int \prod_{\beta=1}^{3} dr_\beta \prod_{\alpha=1}^{4} \delta(r_\alpha^2 - m^2) \, F(r_\beta, C) \, e^{-i\lambda \sum_{\alpha=1}^{4}(\xi_\alpha, r_\alpha)} , \qquad (16.20)$$

with $r_4 = -C - \sum_{\beta=1}^{3} r_\beta$, $\xi_\alpha = \rho_\alpha - N = c_\alpha R_\alpha$, and $F$ a smooth function. We introduce $t_\alpha = r_\alpha^2$ and eight other regular functions $u_1, \ldots, u_8$, of $r_\beta$ as new variables of integration. The $u_h$ can be chosen such that the Jacobian

$$J(t_\alpha, u_h, C) = \frac{\partial(r_1, r_2, r_3)}{\partial(t_\alpha, u_i)}$$

is regular, and that the $u_h$ vanish at $r_\alpha = R_\alpha$. In particular, we might choose the $u_h$ to be eight components of the 3-vectors $\vec{r}_\beta - \vec{R}_\beta$, $\beta = 1, 2, 3$. We obtain

$$I_R(\lambda) \sim \int d^8 u \, F(r_\beta, C) \, J(m^2, u, C) \, e^{-i\lambda \eta(u)} , \qquad (16.21)$$

where the $r_\beta$ in the argument of $F$ are regular functions of $u = \{u_1, \ldots, u_8\}$. $J(m^2, u, C)$ will henceforth be denoted $J(u, C)$. $\eta$ is given by

$$\eta(u) = \sum_\alpha c_\alpha \big(R_\alpha, r_\alpha(u)\big) . \qquad (16.22)$$

$r_\alpha(u)$ and therefore $\eta(u)$ can be expanded into a power series in $u$. The constant term gives rise to a factor $e^{-i\lambda \eta_0}$, $\eta_0 = \eta(0) = m^2 \sum c_\alpha$, which can be drawn in front of the integral. The term of first order in $\eta$ vanishes because the $R_\alpha$ were defined such that at $u = 0$ the gradient of $\eta$ is perpendicular to $\mathcal{M} = \{t_\alpha = m^2\}$, hence the restriction of $\eta$ to $\mathcal{M}$ is stationary at $u = 0$.

The quadratic form ${}^2\eta(u)$ consisting of the terms of second order in $\eta(u)$ is positive definite. Take $(R_\alpha, r_\alpha(u))$ for a given $\alpha$, which index will be dropped for the moment. We have

$$r(u) = R + {}^1r(u) + {}^2r(u) + \cdots ,$$

${}^n r(u)$ being the term of order $n$ in the Taylor expansion of $r(u)$, $r(u) \in \mathcal{M}$. On $\mathcal{M}$ we have $(R, r) = |R_0| w(\vec{r}) - (\vec{R}, \vec{r})$. The terms of second order are obtained from

$$w(\vec{r}) = \sqrt{m^2 + (\vec{R} + {}^1\vec{r} + {}^2\vec{r} + \cdots)^2}$$

$$= w_0 \left( 1 + \frac{(\vec{R}, {}^1\vec{r})}{w_0^2} + \frac{(\vec{R}, {}^2\vec{r})}{w_0^2} + \frac{1}{2} \frac{({}^1\vec{r}, {}^1\vec{r})}{w_0^2} - \frac{1}{2} \frac{(\vec{R}, {}^1\vec{r})^2}{w_0^4} + \cdots \right) ,$$

$\omega_0 = |R_0|$, as $^2(R,r) = \frac{1}{2}\left((^1\vec{r},^1\vec{r}) - \frac{(\vec{R},^1\vec{r})^2}{|R_0|^2}\right) > 0$, except for $^1\vec{r} = 0$. But no matter in which direction we move away on $\mathcal{M}$ from $\{R_\alpha\}$, there is at least one $\alpha$ for which $^1\vec{r}_\alpha \neq 0$.

Because of the positivity of $^2\eta$ we can choose the coordinates $u_h$ of $\mathcal{M}$ such that

$$^2\eta(u) = \sum_{h=1}^{8} u_h^2 . \tag{16.23}$$

The higher terms in the $\eta$-expansion do not influence the $\lambda$-asymptotics since they are negligible with respect to $^2\eta$ in the relevant region $|u| \to 0$.[2] We can therefore replace (16.21) by

$$I_R(\lambda) \sim e^{-i\lambda E_0} F(R_\beta, C) J(0, C) \prod_{h=1}^{8} \int du_h e^{-i\lambda u_h^2} , \tag{16.24}$$

because the dominant asymptotic term is determined by the value of $FJ$ at the critical point $u = 0$. The integral in this expression has the value (see [GR], formula 3.381) $\frac{\sqrt{\pi}}{\sqrt{\lambda}} e^{-i\frac{\pi}{4}}$. Thus we arrive at the final result

$$I_R(\lambda) \sim e^{-i\lambda E_0} \lambda^{-4} \pi^4 F(R_\beta, C) \hat{J}(C) , \tag{16.25}$$

with

$$\hat{J}(C) = J(t_\alpha, u_h, C)\big|_{t_\alpha = m^2, u_h = 0} .$$

For variables $u_h$ not satisfying the condition (16.23) we have $^2\eta(u) = \sum_{r,s} u_r M_{rs} u_s$. In this case there appears in the result (16.25) an additional factor $(\det M)^{-\frac{1}{2}}$.

The asymptotically leading term of $E_G(\lambda)$ defined in (16.9) is found by using the above results for $E_R$ twice, once for the unprimed, once for the primed variables, in the latter case with the obvious change of $i$ to $-i$ in the poles and exponentials. The factors $e^{i\lambda NC}$ and $e^{-i\lambda E_0}$ that arose at various stages cancel between the two parts, and we obtain

$$E_G(\lambda) \sim \frac{\pi^8}{\lambda^8} \int dC \, \hat{J}(C)^2 \, w(r'_1, \ldots, r_3; C) \, F(r'_1, \ldots, r_3, C)\big|_{r_\beta = r'_\beta = R_\beta} . \tag{16.26}$$

Like in the 1-particle case, the $\lambda^{-8}$-decrease agrees with geometrical expectation. In the elastic case the effects of momentum conservation must not be forgotten.

---

[2] It is easily checked that applying the following arguments to the function (14.3) in Lemma 14.2, replacing $\omega(\vec{p})$ in the exponent by its expansion $m + (\vec{p}^{\,2}/2m)$ up to second order, yields the same result as the more complete derivation given there. For more details, see [Ste 68].

An expression which is symmetric in the four $R_\alpha$ can be obtained by replacing $C$ by $N$ as variable of integration. This is possible because the connection between $N$ and $C$ is 1–1 under our assumptions. We find

$$E_G(\lambda) = \frac{\pi^8}{\lambda^8} \int dN \, \frac{\partial(C)}{\partial(N)} \, \hat{J}(C(N))^2 \, w(R_1(N), \ldots, R_4(N))$$
$$\times F(R_1(N), \ldots, R_4(N)) \qquad (16.27)$$

in obvious notation.

Returning to the original notation which distinguishes between incoming and outgoing particles, we define

$$P_1 = -P_1' = R_3, \quad P_2 = -P_2' = R_4, \quad Q_1 = -Q_1' = R_1, \quad Q_2 = -Q_2' = R_2. \qquad (16.28)$$

The $F$-factor in (16.26) becomes

$$F = \prod_{j=1}^{2} |\tilde{f}_j(Q_j)|^2 \prod_{i=2}^{2} |\tilde{g}_i(-P_i)|^2 ; \qquad (16.29)$$

the $w$-factor is (see (16.8))

$$w = (2\pi)^8 \prod_i (P_i^2 - m^2)^2 \prod_j (Q_j^2 - m^2)^2$$
$$\times \tilde{\mathcal{W}}'_G(-P_1, -P_2, -Q_1, -Q_2, -|P_1, P_2, Q_1, Q_2, +) , \qquad (16.30)$$

the prime in $\tilde{\mathcal{W}}'_G$ denoting that the $\delta^4$ of momentum conservation is omitted. This factor $w$ contains the information on the interaction, while $\hat{J}^2 F$ depends on the kinematics of the process and on the characteristics of the sources and probes. $|\tilde{f}_j|^2$ and $|\tilde{g}_i|^2$ can be determined experimentally from measurements on 1-particle states. Note that $w$ is the restriction to the mass shell of the fully amputated 8-point function $\tilde{\mathcal{W}}'_G$. The formula (16.30) formally summed over all graphs $G$ is a generalization to inclusive processes of the LSZ reduction formula. It is customarily derived by inserting a complete set of free out-states between the $T^-$- and the $T^+$-factors and using the original reduction formula.[3] But this derivation cannot be generalized to QED, because there the LSZ asymptotic conditions do not hold.

$w$ is a measure of the probability that two particles with momenta $-P_1$ and $-P_2$ when hitting each other produce two particles of momenta $Q_1$ and $Q_2$ plus an unspecified number of unobserved particles. It is related to the more familiar inclusive cross section (see (13.13))

$$d\sigma = \frac{\pi^2}{\sqrt{(P_1, P_2)^2 - m^4}} \, w \prod_j \frac{d^3 Q_j}{2\omega(\vec{Q}_j)} . \qquad (16.31)$$

---

[3] For details see Sect. 17.1.

The cross section is the relevant quantity for the analysis of realistic scattering experiments, with which our arrangement of sources and probes that are active only during brief intervals of time apparently has but a limited similarity.

But this appearance is deceptive. True, the current density in our state $\Phi$ is time-dependent. But for increasing $\lambda$ this time dependence becomes weaker and weaker due to the "spreading of the wave packet". Hence in a small neighbourhood of the reaction point $N$ and in the 4-dimensional localization regions of the probes, whose size does *not* increase with $\lambda$, the current densities are sensibly constant. If $G$ contains at least two cross lines, then $w$ is continuous and slowly varying. For sufficiently small supports of $\tilde{g}_i$, $\tilde{f}_j$, the relevant $C$-region is small, and $w$ can be considered a constant and drawn in front of the integral, taking it at an arbitrarily chosen value $\bar{C}$ in the relevant region. We obtain in an excellent approximation

$$E_G(\lambda) \sim \frac{\pi^8}{\lambda^8} w\big(P_1(\bar{C}),\ldots,Q_2(\bar{C})\big)$$
$$\times \int dC\, \hat{J}(C)^2 \prod |\tilde{g}_i(-P_i(C))|^2 \prod |\tilde{f}_j(Q_j(C))|^2 \ . \tag{16.32}$$

$w$ occurring linearly in this expression, we can sum over all relevant graphs up to a given perturbative order, or formally even over all orders, obtaining an expression of the same form. The $C$-integral is a kinematical factor not depending on the form and strength of the interaction. It may be considered a generalized phase-space factor.

If $G$ contains exactly one cross line, we use (16.10) to write

$$w(\cdots;C) = \delta_+(C)\, w_L(Q_1,Q_2,P_1;C)\, w_R(Q_1,Q_2,P_1;C) \ .$$

Summing over all graphs $G$ of this type of a given order $\sigma$, we find that the products $w_L w_R$ sum to the term of order $\sigma$ of $|w_R(\cdots)|^2$, since the formal sum over all orders of $w_L$ is the complex conjugate of the summed $w_R$. The factor $|w_R|^2_\sigma$ is continuous and slowly varying and can be drawn in front of the integral like $w$ in the previous case, yielding

$$E_\sigma(\lambda) \sim \frac{\pi^8}{\lambda^8} |w_R(Q_1,Q_2,P_1;\bar{C})|^2_\sigma \int dC\, \delta_+(C)\, |\hat{J}(C)|^2$$
$$\times |\tilde{g}_1(-P_1(C))|^2 \cdots |\tilde{f}_2(Q_2(C))|^2 \ . \tag{16.33}$$

Again, the $C$-integral is a generalized phase-space factor.

The graphs with no cross lines describe an exclusive process: elastic scattering of two particles, both of whom are observed. From (16.11) we obtain in the same way as above

$$E_\sigma(\lambda) \sim \frac{\pi^8}{\lambda^8} |w_R(Q_1,Q_2,P_1)|^2_\sigma\, |\hat{J}(0)|^2\, |\tilde{g}_1(-P_1)|^2 \cdots |\tilde{f}_2(Q_2)|^2 \ . \tag{16.34}$$

The $P_i$, $Q_j$, lying on the negative and positive mass shell, respectively, are determined such that $\sum P_i + \sum Q_j = 0$ and that a $N \in R^4$ exists with $N - \beta_i = -c_i P_i$, $\nu_j - N = d_j Q_j$, $c_i$ and $d_j$ positive.

The generalization to the case of an arbitrary number $n$ of probes at the spacelike separated positions $a_j = \lambda \nu_j$; $j = 1, \ldots, n$, is immediate. An expression of the form (16.26) is again obtained, but with the factor $(\pi/\lambda)^{2(n+2)}$ in front and with $w$ given by

$$w = (2\pi)^{2(n+2)} \prod_{i=1}^{2}(P_i{}^2 - m^2)^2 \prod_{j=1}^{n}(Q_j{}^2 - m^2)^2$$
$$\times \tilde{W}'_G(-P_1, -P_2, -Q_1, \ldots, -Q_n, -|P_1, \ldots, Q_n, +)\big|_{\sum P_i + \sum Q_j = -C} .$$
(16.35)

The $j$-product from (16.29) extends over $n$ factors, and $\hat{J}(C)$ is an obvious higher-dimensional version of the former one. The same simplifications depending on the number of cross lines occur as for $n = 2$.

## 16.2 Reactions in QED

The scattering formalism explained in the preceding section for a massive scalar theory must now be generalized to QED. Like in the scalar example, we consider the typical case of two incoming particles and two observed outgoing particles. But the sources and probes are those of QED as introduced in Chap. 15. We use the methods and results established in the massive case with the necessary changes. The complications due to spin and the presence of several fundamental fields are trivial. But there are some less trivial problems to be solved, which are due to the non-locality of the physical Fermi fields $\Psi$, $\bar{\Psi}$, to the vanishing mass of the observed photons, and to the IR problems in reactions involving charged particles.

The results to be established apply equally to the Källén gauge and to the physical states of Chap. 12. The only point where the non-locality of $\Psi$ might cause difficulties is the replacement of ordinary field products by $T^{\pm}$-products in passing from (16.3) to (16.4). As has been noted in connection with (15.2), this replacement is still asymptotically valid if applied to a product of a probe factor and a source factor. The argument is based (see (14.26)) on relations of the type

$$\Psi(x)\Psi(y) - T^+(x,y) = \theta(y^0 - x^0)\{\Psi(x), \Psi(y)\} ,$$

which shows that any matrix element of this difference, smoothed over a test function, vanishes strongly for $x^0 - y^0 \to \infty$. It also vanishes strongly for $|\vec{x} - \vec{y}| \to \infty$ but $x^0 - y^0 = \text{const}$ for local fields, but not in our case of non-local fields. Nevertheless, the replacement is allowed if $\{\Psi(x), \Psi(y)\}$ tends to zero in equal-time directions in an operator sense, i.e. even if sandwiched between states for which the terms on the left-hand side do not tend to zero individually. That this is actually the case is heuristically plausible, starting from the formal definition (12.9) in terms of the local GB fields. Calling the exponential in this definition $1 + E(x)$, we can write $\{\Psi(x), \Psi(y)\}$ as a sum of terms, each of which contains one of the factors $\{\psi(x), \psi(y)\}$, $[\psi(x), E(y)]$, $[E(x), \psi(y)]$, or $[E(x), E(y)]$. The first of these expressions vanishes for spacelike $x - y$. The second one depends on the commutator $[\psi(x), A_\rho(u)]$ at $u^0 = x^0$, $|\vec{x} - \vec{u}| < |x^0 - y^0|$. For $|\vec{x} - \vec{y}| \to \infty$ in the critical directions $x^0 - y^0 = $ const, this commutator is $\neq 0$ only for $\vec{u} \sim \vec{x}$, hence for large $|\vec{u} - \vec{y}|$, where the factor $\hat{r}^\rho(\vec{y} - \vec{u})$ in the exponent is small. A similar argument applies to the third expression. The fourth contains a factor $[A_\rho(u), A_\sigma(v)]$ at $u^0 = x^0$, $v^0 = y^0$, $|\vec{u} - \vec{v}| \leq |x^0 - y^0|$, and there either $\hat{r}^\rho(\vec{x} - \vec{u})$ or $\hat{r}^\sigma(\vec{y} - \vec{v})$ or both are small for large $|\vec{x} - \vec{y}|$.

This non-locality problem could be avoided by working with more realistic probes which are local and gauge invariant. But this would entail substantial additive complications in our derivations, which more than outweigh the advantage of such an alternative. We will therefore proceed under the assumption that the introduction of $T^\pm$-products like in (16.4) is justified despite the non-locality of the physical fields. A more convincing argument that this leads indeed to a correct result will be given later on at the end of Sect. 16.4. Like in Chap. 15 we use the local $F_{\mu\nu}$ as physical photon fields, not the potentials $\mathcal{A}_\mu$. Similar problems therefore do not occur for the photons.

Let $D^\pm$, $D^\gamma$, be the probes introduced in Chaps. 14 and 15. Consider the putative 2-particle state

$$\Phi(b_1, b_2) = \varphi_1(g_1^\#, b_1)\, \varphi_2(g_2^\#, b_2)\, \Omega\;, \tag{16.36}$$

with $\varphi_i$ any of the fields $\Psi$, $\bar{\Psi}$, $\vec{E}$, and $g_i^\#$ a corresponding test spinor or vector $\bar{g}_i$, $g_i^p$, or $\vec{g}^*$, as the case may be. The source positions have the same meaning as in (16.1). Let $D_1(a_1)$, $D_2(a_2)$, defined as

$$D_j(a_j) = \varphi_j(f_{j,a_j})^* \varphi_j(f_{j,a_j})\;, \tag{16.37}$$

be two probes of arbitrary type. The same notations as in (16.36) are used. We introduce the scaling factor $\lambda$ of (16.2), and we wish to determine the asymptotic behaviour for $\lambda \to \infty$ of

$$\begin{aligned}E(\lambda) &= \bigl(\Phi,\, D_1(a_1)\, D_2(a_2)\, \Phi\bigr) \\ &\sim \Bigl(\Omega, \prod_i \varphi_i(g_{i,\lambda\beta_i}^\#)^* \prod_j \varphi_j(f_{j,\lambda\nu_j})^* \prod_j \varphi_j(f_{j,\lambda\nu_j}) \prod_i \varphi_i(g_{i,\lambda\beta_i}^\#)\, \Omega\Bigr)\;,\end{aligned} \tag{16.38}$$

with $\nu_i$, $\beta_j$, being fixed in the same geometrical configuration as in the scalar example. The asymptotic quasi-locality described before has been used.

Equation (16.4) still holds with the following changes. $\overline{\tilde{g}(p')}$ is replaced by $\tilde{g}^{\#}(-p')$, and similarly for $\overline{\tilde{f}}$. The variables $p'_1, \ldots, q_2$, are variables of the appropriate field type. The ordering of variables in $\tilde{\mathcal{W}}$ must be changed to $p'_1, \ldots, q'_2 | q_2, q_1, p_2, p_1$. Summation over indices in the test spinors and vectors and in the corresponding fields is understood.

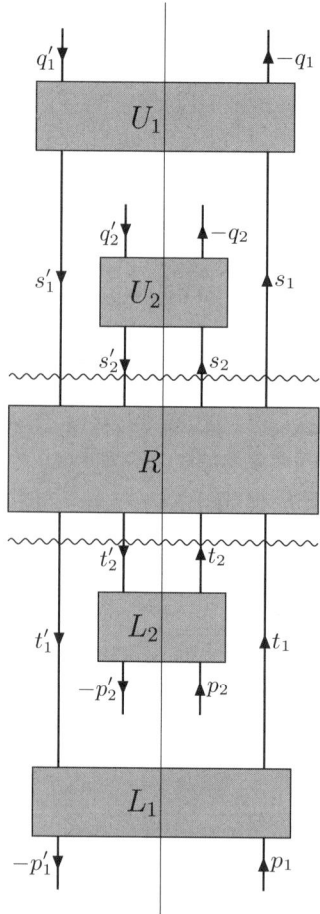

**Fig. 16.4.** Relevant graphs of $\tilde{\mathcal{W}}(p'_1, \ldots, -|\ldots, p_1, +)$. The wavy lines denote cuts

The asymptotically relevant graphs contributing to $\tilde{\mathcal{W}}$ are of the form shown in Fig. 16.4. In contrast to the 1-particle case of Chap. 15, there are now two cuts through the graph designating the asymptotically relevant singularities, one on the incoming side, one on the outgoing side of the $R$-blob representing the scattering event. The lines are of arbitrary type, except that the variables of a quartet $(q_j, s_j, q'_j, s'_j)$ or $(p_i, t_i, p'_i, t'_i)$ belong to lines of the same type. To facilitate the transfer of the results of the previous section to the new situation, the direction of the lines is chosen like in Fig. 16.1. For positron lines this choice clashes with the earlier convention of correlating the flow of momentum with the flow of charge. This necessitates some obvious but tiresome changes in the graph rules. In order to avoid this problem without fundamental significance, I will take the liberty of never explicitly considering processes involving observed positrons. The $U_i$- and $L_j$-parts are the same as in the 1-particle case of Chap. 15. They contain only photonic cross lines. In Fig. 16.1 they are not present, because under our assumptions they do not exist in a purely massive theory, as has been noted before. The factorization of the $U$- and $L$-parts of the graph is based on the same sort of consideration as used in the massive case for the discussion of the possible influence of threshold singularities. Singularities in the relevant variables $(\nu_j, q_j + q'_j)$ can only originate from $U_j$-factors of the type shown.

The considerations of Sect. 16.1 concerning no-interaction events in the forward direction generalize immediately to the new situation.

For studying reactions in the non-forward direction we follow the procedure of Sect. 16.1. The first problem encountered is that of the threshold singularities. Subgraphs of the form of Fig. 16.2 cannot be excluded by conditions on the $\tilde{f}$-$\tilde{g}$-supports if the $k$-line is photonic. If $q'_1$ is also a photon variable, we expect a weak singularity at $(p'_1 + p'_2 + q'_2)^2 = 0$ coming from the points at which the cut lines are photonic and their momenta are parallel and lightlike. But this singularity is asymptotically irrelevant, as is shown by a slight extension of the handling of the photonic IR problem in Sect. 15.3 (see (15.87)ff). The same procedure as there is started from the 4-photon blob at the top of the graph. If $p'_i$ and $q'_2$ are not all photonic, the WT identity for the lower blob is non-trivial, i.e. not of the simple form (15.88). If, for example, $q'_2$ is a fermion variable, there is on the right-hand side of the WT identity a term associated to this line containing a factor $(\slashed{q}'_2 - m)$. Multiplied with the propagator factor $(\slashed{q}'_2 - m)^{-1}$ it removes the singularity of that propagator, which renders the graph asymptotically irrelevant. If $q'_1$ is a fermion variable, then there is a weak singularity due to the possible cross momentum $k = 0$. This is part of the fermionic IR problem which will be solved in Sect. 16.4. That subgraphs of the form of Fig. 15.10, with $q', -q$ replaced by $q'_i, -q_i$ or $s'_i, s_i$, are IR harmless follows from the results of Sect. 15.3. Similar subgraphs with $q', -q$ replaced by $s'_i, s_j$ or $s'_i, s'_j$ with $i \neq j$ are not dangerous because the momenta involved are never parallel.

Pursuing further the path which has lead to (16.26), we meet the two non-trivial complications already referred to, caused by the vanishing photon mass and the IR problem. They are solved in the next and the next but one section, respectively.

## 16.3 Photons

In this section we study the necessary changes in the considerations of Sect. 16.1 if observed particles of mass zero are involved. The IR problems are still ignored, i.e. the blob functions $R$, $L_i$, $U_j$, in Fig. 16.4 are assumed to be smooth. Spin indices are suppressed. The presence of the source and probe factors $L_i$ and $U_j$ is no essential complication relative to the massive case. It simply amounts to redefining the function $F$ occurring there as the product of these factors integrated over the test functions $\tilde{g}_i$, $\tilde{f}_j$, in the external variables. And $R$ assumes the place of $w$. The role of the external variables $p_1, \ldots$ is taken over by the cut variables $s_j, s'_j, t_i, t'_i$. Notice that this entails some changes of sign. We retain the conventions of Sect. 16.1 by defining

$$r_1 = -s_1, \quad r_2 = -s_2, \quad r_3 = t_1, \quad r_4 = t_2, \\ r'_1 = -s'_1, \quad r'_2 = -s'_2, \quad r'_3 = t'_1, \quad r'_4 = t'_2, \tag{16.39}$$

and taking over (16.7) unchanged. In (16.8)ff the mass parameters now carry indices, $m_\alpha$ being the electron mass $m$ or the photon mass 0 as the case may be.

The next necessary adaptation is a change in the proof that only positive $c_\alpha$ are relevant in the condition (16.16), the general idea remaining the same. We use a frame of reference in which $\vec{R}_\alpha \neq 0$ for all $\alpha$. The decomposition $r_\alpha = \tau_\alpha R_\alpha + \delta_\alpha$ is meaningless if $m_\alpha = 0$. Instead of $\tau_\alpha$ we use directly $t_\alpha = r_\alpha^2$ as independent variable. In the relevant region $r_\alpha \sim R_\alpha$ we have

$$r^{-i\lambda c_\alpha(R_\alpha, r_\alpha)} = e^{-i\lambda c_\alpha |R_\alpha^0| \sqrt{\vec{r}_\alpha^2 + t_\alpha}} e^{-i\lambda c_\alpha(\vec{R}_\alpha, \vec{r}_\alpha)},$$

the positive square root being taken. The corresponding singular factor is $(t_\alpha + i\varepsilon)^{-1}$. Again we shift the $t_\alpha$-integration into the upper half-plane: $t_\alpha \to t_\alpha + it'_\alpha$, $t'_\alpha > 0$, in a neighbourhood of the critical point. If this neighbourhood is sufficiently small, then $|\vec{r}|^2 + t_\alpha$ is positive in it and $I = \Im \sqrt{|\vec{r}_\alpha|^2 + t_\alpha + it'_\alpha} > 0$. Hence the exponential contains the factor $e^{\lambda c_\alpha |R_\alpha^0| I}$ which decreases rapidly for $\lambda \to \infty$ if $c_\alpha < 0$.

The next change concerns Lemma 16.1 and its proof. The uniqueness of the scattering point $N$ for small $\tilde{f}$-$\tilde{g}$-supports remains true, but this does not in all cases imply the uniqueness of the critical momenta $R_\alpha$. We prove the following generalized lemma.

**Lemma 16.2.** *The set $S_C$ of points $N$ for which $\rho_\alpha - N = c_\alpha R_\alpha$ with $c_\alpha > 0$ solves condition (16.16) is discrete. A given solution $N$ determines the momenta $R_\alpha$ uniquely, except if all four particles involved are massless and $C$ lies in the 3-plane $E_3$ spanned by $\rho_1 - \rho_4$, $\rho_2 - \rho_4$, $\rho_3 - \rho_4$.*

The exceptional case is realized, in particular, in elastic $\gamma$-$\gamma$-scattering where $C = 0$. It is also realized in inclusive processes with only photons being observed. But there it is of little consequence, because it concerns only a lower-dimensional subset of the integration variable $C$, inducing a discontinuity in the $C$-dependence of the integrand which does not matter any more than the other singularities in this variable.

**Proof of Lemma 16.2.** We first prove the essential uniqueness of $N$. Assume at first that at least one of the four particles is massive. In this case the proof is a simple generalization of the proof of Lemma 16.1. Let $M$ be the non-empty set of indices $\alpha$ belonging to massive particles, $M'$ its complement. Define

$$Z(N) = m \sum_{\alpha \in M} \sqrt{(\rho_\alpha - N)^2} + N C \tag{16.40}$$

and look for the stationary points of this function under the subsidiary conditions $(\rho_\beta - N)^2 = 0$ for $\beta \in M'$. Introducing Lagrange multipliers $\ell_\beta$, we find the condition of stationarity

$$\operatorname{grad}\left(Z + \sum_{M'} \ell_\beta (\rho_\beta - N)^2\right)$$
$$= m \sum_M \frac{\rho_\alpha - N}{\sqrt{(\rho_\alpha - N)^2}} + 2 \sum_{M'} \ell_\beta (\rho_\beta - N) + C = 0 \ . \tag{16.41}$$

This gives a solution $N(\ell_\gamma)$. The values of $\ell_\gamma$ are determined from $(\rho_\beta - N(\ell_\gamma))^2 = 0 \ \forall \beta \in M'$. Condition (16.41) and the subsidiary conditions are satisfied if $N$ is a solution of the geometrical problem (16.16) and $\ell_\beta = \frac{1}{2c_\beta}$. Then (16.41) is the momentum conservation law $\sum_\alpha R_\alpha + C = 0$. Hence, exactly like in the purely massive case, $Z$ is stationary at the solutions of the geometrical problem. That these points of stationarity are isolated maxima is proved like in the massive case. A problem might arise if only one particle is massive, e.g. for $\alpha = 1$. Then

$$Z_2(\Delta) = \frac{1}{2} \frac{1}{\sqrt{\xi_1^2}} \left(\Delta^2 - \frac{1}{m^2}(R_1, \Delta)\right)^2 = 0$$

if $\Delta$ is parallel to $R_1$. But $R_1$ being timelike, the subsidiary conditions $(\rho_\beta - N' - \Delta)^2 = 0$ would then be violated for $\beta \in M'$ if $(\rho_\beta - N')^2 = 0$. (We use the notation of the proof of Lemma 16.1.) There remains the 4-photon case. The conditions $(\rho_\alpha - N)^2 = 0$ must be satisfied for all $\alpha$. We set $\rho_4 = 0$. From $\alpha = 4$ we obtain $N^2 = 0$. The remaining three conditions $(\rho_\beta - N)^2 = 2(\rho_\beta, N) + \rho_\beta^2 = 0$ determine the component of $N$ in the timelike 3-plane $E_3$ spanned by $\rho_1, \rho_2, \rho_3$, uniquely. The component $n^\perp$ of $N$ orthogonal to $E_3$ is then determined from $N^2 = 0$ up to a sign. But the condition $\sum_\alpha \frac{\rho_\alpha - N}{c_\alpha} + C = 0$ yields $-N^\perp \sum \frac{1}{c_\alpha} = C^\perp$, and this determines the sign of $N^\perp$ uniquely if $c_\alpha > 0$.

Now we address the question of the uniqueness of the $R_\alpha$. For $\alpha \in M$ we obtain the unambiguous result $R_\alpha = m \frac{\rho_\alpha - N}{\sqrt{(\rho_\alpha - N)^2}}$. For $\beta \in M'$ the direction of $R_\beta$ is fixed to be the direction of $\rho_\alpha - N$. Moreover, the sum $\sum_{M'} R_\beta = -C - \sum_M R_\alpha$ is uniquely determined. If $M'$ contains at most three indices (but at least one), then the $\rho_\beta - N$, $\beta \in M'$, are linearly independent. This is also true if $M'$ contains all four indices and $C$ does not lie in $E_3$. Then $N$ does not lie in $E_3$ and the four $\rho_\beta - N$ are linearly independent. In these cases the decomposition of $-C - \sum_M R_\alpha$ with respect to the $M'$-directions is unique, yielding unique $R_\beta$'s. In the remaining case of four massless particles and $C \in E_3$, $N$ lies in $E_3$, and therefore the four $\rho_\beta - N$ lie in the 3-plane $E_3$, hence are linearly dependent. But any three of them are still independent, so that the decomposition $-C = \sum_\beta R_\beta$ with respect to the four $R_\beta$-directions yields components $R_\beta$ depending on one free parameter. In the case of elastic photon–photon scattering, where $C = 0$, this free parameter is a scaling parameter common to all $R_\alpha$. This ends the proof of the Lemma.

The form (16.20) of $I_R(\lambda)$ is taken over as it stands, except that the mass parameters acquire an index: $m \to m_\alpha$. We at first exclude the exceptional case of elastic $\gamma$–$\gamma$ scattering. In all other cases we introduce the same variables $t_\alpha$ and $u_j$ as in the massive case. For inclusive $\gamma$–$\gamma$ scattering the Jacobian $J(t_\alpha, u_h, C)$ may then be singular in $C$ for $C \in E_3$. But this singularity still defines a distribution in $C$ and is therefore acceptable, as we know from the massive case. Thus we arrive again at (16.21). There is a slight complication in the proof of the positivity of $^2\eta(u)$: If $r_\alpha$ is photonic, then $^2(R_\alpha, r_\alpha) = 0$ if $^1\vec{r}_\alpha$ and $\vec{R}_\alpha$ are parallel, not only if $^1\vec{r}_\alpha = 0$. But using the condition $\sum_\alpha {}^1\vec{r}_\alpha = 0$ we find that $^1\vec{r}_\alpha$ and $\vec{R}_\alpha$ cannot be parallel for all $\alpha$, except for the case of $\gamma$–$\gamma$ scattering with $C \in E_3$, which exception is of no consequence. From there on the arguments of the massive case go through without a hitch, leading to the result (16.26), where of course the factors $w$ and $F$ are those of QED.

In bringing this result into contact with experiment, we must remember that the $r_\alpha$ are not external variables like in (16.6). They are defined by (16.39) in terms of the internal cut variables $s_1, \ldots, t'_2$. Hence (16.28) is replaced by

$$S_1 = S'_1 = -R_1, \quad S_2 = S'_2 = -R_2, \quad T_1 = T'_1 = R_3, \quad T_2 = T'_2 = R_4, \tag{16.42}$$

and the expression (16.30) must be changed accordingly, not forgetting to set $m = 0$ in the amputations referring to photons. Equation (16.29) is replaced by

$$F = \prod_{j=1}^{2} U_j(-S_j, -S_j) \prod_{i=1}^{2} L_i(T_i, T_i). \tag{16.43}$$

As a result, the $w$-factor drawn in front of the integral in (16.32) is at first expressed in terms of $S_j(\bar{C})$, $T_i(\bar{C})$. But for fermionic variables we know that only very little momentum is transferred in the $U$–$L$-parts across the sector boundary, so that $S_j$ is close to $-Q_j$, $T_i$ to $P_i$, and we may replace the $S_j, T_i$ in the slowly varying $w$ by the corresponding external momenta. For the photon variables it is known from Sect. 15.3 that $s_j$ and $-q_j$ are almost parallel, but they may differ considerably in length. But, as has been argued there, we are justified in considering $S_j$, $T_i$, as the true momenta of the particles observed in a realistic experiment, which is not described adequately by our simplified arrangement. Therefore, $w$ expressed in terms of $S_j$, $T_i$, *is* a measure of the scattering probability of particles with these momenta, and the connection (16.31) with cross sections is to be written in terms of these variables.

The simplifications applying to graphs with one or no cross line hold here as well.

Another essential point in applying these results to reality is the correct handling of spin, which has been neglected hitherto. In Gupta–Bleuler states this is accomplished by introducing vector indices. A photonic factor $U_j$ becomes $U_j^{\sigma_j' \sigma_j}$, a photonic $L_i$ becomes $L_i^{\tau_i' \tau_i}$, and $w$ acquires indices $\tau_i', \sigma_j', \sigma_j, \tau_i$, associated with the photon fields in $\mathcal{W}$, even with those whose variables $T_2, T_2'$, do not occur explicitly in the argument of $w$ because of their redundancy. The indices in the $U$–$L$- and $w$-factors pertaining to the same cut line are contracted. For example, if $S_1$ is a photon variable, we form in analogy to (15.86)

$$U_1^{\sigma_1' \sigma_1} g_{\sigma_1' \rho_1'} g_{\sigma_1 \rho_1} w^{\cdots \rho_1' \rho_1 \cdots}(\cdots), \qquad (16.44)$$

$\rho_1'$ and $\rho_1$ in $w$ belonging to the variables $-S_1' = -S_1$. The restriction to the physical degrees of freedom is achieved as described in the 1-particle case in (15.90). Let $w^{\mathrm{ph}}$ be formed with the physical photon fields $F^{0v}$ instead of the unphysical $A^\sigma$, and analogously for $U_j$, $L_i$, e.g. $U_1^{u'u}$, $u, u' = 1, 2, 3$, if $U_1$ is a photonic factor. The contraction (16.44) is then carried out with factors $\delta_{v'w'} \delta_{vw}$, and a factor $|\vec{\xi}_1|^{-2}$ is added both to $U_1$ and to $w$. The fermionic variables do not concern us at the moment.

For elastic $\gamma$–$\gamma$ scattering the argument is altered by the scaling freedom of the $R_\alpha$. Instead of the critical point $\{R_\alpha\} \in \mathcal{M}$ there exists now a critical 1-dimensional submanifold $\mathcal{M}_C$, whose points are of the form $R_\alpha = \rho \hat{\xi}_\alpha$, $\hat{\xi}_\alpha = \frac{\xi_\alpha}{|\xi_\alpha|}$, with $\rho$ a real parameter varying over a small positive interval. We extend this parameter to the relevant part of $\mathcal{M}$ by introducing submanifolds $\mathcal{M}_\rho \subset \mathcal{M}$ which intersect $\mathcal{M}_C$ at the $\rho$-point under a finite angle, and with $\mathcal{M}_\rho \cap \mathcal{M}_{\rho'} = \emptyset$ for $\rho \neq \rho'$. We may then introduce $t_\alpha = r_\alpha^2$, $\rho$, and seven regular functions $u_1, \ldots, u_7$, of the $r_\alpha$ as new variables, such that $J(t_\alpha, \rho, u_h) = \frac{\partial(r_1, r_2, r_3)}{\partial(t_\alpha, \rho, u_i)}$ is regular and the $u_h$ vanish on $\mathcal{M}_C$. From there on we proceed in close analogy to the massive case. Equation (16.21) becomes

$$I_R(\lambda) = \int d\rho \, d^7 u \, F(r_\beta) \, J(\rho, u) \, e^{-i\lambda \eta(\rho, u)}, \qquad (16.45)$$

with $\eta(\rho, u) = \sum_\alpha c_\alpha \left(R_\alpha, r_\alpha(\rho, u)\right)$ then being expanded into a power series in $u$. Its term $^2\eta$ of second order is, for constant $\rho$, again a positive-definite quadratic form in $u$, because the dangerous case of $^1 r_\alpha$ being parallel to $R_\alpha$

for all $\alpha$ occurs only in the direction tangential to $\mathcal{M}_C$. The zero-order term is $\eta_0 = \sum c_\alpha(R_\alpha, R_\alpha) = 0$. The $u_h$ can be chosen such that $^2\eta(u) = \sum_{h=1}^{7} u_h{}^2$ for all $\rho$. Then (16.24) and (16.25) become

$$I_R(\lambda) \sim \int d\rho\, F(R_\beta(\rho))\, \hat{J}(\rho) \prod_1^7 \left\{ \int du_h e^{-i\lambda u_h{}^2} \right\}$$

$$\sim \lambda^{-\frac{7}{2}} \pi^{\frac{7}{2}} e^{-i\frac{7\pi}{4}} \int d\rho\, F(R_\beta(\rho))\, \hat{J}(\rho)\,, \qquad (16.46)$$

with $\hat{J}(\rho) = J(t_\alpha, \rho, u_h)\big|_{t_\alpha=0,\, u_h=0}$. And (16.34) becomes, defining $S_j(\rho)$, $T_i(\rho)$ by (16.42) and using in the decomposition (16.11) the easily derived relation $w_L = w_R{}^*$,

$$E_G(\lambda) \sim \frac{\pi^7}{\lambda^7} \int d\rho'\, d\rho\, \hat{J}(\rho')\, \hat{J}(\rho)\, w_R\big(-S_1(\rho'), -S_2(\rho'), T_1(\rho')\big)^*$$

$$\times w_R\big(-S_1(\rho), -S_2(\rho), T_1(\rho)\big) \prod_j U_j\big(-S_j(\rho'), -S_j(\rho)\big)$$

$$\times \prod_i L_i\big(T_i(\rho'), T_i(\rho)\big)\,. \qquad (16.47)$$

Concerning the connection of this result to realistic experiments we refer to the remarks made in Sect. 15.3. For sufficiently small supports of the $\tilde{f}$ and $\tilde{g}$ (in fact, it suffices that one of these supports is small) only values $\rho \sim \rho'$ contribute, and we may therefore replace $\rho'$ by $\rho$ in the factor $w_R{}^*$ in a sufficient approximation. Moreover, we may interpret the contribution of a given small $\rho$-interval as pertaining to a realistic experiment in which the energies of all four photons are precisely measured and are given by $-T_i^0(\rho)$ and $-S_j^0(\rho)$ for the incoming and outgoing particles, respectively, so that $\left|w_R^{v_1 v_2 u_1 u_2}\big(-S_1(\rho), \ldots, T_1(\rho)\big)\right|^2$ is a measure of the probability of such a scattering event, the $v_i$, $u_j$ indicating the polarizations of the photons. They are the physical spin parameters discussed in Sect. 16.2.

From the 1-particle results of Sect. 15.3 one might at first have expected independent scaling parameters $\rho_\alpha$ for each of the photons. That there is actually only one such parameter in the elastic case and none in the other cases is due to the stringent requirement that the positions of the sources and probes must be chosen such that there actually exists a scattering point $N$ satisfying condition (16.16).

As in the massive case, the generalization to any number of photon probes is straightforward.

## 16.4 Electrons

The IR problems associated with charged observed particles must yet be solved. Consider a 2-sector graph $G$ of the form shown in Fig. 16.4 containing charged particles. To fix the ideas, we assume that all four particles involved are electrons. The graph is divided by two cuts into three parts: the incoming $L$-part, the outgoing $U$-part, and the middle $R$-part describing the scattering event. The asymptotic behaviour of its contribution $E_G(\lambda)$ to $E(\lambda)$ (see (16.38) and (16.9)) is determined by the singularities in the cut variables $s_j, s'_j, t_i, t'_i$. The role of the product $wF$ in the massive case (16.9) is now played by the product of the $U$-, $L$-, and $R$-parts. These factors are not smooth in the cut variables, so that the relevant singularities are not just the poles from the cut lines. We must therefore determine the singularities of the various parts and show that in summing over graphs their dangerous singularities cancel. The fact that the cut variables are not independent for a given cross momentum $C$ must be kept in mind, though it does fortunately not seriously influence the determination of the singularities.

These singularities we find with the help of the methods established in Sect. 15.2 for the 1-particle case. For the reason explained there we use for the clothed propagators containing the cut lines the decomposition

$$t^+(s_1) \frac{1}{t^+(s_1)} t^+(s_1) \qquad (16.48)$$

etc. The factors $1/t^+(s_1)$, etc., have the form explained in Sect. 15.2 and need no further discussion. The denominator factor $t^+(s_1)$ is written as $\sqrt{t^+(s_1)}\sqrt{t^+(s_1)}$, one root being assigned to the $U$-part, the other one to the $R$-part, and similarly for the other cut lines. See (15.34) for the meaning of these square roots. We may then determine the singularities of the three parts separately.

The $U$-part, the part of $G$ above the upper cut in Fig. 16.4, is the product of two 1-particle parts $U_1$ and $U_2$ of the form found in Sect. 15.2. The functions $U'_j, \hat{U}_j, U_j, N_j^U$, and $U_{MSj}$, are defined like there. The relevant variables are $\Delta_j, \Delta'_j, s_j, s'_j, \nu_j$. We define $N^U = N_1^U N_2^U$. The $L$-part and $N^L$ are written as products in the same way.

There remains the $R$-part, the part of $G$ between the two cuts, including the cut lines. In contradistinction to $U$ and $L$, it is in general a connected subgraph of $G$. We must determine its dominant singularities at the mass shell in the cut variables $s_j, s'_j, t_i, t'_i$. In analogy to (15.24) we introduce the notations

$$\begin{aligned} s_j^+ &= s_{j0} + \omega_j, & s'^+_j &= s'_{j0} + \omega'_j, \\ \hat{s}_j &= (-\omega_j, \vec{s}_j), & \hat{s}'_j &= (-\omega'_j, \vec{s}'_j), \\ \Delta_j^U &= 2\omega_j s_j^+, & \Delta'^U_j &= 2\omega'_j s'^+_j. \end{aligned} \qquad (16.49)$$

$t_i^+, \hat{t}_i, \Delta_i^L$, and the corresponding primed variables are defined from $t_i, t'_i$ in the same way. Then we find by the methods of Sect. 16.1 that the asymptotically leading term in $E_G(\lambda)$ depends on the singularity of $R$ (and of course of $U$ and $L$) in the variables $\Delta_h^\rho$ ($\rho{=}U$ or $L$), at $\Delta_h^\rho = \Delta_h'^{\,\rho} = 0$, taken at

$$\hat{s}_j = \hat{s}'_j = -m\,\frac{\nu_j - N}{\sqrt{(\nu_j - N)^2}} =: S_j \;,$$
$$\hat{t}_i = \hat{t}'_i = m\,\frac{\beta_i - N}{\sqrt{(\beta_i - N)^2}} =: T_i \;, \qquad (16.50)$$

where $\nu_j$, $\beta_i$, $N$, are defined like in Sect. 16.1. We call the variables $s_j$, $s'_j$, with the same index $j$ a *pair*; also $t_i$ and $t'_i$ with the same $i$. The lines associated with a pair of variables are called a pair too. The $\Delta_h^\rho$ and $\Delta_h'^{\,\rho}$ take over the role of the variables $t_\alpha$ of the massive case.

The pole singularities of the cut lines are in general modified by IR singularities due to soft photon lines. The simplest examples of this effect are created by lines directly connecting two cut lines on the $R$-side of the cut. If these lines belong to the same pair, we are back in the 1-particle situation discussed in Sect. 15.2. The IR cancellation found there between the different possibilities of such connections inside a pair occurs here as well.

A similar cancellation happens between the various possibilities of connecting the cut lines of two different pairs. Let us consider an example of this in order $e^6$, the lowest order in which IR problems exist in reactions. Take a line connecting the $s_1$-line to the $s'_2$-line. It gives rise to the integral

$$\int dk\,\delta_-(k)\,\frac{1}{(s_1+k)^2 - m^2 + i\varepsilon}\,\frac{1}{(s'_2+k)^2 - m^2 - i\varepsilon}\,F(s_1+k, s'_1, s_2, s'_2+k)\;.$$

The factor $F$ is due to the rest of the graph integrated over all variables except those of the two $s$-pairs. It is a smooth function. The $k$-integral is IR divergent at the origin $k = 0$ on the mass shell $s_1^2 = s_2'^{\,2} = m^2$. Notice that only the value $F = F(s_1, \ldots, s'_2)$ at $k = 0$ appears in this singularity: as far as the IR problem is concerned, we can draw this factor in front of the integral. There are three more ways of connecting a cut 1-line with a cut 2-line by a photon line, all of them leading to IR divergences at $\Delta_1^U = \Delta_2^U = \Delta_1'^{\,U} = \Delta_2'^{\,U} = 0$. We isolate the IR relevant parts of the four integrals like in the $e^2$-example of Sect. 15.2 (see (15.10)), and we use that we are only interested in the values of the integrals at $\hat{s}_j = \hat{s}'_j = S_j$. Omitting factors common to the four terms, including $F$ and $\prod_j (\Delta_j^U \Delta_j'^{\,U})^{-1}$, and dropping the index $U$, they sum to

$$\Sigma(\Delta_j, \Delta'_j) = \frac{i}{2\pi}\int dk\,\frac{1}{k^2 + i\varepsilon}\,\frac{1}{-\Delta_1 - K_1 + i\varepsilon}\,\frac{1}{-\Delta_2 + K_2 + i\varepsilon}$$
$$-\frac{i}{2\pi}\int dk\,\frac{1}{k^2 - i\varepsilon}\,\frac{1}{-\Delta'_1 + K_1 - i\varepsilon}\,\frac{1}{-\Delta'_2 - K_2 - i\varepsilon}$$
$$+\int dk\,\delta_-(k)\,\frac{1}{-\Delta'_1 + K_1 - i\varepsilon}\,\frac{1}{-\Delta_2 + K_2 + i\varepsilon}$$
$$+\int dk\,\delta_+(k)\,\frac{1}{-\Delta_1 - K_1 + i\varepsilon}\,\frac{1}{-\Delta'_2 - K_2 - i\varepsilon}\;. \qquad (16.51)$$

The $k$-integrations extend over $\mathcal{K}$, and we have defined $K_j = 2(S_j, k)$. We notice at once that the four integrands sum to zero at the critical point

$\Delta_j = \Delta'_j = 0$. This suggests cancellation of the IR singularities present in the individual terms. But this appearance is deceptive, because $\Sigma$ is not continuous at the critical point. The first term is logarithmically divergent at $\Delta_1 = \Delta_2 = 0$, and similarly for the other terms. However, if we split $\frac{1}{k^2 \pm i\varepsilon} = \frac{\mathcal{P}}{k^2} \mp i\pi\big(\delta_+(k) + \delta_-(k)\big)$ and consider the principal-value contributions, we find them to sum to

$$-\frac{i}{2\pi} \int dk \, \frac{\mathcal{P}}{k^2} \, \frac{[\Delta_1 \Delta_2 - \Delta'_1 \Delta'_2] + [(\Delta_2 + \Delta'_2) K_1 - (\Delta_1 + \Delta'_1) K_2]}{(\Delta_1 + K_1 - i\varepsilon)(-\Delta_2 + K_2 + i\varepsilon) - (-\Delta'_1 + K_1 - i\varepsilon)(\Delta'_2 + K_2 + i\varepsilon)} \, . \tag{16.52}$$

The first square bracket multiplies a $k$-integral which is IR divergent of second order at

$$D := \sqrt{\Delta_1{}^2 + \Delta_2{}^2 + \Delta'_1{}^2 + \Delta'_2{}^2} = 0 \, . \tag{16.53}$$

This means that this first term has no logarithmic singularity at $D = 0$ but only a bounded discontinuity. Its $\lim_{D \to 0}$ exists if taken along a fixed direction in $\Delta$-space. The same is true for the terms coming from the second square bracket, hence the $\frac{\mathcal{P}}{k^2}$-contribution to $\Sigma$ has this property. The same can be shown for the $\delta_\pm$-contributions, so that it holds for $\Sigma$: $\lim_{D \to 0} \Sigma(\Delta_j, \Delta'_j)$ exists if taken along a fixed direction in $\Delta$-space.

The remarks made in Chap. 15 about the analogous problem in the 1-particle case (see (15.21) and its discussion) apply here as well. If the terms in $\Sigma$ are made sense of individually by introducing an IR regularization, e.g. a small photon mass, then $\Sigma$ *is* continuous around $\Delta_j = \Delta'_j = 0$ and vanishes at this critical point. Hence the conventional method of introducing such a regularization, then restricting the expression to the mass shell, then removing the regularization, misses the contribution (16.52) and the question of the legitimacy of the regulator method arises. It will be shown in Sect. 17.2 that as a matter of fact the additional term does not contribute to the leading term of $E(\lambda)$. This result does, however, not generalize to higher orders, where terms of the type (16.52) become essential. Nor does the result apply to the 1-particle case of Chap. 15, even in the lowest non-trivial order. And this 1-particle problem is present in our 2-particle case as well, where it is caused by soft photon lines connecting trajectories of the same pair.

The generalization to arbitrary orders follows the procedure used for the 1-particle case in Sect. 15.2. The singularities in the cut variables are determined by decomposing the $R$-graph into a core and a web, the latter being of a simple structure and containing the IR problem in full. The $R$-web contains eight semitrajectories associated with the variables $s_j$, $s'_j$, $t_i$, $t'_i$. The trajectories with the same $j$ or the same $i$ form a "pair". They are associated with the same observed particle.

The semitrajectories start inward from the cut lines. We must make sure that they remain separated, that they do not get entangled. This we achieve by choosing the size $R$ of the soft region $\mathcal{K}$ sufficiently small, depending on the $\tilde{f}$-$\tilde{g}$-supports and the order of perturbation theory, such that there are necessarily hard photon lines present, i.e. a core which is met by any subtrajectory before it runs into another one. The dependence of $R$ on the perturbative order must be kept in mind if using the summed version $W_R$ of the web to be described presently. This summed form makes sense only as a generating functional for its finite-order terms. For any chosen value of $R$, these finite-order terms are given accurately up to a sufficiently low order.

The photon lines of the web connect any two of the semitrajectories in accordance with the rules layed down in Chap. 15. We order the pairs as follows: $s_1$, $s_2$, $t_1$, $t_2$, and let photon lines between different pairs point from the earlier to the later pair, while for lines staying inside a pair the rules of the 1-particle case apply. The web rules are as given in Sect. 15.2, with the variables appearing in the vertex factors and the fermion propagators being those of the semitrajectory to which they belong.

Adding all possible webs, we arrive at the equivalent of (15.60). It reads

$$W_R(\Delta_j^U, \ldots, \Delta_j'^L)$$
$$= \frac{1}{(2\pi)^8} \prod_j \left[ \int_0^\infty d\sigma_j^U\, e^{-i\sigma_j^U(\Delta_j^U - i\varepsilon)} \int_0^\infty d\sigma_j'^U\, e^{i\sigma_j'^U(\Delta_j'^U + i\varepsilon)} \right]$$
$$\times \prod_i \left[ \int_0^\infty d\sigma_i^L\, e^{-i\sigma_i^L(\Delta_i^L - i\varepsilon)} \int_0^\infty d\sigma_i'^L\, e^{i\sigma_i'^L(\Delta_i'^L + i\varepsilon)} \right]$$
$$\times \exp\left\{ \sum_j H_j^U + \sum_i H_i^L + H_{12}^U + H_{12}^L + H_{12}^{UL} + H_{21}^{UL} \right\}. \quad (16.54)$$

$H_j^U$ has the form of the first exponent in (15.60) with $\sigma$ replaced by $\sigma_j^U$ and analogously for $\sigma_j'$, $\Delta_j$, $\Delta_j'$. $K$ becomes $K_j^U = 2(S_j, k)$, and the $\delta_+$ in the third term becomes a $\delta_-$. This last change is necessary because we are now concerned with semitrajectories below the cut, not above it like in (15.60). $S_j$ is the critical value of $s_j$, $s_j'$, defined in (16.42). $H_i^L$ has again the form (15.60), this time with a $\delta_+$ remaining in the third term and with the variables $\sigma_i^L$, $\sigma_i'^L$, $\Delta_i^L$, $\Delta_i'^L$, and $K_i^L = 2(T_i, k)$. The remaining terms describe the contributions of photon lines connecting different pairs. $H_{12}^U$ is given by

$$H_{12}^U = \frac{e^2(S_1, S_2)}{2\pi^3} \left[ \frac{i}{2\pi} \int_K \frac{dk}{k^2 + i\varepsilon} \frac{e^{i\sigma_1^U K_1^U} - 1}{K_1^U} \frac{e^{-i\sigma_2^U K_2^U} - 1}{K_2^U} \right.$$
$$- \frac{i}{2\pi} \int_K \frac{dk}{k^2 - i\varepsilon} \frac{e^{i\sigma_1'^U K_1^U} - 1}{K_1^U} \frac{e^{-i\sigma_2'^U K_2^U} - 1}{K_2^U}$$
$$- \int_K dk\, \delta_+(k) \frac{e^{i\sigma_1'^U K_1^U} - 1}{K_1^U} \frac{e^{-i\sigma_2^U K_2^U} - 1}{K_2^U}$$
$$\left. - \int_K dk\, \delta_-(k) \frac{e^{i\sigma_1^U K_1^U} - 1}{K_1^U} \frac{e^{-i\sigma_2'^U K_2^U} - 1}{K_2^U} \right]. \quad (16.55)$$

## 16.4 Electrons

The coefficients $\pm\frac{i}{2\pi}$ of the one-sector terms differ from the corresponding $\pm\frac{i}{4\pi}$ of $H_j^U$ because the latter describe photon lines connecting a trajectory to itself and contain a combinatorial factor $\frac{1}{2}$ from the invariance of the graph under a reversal of direction of such a line.

$H_{12}^L$ has the same form, except that the variables are $\sigma_i^L, \ldots,$ and $T_i$ instead of $S_j$, and all $\sigma$-exponents change their sign. $H_{ji}^{UL}$ is

$$H_{ji}^{UL} = \frac{e^2(S_j, T_i)}{2\pi^3}\left[-\frac{i}{2\pi}\int_{\mathcal{K}}\frac{dk}{k^2 + i\varepsilon}\frac{e^{i\sigma_j^U K_j^U} - 1}{K_j^U}\frac{e^{i\sigma_i^L K_i^L} - 1}{K_i^L}\right.$$

$$+\frac{i}{2\pi}\int_{\mathcal{K}}\frac{dk}{k^2 - i\varepsilon}\frac{e^{i\sigma_j'^U K_j^U} - 1}{K_j^U}\frac{e^{i\sigma_i'^L K_i^L} - 1}{K_i^L}$$

$$+\int_{\mathcal{K}} dk\, \delta_-(k)\frac{e^{i\sigma_j^U K_j^U} - 1}{K_j^U}\frac{e^{i\sigma_i'^L K_i^L} - 1}{K_i^L}$$

$$\left.+\int_{\mathcal{K}} dk\, \delta_+(k)\frac{e^{i\sigma_j'^U K_j^U} - 1}{K_j^U}\frac{e^{i\sigma_i^L K_i^L} - 1}{K_i^L}\right]. \tag{16.56}$$

Terms corresponding to the second exponential in (15.60) do not occur because the $R$-part contains no external vertices.

The possible presence of observed positrons brings no essential complication. The only change in the web rules is a change of sign of the vertex factors on positron semitrajectories. This turns the factor $(S_1, S_2)$ into $-(S_1, S_2)$ in (16.55), $(S_j, T_i)$ into $-(S_j, T_i)$ in (16.56), if the two variables involved refer to particles of opposite charge.

In the foregoing considerations a new source of IR problems existing in the $R$-part has been ignored. There may exist fermionic cross lines associated with unobserved charged particles produced in the reaction. Their momenta are by definition restricted to the mass shell. Hence a soft photon line connecting such a cross line to itself like in Fig. 10.4 or to another fermionic cross line like in Fig. 11.1 creates an IR divergence. But this is the mild IR problem that has been solved in Chap. 11 and therefore needs not concern us now. However, another problem does concern us: soft photon lines connecting a fermionic cross line to one of the cut lines of $G$, or more generally to one of the semitrajectories of the $R$-part, produce logarithmic singularities in the corresponding variable $\Delta$ or $\Delta'$. But we show that these singularities cancel in $W_R$, so that they have no influence on the asymptotic behaviour of $E(\lambda)$.

The $\Delta$-singularities can again be isolated in the web of a web–core splitting of the $R$-part. The enlarged web contains also the fermionic cross lines and their propagators $\pm\delta_\pm(c)$, where $c$ is a cross momentum. To a $c$-line we associate two new semitrajectories starting from it in both directions. This means that we must be more specific about the graphs $G$ that we consider. Instead of characterizing them simply by the number and nature of the cut pairs of lines, we must also specify the number and direction of the fermionic cross lines. The construction of the web follows the by now well-trodden path. The summation over all possibilities of attaching a given number of photonic web lines to a cross-semitrajectory is simpler than for the trajectories starting from cut lines, because the momentum $c$ of the cross line is restricted to the mass shell. Take e.g. the case of $c$ flowing from the $T^-$- to the $T^+$-sector, and the adjacent $T^+$-subtrajectory. Let $\{\ell_j\}$ be the momenta of the photon lines which connect that subtrajectory to itself, $\{k_i\}$ the photon momenta entering the trajectory coming from other trajectories. Then the sum (15.58) is replaced by the simple expression

$$\delta_+(c) \prod_j \left(-[\ell_j]^2\right) \prod_i [k_i] ,  \tag{16.57}$$

with $[p] := \frac{i}{2\pi} \frac{1}{2(c,p)+i\varepsilon}$. In this result the web part of the mass renormalization is duly included. We will not give its derivation, which is not difficult.[4] If no $\ell_j$-lines are present, the result is easily derived by induction with respect to the number of $k_i$'s. And the $k_i$-factors are the ones that really interest us, the $\ell_j$-factors being part of the mild IR problem that has already been solved (they involve no $\Delta$-semitrajectory).

From here on we proceed as before. Summing over all orders, we still find a nice separation of the contributions from photons connecting two pairs of semitrajectories. If at least one of these pairs belongs to a cross line, the corresponding contribution is trivial. Let us take by way of example a cross line with momentum $c$ flowing from $T^+$ to $T^-$, and a $T_i$ pair. Dropping the index $i$ we get a new factor of the general type (16.56) in the integrand of the $\tau$–$\tau'$-integral defining the asymptotically relevant part of $W_R$. The new factor reads

$$-\int dc\,\delta_-(c)\, \exp\left\{ \frac{e^2(c,T)}{2\pi^3} \int \frac{dy}{(Y_c+i\varepsilon)(Y_T+i\varepsilon)} \left[ -\frac{1}{2\pi} \frac{e^{i\tau_T Y_T}-1}{y^2+i\varepsilon} \right.\right.$$
$$\left.\left. -\frac{1}{2\pi} \frac{e^{i\tau'_T Y_T}-1}{y^2-i\varepsilon} - i\,\delta_+(y)\left(e^{i\tau_T Y_T}-1\right) \right.\right.$$
$$\left.\left. + i\,\delta_-(y)\left(e^{i\tau'_T Y_T}-1\right) \right]\right\} \tag{16.58}$$

in obvious notation. In the first term of the $y$-integral we integrate first over $y^0$, closing the path of integration in the lower half-plane. The only pole included in the contour is the one at $y^0 = |\vec{y}| - i\varepsilon$, and its contribution precisely cancels the $\delta_+(y)$-term. The other two terms of the $y$-integral cancel in the same way. Hence $\exp\{\cdots\} = 1$ in (16.58), which shows that the $c$–$T$-singularities are irrelevant to the $\lambda$-asymptotics. The same holds for the other combinations of a cross line and a cut pair. Hence (16.54) gives the correct form of the asymptotically relevant $R$-singularity.

---

[4]See Lemma 17.1 for a proof.

We must study the singularity of $W_R$ at $\Delta_h^\rho = \Delta'^\rho_h = 0$ for all $\rho$, $h$, where $\rho = U$ or $L$, and $h = 1$ or $2$. We define

$$D = \sqrt{\sum_{h,\rho}\left[\left(\Delta_h^\rho\right)^2 + \left(\Delta'^\rho_h\right)^2\right]}, \tag{16.59}$$

and introduce the scaled variables

$$\Delta_h^\rho = \alpha_h^\rho D, \quad \Delta'^\rho_h = \beta_h^\rho D, \quad \tau_h^\rho = \sigma_h^\rho D, \quad \tau'^\rho_h = \sigma'^\rho_h D, \quad k = yD. \tag{16.60}$$

Note that $\sum_{h\rho}\left((\alpha_h^\rho)^2 + (\beta_h^\rho)^2\right) = 1$. Defining $Y_j^U = 2(S_j, y)$, $Y_i^L = 2(T_i, y)$, $\mathcal{K}_D = \{y : |\vec{y}| \leq \frac{R}{D}\}$, we find as a generalization of (15.63):

$$W_R(D) = \frac{1}{(2\pi D)^8} \prod_{h,\rho}\left(\int_0^\infty d\tau_h^\rho\, e^{-i\tau_h^\rho(\alpha_h^\rho - i\varepsilon)} \int_0^\infty d\tau'^\rho_h\, e^{i\tau'^\rho_h(\beta_h^\rho + i\varepsilon)}\right)$$

$$\times \exp\left\{\sum_{h\rho} M_h^\rho + \sum_\rho M_{12}^\rho + M_{12}^{UL} + M_{21}^{UL}\right\}. \tag{16.61}$$

The $M_h^\rho$ etc. are defined like the $H_h^\rho$ etc. in (16.54), except that all variables are replaced by their rescaled versions according to (16.60), $K_h^\rho$ by $Y_h^\rho$, and the $y$-integral replacing the $k$-integral extends over $\mathcal{K}_D$. Note that, apart from the factor $D^{-8}$ in front, the expression (16.61) depends on $D$ exclusively through the domain of integration $\mathcal{K}_D$. The 1-particle terms $M_h^\rho$ are of the form studied in Chap. 15 in connection with (15.63), and we can take over the results obtained there for the $D \to 0$ behaviour of these terms.

All the 2-particle terms are of a similar structure. It suffices therefore to discuss one of them in detail. We choose $M_{12}^L$. Since it depends only on the $L$-variables, we drop the index $L$ for the time being. The $\delta_+$-term

$$-\int dy\, \delta_+(y) \frac{e^{-i\tau'_1 Y_1} - 1}{Y_1} \frac{e^{i\tau_2 Y_2} - 1}{Y_2}$$

in $M_{12}$ (see (16.55)) can be written as

$$Z_3 = -X_+(-\tau'_1, 0) - X_+(0, \tau_2) + X_+(-\tau'_1, \tau_2), \tag{16.62}$$

with

$$X_\pm(u_1, u_2) = \int dy\, \delta_\pm(y) \frac{1 - e^{i(u_1 Y_1 + u_2 Y_2)}}{Y_1 Y_2}.$$

Note that this integral converges at $y = 0$ despite the vanishing there of the $Y_j$ because the numerator vanishes too. Analogously, the $\delta_-$-term is

$$Z_4 = -X_-(-\tau_1, 0) - X_-(0, \tau'_2) + X_-(-\tau_1, \tau'_2). \tag{16.63}$$

The $(y^2 + i\varepsilon)^{-1}$-term is

$$Z_1 = \frac{i}{2\pi} \int \frac{dy}{y^2 + i\varepsilon} \frac{(e^{-i\tau_1 Y_1} - 1)(e^{i\tau_2 Y_2} - 1)}{Y_1 Y_2} .$$

The zeros of the $Y_j$ are no poles of the quotient, the numerator vanishing for $Y_j = 0$. Therefore we may introduce an arbitrary $i\varepsilon$-prescription for these would-be poles, obtaining existent integrals after the numerator has been multiplied out and the resulting terms separated. We define $Y_1 Y_2 = (Y_1 + i\varepsilon)(Y_2 - i\varepsilon)$. The $y^0$ integration path passes then below the $Y_1$-pole, and above the $Y_2$-pole. We decompose the numerator as

$$(1 - e^{i\tau_2 Y_2}) + (1 - e^{-i\tau_1 Y_1}) - (1 - e^{i(\tau_2 Y_2 - \tau_1 Y_1)}) . \tag{16.64}$$

In the first contribution we may close the contour of the $y^0$-integration by an infinite semicircle in the lower half-plane. The putative $Y_2 = 0$ pole enclosed by this contour does not actually exist because its residue vanishes. The only remaining contribution is that from the pole at $y^0 = |\vec{y}| - i\varepsilon$. The result is

$$\int dy\, \delta_+(y) \frac{1 - e^{i\tau_2 Y_2}}{Y_1 Y_2} = X_+(0, \tau_2) ,$$

which cancels a term in $Z_3$. Analogously, the second term in (16.64) yields $X_-(-\tau_1, 0)$, cancelling a term in $Z_4$. The third contribution remains. The $1/(y^2 - i\varepsilon)$-term $Z_2$ is handled in the same way, leading to a further cancellation of terms in $Z_3$ and $Z_4$. What remains in the exponent of (16.61) is

$$\int_{\mathcal{K}_D} \frac{dy}{(Y_1 + i\varepsilon)(Y_2 - i\varepsilon)} \left[ -\frac{i}{2\pi} \frac{1}{y^2 + i\varepsilon} \left(1 - e^{i(-\tau_1 Y_1 + \tau_2 Y_2)}\right) \right.$$
$$+ \frac{i}{2\pi} \frac{1}{y^2 - i\varepsilon} \left(1 - e^{i(-\tau'_1 Y_1 + \tau'_2 Y_2)}\right) + \delta_+(y) \left(1 - e^{i(-\tau'_1 Y_1 + \tau_2 Y_2)}\right)$$
$$\left. + \delta_-(y) \left(1 - e^{i(-\tau_1 Y_1 + \tau'_2 Y_2)}\right) \right] . \tag{16.65}$$

We notice first that the unit terms in the round brackets sum to zero and can therefore be omitted. Second, the integral exists as a continuous function of $\tau_j$, $\tau'_j$. The singular factors in the integrand are defined as distributions. A problem concerning their multiplication might arise at $y = 0$, where they are all singular and their product is not defined as a distribution. But the singularity is of order $|y|^{-4}$, it is almost integrable, and the regular round brackets vanish at $y = 0$, which kills the divergence. Third, the limit $D \to 0$ of the integral exists as a locally integrable function of $\tau_j$, $\tau'_j$, which is continuous almost everywhere. Taking this limit amounts to replacing $\mathcal{K}_D$ by $R^4$ as domain of integration. In view of the definition of $\mathcal{K}_D$ it makes sense to define this integral by first integrating over $y^0$, then over $\vec{y}$. The $y^0$-integral clearly exists and yields a function whose absolute value decreases like $|\vec{y}|^{-3}$ for

$|\vec{y}| \to \infty$. Moreover, the exponentials in the various terms contain factors like e.g. $e^{-2i(\vec{y},\tau_1 \vec{S}_1 - \tau_2 \vec{S}_2)}$, which is unchanged by the $y^0$-integration and is oscillatory in almost all directions except if $\tau_1 = \tau_2 = 0$, and similar conditions for the other terms. At these exceptional points the limit is logarithmically divergent, while it exists elsewhere. Like in the discussion of $c_2$ in (15.74) we find that $M_{12}^L$ is for $D=0$ a locally integrable function without a strong oscillatory behaviour, which is continuous almost everywhere and has at most weak singularities where certain combinations of at least two $\tau$–$\tau'$-variables vanish, and it increases at most weakly at infinity. The same is then true for every term of finite order in the perturbative expansion of $\exp(M_{12}^L)$. The same result holds for the other 2-particle terms in (16.61). As in the 1-particle case we conclude that $\lim_{D \to 0} D^8 W_R(D)$ exists as a function of the directional parameters $\alpha_h^\rho$, $\beta_h^\rho$ in $\Delta$-space for almost all values of these parameters.

The leading singularity at $D=0$ of $W_R(D)$ contains exponentials of the type $e^{C_1 \log D}$ of (15.75) coming from the 1-particle terms $M_h^\rho$. Like in the 1-particle case, these singularities cancel against corresponding singularities in $W_U$ and $W_L$ and the numerators $(W_+^\rho W_-^\rho)^{1/2}$. The 2-particle terms do not contribute such undesirable logarithms. Effectively, $W_R(D)$ behaves then for $D \to 0$ as

$$W_R(D) \simeq \frac{1}{(2\pi D)^8} \prod_{h\rho} \left( \int_0^\infty d\tau_h^\rho \, e^{-i\tau_h^\rho \alpha_h^\rho} \int_0^\infty d\tau_h'^\rho \, e^{i\tau_h'^\rho \beta_h^\rho} \right) e^{\bar{M}_1 + \bar{M}_2}. \quad (16.66)$$

The sign $\simeq$ denotes equality of the leading singularities at $D=0$ apart from the log $D$-factors which have been omitted. The $y$-integral extends over $R^4$, the $i\varepsilon$ in $\alpha_h^\rho$, $\beta_h^\rho$, and $Y_h^\rho$ have been suppressed. $\bar{M}_1$ contains the $C_2$-terms according to (15.75) of the $M_h^\rho$, and $\bar{M}_2$ is the sum over the 2-particle $M$'s in (16.61) taken at $D=0$, i.e. with the $y$-integrals extending over the full space $R^4$.

$W_R$ can be rewritten in terms of the original variables $\Delta_h^\rho$, $\Delta_h'^\rho$, $S_j$, $T_i$. Its leading singularity, called $N_R(\Delta_j^U, \ldots, T_i)$ is Lorentz invariant for the same reason as its 1-particle analogue $N_U$ of Chap. 15: it depends on $S_j$, $T_i$, only through the invariants $(S_{j_1}, S_{j_2})$ etc. But, contrary to the 1-particle case, there exists as yet no explicit expression for $N_R$ as a function of these variables only. We define therefore $N^R$ as a function still depending on the 4-vectors $S_j$, $T_i$ by

$$N^R(\Delta_1^U, \ldots) = \lim_{\kappa \to 0} \kappa^2 \frac{W_R(\kappa \Delta_1^U, \ldots)}{\prod_j \sqrt{N_+(\kappa \Delta_j^U) N_-(\kappa \Delta_j'^U)} \prod_j \sqrt{N_+(\kappa \Delta_i^L) N_-(\kappa \Delta_i'^L)}} . \quad (16.67)$$

As a function of the $\Delta$-variables, $N^U$ and $N^L$ are homogeneous of order $-1$, $N^R$ of order $-2$.

Let $U(\Delta_j^U, \Delta_j'^U)$ be the summed $U$-parts of the graphs $G$, including their shares of the clothed $S_j$-propagators, integrated over the test spinors $\tilde{f}_j$. And let $L(\Delta_i^L, \Delta_i'^L)$ and $R(\Delta_\alpha, \Delta_\alpha')$ be defined analogously. In $R$ we have introduced the notations (see (16.42))

$$\Delta_1 = \Delta_1^U, \quad \Delta_2 = \Delta_2^U, \quad \Delta_3 = \Delta_1^L, \quad \Delta_4 = \Delta_2^L, \tag{16.68}$$

and the same for the $\Delta_\alpha'$. We will also use the $R_\alpha$ defined in (16.42). In analogy to Theorem 15.3 we define

$$\begin{aligned} U_{\mathrm{MS}} &= \lim_{D \to 0} \frac{U(\Delta_j^U, \ldots)}{N^U(\Delta_j^U, \ldots)}, \\ L_{\mathrm{MS}} &= \lim_{D \to 0} \frac{L(\Delta_i^L, \ldots)}{N^L(\Delta_i^L, \ldots)}, \\ R_{\mathrm{MS}} &= \lim_{D \to 0} \frac{R(\Delta_\alpha, \ldots)}{N^R(\Delta_\alpha, \ldots)}. \end{aligned} \tag{16.69}$$

Remember the dependence of these expressions on the $R_\alpha$. The limits $D \to 0$ exist by our previous results. They generalize the restriction to the mass shell of the amputated $U$, $L$, $R$, that are known from the conventional formalism.

Writing down the QED analogue of (16.9) and proceeding along the lines established in the massive theory we arrive at

$$\begin{aligned} E(\lambda) &\sim \int dC \, U_{\mathrm{MS}}(C, R_\alpha) \, R_{\mathrm{MS}}(C, R_\alpha) \, L_{\mathrm{MS}}(C, R_\alpha) \\ &\times \int \prod_{\beta=1}^{3} \{dr_\beta dr_\beta'\} \, N(\Delta_\alpha, \Delta_\alpha', R_\alpha) \prod_{1}^{4} e^{-i\lambda(\xi_\alpha, r_\alpha - r_\alpha')}, \end{aligned} \tag{16.70}$$

with

$$N = N^U N^R N^L. \tag{16.71}$$

The behaviour of the $r_\beta$–$r_\beta'$-integral $A$ for $\lambda \to \infty$ is determined by a slight generalization of the methods used in the massive case (see also the 1-particle analogue (15.25)-(15.31)). Notice that we talk about the full integral, not splitting it into a primed and an unprimed part. As new variables of integration we introduce $\Delta_\alpha$, $\Delta_\alpha'$, (replacing the $t_\alpha$, $t_\alpha'$, of the massive case) and regular functions $u_1, \ldots, u_8; u_1', \ldots, u_8'$, of the $r_\beta; r_\beta'$, respectively, which vanish at $\vec{r}_\beta = \vec{r}_\beta' = \vec{R}_\beta$. As mentioned after (16.20), they may be chosen as a suitable subset of the components of $\vec{r}_\beta - \vec{R}_\beta$, $\vec{r}_\beta' - \vec{R}_\beta$. The Jacobian $J(\Delta_\alpha, u_h, C) = \frac{\partial(r_1, r_2, r_3)}{\partial(\Delta, u)}$ is regular. Since only an arbitrarily small neighbourhood of the critical point $\Delta_\alpha = 0 \,\forall\, \alpha$ is relevant for our purposes, we may replace $J$ in our expressions by its value $J(u, C)$ at $\Delta = 0$.

Like in (15.25) we split $(\xi_\alpha, r_\alpha) = (\xi_\alpha, \hat{r}_\alpha) + \xi_\alpha^0 r_\alpha^+$ and replace $r_\alpha^+$ by $\frac{\Delta_\alpha}{2\omega(\vec{r}_\alpha)}$, where the denominator may be evaluated at $\vec{R}_\alpha$ instead of $\vec{r}_\alpha$ without affecting the $\lambda$-asymptotics. The sum $\sum(\xi_\alpha, \hat{r}_\alpha)$ is then expressed as a function of the $u_i$. It depends only on the $\vec{r}_\alpha$, not on the $\Delta_\alpha$. We obtain

$$A(\lambda) = \int d^8u\, d^8u'\, J(u,C)\, J(u',C)\, e^{-i\lambda(\eta(u)-\eta(u'))}$$
$$\times \int d^4\Delta\, d^4\Delta'\, N(\Delta,\Delta')\, e^{i\lambda \sum_\alpha |n_\alpha|(\Delta_\alpha-\Delta'_\alpha)} \quad (16.72)$$

with $n_\alpha = \frac{\xi^0_\alpha}{2\omega(\vec{R}_\alpha)}$. $\eta(u)$ is the function defined in (16.21) and possesses the properties established there. In particular, the $u$–$u'$-integral has the value $\pi^8 \lambda^{-8} J(0,C)^2 (\det M)^{-1}$ found there. That the various approximations used ($\Delta_\alpha = 0$ in $J$, $\vec{r}_\alpha = \vec{R}_\alpha$ in $n$) are of no consequence is seen by remembering that only the critical values $r_\alpha = r'_\alpha = R_\alpha$, $\Delta_\alpha = \Delta'_\alpha = 0$ are of relevance. Hence we may introduce, without changing the asymptotics, a test function in $r_\alpha, \ldots, \Delta'_\alpha$, into our expression, which is $\equiv 1$ in an arbitrarily small neighbourhood of the critical point and vanishes outside a larger but still arbitrarily small neighbourhood. Going over to the scaled variables $\delta = \lambda\Delta$, the $\Delta$–$\Delta'$-integral in (16.72) is found to be

$$I = \int d^4\delta\, d^4\delta'\, N(\delta,\delta')\, e^{-i \sum_\alpha n_\alpha(\delta_\alpha - \delta'_\alpha)}, \quad (16.73)$$

i.e. it does not depend on $\lambda$. We have used that $N$ is homogeneous of order $-8$. This fact tells us also that $N$ decreases at infinity of $8^{\text{th}}$ order, which together with the oscillatory factor renders the integral conditionally convergent. Of course $N$ depends also on the $R_\alpha$. $I$ cannot be apportioned in a natural way to $U$, $L$, $R$.

We have arrived at the analogue of (16.26),

$$E(\lambda) \sim \frac{\pi^8}{\lambda^8} \int dC\, J(0,C)^2 (\det M)^{-1}\, U_{\text{MS}}(C)\, R_{\text{MS}}(C)\, L_{\text{MS}}(C)\, I(C) \ . \quad (16.74)$$

Again, the dependence on $R_\alpha$, i.e. on $P_i$ and $Q_j$, of the terms in this expression is understood. For the definition of $M$ see the remark after (16.25). The physical interpretation of this expression will be discussed in Chap. 17.

The simplification leading in the massive case to (16.32) is generally possible in QED if at least one of the observed incoming or outgoing particles is charged. This is so because the result (16.74) is obtained by a summation over graphs with an arbitrary number of photonic cross lines. Graphs with a given finite number of cross lines, especially zero or one, give a vanishing contribution. This means that the probability of a given fixed number of soft photons being emitted is zero, a well-known result. Let us discuss this for the case of no cross line, the other cases being easy generalizations of this. In this special case, the integrand $URL$, including the webs, factorizes into a $T^+$- and a $T^-$-term. We need only consider the $T^+$-factor of the web. The corresponding $U^+$-web $W_U^+$ is given by a simplified version of (16.66) containing only the factors referring exclusively to unprimed variables, however including also the corresponding terms coming from the second exponential in (15.60) describing the influence of the composite external vertices. This "physical" exponent summed over our restricted set of graphs does not vanish. The terms involving both pairs of trajectories do not affect the strength of the critical singularity at $D = 0$, as has been shown before. But from the 1-particle contributions we expect in the integrand factors of the form $\exp[C_1(\tilde{s}) \log D]$ as indicated in (15.75). These factors vanish for $D \to 0$ like $D^{C_1(\tilde{s})}$ if $C_1$ is positive, thus weakening the total singularity. That this really happens depends crucially on working in a physical state space. If we retain only the Gupta–Bleuler factors in (15.60), the exponent reduces to the $G_F$ defined in (15.64), up to a positive coefficient. From our analysis of (15.64) we know that the logarithmic divergence of $G_F$ for $D \to 0$ is due to the constant terms in the numerator and is given by the logarithmic divergence of $\int_{K_D} dy\, \delta_+(y) \frac{1}{Y^2}$ which is caused by the insufficiently fast decrease of the integrand at infinity. Since the integrand is positive the divergence is positive, implying a negative $C_1$, $\log D$ being negative for small $D$. But this changes if the $\Psi_n$-contributions are included. Since the web is free of UV problems we can achieve this by going over to the B-version of the graph rules given in Sect. 12.2. This means replacing the factor $g_{\mu\nu}$ in the photon propagators by the expression (12.23),

$$-M_{\mu\nu}(k) = \left(\delta_\mu^\kappa + i\, k_\mu \tilde{r}^\kappa(k)\right) g_{\kappa\lambda} \left(\delta_\nu^\lambda - i\, k_\nu \tilde{r}^\lambda(-k)\right),$$

and dropping the $\Psi_n$-contributions. This change of rules only in the web seems at first inadmissible. A semitrajectory of a web does in general not end in an external point, so that the proof of the equivalence of the two sets of rules does not apply. But this can be amended by including in the web soft photon lines with both ends on one of the basic trajectories, even if one or both ends lie in the core, or if the line crosses the cut, i.e. belongs to an amalgamated $U$–$R$ or $R$–$L$ web without a 2-particle cut. These latter contributions are soft-particle contributions from graphs without a corresponding 2-line cut, which are therefore asymptotically irrelevant. These enlargements of the web are legitimate because they do not change the strength of the web singularity at $D = 0$. Nor do they introduce UV problems. After the changeover from the A-rules to the B-rules has been effected, the webs containing irrelevant additional lines can be dropped again, resulting in webs $W_U^p, W_L^p, W_R^p$ of the old form but with the physical photon propagators. The earlier Gupta–Bleuler derivation of $C_1$ can be applied to the new situation. Remember that $\tilde{r}(k)$ does not depend on $k_0$, so that the method of residues can still be used for calculating the $k_0$-integral. The only change in the result is the replacement of the positive factor $(\hat{s}, \hat{s}) = m^2$ by $-(\hat{s}, M\hat{s})$ which has been shown to be negative in (12.34) and (12.35). Hence $C_1$ changes its sign relative to the Gupta–Bleuler case. As a result, the $D$-singularity of the $U^+$-part is weakened relative to the fully summed one and is therefore asymptotically irrelevant. This weakening is partly, but not completely, offset by a corresponding but weaker effect from the division by a $\sqrt{W^+}$ for each semitrajectory. The B-rules must be applied to these denominators too. Similar results hold for $W^R$ and $W^L$. This result and its generalization to finitely many cross lines prove the contention that in reactions involving charged particles infinitely many photons are necessarily emitted. The contribution to the inclusive cross section of processes involving only finitely many photons is zero.

It must be pointed out that this argumentation has to be taken with a grain of salt. It ignores the proviso mentioned in connection with $W_R(D)$, that the summed exponentials are not completely reliable representations of the true form of the IR singularities. But they are not likely to paint a completely wrong picture; they are expected to render the essential features correctly, if not the details. The non-occurrence of exclusive processes involving charged particles and only finitely many photons is confirmed by other methods of approximation which are better adapted than perturbation theory to the subtler aspects of the IR problem, e.g. models of the Bloch–Nordsieck type or the use of coherent states as asymptotic states (see [JR], Sect. 16-1 and Suppl. S4). That finite-order perturbation theory cannot be expected to lead to a full understanding of the IR situation is seen from the very different nature of the IR singularities of the $\mathcal{W}$-functions which has been noted in Sects. 6.3 and 10.1: the perturbative singularities are stronger than the free ones, while for physical fields the exact ones must be weaker. This insensitivity of perturbation theory to the finer points of the IR problem has other consequences of importance in the present context. Calculating the cross section for elastic $e^-$–$e^-$ scattering, we find in order $e^4$ a finite, non-vanishing result, in apparent contradiction to what has just been stated. But the contributions of higher order to this process are IR divergent, an indication that the process does not exist in nature. Hence even the apparently exclusive $e^4$-term is meaningful only as the lowest non-vanishing term in the expansion of an inclusive cross section. Another point to be noted is that the sign difference between Gupta–Bleuler and physical gauges which was crucial to our argument, is irrelevant for the IR cancellations in finite-order perturbation theory. The expressions for reaction cross sections to be presented in Sect. 16.5 do not depend on whether they are calculated from Gupta–Bleuler fields or physical fields, even though the derivation of these expressions, especially the use of the approximation corresponding to (16.32), is rather more convincing for the physical fields. Of course, the equality of the two results may serve as an a posteriori justification of the Gupta–Bleuler derivation.

From the foregoing arguments it is clear that the necessity of including arbitrarily high numbers of cross lines holds for the $L$-, $U$-, and $R$-part separately. This implies that $R_{\mathrm{MS}}$ and $I$ are smooth and slowly varying as a function of $C$. $U_{\mathrm{MS}}$ and $L_{\mathrm{MS}}$ are smooth too, and their product is non-vanishing only in a small $C$-region determined by the supports of the $\tilde{f}_j$ and $\tilde{g}_i$. If $\bar{C}$ is a value in this $C$-region, we obtain in an excellent approximation

$$E(\lambda) \sim \frac{\pi^8}{\lambda^8} R_{\mathrm{MS}}(\bar{C}) I(\bar{C}) \int dC\, J(0,C)^2 (\det M)^{-1} U_{\mathrm{MS}}(C)\, L_{\mathrm{MS}}(C)\ ,\quad(16.75)$$

with $P_1+\cdots+Q_2 = -\bar{C}$, if the $\tilde{f}$–$\tilde{g}$-supports are sufficiently small. This is the equivalent to (16.32) of the massive theory. The factor $R_{\mathrm{MS}} I$ is the physically interesting one, specifying the cross section for the inclusive process under consideration. The other factors are partly kinematical, partly describing the characteristics of probes and sources.

Like in the massive case, (16.30), $R_{\mathrm{MS}}$ is a mass-shell restriction of $\tilde{\mathcal{W}}^{\mathrm{amp}\prime}(-P_1,\ldots,-Q_2,-|Q_2,\ldots,P_1,+)$, but amputation is not achieved by multiplication with $(P_i^2-m^2)$ etc. but by the more complicated prescription defined in (16.69). For the possibility of expressing the 8-pair function $\tilde{\mathcal{W}}(\cdots|\cdots)$ in terms of the physical fields $\Psi$, $\bar{\Psi}$, instead of the unphysical $\psi$, $\bar{\psi}$, we refer to the remarks made for the 1-particle case after (15.42).

Finally, we return to a problem introduced at the beginning of Sect. 16.2, that of justifying the replacement of an ordinary product of physical Fermi fields by their $T^+$- or $T^-$-products. We indicate briefly a possible way of proving the admissibility of these replacements for our purposes. For the sources the problem can be solved by the simple observation that it is not necessary to locate the two sources at equal times. If $b_1$, $b_2$, are their positions, we may consider the case (always realized in a suitable reference frame) $b_2^0 < b_1^0 < a_j^0$ and $(b_1-b_2)^2 < 0$. In this case the source fields are already correctly time ordered or anti-time ordered in both sectors, and the problem is solved like that of the time ordering of the probe positions $a_j$ relative to the $b_i$ (see at the beginning of Sect. 16.2). The real problem concerns the probes, where this trick does not work. The graph points of the probe fields lie in the $U$-part, so we must consider this part. We note that in principle the introduction of time orderings is not necessary. It is certainly possible to derive our results working directly with the ordinary products. Such a derivation would involve graphs with eight sectors instead of only two. It is to be expected that a web–core splitting would again do the trick, but this splitting would be vastly more complicated than what we have done, and it has not been worked out explicitly.

Consider the $U$-part of such a graph. We alternate $T^+$- and $T^-$-sectors, in order to avoid internal sectors. The IR problems connected with composite external vertices are expected to come from soft photon lines connecting one of the basic trajectories, let us say the right-hand 1-trajectory, to a composite vertex at the end of another (or the same) basic trajectory, e.g. of either of the 2-trajectories. The propagators of the line in question are identical in the two cases, and so is the rest of the graphs except that a factor $i$ in the end-vertex of one 2-trajectory becomes $-i$ in the other 2-trajectory. Hence the two contributions cancel. In the case of such a photon line connecting a 1-trajectory to its own end point or to that of the other 1-trajectory, we find the same mechanism of cancellation or the slightly more complicated one explained in connection with (15.60): a $\frac{-i}{2\pi} \frac{1}{k^2+i\varepsilon}$ is changed into a $-\delta_+(k)$ by integration over $k_0$. Hence the $\Psi_n$- and $\bar{\Psi}_n$-vertices have no IR consequences. A dependence on them survives only in the cores, and these factorize with respect to pairs of trajectories, i.e. to particles. But inside pairs we do not want to time-order, so there is no problem. Apart from the factorizing cores, the graphs (or rather their asymptotically relevant parts) have exactly the same form as in the Källén gauge. This is true especially for the physically interesting $R$-part. But in the local Källén gauge the introduction of $T^\pm$-products is clearly legitimate, like in the massive theory of Sect. 16.1. Hence the asymptotically relevant singularities are the same irrespective of whether they are calculated from 8-sector graphs or 2-sector graphs. The final result, the asymptotic expression for $E(\lambda)$, does not depend on whether the $T^\pm$-trick has been used or not.

# Chapter 17

# Cross Sections

There remains the question of relating the results of the preceding chapters to experiments. Conventionally, the result of a scattering experiment is expressed as a "scattering cross section". There arises the problem of how to extract such an expression from our results. In Sect. 17.1 this problem is investigated and found to be non-trivial, especially in view of the discrepancies noted in the previous chapters between our approach and the traditional one working with IR regularizations. A possible solution is presented and discussed in Sect. 17.2. The results obtained are then illustrated in Sect. 17.3 with the help of a concrete example: Compton scattering in low orders.

## 17.1 Cross Sections: the Problem

In Chap. 16 the scattering events of QED have been studied on the basis of a particle notion which has been suggested and elaborated in the preceding chapters. According to this notion, the particles of a relativistic quantum field theory are not fundamental objects of the theory, but merely convenient constructs useful for the phenomenological description of those of its states which are of the greatest physical interest. A particlelike behaviour of these states emerges from the theory without having been put in beforehand.

The result of this analysis is the expression (16.75). In deriving it we have worked with simple representations of scattering states and probes which have little resemblance with actual sources and detectors. Real sources and detectors are vastly more complicated and in general not accessible to a detailed theoretical description, especially a perturbative one. To make a comparison between theory and experiment possible, any reference to the specific structure of sources and detectors must be removed from the theoretical expression: we must try to isolate that part of the expression which describes the reaction in itself. This part is usually written as a "cross section" of the general form of (16.31):[1]

$$d\sigma = \frac{\pi^2 (2m)^f}{\sqrt{(P_1, P_2)^2 - m_1^2 m_2^2}} \Sigma(-P_1, -P_2; Q_1, \ldots, Q_n) \prod_{j=1}^{n} \frac{d^3 Q_j}{2\omega(\vec{Q}_j)} . \qquad (17.1)$$

Here we consider an in general inclusive process with two incoming particles of momenta $-P_1, -P_2$, and masses $m_1$, $m_2$, and $n$ observed outgoing particles with momenta $Q_1, \ldots, Q_n$. $f$ is the number of observed fermions. For the special case of exclusive processes (no unobserved particles), see (13.13). The explicit factors in $d\sigma$ are kinematical. Dynamics enters through the function $\Sigma$. Furnishing an expression for this function is one of the primary duties of field theory.

The problem of finding an expression for the "scattering function" $\Sigma$ in QED will occupy us in the present and the following section. The results are unexpected. As announced at the beginning of Chap. 16, charged particles appear to be complex objects which cannot be defined without reference to the way in which they are observed. But this reference can be cast in a form which appeals only to general, universal properties of sources and detectors, not to their detailed structure.

Let us address the problem first in a theory without IR problems, in which the situation is clear: the massive scalar theory of Sect. 16.1. To this theory the LSZ formalism introduced in Sect. 6.3 and explained in more detail in Sect. 13.1 is applicable. For exclusive processes the cross section is given by (13.13) and (13.14) and is expressible in terms of field theoretical quantities by the reduction formula (13.7) and (13.8). We find

$$\Sigma_{\text{excl}}(-P_1, \ldots, Q_n) = \tilde{\tau}'^{-}(-P_1, -P_2, \ldots, -Q_j, \ldots) \tilde{\tau}'^{+}(Q_n, \ldots, P_2, P_1)$$
$$\times \delta^4 \left( \sum Q_j + \sum P_i \right) . \qquad (17.2)$$

The following definition has been used:

$$\tilde{\tau}^{\pm}(p_1, \ldots, p_N) = \prod_{j=1}^{N} \left( \pm \frac{i}{2\pi} \frac{1}{p_j^2 - m^2} \right) \tilde{\tau}'^{\pm}(p_1, \ldots, p_N) \delta^4 \left( \sum p_j \right) . \qquad (17.3)$$

$\tilde{\tau}'^{\pm}$ is defined on the manifold $\sum p_j = 0$. $\tilde{\tau}'^{+}$ is related by (13.7) to an $S$-matrix element. From expression (17.2) it is easy to arrive at the corresponding expression for an inclusive process. Consider the case of two incoming and two observed outgoing particles plus an unknown number of unobserved outgoing particles. In order to arrive at the corresponding $\Sigma$ we sum the $\Sigma_{\text{excl}}$ for all possible numbers of out-particles, integrating over the momenta of the unobserved particles. We assume asymptotic completeness, which can be expressed as

$$\sum_{n=0}^{\infty} \frac{1}{n!} \int \prod_{j=1}^{n} \frac{d^3 k}{2\omega(\vec{k})} \prod_j a^*(\vec{k}_j) P_\Omega \prod_j a(\vec{k}_j) = 1 , \qquad (17.4)$$

---
[1] See e.g. [IZ], Sect. 5-1-1.

where the annihilation operator $a$ stands for $a_{\text{out}}$ and $P_\Omega$ is the projection onto the vacuum. Using this we find, starting from (13.6),

$$\Sigma(-P_i; Q_j) = \left( \Omega, \prod_{i=1}^{2} a_{\text{in}}(-\vec{P}_i) \prod_{j=1}^{n} a_{\text{out}}^*(\vec{Q}_j) \prod_{j} a_{\text{out}}(\vec{Q}_j) \prod_{i} a_{\text{in}}^*(-\vec{P}_i) \Omega \right). \tag{17.5}$$

Applying the LSZ asymptotic condition and reduction procedure one arrives at the generalized reduction formula

$$\Sigma = (2\pi)^8 \prod_i [(p_i^2 - m^2)(p_i'^2 - m^2)] \prod_j [(q_j^2 - m^2)(q_j'^2 - m^2)]$$

$$\times \tilde{\mathcal{W}}'(-p_i', q_j', -|-q_j, p_i, +)\Big|_{p_i = p_i' = P_i,\, q_j = q_j' = -Q_j}. \tag{17.6}$$

$\tilde{\mathcal{W}}'$ is $\tilde{\mathcal{W}}$ without the $\delta^4$ of momentum conservation. Notice that $\tilde{\mathcal{W}}'$ possesses poles at the mass shell which are cancelled by the amputation factors in front. Hence the product in (17.6) must be defined by a limiting procedure in which $p_i$, $p_i'$ tend to $P_i$, etc., from off-shell values, preferably with $p_i^2$, $p_i'^2 < m^2$.

Comparison of (16.31) and (16.32) shows that the desired function $\Sigma$ is essentially the $\lambda^{-8}$-coefficient in $E(\lambda)$ without the $C$-integral, which latter obviously refers to sources and probes. $\hat{J}(C)$ is a calculable, purely kinematical factor. That for extracting $\Sigma$ from $E(\lambda)$ we have to rely on a comparison with the LSZ result (16.31) is clearly unsatisfactory. The problem is due to the fact that the experimental meaning of $E(\lambda)$ is not clear. The probe operator $D(a)$ is not defined operationally, so that we do not know how to measure its expectation value. But for the moment we ignore this problem. We will return to it later on.

The foregoing considerations can be generalized without difficulty to massive QED. This is of interest for the comparison between the traditional approach to scattering and ours. Under "massive QED" we understand here simply the normal QED regularized with the help of a positive photon mass $\mu > 0$. This means that the singular factors of the photon propagators are $\delta_\pm^\mu(k)$, $(k^2 - \mu^2 \pm i\varepsilon)^{-1}$, but otherwise the graph rules are unchanged. This does not really define true massive QED, it is not an acceptable theory of a massive vector field. A correct formulation of such a theory can be found in Sect. 3-2-3 of [IZ]. It differs from our truncated version only by gauge terms which do not contribute to observable quantities. In perturbation theory our version is perfectly unproblematic.

As long as we consider a fixed positive value of $\mu$, the results found for the massive scalar theory carry over in an obvious way. But for studying what happens for $\mu$ tending to zero, we must introduce two changes already for positive $\mu$. Firstly, we must no longer demand that the supports of the test functions $\tilde{f}_j$, $\tilde{g}_i$, do not overlap the continuous spectrum. Secondly, the condition that the residue of the mass-shell pole of the fermionic 2-point function have the free value must no longer be used. But these are complications with which we are familiar from massless QED, and they are handled like there. The more general $\tilde{f}$-$\tilde{g}$-supports lead to the emergence of non-trivial $U$- and $L$-parts containing cross lines, which produce however no IR singularities. The $\psi$-normalization is handled as we handled it in true QED. UV finiteness is achieved by subtractions at the origin as explained in Sect. 10.2. The proper position of the 1-particle pole in the fermionic 2-point function is then assured by a finite mass renormalization, but no similar action is taken with respect to its residue. This leads to the presence of non-trivial SEPs on the basic trajectories as shown in Fig. 15.7. But like the cross lines in the $U$- and $L$-parts, these SEPs produce no IR problems. They do not change the $\lambda^{-8}$-behaviour of $E_G(\lambda)$, but they affect the coefficient of this leading term, making it diverge in the limit $\mu \to 0$.

More in detail: as in Chap. 16, consider the example of $e^-$–$e^-$ scattering. Take graphs $G$ of the form shown in Fig. 16.4. The clothed cut propagators are decomposed like in (16.48) and the denominator shared evenly between the adjacent subgraphs of $G$. The contributions $U'(s_j, s'_j)$, $L'(t_i, t'_i)$, and $R'(s_j, \ldots, t'_i)$, of the various subgraphs without the cut lines are smooth at the mass shell $s_j^2 = \cdots = t_i'^2 = m^2$ in the relevant region of these variables. The same holds for $\Delta_j^U t^+(s_j)$ etc. Hence

$$U_{MS}(S_j) = \left. \frac{U'(s_j, s'_j)}{(2\pi)^2 \prod_j \sqrt{\Delta_j^U t'^+(s_j) \Delta_j'^U t'^-(s'_j)}} \right|_{s_j = s'_j = S_j} \quad (17.7)$$

and the similarly defined $L_{MS}$ and $R_{MS}$ exist. $t'^\pm$ is the relevant part of $t^\pm$ as defined in (15.34). The asymptotic behaviour of $E(\lambda)$ is given by the formula (16.75) without the factor $I$. In analogy to the scalar theory, the function $\Sigma$ in the cross section (17.1) is given by the factor $\pi^8 R_{MS}$ of (16.75), the factors $U_{MS}$ and $L_{MS}$ being associated with the probes and sources.

But the $U_{MS}, \ldots$ defined in (17.7) diverge for $\mu \to 0$. More exactly, the contribution of an individual graph $G$ to $U_{MS}$, etc., diverges in this limit. As it turns out, the divergences cancel in the sum over the relevant graphs $G$ of a given order. This is shown as follows. In generalization of the second-order example of Sect. 15.2[2] we can isolate in a simple way the parts of $G$ producing the divergence, exactly as we did in massless QED. The potential divergence of the graph $G$ is contained in a web defined like in true QED, except that its photon propagators are the massive ones. Faithfully following the procedure used in the massless case, we arrive again at (16.75), but this time with $U_{MS}, L_{MS}, R_{MS}$, defined by (16.69) and with a factor $I$ present. However, these changes relative to the previous result (we are still in the massive case!) are only apparent, if the usual sum over graphs is considered. This is so

---
[2]See the discussion surrounding (15.21).

## 17.1 Cross Sections: the Problem

because, firstly, for $\mu > 0$ the webs $W_U, \ldots$ without the cut propagators $(\Delta_j^U)^{-1}, \ldots$ are continuous in the $\Delta$'s. This is evident from the definition of the webs. Secondly, we will presently prove that the function

$$\hat{W}^U(\Delta_j^U, \Delta_j'^{\,U}) = (2\pi)^4 \prod_j (\Delta_j^U \Delta_j'^{\,U}) \, W^U(\Delta_j^U, \Delta_j'^{\,U}) \qquad (17.8)$$

and the analogously defined $\hat{W}^L$, $\hat{W}^R$, are equal to 1 at $\Delta_h^\rho = 0 \; \forall \, \rho, h$. This means that the leading singularities of the various webs are given by their zero-order contributions. These results imply that the seemingly different definitions of $U_{MS}$ etc. given above for $\mu > 0$ in fact agree, and that $I(C) = 1$. Hence there is no discrepancy between the two forms of $E(\lambda)$ derived with or without introducing a web–core splitting.

There follows the promised proof of these claims.

**Lemma 17.1.** *Let $\hat{W}^\alpha(\Delta)$, $\alpha = U$, $L$ or $R$, be a web function defined as in (17.8), with $\Delta$ the set of cut variables on which it depends. Then*

$$\hat{W}^\alpha(0) = 1 \,. \qquad (17.9)$$

**Proof.** Equation (17.9) is most easily verified in a summed form analogous to (15.63) and (16.54), but taken at the point $\Delta = 0$ of $\Delta$-space. This restriction to the origin allows substantial simplifications in deriving the sums over graphs. We start with $\hat{W}_U$ defined from $W_U$ like in (17.8). Because $W_U$ factorizes into two 1-particle terms, it suffices to consider the 1-particle case treated in Chap. 15. Essentially we follow the procedure used there (see the subsection "Proof of Theorems 15.1–3" of Sect. 15.2). Consider a web graph $G$ and in it one of its two semitrajectories, connected with the variables $\Delta$ and $\hat{s} = S$. Let $N$ be the number of vertices on this semitrajectory, and

$$A_N = F_N \prod_{j=1}^{N} \left( \sum_{i=1}^{j} K_i \right)^{-1} \qquad (17.10)$$

the product of its fermion propagators at $\Delta = 0$, with the exclusion of the first one belonging to the cut line. The notation is that of (15.54) and (15.56).[3] The coefficient $F_N$ is

$$F_N = \left( e \sqrt{\frac{2}{\pi}} \right)^N \left( \frac{1}{2\pi} \right)^N \prod_{i=1}^{N} S^{\mu_i} \,, \qquad (17.11)$$

and $K_i = 2(S, k_i)$ where $k_1, \ldots, k_N$ are the photon momenta flowing into the trajectory, numbered consecutively starting from the cut line. We need

---
[3] Note that the original $A_N$ of (15.55) without the factor $j = 0$ is continuous in $\Delta$ as a distribution in the variables $k_i$. Its restriction to $\Delta = 0$ is therefore legitimate.

to calculate the sum $F_N \Sigma_N$ of the $A_N$ over all different orderings of the $N$ vertices. Two orderings are not considered different if they are related through the exchange of the two end points of one and the same photon line. This convention differs from the one used in Chap. 15. We distinguish between photon lines connecting the $s$-trajectory to the $s'$-trajectory (more generally: to a different trajectory) and those starting as well as ending at the $s$-trajectory. Let $\{\ell_j\}$ be the set of momenta of the latter, while now $\{k_i\}$ denotes the momenta of the former. $K_i$ and $L_j$ are defined accordingly. The $\ell$-lines are directed against the direction of the trajectory. We must also remember to apply a "mass renormalization" to SEPs in order to prevent the build-up of singularities of higher orders which might even forbid the restriction to $\Delta = 0$. It will turn out that SEPs of higher than second order cancel at $\Delta = 0$, so that they need not be renormalized. Let $\ell$ be the photon momentum of a SEP of second order, $v$ the total momentum flowing into the trajectory before this SEP. Then the fermion propagator of the SEP is up to a constant factor, for the moment leaving $\Delta$ unrestrained, $(\Delta + 2(S, v) + i\varepsilon)^{-1}$. Mass renormalization replaces it by

$$\frac{1}{\Delta + 2(S, v + \ell)} - \frac{1}{2(S, \ell)} = -\frac{\Delta + V}{L(V + L + \Delta)},$$

in obvious notation. The numerator cancels the denominator of the following line, thus removing the unwanted singularity. In the denominator we may again set $\Delta = 0$.

With these prescriptions we claim that

$$\Sigma_N = \prod_j \left(-\frac{1}{(L_j + i\varepsilon)^2}\right) \prod_i \frac{1}{K_i + i\varepsilon}. \tag{17.12}$$

We prove this first for the case (case A) that $\{\ell_j\}$ is empty, so that

$$\Sigma_N = \sum \frac{1}{K_{i_1}} \frac{1}{K_{i_1} + K_{i_2}} \cdots \frac{1}{K_{i_1} + \cdots + K_{i_N}},$$

the sum extending over all permutations $(i_1, \ldots, i_N)$ of the indices $(1, \ldots, N)$. We use induction with respect to $N$. Assuming the result to be correct for $N - 1$ vertices, we find

$$\Sigma_N = \sum_{i=1}^{N} \frac{1}{\sum_1^N K_r} \prod_{j \neq i} \frac{1}{K_j}$$
$$= \frac{1}{\sum_1^N K_r} \sum_i K_i \prod_j \frac{1}{K_j} = \prod_j \frac{1}{K_j},$$

as claimed. Next, admit the presence of $\ell$-lines (case B). Consider a particular ordering of the vertices and select the first *starting* point $P$ of an $\ell$-line. The endpoint of this line is then earlier than $P$, as may be other endpoints of $\ell$-lines, but no starting points. Let $\ell$ be the momentum of the distinguished line. The other photon momenta flowing into the trajectory before $P$, be they $k$- or $\ell$-momenta, we call $z_1, \ldots, z_n$. We sum over all permutations of the vertices earlier than $P$, leaving those later than $P$ unchanged. If $n = 0$, we are dealing with a single SEP of second order, and our rules give from it a factor $-(L + i\varepsilon)^{-2}$. The remainder of the trajectory is a $A_{N-2}$, and a summation over the orderings of its vertices yields the desired result (17.12) if we assume this result to be correct for the case $N - 2$. If $n > 0$, we first consider the contributions in which there is at least one $z_i$-vertex between the $\ell$-vertex and $P$. Give the last $z_i$ the number $i = n$ and sum over all orderings of the $(z_1, \ldots, z_{n-1}, \ell)$-vertices. By the established result for case A this sum is

$$\frac{1}{L} \prod_1^{n-1} \frac{1}{Z_i} \frac{1}{\sum_1^n Z_i + L} \frac{1}{\sum_1^n Z_i} \, .$$

The last factor comes from the first line *after* $P$, while the contributions from yet later lines are omitted. There remains the contribution from arrangements where all $z_i$-lines are earlier than the $\ell$-vertex. This implies again the presence of a SEP of second order which must be renormalized. Call the last $z$-variable $z_n$ and sum over all orderings of the earlier ones. The result is

$$-\prod_1^{n-1} \frac{1}{Z_i} \frac{1}{\sum_1^n Z_i} \frac{1}{\sum_1^n Z_i + L} \frac{1}{L} \, ,$$

which exactly cancels the previous expression.

$\hat{W}_U(0)$ is obtained from (17.12) by a summation over $N$ like the one leading to (15.60). The combinatorial factors are the same as there, except that the factor $2^{-L-L'}$ is missing because the two endpoints of an $\ell$-line are now not exchanged. The second exponential in (15.60) having been shown to be trivial, we leave it out from our expression. We find

$$\hat{W}_U(0) = \exp\left\{ \frac{e^2 m^2}{2\pi^3} \left[ \frac{i}{2\pi} \int \frac{dk}{k^2 - \mu^2 + i\varepsilon} \frac{1}{(K + i\varepsilon)^2} \right.\right.$$
$$\left.\left. - \frac{i}{2\pi} \int \frac{dk}{k^2 - \mu^2 - i\varepsilon} \frac{1}{(K - i\varepsilon)^2} - \int dk\, \delta_+^\mu(k)\, \frac{1}{K^2} \right] \right\}, \quad (17.13)$$

with $K = 2(S, k)$. In the second term in the square bracket we change the sign of $-i\varepsilon$ by the substitution $k \to -k$. No $i\varepsilon$-prescription is necessary in the third term because the plane $K = 0$ does not intersect the support of $\delta_+^\mu$ if $\mu > 0$. This term can also be written $-\frac{1}{2} \int dk\, \delta(k^2 - \mu^2) \frac{1}{K^2}$, and $K$ can be replaced in this by $(K + i\varepsilon)$.

The $\hat{W}_\pm$ are calculated in the same way. There is only one trajectory present so that the $K_i$-terms in (17.12) are absent. $\hat{W}^U(0) = \dfrac{\hat{W}_U(0)}{\sqrt{\hat{W}_+(0)\, \hat{W}_-(0)}}$

is thus obtained from (17.13) by replacing the coefficients $\pm \frac{i}{2\pi}$ of the first two terms by $\pm \frac{i}{4\pi}$. This yields

$$\hat{W}^U(0) = \exp\left\{\frac{e^2 m^2}{4\pi^3}\int \frac{dk}{(K+i\varepsilon)^2}\right.$$
$$\left.\times\left[\frac{i}{2\pi}\frac{1}{k^2-\mu^2+i\varepsilon} - \frac{i}{2\pi}\frac{1}{k^2-\mu^2-i\varepsilon} - \delta(k^2-\mu^2)\right]\right\}$$
$$= 1, \qquad (17.14)$$

because the square bracket vanishes.

$\hat{W}^R(0)$ is found by the same method. The result is similar to the expression (16.54) for $W_R$:

$$\hat{W}^R(0) = \exp\left\{\sum_{h\rho} H_{h\rho} + H_{12}^U + H_{12}^L + H_{12}^{UL} + H_{21}^{UL}\right\}. \qquad (17.15)$$

The 1-pair factors $e^{H_{h\rho}}$ are products of two 1-particle factors of the form (17.14), hence they are $=1$. The 2-pair term $H_{12}^L$ is

$$H_{12}^L = \frac{e^2(S_1, S_2)}{2\pi^3}\int dk \frac{1}{(K_1-i\varepsilon)(K_2+i\varepsilon)}$$
$$\times\left[\frac{i}{2\pi}\frac{1}{k^2-\mu^2+i\varepsilon} - \frac{i}{2\pi}\frac{1}{k^2-\mu^2-i\varepsilon} - \delta_+^\mu(k) - \delta_-^\mu(k)\right],$$

which vanishes because the square bracket vanishes. The same is true for the other 2-pair terms. This proves

$$\hat{W}^R(0) = 1, \qquad (17.16)$$

and Lemma 17.1 is proved.

We come finally to the true QED with a vanishing photon mass $\mu = 0$. In our formalism the scattering cross section is extracted from the coefficient of the leading term in $E(\lambda)$. For $\mu = 0$ this coefficient is given by (16.75). It contains the non-trivial factor $I(\tilde{C})$ which is not present in the corresponding expression for $\mu > 0$. Hence we cannot isolate the scattering function $\Sigma$ by simple analogy. Does it include the factor $I$ or does it not?

The problem does not arise in the conventional approach to scattering in QED. This approach relies on the LSZ formalism, making sense of it with the help of an IR regularization, preferentially by introducing a positive photon mass $\mu$. This amounts essentially to treating massless QED as a limit of its massive variant. There is a problem there concerning the observed photons. As is well known, massive vector particles have three independent spin states, massless ones only two: there seems to be a discontinuity at $\mu = 0$. Moreover, in our treatment there exists another, more severe problem of this kind: in elastic $\gamma$–$\gamma$ scattering $E(\lambda)$ decreases like $\lambda^{-8}$ in the massive case, like $\lambda^{-7}$ in the massless case. This makes the discussion of the $\mu$-limit difficult. But the observed photons are hard and therefore IR harmless. The problems connected with them are not the problems that concern us now. In our approach there is no need to start from a massive theory. We want to do this only for investigating the discrepancy between the conventional and our results, which is an IR effect due to soft photons. We will therefore only discuss purely fermionic processes, in particular our standard example of inclusive $e^-$–$e^-$ scattering. Photonic processes may be handled similarly, starting from a hybrid semimassive QED defined by graph rules in which the internal photon propagators of the reaction part are massive, the external cut propagators massless. This means treating $\mu$ purely as a regularization parameter, giving up any pretence of starting from a meaningful massive theory. This is entirely in the spirit of the conventional approach as used in practice. But we will confine ourselves to the electron–electron example.

There is no question that, taken as a field theory, massive QED converges to massless QED for $\mu \to 0$. The functions $\mathcal{W}_\sigma$ and $\tilde{\mathcal{W}}_\sigma$ of Part II, especially the Wightman functions defining the theory, are limits in the sense of distributions of their massive or semimassive counterparts. But this does not guarantee that the same is true for the quantity of interest to us: the coefficient $K$ in (16.75). It is defined as $K = \lim_{\lambda \to \infty} \lambda^8 E(\lambda)$, and this limit need not commute with the limit $\mu \to 0$. As already mentioned, the factor $I$ is not there in the massive case. The $C$-integral is a kinematical expression which does not depend on $\mu$. The only $\mu$-dependent factors of the massive case are $U_{MS}, L_{MS}$, and $R_{MS}$, and these converge for $\mu \to 0$ to their counterparts in (16.75), which are defined by (16.69). As has been discussed above, these definitions hold also in the massive case, even though there the introduction of the web functions is unnecessary since they take their free form at $D = 0$. Consider $U_{MS}$, more exactly one of its 1-particle factors. We start from the decomposition $U' = C_U W_U$ constructed in the proof of Lemma 15.4. Obviously we have

$$\frac{U'}{W_U} = C'_U(s, s') , \qquad (17.17)$$

where $C'_U$ is the core function introduced in (15.53), which is smooth. This holds too for $\mu > 0$, the core function $C'$ in that case being defined exactly like in the massless case, but with the massive webs. The same arguments which are used to show smoothness of $C'_U$ in $s$, $s'$, establish that it is smooth as well as a function of $\mu$ for $\mu \geq 0$, hence its limit for $\mu \to 0$ exists and is the correct value of the massless case. The same arguments apply to $t^\pm$. Moreover, $C_\pm(-\vec{\nu}) \neq 0$, so its occurrence in a denominator is not dangerous. Hence

$$U_{MS} = \left.\frac{C'_U}{\sqrt{C'_+ C'_-}}\right|_{s=s'=-\nu}$$

is continuous in $\mu$ for $\mu \geq 0$. And the 2-particle $U_{MS}$ in (16.75) is the product of the ($\mu = 0$)-values of the corresponding 1-particle expressions, hence the limit of the massive 2-particle $U_{MS}$. The same holds for $L_{MS}$ and $R_{MS}$. Hence $\lim_{\mu \to 0} K(\mu) \neq K(0)$ if $K(0)$ is defined from (16.75), because the factor $I(\bar{C})$ is not present in the $\mu$-limit.[4]

This raises the question of how to compare our results with experiment, in particular how to extract from our asymptotic formula an expression taking the place of the conventional cross section. For this, a physical definition of a cross section which does not rely on the LSZ formalism is needed. Stating that $d\sigma$ in (17.1) denotes the fraction of particles scattered into the momentum region $\prod d^3 Q_j$ does not constitute an operational definition. It refers to the measured momenta of observed particles. And this observation and these measurements bring in the detectors and sources, and therefore the non-separability of their influence on the observed counting rates from the influence of the reaction: it brings in the ominous factor $I$. The problem is connected with a problem mentioned earlier, that of the lack of an operational interpretation of the sources and probes used in our derivations. It is this problem which prevents us even in the massive case from extracting a cross section directly from our results, without having recourse to a comparison with the LSZ result.

## 17.2 A Solution

The preceding section ended with the statement of a problem. The solution of this problem presented in the present section is tentative. It is meant as a proposal, not as a statement of facts. It is for the reader to decide whether he will accept it or not.

The factor $I$ *is* there in our formalism. And it is caused by soft bremsstrahlung produced in the reaction itself as well as in the experimental apparatus. It can therefore not be attributed solely to the one or the other, and it can also not be divided among them in a convincing way. This suggests that the experiment should be considered as a whole, without attempting a clean separation between preparation of the initial state, reaction, and observation of the resulting final state. But the offending factor $I$ has a simple universal form not depending on the exact structure of sources and detectors. It is caused by soft photons emitted by the charged particles after leaving the source and before entering the detector, not by photons produced inside the apparatus. These latter are entirely contained in the cores and thus in the factors $U_{MS}$ and $L_{MS}$. But the incoming particles have necessarily been interacting with their sources, the observed outgoing particles are necessarily interacting with the detectors with which they are observed. Hence they unavoidably produce bremsstrahlung of the IR relevant type described by

---

[4]It must be said that at present the possibility that $I(C) \equiv 1$ for some obscure reason cannot be strictly excluded. However, this seems extremely unlikely.

the web and thus responsible for $I$, irrespective of the exact nature of sources and detectors.

The following assumption is therefore reasonable. Consider again the example of inclusive $e^--e^-$ scattering, but using realistic sources and detectors sharing with our simplified ones the good localization in $x$- and $p$-space. More precisely: the detectors may be large, but they should localize the observed particles in small regions. And they should provide a good momentum discrimination. The same goes for the sources. Let $E(\lambda)$ be the output of the experiment, typically the rate of coincident triggerings of the two detectors, or a related quantity. Then $E(\lambda)$ is given by (16.75), where $R_{MS}$, $I$, and the $C$-integral are the expressions derived in Sect. 16.4.

But $U_{MS}$ and $L_{MS}$ describe the properties of the detectors and sources actually used in the experiment. Each of them is a product of two factors, one for each of the two detectors and sources present. In general it is, however, very difficult if not impossible to find sufficiently accurate theoretical expressions for the detector and source factors. They must be determined empirically. This can be done by analyzing 1-particle states. A possible way of proceeding is the following. Take a given source and a given detector in a geometrical arrangement as discussed in Chap. 15. The regions of momentum sensitivity of the two must overlap and the vector $\nu$ giving the direction of their separation must lie in this overlap. We assume $(\nu, \nu) = m^2$. Then, for large $\lambda$ the output $E(\lambda\nu)$ of the experiment is given by (15.31) and (15.41) as

$$E(\lambda\nu) = \frac{2\pi^3 m}{\lambda^3} U_{MS} L_{MS} I_1 , \qquad (17.18)$$

with

$$I_1 = \int d\delta \, d\delta' \, e^{-in(\delta-\delta')} \, N^U(\delta, \delta') \, N^L(\delta, \delta') \qquad (17.19)$$

for any $n > 0$. From the measured value of $E$ and the calculated value of $I_1$ we obtain the product $U_{MS} L_{MS}$ of detector and source factor. The problem of separating the two remains. It can be solved by at first replacing $U_{MS}$ by a factor which is theoretically understood but still leads to a measurable $E$.

An obvious possibility is that of replacing the probe operator

$$D^-(a) = \bar{\Psi}(f_a)^* \bar{\Psi}(f_a) \qquad (17.20)$$

by a component of the current density

$$j^\mu(a) = e \, N_1 \left( \bar{\psi}(a) \, \gamma^\mu \psi(a) \right) \qquad (17.21)$$

present in the field equations (8.1). $E(\lambda\nu)$ is then the expectation value of $j^\mu$ measured at the point $a = \lambda\nu$ in the state produced by the given source. This expectation value we know how to measure at least in principle, even though this may be technically difficult for certain types of sources. And the expression (17.21) is sufficiently close to $D^-$ to make $E(\lambda\nu)$ asymptotically calculable with essentially the methods developed in Chap. 15. There are, however, some important differences between the expressions (17.20) and (17.21). One of them works in our favour: $j^\mu$ is gauge invariant. It can therefore be written as a functional of the local GB fields $\psi$, $\bar\psi$. The complications due to the non-locality of the physical fields $\Psi$, $\bar\Psi$ do not arise. But there exist also the following less favourable distinctions which necessitate some elaborations.

(a) Whilst the product of the smeared fields $\bar\Psi^*(f_a)$ and $\bar\Psi(f_a)$ is well defined, this is not so for the product of fields at a point occurring in (17.21). Hence the necessity of using the normal-product description $N_1$.

(b) The operator $j^\mu$ does not annihilate the vacuum.

(c) In contrast to the probe $D^-$, the current $j^\mu$ has no momentum selectivity.

The renormalization problem (a) is solved by a combination of results obtained in Sect. 9.4 in connection with (9.23), and near the end of Sect. 10.3 in the discussion of the properties of the renormalized $\mathcal{W}$-functions. In the latter discussion it has been pointed out that the identification of variables in adjacent fields like in $\bar\psi(z)\gamma^\mu\psi(z)$ produces UV divergences in renormalization parts straddling the boundary between the corresponding two sectors. They must be subtracted like the more familiar renormalization parts staying inside a sector. This is the meaning of the $N_1$-prescription. Consider first a graph in which the two $z$-points are directly connected by a fermion line. This line forms a renormalization part of a novel type, which is disconnected from the rest of the graph. It is clearly annihilated by subtracting it at coincident variables since its variables coincide in the first place: these graphs do not contribute to $N_1(\bar\psi(z)\psi(z))$. For the remaining graphs it has been shown in Sect. 9.4 that their integrands summed over the graphs with the same scaffolding coincide as singular functions for the products $\bar\psi(z)\gamma^\mu\psi(z)$ and $-(\gamma^\mu\psi(z))\bar\psi(z)$ and therefore also for $\frac{1}{2}[\bar\psi(z),\gamma^\mu\psi(z)]$. Since the subtractions of Sect. 10.3 are defined in the same way in all these cases, the identity of the three expressions remains valid under renormalization, i.e. we can actually use the simple product form $\bar\psi(z)\gamma^\mu\psi(z)$, properly renormalized as indicated. The $U$-part of the $j$-graphs must be understood in this sense. Only renormalization parts inside $U$ are of interest, because it has been shown in Sect. 15.1 that the subtraction terms of cut renormalization parts are asymptotically irrelevant.

Point (b) in the list of complications is no problem. It is still true that $(\Omega, j^\mu\Omega) = 0$, and this suffices for our purposes. Point (c) is harmless too. It simply implies that the $U$-part is more complicated than in the probe case: it may contain fermionic cross lines. But this is a situation that has already been met in the reaction part $R$ discussed in Sect. 16.4, and it is handled like there.

As a result, the expectation value of $j^\mu(\lambda\nu)$ in the state produced by a particular source is given for large $\lambda$ by an expression of the form (17.18). Call the corresponding detector factor $U^j$. This $E(\lambda \nu)$ is measurable and $U^j$ and $I_1$ are calculable, hence the source factor $L_{MS}$ can be determined. Combining this source with a general detector in a 1-particle experiment gives again an output of the form (17.18), from which $U_{MS}$ can be determined because $I_1$ and $L_{MS}$ are known. This procedure yields the values of $U_{MS}$ and $L_{MS}$ at the momentum $\nu$ parallel to the separation between detector and source. By varying this direction the whole sensitive momentum region can be covered.

If we use the same procedure starting from the conventional form $2\pi^3 m\, U_{MS}^{conv} L_{MS}^{conv}$ of the leading coefficient in $E$, we obtain

$$U^j L_{MS} I_1 = U^j L_{MS}^{conv} \quad \Longrightarrow \quad L_{MS}^{conv} = L_{MS} I_1 \qquad (17.22)$$

and hence

$$U_{MS} L_{MS} I_1 = U_{MS}^{conv} L_{MS}^{conv} \quad \Longrightarrow \quad U_{MS}^{conv} = U_{MS} \;. \qquad (17.23)$$

$U^j$ is the calculable $U_{MS}$ of the current and is therefore the same in our formalism and the conventional one. Remember that $I_1$ is a constant not depending on energy. The asymmetry between sources and detectors in these relations is no reason for concern. Preparation of a state and observation of a state are different matters. They are not related through time reversal or any other symmetry. Note, however, that the relations (17.22) and (17.23) hold only if the suggested method of determining the source and detector factors is used. If other methods of a more theoretical nature are employed, things may be different. For instance, our method is clearly not applicable to the determination of $L_{MS}$ for a fixed target.

Let us sum up.

In massive theories, the results of a scattering experiment can be described by a scattering cross section of the form (17.1). Theoretically we find in our formalism that $\Sigma = R_{MS}$, the factor in the asymptotic form of $E(\lambda)$ pertaining to the reaction part of the graphs contributing to $E$. We consider only inclusive processes, since exclusive reactions involving charged particles do not exist in QED.

In massless QED, the relevant $E$-coefficient contains an additional factor $I$. The product $R_{MS} I$ takes over the role of $R_{MS}$ in the massive case. $R_{MS}$ refers exclusively to the reaction proper, $I$ is an IR correction produced by soft photon processes in the reaction as well as near the experimental apparatus. But it does not depend on the detailed structure of the apparatus, it has an universal form. Nor does it depend on the details of the reaction. If the observed charged particles are not electrons but hadrons, then the effects of the strong interaction are entirely contained in $R_{MS}$, while $I$ retains its pure QED form. The same is true for effects of the weak interaction. It is therefore reasonable to retain the name "cross section" for the factor $R_{MS}$, more exactly for the expression (17.1) with $\Sigma = R_{MS}$.[5] $I$ we call the "IR halo" or simply "halo" of the process.

---

[5] Alternatively, one might define $\Sigma = R_{MS} I$. The choice between the two possibilities is a matter of taste.

334   17. Cross Sections

The conventional method of calculating cross sections in QED misses the halo. But it is a possible way of computing $R_{MS}$. That is, we can first introduce an IR regularization, be it a positive photon mass, a low-energy cutoff, or a long-distance cutoff, then calculate $\Sigma$ from the LSZ reduction formula, then remove the regularization. The result is $R_{MS}$.

Alternatively, $R_{MS}$ may be calculated directly in massless QED as follows. Take the process considered in (17.1). The particles are of arbitrary type and polarization. Start from the 2-sector function $\tilde{\mathcal{W}}(-p'_1, -p'_2, q'_1, \ldots, q'_n, -| - q_n, \ldots, -q_1, p_2, p_1, +)$. The fields belonging to photon variables are the electric field strengths $\vec{E}$. The $p_i$-field is a $\psi$ or a $\bar{\psi}$ if $p_i$ refers to an electron or positron, respectively, and the other way round for $p'_i$. The $q_j$-field is $\bar{\psi}$ or $\psi$ if the $j$-particle is an electron or positron, and the other way round for $q'_j$. Why the unphysical $\psi$, $\bar{\psi}$, are preferable to the physical $\Psi$, $\bar{\Psi}$, has been discussed in Sect. 15.2.[6] Contract the fields with the appropriate polarization factors as defined at the end of Sect. 14.2: a transversal 3-vector $\vec{e}$ for photons, polarization spinors (see(14.43)) $\eta(-q)$, $\eta(-p)^*\gamma_0$, $\eta(-p')$, and $\eta(-q')^*\gamma_0$ for electrons, $\eta(-q)^*\gamma_0$, $\eta(-p)$, $\eta(-q')$, and $\eta(-p')^*\gamma_0$ for positrons. This handling of polarization supposes that the sources and detectors produce and register particles with a given polarization. If polarization is not measured, then a summation over polarizations is necessary in the out-states, an averaging in the in-states.

The scalar function obtained from $\tilde{\mathcal{W}}$ by this contraction must then be divided by $\sqrt{t^+}$ or $\sqrt{t^-}$ for every charged field, or rather by its relevant scalar part as defined in (15.34). The resulting function can be written as

$$S(p'_1, \ldots | \ldots, p_1) = \delta^4\left(\sum_i (p_i - p'_i) - \sum_j (q_j - q'_j)\right) s(p'_1, \ldots, p_1) . \quad (17.24)$$

$s$ is only defined on the support of the $\delta^4$-factor: one of its variables is redundant. To $s$ we apply a generalized amputation. If the pair $(p_i, p'_i)$ or $(q_j, q'_j)$ refers to a photon, multiply $s$ with $(2\pi)^2 p_i^2 p'^2_i$ or $(2\pi)^2 q_j^2 q'^2_j$, respectively. Divide the result by $N^R$, a straightforward generalization to any number of charged particles of the $N^R$ defined in (16.67). It depends on the $P_i$ and $Q_j$ pertaining to charged particles and on a pair of variables $\Delta_j = 2\omega(q_{j0} + \omega_j)$, $\Delta'_j = 2\omega_j(q'_{j0} + \omega_j)$, with $\omega_j = \omega(\vec{Q}_j)$, for each outgoing charged particle and similarly for incoming charged particles. Finally, in the resulting expression let $p_i$ and $p'_i$ tend to $P_i$, $q_j$ and $q'_j$ to $-Q_j$, preferably in spacelike directions, i.e. from $p_i^2 \sim p'^2_i < m^2$ etc. (Remember that $P_i$ and $Q_j$ lie on the mass shell.) The result of this procedure is the desired function $R_{MS}$.

The results obtained here differ from those normally used in the literature, and this is clearly a serious matter. I am called upon to defend my point of view against this objection and to explain why I do not consider the objection fatal. At least one of the two results, the traditional one or the probe one,

---
[6] The result is the same for both choices.

must be wrong. It might be the latter. But I know of no proof that it is not the traditional one. The recipe used in that approach is introduced ad hoc, without a convincing justification beside the fact that it yields finite results. It involves the exchange of two limits, that of going to the mass shell and that of letting the photon mass tend to zero, and such an exchange is known to be dangerous and in need of justification. No such justification exists in field theory, as far as I am aware. A justification of sorts may be found in $S$-matrix theory, which claims that the $S$-matrix contains all the physically relevant (= observable) information that a field theory is capable of furnishing. What happens off-shell is irrelevant. If that is true, then it is legitimate to go to the mass shell first – one really ought to have started from there in the beginning – and vary the parameters of the theory, like the photon mass, afterwards. But this extreme thesis is hardly acceptable. An objection to it has already been formulated at the end of Sect. 13.1: $S$-matrix theory is incomplete because it is incapable of estimating the error made by measuring asymptotic observables at finite times. It cannot delimit the boundaries within which it is applicable. Moreover, in gauge theories like QED the $S$-matrix in its standard definition does not even exist due to the IR problems. And how can a non-existent object be the central object of a theory?

The method that has been used in this book, that of characterizing particles through their observable localization behaviour rather than by stipulating asymptotic conditions, avoids the legitimation problem of the conventional approach by working exclusively with the correct theory, QED with massless photons, not making a detour through theories with unphysical values of an essential parameter. This, of course, does not guarantee the correctness of the method. But I wish to emphasize that the problem of an insufficient legitimation of the traditional method exists independent of whether or not a convincing alternative is available.

In the last resort, the question must be decided by experiment. It would therefore be important to compute the numerical consequences of the discrepancy. I have not tried to do so. But let me indicate qualitatively what sort of effects are expected in low orders of perturbation theory. The results to be stated will be demonstrated in the next section for the example of Compton scattering. The methods used in this case carry over without difficulty to other processes.

The offending additional terms necessarily involve soft web photons. Since the web alone cannot produce scattering in a non-forward direction, there must also be a core present. This means that the two methods give identical results in the lowest non-vanishing order of perturbation theory, i.e. in the tree approximation (the sectors are trees disconnected from each other). In the next order, the one-loop order, the difference between the two results is the tree-order result multiplied by $nI_1^{(2)}$. $I_1^{(2)}$ is the $e^2$-term in $I_1$, $n$ is in the worst case the number of charged observable particles. If the relation (17.22) holds, then the incoming particles do not contribute to the discrepancy. $I_1^{(2)}$ is a momentum-independent constant: the dependence of the discrepancy on energy and scattering angles is that of the tree approximation. Only in higher orders becomes the halo fully effective.

## 17.3 Compton Scattering, an Example

As an example of how our results work in practice, let us consider Compton scattering in the two lowest non-vanishing orders of perturbation theory. This should also help to clarify some important details that may not have found sufficient emphasis in the lengthy calculations of the preceding chapters. Let $-P_e$ and $-P_\gamma$ be the 4-momenta of the incoming particles, $Q_e$ and $Q_\gamma$ of the outgoing particles, so that $P_e^2 = Q_e^2 = m^2$, $P_\gamma^2 = Q_\gamma^2 = 0$. We wish to calculate the scattering function $\Sigma(-P_e, -P_\gamma; Q_e, Q_\gamma) = R_{MS}$ in the cross section (17.1), and the halo $I(P_e, Q_e)$. These two functions describe the reaction itself. The source and detector functions $L_{MS}$ and $U_{MS}$ do not concern us now.

First we repeat the rules for finding $R_{MS}$ and $I$ in general orders. Until further notice, $R_{MS}$, $I$, and other functions denote the full sum over all relevant graphs, considered as formal power series in the coupling constant $e$. $R_{MS}$ is derived from the reaction parts of graphs of the form of Fig. 16.4, but with the cut variables renamed as shown in Fig. 17.1. The indices $i$ and $j$ in $p_i$, $q_j$, take the values $e$ and $\gamma$ instead of 1 and 2 as in Chap. 16. The cut variables $p_e, \ldots$ refer to the observed particles.

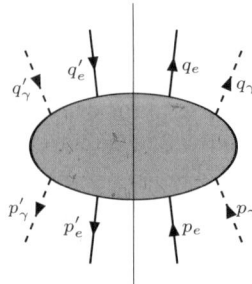

**Fig. 17.1.** The reaction part of a Compton graph

We formulate now the prescription for calculating $R_{MS}$ as distilled from Chaps. 15–17. The observed photons are hard, hence IR harmless: the $R$-part of an individual graph depends smoothly on $p_\gamma, \ldots, q'_\gamma$. The corresponding cut lines can be amputated conventionally like in the massive example of Sect. 17.1, simply by omitting their propagators. Adapting the procedures and notations of Chaps. 15 and 16 to the Compton case, we define $R'$ as the

## 17.3 Compton Scattering, an Example

(summed) $R$-part of our graphs, including the propagators of the cut electron lines but not those of the cut photon lines. Next we renormalize the electron field by defining

$$\hat{R}(p_e,\ldots) = \frac{1}{\sqrt{t'^{+}(p_e)}} \frac{1}{\sqrt{t'^{+}(q_e)}} \frac{1}{\sqrt{t'^{-}(p'_e)}} \frac{1}{\sqrt{t'^{-}(q'_e)}} R'(p_e,\ldots), \quad (17.25)$$

with $t'^{\pm}$ the IR relevant parts of the 2-point functions as defined in (15.34). The singularity function $N^R(\Delta^U,\ldots,\Delta'^L)$ is defined like in (16.67) as a homogenized version of

$$N^R = \frac{W_R(\Delta^U,\ldots,\Delta'^L)}{\prod_\rho \left(\sqrt{N_+(\Delta^\rho)\, N_-(\Delta'^\rho)}\right)}. \quad (17.26)$$

The index $\rho$ takes the values $U$, $L$, also in what follows. $N^R$ depends on the variables $\Delta^\rho$, $\Delta'^\rho$, associated with the cut electron lines. $W_R$ is constructed from webs containing the corresponding four semitrajectories, not eight like in (16.67). Finally we define

$$R(\Delta^\rho, \Delta'^\rho) = \hat{R}\Big|_{\vec{p}_e = \vec{p}'_e = \vec{P}_e,\, \vec{q}_e = \vec{q}'_e = -\vec{Q}_e} \quad (17.27)$$

and

$$R_{MS} = \lim_{D \to 0} \frac{R}{N^R}, \quad (17.28)$$

with $D^2 = \sum_\rho \left((\Delta^\rho)^2 + (\Delta'^\rho)^2\right)$. $R_{MS}$ is the value of $\hat{R}/N^R$ at $p_e = p'_e = P_e$, $q_e = q'_e = -Q_e$, this value being defined by the limiting procedures (17.27) and (17.28), carried out in this order.

$N^U(\Delta^U, \Delta'^U)$ and $N^L(\Delta^L, \Delta'^L)$ are the 1-particle functions of Chap. 15. The halo $I(P_e, Q_e)$ is given by

$$I = \int \prod_\rho (d\Delta^\rho d\Delta'^\rho)\, N^U N^R N^L e^{i \sum_\rho n_\rho (\Delta^\rho\ \Delta'^\rho)} \quad (17.29)$$

with $n_U = \frac{a^0 - N^0}{2Q_e^0}$, $n_L = \frac{b^0 - N_0}{2P_e^0}$. $a$ and $b$ are the positions of the electron detector and source and $N$ is the scattering center defined in Lemma 16.2. Both $n_\rho$ are positive.

The singularity functions $N^\alpha$ are only defined up to a constant factor. It is convenient to normalize them such that their order-zero terms are

$$N^U_{(0)} = \frac{1}{2\pi\sqrt{\Delta^U \Delta'^U}}, \qquad N^L_{(0)} = \frac{1}{2\pi\sqrt{\Delta_L \Delta'^L}},$$

$$N^R_{(0)} = N^U_{(0)} N^L_{(0)}. \quad (17.30)$$

This yields the halo of order zero,

$$I_{(0)} = \prod_\rho \left[ \int \frac{d\Delta^\rho}{2\pi(\Delta^\rho - i\varepsilon)} e^{in_\rho \Delta^\rho} \int \frac{d\Delta'^\rho}{2\pi(\Delta'^\rho + i\varepsilon)} e^{-in_\rho \Delta'^\rho} \right] = 1. \quad (17.31)$$

In non-forward directions the lowest non-vanishing term in $R'$ is the term of order $e^4$. It must be divided by the terms of order zero of $\sqrt{t'^+(p_e)}$ etc. as stated in (17.25). We have

$$t^+_{(0)}(p) = \frac{i}{2\pi} \frac{\slashed{p} + m}{p^2 - m^2 + i\varepsilon}.$$

Its relevant singular contribution is $N^{(0)}_+(\slashed{p}+m)$ with $N^{(0)}_+ = W^{(0)}_+ = \frac{i}{2\pi(-\Delta+i\varepsilon)}$: the free term in the web function $W_+$. From this we obtain

$$t'^+(p) = 2m\, N^{(0)}_+. \quad (17.32)$$

The factor $N^{(0)}_+$ is cancelled in (17.28) by a corresponding factor in $N^R$, and we obtain

$$R^{(4)}_{MS} = \frac{1}{(2m)^2} R^{\mathrm{amp}}_{(4)} \Big|_{p_e=p'_e=P_e,\, q_e=q'_e=-Q_e}. \quad (17.33)$$

$R^{\mathrm{amp}}$ is the fully amputated 4-point function

$$R^{\mathrm{amp}} = (2\pi)^8 \prod_i \left[(p_i^2 - m_i^2)(p'^2_i - m_i^2)\right] \prod_j \left[(q_j^2 - m_j^2)(q'^2_j - m_j^2)\right]$$

$$\times \tilde{\mathcal{W}}'(-p'_i, q'_j, -|-q_j, p_i, +). \quad (17.34)$$

A typical graph contributing to the 4-point function $\tilde{\mathcal{W}}'_{(4)}$ of (17.34) is shown in Fig. 17.2, where the dotted photon lines must be ignored. The two types of sectors in this figure with crossed or uncrossed photon lines respectively can be combined arbitrarily, so that there are four graphs contributing to $\tilde{\mathcal{W}}'_{(4)}$. Since they are all of similar structure, it suffices to consider one of them, e.g. the one shown explicitly. Its contribution to $R^{\mathrm{amp}}_{(4)}$ is the product of a factor

$$\frac{i}{(2\pi)^2}(\slashed{p}_e + m)\gamma^h \frac{q^0_\gamma}{|\vec{q}_\gamma|} \frac{1}{\slashed{p}_e - \slashed{q}_\gamma - m} \gamma^j \frac{-p^0_\gamma}{|\vec{p}_\gamma|}(\slashed{q}_c + m) \quad (17.35\mathrm{a})$$

from the $T^+$-sector, and

$$\frac{-i}{(2\pi)^2}(\slashed{q}'_e + m)\gamma^{h'} \frac{-q'^0_\gamma}{|\vec{q}'_\gamma|} \frac{1}{\slashed{p}'_e + \slashed{p}'_\gamma - m} \gamma^{j'} \frac{p'^0_\gamma}{|\vec{p}'_\gamma|}(\slashed{p}_e + m) \quad (17.35\mathrm{b})$$

from the $T^-$-sector. The external electron fields are by our conventions components $E^j, \ldots$ of the electric field strength. The corresponding vertex

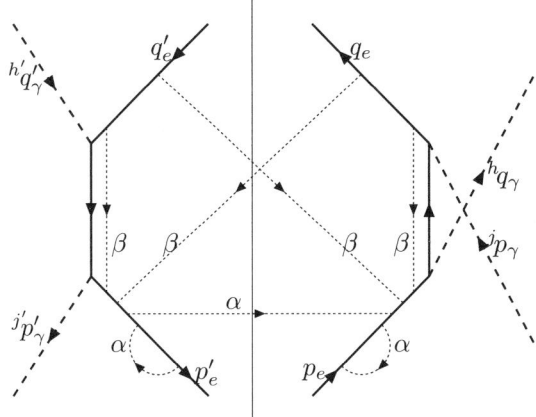

**Fig. 17.2.** Graphs contributing to $R_{MS}$ in orders $e^4$ and $e^6$.
The dotted lines signify alternative positions of internal photon lines in order $e^6$

factors have been explained in (15.90). The restrictions of the expressions (17.35) to the mass shell values $p_e = p'_e = P_e$, $q_e = q'_e = -Q_e \neq 0$, clearly exist. Adding the corresponding products from the other three graphs of order $e^4$, multiplying in the polarization factors explained at the end of Chap. 14,[7] and inserting the result as the scattering function $\Sigma$ into (17.1), we recover the result of the traditional formalism as stated e.g. in [IZ], Sect. 5-2-1.

Since $R_{MS}^{(4)}$ is the lowest non-vanishing term in $R_{MS}$, it can in order $e^4$ only be multiplied by the halo $I_{(0)} = 1$ of order zero: there is no halo effect in this order.

The next non-vanishing order is $e^6$. Using $I_{(0)} = 1$ we have

$$(R_{MS}I)_{(6)} = R_{MS}^{(6)} + R_{MS}^{(4)}I_{(2)} \,. \tag{17.36}$$

Let us first calculate the halo-free term $R_{MS}^{(6)}$. According to prescription we define $R'_{(6)}$ as a sum over 2-sector graphs of order $e^6$ with the appropriate external lines. The propagators of external electron lines are included but not those of external photon lines. The relevant graphs are generated from the double tree graphs of order $e^4$ by adding an internal photon line. Possible positions of this additional line are shown in Fig. 17.2 as dotted lines. Not

---

[7]Using the relations (14.41)–(14.43) we find

$$(\not{p} + m) H_\pm (\not{p} + m) = H_\pm (\not{p} + m)^2 = 2m (\not{p} + m) H_\pm$$

for $p^2 = m^2$, hence $\eta_\pm(p)(\not{p} + m) = 2m\, \eta_+(p)$.

shown are lines with at least one end on the internal, vertically drawn electron lines. These cause no IR problems. They belong to the core, leading to a trivial web. These graphs are handled exactly like the $e^4$-graphs from before and give their conventional contributions to $\Sigma$. Also not shown are photon lines connecting lines of the out-pair $(q_e, q'_e)$ among themselves. These graphs are handled like the graphs in which the incoming lines are connected by lines labelled $\alpha$ in Fig. 17.2.

Next we need $\hat{R}^{(6)}$ as defined in (17.25). For that we must find the expansion up to order $e^2$ of $\frac{1}{\sqrt{t'^+(s)}}$. From (15.34) and (17.32) we find

$$t'^+(s) = \frac{2m}{-\Delta + i\varepsilon}\left(A_{(0)}(s^2) + e^2 A_{(2)}(s^2) + \cdots\right)$$
$$= 2m\, N_+^{(0)}\left(1 + e^2 \frac{A_{(2)}}{A_{(0)}} + \cdots\right),$$

whence

$$\left(\frac{1}{\sqrt{t'^+}}\right)_{(2)} = -\frac{1}{2\sqrt{2m}\sqrt{N_+^{(0)}}} \cdot \frac{\frac{A_{(2)}}{-\Delta}}{\frac{A_{(0)}}{-\Delta}}. \tag{17.37}$$

The numerator of the last fraction can be replaced by

$$\frac{A_{(2)}}{-\Delta} \sim \frac{A_{(2)} + (\slashed{s} + m)B_{(2)}}{s^2 - m^2} = \frac{t^+_{(2)}}{\slashed{s} + m}. \tag{17.38}$$

The symbol $a \sim b$ means here: "$a$ is replaceable by $b$ for our purposes". In the present instance it means that the terms on both sides have the same leading singularity at $\Delta = 0$. Remember that $t^+$ is the clothed propagator. The expression (17.37), hence (17.38), occurs in a graph multiplied by a factor $(\slashed{s} + m) \sim \frac{1}{2m}(\slashed{s} + m)^2$, hence (17.38) can be replaced by

$$\frac{A_{(2)}}{-\Delta} \sim \frac{t^+_{(2)}}{2m}. \tag{17.39}$$

From this we obtain

$$\left(\frac{1}{\sqrt{t'^+(s)}}\right)_{(2)} \sim -\frac{1}{2\sqrt{2m}\sqrt{N_+^{(0)}(s)}} \cdot 2\pi i \Delta \frac{t^+_{(2)}(s)}{2m} \tag{17.40}$$

and

$$\hat{R}^{(6)} = \frac{1}{(2m)^2} C \left[R'_{(6)} - \frac{1}{2}R'_{(4)}\left(2\pi i \Delta^L \frac{t^+_{(2)}(p_e)}{2m} + \cdots\right)\right] \tag{17.41}$$

with

$$C = \frac{1}{\prod_\rho \sqrt{N_+^{(0)}(\Delta^\rho)\, N_-^{(0)}(\Delta'^\rho)}} = (2\pi)^2 \prod_\rho \sqrt{(\Delta^\rho - i\varepsilon)(\Delta'^\rho + i\varepsilon)}. \tag{17.42}$$

The dots in the round brackets of (17.41) denote terms like the first one, coming from the other three external electron lines. $\rho$ takes the values $U$ and $L$.

The $R'_{(6)}$-graphs with SEPs in the external electron lines are, up to irrelevant terms, of the form of the subtraction term in the square bracket of (17.41), except that the factor $-1/2$ is replaced by 1. This means that this square bracket, which we call $R''$, can be replaced for our purposes by a sum over $R'_{(6)}$-graphs in which SEPs in external lines are multiplied by $1/2$. (This is just the result of these SEPs being shared equally between the $R$-part and the respective $U$- or $L$-part of the original full graph.) Equation (17.41) becomes

$$\hat{R}^{(6)} = \frac{1}{(2m)^2} \, C \, R''_{(6)} \,. \tag{17.43}$$

$R^{(6)}$ is then defined by (17.27).

Finally we need to calculate

$$R^{(6)}_{MS} = \left(\frac{R}{N^R}\right)_{(6)} = \frac{R^{(6)}}{N^R_{(0)}} + R^{(4)} \left(\frac{1}{N^R}\right)_{(2)} \,, \tag{17.44}$$

all terms taken at $p_e = p'_e = P_e$, $q_e = q'_e = -Q_e$. $N^R_{(0)}$ is defined by (17.30) to be

$$N^R_{(0)} = C^{-1} \,. \tag{17.45}$$

This gives the first term in the last expression of (17.44) as

$$\frac{1}{(2m)^2} \, R''^{\text{amp}}_{(6)} \bigg|_{\Delta^\rho = \Delta'^\rho = 0} \,. \tag{17.46}$$

In the second term we write $N^R \sim C W_R$ with $W_R$ the web function defined in Sect. 16.4. The $\kappa$-limit in the correct definition (16.67) of $N^R$ need not be performed since we are at present only concerned with the singularity of $N^R$ at $\Delta^\rho = \Delta'^\rho = 0$, not with its homogeneity. $N^R_{(2)}$ is found by a calculation similar to the one leading to (17.41). The result is

$$N^R_{(2)} = C \left[ W^{(2)}_R - \frac{1}{2} N^{(0)}_R \sum_\rho \left( \hat{N}^{(2)}_+(\Delta^\rho) + \hat{N}^{(2)}_-(\Delta'^\rho) \right) \right] \,. \tag{17.47}$$

The $\hat{N}_\pm$ are the amputated singularity functions defined analogously to (17.8). The computation of $(1/N^R)_{(2)}$ is then easy, and from (17.44) we find after some calculation

$$R^{(6)}_{MS} = \frac{1}{(2m)^2} \left[ R''^{\text{amp}}_{(6)} - R^{\text{amp}}_{(4)} \hat{W}^R_{(2)} \right] \bigg|_{p_e = p'_e = P_e, \, q_e = q'_e = -Q_e} \,. \tag{17.48}$$

$R''^{\text{amp}}_{(6)}$ is defined like in (17.34), except that $\tilde{\mathcal{W}}'_{(6)}$ is replaced by $\tilde{\mathcal{W}}''_{(6)}$, which is distinguished from the former by factors $1/2$ multiplying SEPs in external electron lines. $\hat{W}^R_{(2)}$ is the web function of second order without the cut propagators of the trajectories, and with SEPs multiplied by $1/2$. It is a sum of ten terms, one for each IR divergent graph, seven of which are shown in Fig. 17.2. An example with a photon line connecting different external lines is the case $q'_e-p_e$. Its contribution to $\hat{W}^R_{(2)}$ is

$$\frac{1}{2\pi^3}(P_e, Q_e) \int_K dk\, \delta_+(k) \frac{1}{-\Delta^L + 2(P_e,k) + i\varepsilon} \frac{1}{-\Delta'^U + 2(Q_e,k) - i\varepsilon}. \quad (17.49)$$

The contribution of an SEP in the $p_e$-line is

$$\frac{im^2}{8\pi^4} \int_K \frac{dk}{k^2 + i\varepsilon} \frac{1}{-\Delta^L + 2(P_e,k) + i\varepsilon} \frac{1}{2(P_e,k) + i\varepsilon}. \quad (17.50)$$

The other contributions are analogous.

The expression (17.48) exists and coincides with the conventional scattering function $\Sigma$ obtained with the help of a regularizing photon mass $\mu$. More precisely, the following can be shown. Keep $\vec{p}_e = \vec{p}'_e = \vec{P}_e$, $\vec{q}_e = \vec{q}'_e = -\vec{Q}_e$ and the external photon momenta $-P_\gamma$, $Q_\gamma$ fixed and consider the square bracket $K$ in (17.48) as a function of $\Delta^\rho$, $\Delta'^\rho$, and a photon mass $\mu$ introduced as in the semimassive QED of Sect. 17.1. Then $K$ is continuous for $\mu \geq 0$, and $\Delta^\rho$, $\Delta'^\rho$ in a small neighbourhood of the mass shell $\Delta^\rho = \Delta'^\rho = 0$. The value of $K$ at $\Delta^\rho = \Delta'^\rho = \mu = 0$ gives our $R^{(6)}_{MS}$, and it is also the value obtained by the conventional procedure. We do not prove these claims in detail but only indicate the essential points of the argument.

The claimed continuity holds for the contribution to $K$ from any individual graph. As an example, let us consider the graph of Fig. 17.2 with a photon line connecting the $p_e$- and $q_e$-lines. Let $k$ be the momentum of this line. The contribution to $R''^{\mathrm{amp}}_{(6)}$ is of the form

$$K^1 = \int \frac{dk}{k^2 - \mu^2 + i\varepsilon} \frac{1}{(p_e + k)^2 - m^2 + i\varepsilon} \frac{1}{(q_e + k)^2 - m^2 + i\varepsilon}$$
$$\times F(P_i, Q_j, \Delta^\rho, \Delta'^\rho, k). \quad (17.51)$$

$F$ contains the propagators of the IR irrelevant vertical electron lines, the numerators of the other electron propagators, and the vertex factors. It does not depend on $\mu$ and is $C^\infty$ in $\Delta^\rho, \Delta'^\rho, k$ in the interesting region. The explicit propagators in (17.51) must be expressed as functions of $\vec{P}_e, \vec{Q}_e, \Delta^\rho, \Delta'^\rho, k$. $K^1$ is continuous except at $\Delta^\rho = \Delta'^\rho = \mu = 0$, where the $k$-integrand has a singularity of fourth order at $k = 0$ which is not defined as a distribution. Hence $K^1$ possesses there a logarithmic singularity unless $F$ happens to vanish at the critical point. The terms of first order in an expansion of $F$ to this order in the critical variables have this fortunate property, hence their contributions have the claimed continuity: a factor $k$ makes the integral converge, a factor $\Delta^\rho$ or $\Delta'^\rho$ knocks out the logarithm at the origin. Only the contribution from $F(P_i, Q_j, 0, 0, 0)$ need be considered further. This factor can be drawn in front of the integral. The contribution of the graph to the factor $R^{\text{amp}}_{(4)}$ of the subtraction term in (17.48) has the same smoothness properties as $F$ and can be restricted at once to $\Delta^\rho = \Delta'^\rho = 0$. Some tedious calculations of the form used in the construction of the web (see especially (15.49)) show that this restriction agrees with the $F$-factor of the first term. We need then only consider the difference of the $k$-integrals, dropping the index $e$ in $p_e$, $q_e$,

$$\int_K \frac{dk}{k^2 - \mu^2 + i\varepsilon} \left[ \frac{1}{p^+p^- + 2(p,k) + k^2 + i\varepsilon} \frac{1}{q^+q^- + 2(q,k) + k^2 + i\varepsilon} - \frac{1}{-\Delta^U + 2(P,k) + i\varepsilon} \frac{1}{-\Delta^L - 2(Q,k) + i\varepsilon} \right]. \tag{17.52}$$

The notation is that of (15.24): $p^\pm = p^0 \pm \omega(\vec{p})$, $\Delta^L = 2\omega p^+$, and analogously for $q^\pm$ and $\Delta^U$. The subtraction in the square bracket can be achieved in two steps: first subtracting the first factor, then the second. This gives in the first step

$$\int_K \frac{dk}{k^2 - \mu^2 + i\varepsilon} \frac{-p^{+2} + 2(P-p, k) - k^2}{(p^+p^- + 2(p,k) + k^2)(-2\omega_p p^+ + 2(P,k))} \frac{1}{q^+q^- + 2(q,k) + k^2}. \tag{17.53}$$

The dangerous point $\Delta^U = \Delta^L = 0$ corresponds to $p^+ = q^- = 0$. And remember $\vec{p} = \vec{P}$, $\vec{q} = -\vec{Q}$. For $p^+ = q^- = \mu = 0$ the denominator has at $k = 0$ a zero of fifth order. The term $k^2$ in the numerator reduces this to an integrable $|k|^{-3}$, hence it produces a continuous contribution. The factor $k$ in $(P-p, k)$ leads to a logarithmic singularity which is killed by the factor $P - p = (-p^+, \vec{0})$. The $k$-integral multiplying $p^{+2}$ is linearly divergent at $p^+ = q^- = \mu = 0$, which is not sufficient to overcome the vanishing of $p^{+2}$. The same sort of arguments applies to the second subtraction involving the $q$-propagators. And a similar reasoning applies to the other graphs of Fig. 17.2. In the case of SEPs one must not forget to apply mass renormalization.

Summing over all graphs we find that $R^{(6)}_{MS} = \frac{1}{(2m)^2} K|_{\Delta^\rho = \Delta'^\rho = \mu = 0}$ is well defined. That it agrees with the conventional result is easily seen. The conventional method starts from the expression (17.48) *without* the subtraction term, taken at $\mu > 0$. This expression is then restricted to the mass shell $\Delta^\rho = \Delta'^\rho = 0$, and after that the limit $\mu \to 0$ is taken. This is a possible way of arriving at the relevant point $\Delta^\rho = \Delta'^\rho = \mu = 0$ of the full expression as prescribed in our method. But we know from Lemma 17.1 that $\hat{W}^R_{(2)}$ vanishes at $\Delta^\rho = \Delta'^\rho = 0$ for $\mu > 0$, so that in this particular manner of reaching the desired limit point it does not matter whether the subtraction term is there or not.

The equality of our result with the conventional one does not hold for the individual graphs, for which the conventional method diverges. But it does hold for sums over suitable subsets of graphs. Two such subsets are indicated in Fig. 17.2 by the labels $\alpha$ and $\beta$ marking the internal photon lines.

There remains the calculation of the halo term $R_{MS}^{(4)} I_{(2)}$ of the expression (17.36). $R_{MS}^{(4)}$ we know already. $I_{(2)}$ is

$$I_{(2)} = \frac{1}{(2\pi)^4} \int \prod_\rho \frac{d\Delta^\rho}{-\Delta^\rho + i\varepsilon} \frac{d\Delta'^\rho}{\Delta'^\rho + i\varepsilon} \exp\left(i\sum_\rho n_\rho(\Delta^\rho - \Delta'^\rho)\right),$$

$$\times \left[\hat{N}_{(2)}^R + \hat{N}_{(2)}^U + \hat{N}_{(2)}^L\right], \quad (17.54)$$

with $\hat{N}^\alpha$ defined from $N^\alpha$ like in (17.8). The simple additive structure is a peculiarity of the order $e^2$, the lowest non-trivial order. It holds even within the various terms. $\hat{N}_{(2)}^R$ is like $\hat{W}_{(2)}^R$ a sum of terms of the general forms (17.49) and (17.50), but with the $k$-integrations extending over $R^4$ instead of $\mathcal{K}$. $\hat{N}_{(2)}^{U,L}$ are sums of three such terms. The changed domain of integration makes the individual terms diverge, but the divergences cancel in the sum over suitable terms. In the terminology of Sect. 16.4, such a cancelling subset contains the graphs with internal photon lines connecting a fixed pair of electron lines with itself, or with another fixed pair. Two such sets are specified in Fig. 17.2 by the labels $\alpha$ and $\beta$. The contributions of these subsets to (17.54) can be evaluated separately.

Let us start with the set $\beta$. Its contribution to the square bracket in (17.54) is

$$\frac{(P_e, Q_e)}{2\pi^3} \int dk \left\{ \frac{i}{2\pi} \frac{1}{k^2 + i\varepsilon} \frac{1}{-\Delta^U - 2(Q_e, k) + i\varepsilon} \frac{1}{-\Delta^L + 2(P_e, k) + i\varepsilon} \right.$$

$$- \frac{i}{2\pi} \frac{1}{k^2 - i\varepsilon} \frac{1}{-\Delta'^U + 2(Q_e, k) - i\varepsilon} \frac{1}{-\Delta'^L - 2(P_e, k) - i\varepsilon}$$

$$+ \delta_-(k) \frac{1}{-\Delta^U - 2(Q_e, k) + i\varepsilon} \frac{1}{-\Delta'^L - 2(P_e, k) - i\varepsilon}$$

$$\left. + \delta_+(k) \frac{1}{-\Delta'^U + 2(Q_e, k) - i\varepsilon} \frac{1}{-\Delta^L + 2(P_e, k) + i\varepsilon} \right\}. \quad (17.55)$$

The part of the integrand of (17.54) standing in the first line is invariant under the exchange $\Delta^U \leftrightarrow -\Delta'^U$ and under $\Delta^L \leftrightarrow -\Delta'^L$. Applying one or both of these substitutions to a term in the curly bracket of (17.55) does not change its contribution to (17.54). In this way we can replace the curly bracket by

$$\frac{1}{\Delta^U + 2(Q_e, k) - i\varepsilon} \frac{1}{\Delta^L - 2(P_e, k) - i\varepsilon} \left(\frac{i}{2\pi} \frac{1}{k^2 + i\varepsilon} - \frac{i}{2\pi} \frac{1}{k^2 - i\varepsilon} - \delta(k^2)\right) = 0$$

without changing $I_{(2)}$: the $\beta$-graphs give a vanishing contribution to $I_{(2)}$!

This pleasing result depends on the special structure of the 2-pair contribution (17.55). It generalizes to the geometry-dependent 2-pair contributions of second order in any QED reaction, like electron–electron scattering, pair annihilation, or more general reactions with any number of observed outgoing particles. But it does not generalize to higher orders, where the crucial additivity is lost. And it does not generalize to the 1-pair contributions of second order like the contributions of the $\alpha$-graphs in Fig. 17.2. This $\alpha$-contribution can be combined with the term $\hat{N}^L_{(2)}$ which gives an almost identical contribution: only, the $\delta_+$ in the cross term becomes a $\delta_-$. Together, the $\alpha$-graphs and $\hat{N}^L_{(2)}$ contribute to $I_{(2)}$ the expression

$$I^{(2)}_1 = -\frac{m^2}{8\pi^5} \int \frac{d\Delta}{-\Delta + i\varepsilon} \frac{d\Delta'}{\Delta' + i\varepsilon} e^{in(\Delta - \Delta')}$$
$$\times \int dk \left\{ \frac{i}{2\pi} \frac{1}{k^2 + i\varepsilon} \frac{1}{-\Delta + 2(P_e, k) + i\varepsilon} \frac{1}{2(P_e, k) + i\varepsilon} \right.$$
$$- \frac{i}{2\pi} \frac{1}{k^2 - i\varepsilon} \frac{1}{\Delta' + 2(P_e, k) + i\varepsilon} \frac{1}{2(P_e, k) + i\varepsilon}$$
$$\left. -\delta(k^2) \frac{1}{-\Delta + 2(P_e, k) + i\varepsilon} \frac{1}{-\Delta' + 2(P_e, k) - i\varepsilon} \right\}. \quad (17.56)$$

The index $L$ in $\Delta^L$ and $\Delta'^L$ has been dropped. The $\Delta^U$-$\Delta'^U$-dependent factors in (17.54) integrate to 1. The ($\Delta \leftrightarrow -\Delta'$)-symmetry of the $k$-independent factors does this time not imply the vanishing of the expression. Notice that the $k$-integral is Lorentz invariant: it does not depend on $P_e$.

An identical contribution $I^{(2)}_1$ to $I_{(2)}$ comes from the $R$-graphs with photon lines connecting the pair $(q_e, q'_e)$, together with the $\hat{N}^U_{(2)}$-part of (17.54). Hence we find finally that

$$I_{(2)} = 2\, I^{(2)}_1, \quad (17.57)$$

as announced at the end of Sect. 17.2. Remember from there, however, that in an actual experiment the factor 2 in the discrepancy term $2\, R^{(4)}_{MS} I^{(2)}_1$ disappears if the relations (17.22) hold, i.e. if $L_{MS}$ and $U_{MS}$ are determined from 1-particle measurements as suggested in Sect. 17.2.

# References

There is no intention to give an extensive bibliography. Only the works referred to in the text are listed.

## Books

[BD] J. D. Bjorken and S. D. Drell. *Relativistic quantum fields.* New York: McGraw-Hill 1968.

[BLOT] N. N. Bogolubov, A. A. Logunov, A. I. Oksak and I. T. Todorov. *General principles of quantum field theory.* Dordrecht: Kluwer 1990.

[C] F. Constantinescu. *Distributions and their applications in physics.* Oxford: Pergamon 1980.

[ED] J. P. Elliott and P. Dawber. *Symmetry in physics.* New York, London: Macmillan 1979.

[EMOT] A. Erdélyi, W. Magnus, F. Oberhettinger and F. G. Tricomi. *Higher transcendental functions.* New York: McGraw-Hill 1953.

[GR] I. S. Gradstein and I. M. Ryshik. *Tables of series, products, and integrals.* Thun, Frankfurt: Harri Deutsch 1981.

[GS] I. M. Gel'fand and G. E. Shilov. *Generalized functions, vol. 1.* New York, London: Academic Press 1964.

[GW] G. L. Goldberger and K. M. Watson. *Collision theory.* New York: John Wiley 1964.

[H] R. Haag. *Local quantum physics.* Heidelberg, Berlin: Springer 1993.

[IZ] C. Itzykson and J.-B. Zuber. *Quantum field theory.* New York: McGraw-Hill 1980.

[J] R. Jost. *The general theory of quantized fields.* Providence RI: Am. Math. Soc. 1965.

[JR] J. M. Jauch and F. Rohrlich. *The theory of photons and electrons ($2^{nd}$ edition).* New York, Heidelberg, Berlin: Springer 1976.

[K] G. Källén. *Quantum electrodynamics.* New York: Springer 1972.

[N] M. A. Naimark. *Linear representations of the Lorentz group.* New York: Macmillan 1964.

[Sch] L. Schwartz. *Théorie des distributions.* Paris: Hermann 1966.

[StW] R. F. Streater and A. S. Wightman. *PCT, spin and statistics, and all that ($2^{nd}$ edition).* Reading MA: Benjamin/Cummings 1978.

[V] A. H. Völkel. *Fields, particles and currents.* Heidelberg, Berlin: Springer 1977.

[W] S. Weinberg. *The quantum theory of fields, vol.1.* Cambridge: Cambridge U. Press 1995.

## Articles

[AH 67] H. Araki and R. Haag. *Collision cross sections in terms of local observables.* Commun. Math. Phys. 4, 77, 1967.

[Bl 50] K. Bleuler. *Eine neue Methode zur Behandlung der longitudinalen und skalaren Photonen.* Helv. Phys. Acta 23, 567, 1950.

[BN 37] F. Bloch and A. Nordsieck. *Note on the radiation field of the electron.* Phys. Rev. 52, 54, 1937.

[Br 69] R. A. Brandt. *Gauge invariance in quantum electrodynamics.* Ann. Phys. (NY) 52, 122, 69.

[Br 70] —. *Field equations in quantum electrodynamics.* Fortsch. Phys. 18, 249, 1970.

[BS 75] *Green's functions for theories with massless particles.* Ann. Inst. H. Poincaré 23, 147, 1975.

[Bu 75] D. Buchholz. *Collision theory for massless fermions.* Commun. Math. Phys. 42, 269, 1975.

[Bu 77] —. *Collision theory for massless bosons.* Commun. Math. Phys. 52, 147, 1977.

[Bu 82] —. *The physical state space of quantum electrodynamics.* Commun. Math. Phys. 85,49,1982.

[Bu 94] —. *On the manifestations of particles.* Mathematical physics towards the 21st century, ed. R. N. Sen and A. Gersten. Beersheva: Ben Gurion University Press 1994.

[Do 64] D. Dollard. *Asymptotic convergence and the Coulomb interaction.* J. Math. Phys. 5, 729, 1964.

[EE 79]   J.-P. Eckmann and H. Epstein. *Time ordered products and Schwinger functions.* Commun. Math. Phys. 64, 95, 1979.

[EEF 76]  J.-P. Eckmann, H. Epstein and J. Fröhlich. *Asymptotic perturbation expansion for the S-matrix and the definition of time ordered functions in relativistic quantum field models.* Ann. Inst. H. Poincaré 25, 1, 1976.

[FJ 60]   P. G. Federbush and K. A. Johnson. *Uniqueness property of the twofold vacuum expectation value.* Phys. Rev. 120, 1926, 1960.

[FPS 74]  R. Ferrari, L. Picasso and F. Strocchi. *Some remarks on local operators in quantum electrodynamics.* Commun. Mat. Phys. 35, 25, 1974.

[Fr 79]   J. Fröhlich. *The charged sectors of quantum electrodynamics in a framework of local observables.* Commun. Math. Phys. 66, 223, 1979.

[Gu 50]   S. Gupta. *Theory of longitudinal photons.* Proc. Phys. Soc. London, Sect. A63, 681,1950.

[Hb 71]   I. Herbst. *One-particle operators and local internal symmetries.* J. Math. Phys. 12, 2480, 1971.

[He 79]   D. Heckathorn. *Dimensional regularization and renormalization of Coulomb gauge quantum electrodynamics.* Nucl. Phys. B156, 328, 1979.

[HK 64]   R. Haag and D. Kastler. *An algebraic approach to quantum field theory.* J. Math. Phys. 5, 848, 1964.

[Hp 65]   K. Hepp. *On the connection between the LSZ and Wightman quantum field theory.* Commun. Math. Phys. 1, 95, 1965.

[Hp 66]   —. *Proof of the Bogoliubov–Parasiuk theorem on renormalization.* Commun. Math. Phys. 2, 301, 1966.

[J 61]    R. Jost. *Properties of Wightman functions.* Lectures on field theory and the many-body problem, ed. E. R. Caianiello. New York: Academic Press 1961.

[Kä 52]   G. Källén. *On the definition of the renormalizations constants in quantum electrodynamics.* Helv. Phys. Acta 25, 417, 1952.

[KF 70]   P. P. Kulish and L. D. Faddeev. *Asymptotic conditions and infrared divergences in quantum electrodynamics.* Theor. and Math. Phys. 4, 745, 1970.

[Ki 68]   T. W. B. Kibble. *Coherent soft-photon states and infrared divergences.* J. Math. Phys. 9, 315, 1968; Phys. Rev. 173, 1527, 1968; Phys. Rev. 174, 1882, 1968; Phys. Rev. 175, 1624, 1968.

[MSt 80]  G. Morchio and F. Strocchi. *Infrared singularities, vacuum structure and pure phases in local quantum field theory.* Ann. Inst. H. Poincaré 33, 251, 1980.

[MSt 83]  —. *A non-perturbative approach to the infrared problem in QED.* Nucl. Phys. B211, 471, 1883.

[Na 74]  N. Nakanishi. *The Lehmann–Symanzik–Zimmermann formalism for manifestly covariant quantum electrodynamics.* Prog. Theor. Phys. 52, 1929, 1974.

[Os 84]  A. Ostendorf. *Feynman rules for Wightman functions.* Ann. Inst. H. Poincaré 40, 273, 1984.

[P 69]  K. Pohlmeyer. *The Jost–Schroer theorem for zero-mass fields.* Commun. Math. Phys. 12, 204, 1969.

[Sc 63]  B. Schroer. *Infrateilchen in der Quantenfeldtheorie.* Fortschr. Phys. 11, 1, 1963.

[Ste 68]  O. Steinmann. *Particle localization in field theory.* Commun. Math. Phys. 7, 112, 1968.

[Ste 70]  —. *Scattering formalism for non-localizable fields.* Commun. Math. Phys. 18, 179, 1970.

[Ste 93]  —. *Perturbation theory of Wightman functions.* Commun. Math. Phys. 152, 627, 1993.

[Str 70]  F. Strocchi. *Gauge problems in quantum field theory III.* Phys. Rev. D2, 2334, 1970.

[StrW 74]  F. Strocchi and A. S. Wightman. *Proof of the charge superselection rule in local relativistic quantum field theory.* J. Math. Phys. 15, 21, 1974.

[YFS 61]  D. R. Yennie, S. C. Frautschi and H. Suura. *The infrared divergence phenomena and high-order processes.* Ann. Phys. (NY) 13, 379, 1961.

[Zi 67]  W. Zimmermann. *Local field equations for $A^4$-coupling in renormalized perturbation theory.* Commun. Math. Phys. 6, 161, 1967.

[Zi 69]  —. *Convergence of Bogoliubov's method of renormalization.* Commun. Math. Phys. 15, 208, 1969.

[Zi 71]  —. *Local operator products and renormalization in quantum field theory.* Lectures on elementary particles and quantum field theory, ed. S. Deser, M. Grisaru and H. Pendleton. Cambridge MA: MIT Press 1971.

[Zw 75]  D. Zwanziger. *Scattering theory for quantum electrodynamics.* Phys. Rev. D 11, 3481 and 3504, 1975.

# Index

Amputation (of external line) 151, 294
Annihilation operator 46
Anti-time-ordered function 106
Anti-time-ordered product 104
Asymptotic completeness 215, 322
Asymptotic conditions 75, 77, 219
    LSZ condition 77
Backward tube 62
Bloch–Nordsieck model 219f
Bose field 31
BPHZ method 82, 151ff
Bremsstrahlung 243, 330
Canonical commutation relations 36, 45, 74, 142
Canonical formalism 45, 56
Charge, electric 20
    of a quantum field 68
Charge conjugation 52
Charge identity 89, 132ff, 149, 163, 178ff, 204
Charge operator 53, 67
    different definitions 72, 85ff
Charge symmetry 52
Classical field theory 19ff
    electrodynamics as 20ff
        half-classical electrodynamics 20ff
    Lagrangian formulation 19
Classical particle interpretation 232, 233f, 243
Cluster (in perturbation theory) 169
    X-cluster 169
Cluster expansion 41, 50, 51, 101
Clustering 27
    weak clustering 27
Cluster property 36
    *see also*: Weak cluster property
Composite field 151, 190
Compton scattering 336ff

Continuity equation 20, 67
Core 257, 307
Coulomb gauge 70, 192
Coulomb scattering, non-relativistic 217ff
Coupling constant 76
    renormalized (physical) 76
    unrenormalized (bare) 76
Covariant fields 16f, 27
    spinor fields 16
    tensor fields 16
Creation operator 46
Critical point (of reaction graph) 290
Cross line 116
Cross section 216, 221, 321f, 330, 333
    different from conventional form 334ff, 344
    differential 217
    inclusive 294
Current, current density 20, 48, 331
    explicit expression 50f, 52, 69
    as probe 331ff
Current charge 72
Current conservation 131f, 163, 178, 204
Current vertex 131
$\delta$-reduced rules 122
Detectors 239, 321, 330f
    as local observables 221
Distributions 28
Dirac ansatz 190
Dirac equation 22, 69, 195f, 198ff
    free 47
Dirac matrices 15
Dirac operators 96
Dirac spinor 14, 68
    free Dirac spinors 47ff
Elastic scattering 286
Electromagnetic field 20

quantized 53ff
Electron 213
Electron mass 77
   unrenormalized 77, 79
   renormalized 79
Elementary charge 76
Energy-momentum operators 30
Equal-time (anti-)commutator 45, 50
Equations of motion, see Field equations
Factorization of leading coefficient 254
Fell's theorem 188
Fermi field 31
Feynman propagator 64, 107, 108
Field
   operator-valued 27
   distribution-valued 28
   physical 189
Field equations 20, 73ff
   of classical electrodynamics 21
   renormalized 81, 95
   subsidiary conditions 74
   verification 123f, 162f, 194ff
Field monomial 85
Field renormalization 44, 76
Field theory 211, 335
   axiomatic field theory 25ff
Fock space 46, 213
Forest 154
Forest formula 154
Forward direction 218, 285
Fourier transform
   of field 29
   of test function 29
Free field 39ff, 213
Furry's theorem 120
Gauge charge 72
Gauge conditions 22
   Coulomb condition 22
   Lorentz condition 22
Gauge invariance 69, 70
Gauge transformations 22, 69, 189
   global vs. local 72
   Gupta–Bleuler to physical 190
   localized 189, 190
Graph, see Sector graph
Graph rules 116
   in physical gauge 193

$p$-space 120ff, 153f, 196
   A-rules 196f
   B-rules 196, 200
   $x$-space 117ff, 193f
Gupta–Bleuler formalism 54, 70, 188
Haag's theorem 74f
Halo 333, 336
   for Compton scattering 337f, 344f
Helicity 235
Hermiticity 33
H-integrable functions 222f
H-type 223
$in$-field 214f
Indefinite metric, see Indefinite scalar product
Indefinite scalar product 56, 61
Infraparticle 221
Infrared (IR) problem 80ff, 243, 244
   extended 136, 165
   mild 165, 309
   hard 165, 243
   for physical space 201ff
   in reactions 304ff
Infrared regularization 220, 250, 307
Integration of graphs in $x$-space 168f
Interaction picture 74
Invariant function 41, 61ff
IR divergences 103, 167
Källén gauge 80, 95
Kinematical singularity 276
Klein–Gordon equation, smooth solutions of 222
$L$-function, $L$-part 242, 243, 256, 305, 314, 324
Lagrangian 19
   of electrodynamics 21
Locality 31, 34
   local (anti-)commutativity 31
Local singularities
   in $p$-space 136, 151
   in $x$-space 155ff
Loop 116
Lorentz condition 22, 55, 56, 71, 96, 187
Lorentz group 7ff
   orthochronous 8
   proper 8
   restricted 9
Lorentz transformation 8

LSZ condition 77ff, 214, 221
LSZ formalism 77ff, 214ff, 217, 322, 330
Massive QED 254, 323
Massive theory 214
Mass operator 47, 214
Mass renormalization 79
  in web 264
Maxwell charge 72
Maxwell equations 20, 67, 187, 190, 194f
  for potentials 69
Minkowski space 7f
Monomial state 29
$n$-field state 59
Non-abelian theories 93
Non-forward direction 215, 218, 285
Normal product 95f, 141
Observable 31, 70
  localization of 31
Observation of a state 330, 333
One-particle irreducible 151
One-particle singularity 78
One-particle state 231, 240, 272, 331
One-photon state 272
Ordering relations 105, 126, 197
*out*-field 214f
Pair (of variables) 306
p-Maxwell equation 70, 188
Parity 9
Particle 47, 77, 211, 213, 321
  charged 190, 304, 330
  notion of 47
Particle probe, *see* Probe
Pauli term 24
Perturbation theory 82, 98ff
Photon 213
  hard 258
  in reactions 300ff
  soft 80, 190, 258, 330
  unobserved 243
Photon probe 272
Physical gauge 61, 70, 192
Physical state 57, 73, 296
  charged 187
Physical state space 57, 73, 90f
Poincaré group 8
Point splitting 82

Polarization
  fermions 235f
  photons 236f
Positivity 36, 42, 204, 205ff
Positron 213, 298, 309
Potentials, electromagnetic 21, 54
Preparation of a state 231, 284, 330, 333
Preswarm 174
Probe 222, 229, 321
  expectation value 231, 233, 240, 284
    asymptotically dominant term 231, 233, 238, 243, 253, 254, 259, 272, 285, 295, 304, 315
    difference to conventional method 250f, 283, 307, 330
  for free particles 229ff
  for interacting particles 239, 284
Propagators 116, 118, 193, 195f
  clothed electron propagator 252
$\Psi_n$ vertex 192ff, 245, 251
$\rho$-term 98, 101
$R$-part, *see* Renormalization part
$R$-part (of reaction graph) 305, 314
Radiation gauge, *see* Coulomb gauge
Reaction 283ff, 330
  exclusive 285
  geometry of 289
  inclusive 285
  reaction point $N$ 289
Reaction cross section, *see* Cross section
Reaction rates 283
Reconstruction theorem 34f, 73, 190
Reduction formula 215
  generalized 294, 323
Regularization 82
Relativistic causality 31
Relativistic invariance in physical gauges 207ff
Renormalization 135ff, 139
  of coupling constant 77, 98
  of electron field 80, 98
  of electron mass 79, 98, 146, 147
  of photon field 79, 98
  of photon mass 79
Renormalization problem 75ff, 143

in $p$-space 153ff
in $x$-space 156ff
Renormalization conditions 97f, 100, 137, 143ff
Renormalization constants 76, 79, 81, 96
Renormalization part 152
   primitively divergent 152
   superficially divergent (classification) 153
Representations of a group 9ff
   infinitesimal generators 14
   Lorentz group 10ff, 26
      spinor representation 12ff
   of the Poincaré group 26
   of the translation group 26
   two-valued 12, 26
Representation of observables 187
Restriction to mass shell 254, 256, 314
$S$-matrix 215, 216, 283, 335
$S$-matrix theory 335
Scaffolding 116
Scalar field 39,
   free hermitian scalar field 39ff
   massive 283, 322
Scaling degree 106
   maximal scaling degree, condition of 107
Scaling of variables 254, 268, 284
Scattering (of particles) 77
Scattering cross section, *see* Cross section
Scattering function 322, 336
   for Compton scattering 336f
Scattering state 213, 321
Sector 116
Sector graph 116
   integrand of
      $\delta$-reduced rules 122
      in $p$-space 121
      in $x$-space 117ff
   lines 116
   points 116
      external 151
   sectors 116
   vertices 116
Sector line 116
Semimassive QED 329

massless limit 329f
Semitrajectory 257, 307
Several-particle state 233
Sign independence 127
Self-energy part 145, 243, 244, 245, 251, 260, 262, 264
SEP, *see* Self-energy part
Singular function 110, 114
Smoothness (of $\tilde{\tau}$-functions) 180ff, 204f
Source (of a particle) 231, 321, 330f
Spin-statistics theorem 32, 45
Subgraph 151
   dimension 152
   proper 151
   superficially convergent 152
Swarm 174
Test functions 28
   Schwartz space $\mathcal{S}$ 28
Threshold singularity 286, 299
Time-ordered function 106
Time-ordered product 104
   permutation symmetry 104, 125
Time-ordering relations, *see* Ordering relations
Time reversal 9, 23
Trajectory 116
Transition amplitude 77, 216
Translation group 8
Truncated cone 222
Two-particle state 233, 284
$U$-function, $U$-part 242, 243, 256, 305, 314, 324
Ultraviolet (UV) problem 73, 75, 135, 137, 139ff, 143, 146, 158, 197ff
   solution by renormalization 81f
Uniqueness of solution 100ff
UV divergences 81, 103, 135ff
   overlapping 149
Vacuum 27
   definition 27
   existence 27
   uniqueness 27
Vector space $\mathcal{V}$ 25f, 187
Vertex 116
   reduced 163
   vertex factor 118, 148
Vertex function 147ff

Vertex part 243, 251, 259f, 262
VP, *see* Vertex part
$W$-function, *see* Wightman function
$\mathcal{W}$-function 105, 222
W-properties 33, 71, 97, 162, 176ff, 197f, 204
    cluster property 34, 130, 178
    covariance 33, 124, 177
    hermiticity 33, 129
    locality 34, 129
    positivity 35
    in perturbation theory 99f
    spectral property 34, 125, 177f
Ward–Takahashi identities 132
Weak cluster property 34, 100, 130, 176
Weakly singular 184, 227
Web 257, 307, 310, 325
    construction of 259ff
    leading singularities 265ff
    web graph 258
    web rules 257ff
Wick product 81

Wightman axioms 25ff, 75, 191
    field theory 27f
    locality 31
    original formulation 32
    positivity 32, 187, 204
    quantum mechanics 25f
    relativistic invariance 26f
    spectral property 30
    vacuum properties 27
Wightman function 32ff, 73, 75, 96, 123
    definition 33
    differential equations 97
        perturbative form 99
    perturbative expansion of 99ff
        unrenormalized 103
        first order 111ff
        second order 113, 115, 135, 139, 142
        renormalized 139
    properties 33f
Wightman postulates, *see* Wightman axioms
WT identities, *see* Ward–Takahashi identities

Location: http://www.springer.de/phys/

## You are one **click** away from a **world of physics** information!

## Come and visit Springer's
# Physics Online Library

### Books
- Search the Springer website catalogue
- Subscribe to our free alerting service for new books
- Look through the book series profiles

You want to order?    Email to: orders@springer.de

### Journals
- Get abstracts, ToC´s free of charge to everyone
- Use our powerful search engine LINK Search
- Subscribe to our free alerting service LINK *Alert*
- Read full-text articles (available only to subscribers of the paper version of a journal)

You want to subscribe?    Email to: subscriptions@springer.de

### Electronic Media
- Get more information on our software and CD-ROMs

You have a question on
an electronic product?    Email to: helpdesk-em@springer.de

● Bookmark now:

# www.springer.de/phys/

 Springer

Springer · Customer Service
Haberstr. 7 · D-69126 Heidelberg, Germany
Tel: +49 6221 345 200 · Fax: +49 6221 300186
d&p · 6437a/MNT/SF · Gha.

Printing: Saladruck, Berlin
Binding: Buchbinderei Lüderitz & Bauer, Berlin